D0204595

INTERPRETING PROTEIN MASS SPECTRA

INTERPRETING PROTEIN MASS SPECTRA

A Comprehensive Resource

A. PETER SNYDER

AMERICAN CHEMICAL SOCIETY
Washington, D.C.

2000

OXFORD
UNIVERSITY PRESS

Oxford New York
Athens Auckland Bangkok Bogotá Buenos Aires Calcutta
Cape Town Chennai Dar es Salaam Delhi Florence Hong Kong Istanbul
Karachi Kuala Lumpur Madrid Melbourne Mexico City Mumbai
Nairobi Paris São Paulo Singapore Taipei Tokyo Toronto Warsaw

and associated companies in
Berlin Ibadan

Developed and distributed in partnership by
the American Chemical Society and Oxford University Press
Published by Oxford University Press, Inc.
198 Madison Avenue, New York, New York 10016

Library of Congress Cataloging-in-Publication Data
Snyder, A. Peter, 1952-
Interpreting protein mass spectra : a comprehensive resource/
A. Peter Snyder,
p. cm.
"Developed and distributed in partnership by
The American Chemical Society and Oxford University Press"–CIP t.p. verso.
ISBN 0-8412-3571-6
1. Proteins–Spectra. 2. Mass spectrometry. I. Title.
QP551.S653 2000
572′/633–dc21 99-27004

3 5 7 9 8 6 4 2

Printed in the United States of America
on acid-free paper

FOREWORD

Molecular mass spectrometry became a useful analytical technique in the 1940s, mainly for quantitative analysis of common volatiles such as light hydrocarbons. Molecular structure characterization became increasingly important in the 1950s and 1960s, but for these three decades electron ionization was the standard method for converting sample molecules into the gaseous ions required for mass spectral (MS) analysis. However, this demands prior sample vaporization, a most critical limitation to the size of molecules analyzed. New methods such as chemical ionization, field desorption, plasma desorption, and fast atom bombardment pushed the molecular-weight limitation from hundreds to thousands, but by the mid-1980s molecular mass spectrometry (MS) was still only a fringe technique serving the explosive growth of biomedical research.

To the total surprise of many of us, all of this changed a decade ago with the discovery of matrix-assisted laser desorption by Franz Hillenkamp and electrospray ionization (ESI) by John Fenn, increasing by orders of magnitude the molecular-weight range of MS. To me, this has been an astounding revolution, making possible the application of all these basic MS analytical principles developed over many decades to this critically important field of research. Without denigrating the outstanding contributions of matrix-assisted laser desorption/ionization (MALDI), the detailed description of this book concerning the accomplishments and potential of ESI convincingly demonstrates its revolutionary effect. Electrospray ionization is the most "gentle" ionization, minimizing concomitant fragmentation of the sample molecule, so that ESI of an impure sample produces ions that represent the component molecules of the mixture. Further, ESI produces multiply charged ions, whose electrostatic repulsion makes possible the subsequent dissociation of the mass-isolated ions that represent large molecules, providing protein sequence information for "purified" ions. These capabilities are described in detail here for proteins.

This great potential for ESI tandem mass spectrometry will only be realized if biomedical researchers have a comprehensive reference to provide detailed theory, instructons, and examples to guide them when applying this to their critical problems. This volume provides an up-to-date answer to this, and its readers should be amply rewarded with exciting possibilities suggested for their own protein research.

September 1998

Fred W. McLafferty
Peter J. W. Debye Professor

PREFACE

Interpretation of protein mass spectra in itself is a challenge. This is so because of the broad array of protein analytes and the many naturally occurring protein derivatives. Many rules and procedures must be kept in mind when tackling the issue of translating a mass spectrum into a protein structure. In addition, deliberate modifications to the protein generate further mass spectra, yet these procedures aid in analyte characterization.

Such was the task of this book—an attempt to codify and centralize many of the major aspects of experimental methods and interpretation of mass spectra from the point of view of protein structure determination. The journal literature is replete with this information, yet no satisfactory repository existed of comprehensive mass spectral explanations and procedures of protein structure determinations, as well as the background tables and figures that present the individual elements of protein structure. Thus, this book was written with that need in mind. In the process, I put myself in the shoes of both novice student and teacher with respect to the conveyance and the practical learning process of the interpretation of protein mass spectra. Basically, I figured that if I could teach myself the rudimentary, fundamental, and state-of-the-art techniques, methods, and procedures that formulate the interpretation of protein mass spectra, then another person interested in this area would have a good chance of following the outline that I have constructed.

Most assuredly, this pleasant journey I undertook was not accomplished in a vacuum. I drew from some of the most learned minds in this science, as well as an extensive amount of literature. These scientists which I would like to acknowledge are Professors Al L. Burlingame, Alvin Fox, Vernon N. Reinhold, and Dr. Richard D. Smith.

The approach to this book is repetition and redundancy—two qualities that are not normally applied to literature articles. For myself, this is the only way to obtain a respectable mastery of a subject . . . just work problem after problem.

The approach and writing of the book mirrors this concept. Thus, this book can most immediately benefit the introductory, first-year graduate student or entry-level industry scientist. This book "talks" to the reader in an analogous fashion as in the daily interaction of colleagues and peers on a subject of scientific concern. This book also has other levels of use. Reference material and information are included which make this book an ideal reference source—that is, a bookshelf reference which provides answers to specific questions when the experimental situation arises.

Also, this book was created with another goal in mind: for it to be used as a course textbook. The overall style of the book provides the impression of "spoon-feeding" when used as a textbook or learning tool of protein mass spectra. The discourse of a certain topic of protein mass spectra provides a gradual examination of the subject from an introductory to an advanced treatment in most of the chapters. Taken together, these treatments become rather comprehensive by design.

Professor Fred McLafferty's book *Interpretation of Mass Spectra* was a source as a model and inspiration for the creation of this book. Professor McLafferty's book serves as a university textbook, short course, bookshelf reference, and as an actively used experimental tool and companion on a daily basis in many laboratories. Such are the roles and models that I used in writing this book.

Note that I have not yet mentioned the subject of electrospray ionization (ESI). Electrospray ionization is a central element in this book, and almost all mass spectra presented herein were obtained by ESI-mass spectrometry. Despite the ubiquity of ESI in this book, this method of sample ionization is a secondary element in comparison with the primary goal of the book: protein mass spectral interpretation. Most of the tools and elements contained herein can also be used for other sample ionization and delivery techniques, such as matrix-assisted laser desorption/ionization (MALDI), fast atom bombardment (FAB), and secondary-ion mass spectrometry (SIMS).

Even if the scientist, practicing or supervisory, does not regularly use mass spectrometry information as part of their protein and biochemical repertoire, the amount of reference information contained herein on the physical aspects of proteins can provide considerable benefits.

Having presented the philosophies as the driving force of this book, I would like to impart some information about the chapters in order to make the reading more enjoyable and rational. Chapters 1–8 provide a basic backdrop or framework for the meat of the book contained in Chapters 9–12. Taken together, Chapters 1–8 and Appendix 1 provide for a proper positioning of the experimental conditions to optimize the signal, reproducibility, and integrity of the protein analyte within a mass spectrum.

Chapter 1 shows the many different types of mass spectrometers that can accept an ESI source, and these include relatively low-cost to very expensive mass spectrometer systems. A number of figures are presented that provide visual representations of different cartoon, pictorial, and schematic insights into the ESI process and physical manifestations of the spray itself. Chapter 2 presents two very important liquid-based, peptide and protein sample delivery systems: reversed-phase, high-pressure liquid chromatography (RP-HPLC) and capillary electrophoresis (CE). Both systems provide a measure of separation and concentration of the protein analyte with various buffer systems. Many examples make use of RP-HPLC in the purification of whole proteins/peptides, as well as utility in the separation of an enzymatic digest of the protein (Chapter 9). Capillary electrophoresis examples constitute a smaller portion than those of RP-HPLC in this text. Appendix 1 provides experimental details of all the chromatography plots and mass spectral figures presented herein, including the analytical parameters for the HPLC and capillary electrophoresis equipment. However, in Chapter 2 a greater emphasis is placed on CE than on RP-HPLC for two reasons. There are more variations of CE technology with respect to RP-HPLC, and details of the practical use of RP-HPLC can be found throughout the text in the explanations of the experimental protocols, in Appendix 1, and in Table 2.1 in Chapter 2.

Proteins and peptides are high-mass entities, and, as such, molecular isotopes play a fundamental role in mass spectral interpretation. However, for reasons different to those of Professor McLafferty's *Interpretation of Mass Spectra*, Chapter 3 provides a comprehensive treatment of this phenomenon. Chapter 4 presents information on the calculation of a basic fundamental parameter of a peptide or protein: the molecular mass. The molecular mass of the protein is usually convoluted in the mass spectrum; hence, deconvolution mathematical procedures must be used. Multiple-charging and multiple mass spectral peaks characterize the presence of a single species.

Chapter 5 presents resolution parameters used to characterize and refine any one of the multiple peaks that represents the molecular mass. Multiple-charging of a mass spectral peak begs the question of how many charges are present on a particular peak. Chapter 4 provides a discourse on how to determine the answer to the question based on a collection of several distinct peaks, while Chapter 6 answers this question by analyzing a specific multiply charged peak.

The presence/absence of a compound in an ESI-mass spectrum is affected significantly by the liquid milieu and physical characteristics of the mass spectrometer operating parameters. In Chapter 7, differences and variations of the major analytical and solution state parameters are shown regarding how good-quality mass spectra can be attained and characterized.

Alkali ions are ubiquitous and their presence can aid (Chapter 12) or usually hinder a mass spectral interpretation. Chapter 8 provides a short discourse on the presence of alkali ions in mass spectral determinations.

The remaining text of the book is the heart and soul of interpreting a protein mass spectrum. Chapter 9 provides a comprehensive overview of the details in interpreting a mass spectrum from small to relatively large proteins. Biochemical modifications, isotope enriched solvents, and certain inherent amino acid residue phenomena which can be brought to bear on a protein analyte are presented in order to more fully evaluate a mass spectrum and hence protein structure, including modifications to amino acid residues. Many tables and schemes are presented as guides, examples, and references on protein backbone and side-group structural details, as well as experimental procedures that augment the protein identification process.

Chapter 10 deals with the specific protein modification of the addition of the phosphate moiety on a protein. Chapter 11 represents an application section that shows the important aspects of Chapters 9 and 10 of which mass spectrometry has opened new doors of opportunity in the understanding of immunology. In particular, pioneering ESI-mass spectrometry work in this area by Professors Donald Hunt, Victor Englehard and Ruedi Aebersold and Dr. Andy Tomlinson are highlighted.

Last, but not least, Chapter 12 presents the important topic of interpreting glycoprotein mass spectra. In fact, one of the reviewers' characterization of Chapter 12 was ". . . the chapter on protein glycosylation is more like a book, itself". This chapter is a long one indeed, primarily because of the difficulty of the material. This difficulty is mainly due to the great many structural and stereoisomeric variations that carbohydrates have to offer, and these permutations are largely transparent to mass spectrometry. Some of the structural aspects can be realized by careful chemical, biochemical, and enzymatic manipulation of the sample. Rest assured, one mass spectrum of the polysaccharide portion of a glycoprotein will only whet the appetite for the great deal of further information to be discovered by further experimentation. This statement could have been made in a negative way; however, it is my belief that such negativity is what is limiting the mass spectrometry research on the salient details (of which there are many!) on the analysis of the glycan portion of glycoproteins. Extensive tables and examples are given on the subject and plenty of problems are provided. However, the state-of-the-art in the art (in contrast to the science) of interpreting glycoprotein mass spectra still makes fundamental use of inference, a small amount of wishful thinking, and relying on prior and similar literature work on solving the polysaccharide analyte structure and stereoisomeric properties. In a number of scholarly circles, it has been stated that a great deal of the mass spectrometry of glycoproteins basically characterized structures (to a limited extent!) that biochemists and molecular biologists have already comprehensively solved using systematic, yet extensive, experimental protocols.

Note that the title of each chapter displays the electrospray ionization term only when required, and this reflects the necessity of the chapter topic with respect to ESI.

The chapters in the book can be approached in a number of ways by the reader. The entire sequence of Chapters 1–8 can serve as a comprehensive course of study in itself, or some of the chapters can be selected for reading, review, or as a refresher.

The problems throughout the chapters can be addressed and solved appropriately, or the reader may learn from the answers directly. In any case, it is strongly recommended that the reader does not dismiss the problems since they supply important input and new ideas (given in the answers sections) on the topic at hand, as well as background material for further concepts.

Any part of Chapters 9–12 can serve as a reference, guide, or as a focus on a particular aspect of protein sequencing through mass spectra determinations. It is possible that source information is already known about the sample protein, yet more is needed. Certain passages, tables, schemes, figures, and so on, in Chapter 9–12 may be able to supply the next step, more information, or further suggestions as to how to proceed with the elucidation of the protein sample. Comprehensive guides and systematics are given in Chapters 9–12 regarding the delineation of protein and peptide structural attributes and their correlation with the mass spectra.

I would like to thank my supervisor, D. G. Parekh, for his continued support and inspiration throughout the writing of this book. Alice I. Vickers has been a steady, reliable, and dependable secretary, assistant, and administrative supervisor on the many facets of this book. The reader will agree that she makes me look good, . . . more than I deserve. A major point worth mentioning is that after the first draft of the 12 chapters was written and entered into the computer, the computer hard drive experienced a crash and all the files were destroyed. Only early incomplete copies of the chapters were on disks. Ms. Vickers retyped all 12 chapters, including the three appendices. The many figures, tables, and schemes in this book were rendered in the capable hands of Diane M. Billings, Aaron M. Thompson, and their supervisor Ralph F. Falcone.

The preparation of this book has been a very pleasant learning experience. As a matter of fact, I am writing a second companion draft to accommodate topics that are not covered in this volume, including noncovalent and enzyme active site–substrate interactions, hydrogen–deuterium exchange, glycolipids, and interpretation of DNA and RNA mass spectra.

I welcome any comments and suggestions on this book; these can include additions, deletions, replacement material, corrections, further insights into certain topics, and better problem sets. I also welcome any thoughts that you may have on topics that pertain to the interpretation of protein (and biomolecules in general) mass spectra for a possible second companion volume.

Please write or send an e-mail to me with your suggestion(s).

A. Peter Snyder
U.S. Army Edgewood Chemical Biological Center (ECBC)
Building E3220, Room 1326
Aberdeen Proving Ground, MD 21010-5424
apsnyder@sbccom.apgea.army.mil

CONTENTS

INTERPRETING PROTEIN MASS SPECTRA

1

Electrospray Ionization as a Vehicle for Protein Mass Spectra

Techniques for the transformation of a bulk liquid volume into a very fine mist have had utility for many decades. Spray techniques are efficient means for the transportation of either an active ingredient or the whole bulk liquid. Products such as perfumes, air fresheners, deodorants, hair sprays, and nasal and oral therapeutic medicines rely on the transformation of a bulk liquid into a fine mist in order to transfer the active ingredient(s) to the intended destination. Simple mechanical and pressure principles effect the transformation and delivery procedures.

Electrospray (ES), also known as electrohydrodynamic atomization or electrostatic spraying, has the same effect as the above mechanical and pressure procedures, except that electrostatic fields are primarily the cause of the spray.[1] The electrostatic fields can be assisted by mechanical/pressure means, and these assists facilitate the dispersion of the liquid stream. Electrospray has been used as a vehicle to produce liquid metal ion sprays and sprays of heterogeneous materials such as solid–liquid slurries (e.g., limestone and water, zirconia and water).[1]

The spray and electrospray examples described here do not result in charged molecules or ions. These processes merely disperse bulk liquid into mists of neutral drops. However, with a sufficient mechanical or pressure component, the spraying of paint (buildings interior and exterior and automobile paint applications), for example, can result in aerosol drops that have a net charge caused by either ion transfer or charge induction. For both spray and ES modes, liquid flow rates are usually in the greater than or equal to milliliters-per-minute regime.[1]

The vehicle of this book—electrospray ionization (ESI)—can be considered an extension of the spray and ES modes of liquid dispersion. An electric field, usually at the very tip or outlet of the spray capillary, imparts charge to the droplet as in ES; however, the flow rate is less than milliliters per minute. Typical ESI flow rates for mass spectrometry (MS) are in the 1–100 μl/min flow regimes. Lower flow rates, in the 1–100 nl/min, have been introduced, and this variation of the ESI process is labeled Nanospray^{TM} or microelectrospray.[2–6] The ES part of ESI for mass spectrometry can be characterized as atomization of a liquid at a very low flow rate. The sample ionization part of ESI occurs after the liquid sprays out of the capillary delivery system. Auxiliary methods and hardware (*vide infra*) are used to aid in the desolvation of the spray into successively smaller droplets until true ions form. The ESI process is of great benefit to MS systems, because there are many compounds of interest in diverse scientific areas that exhibit low to negligible vapor pressure and a significant degree of polarity. Mass spectrometry as a tool can only accept compounds and species which are in the gas phase. Electrospray ionization is a good example of a method that is gentle enough to faithfully transfer small and large analyte molecules, which possess a wide range of physical parameters and structural moieties, from a bulk liquid to individual ions.

Figure 1.1 shows a conceptual picture of the ESI liquid mist plume.[7] Relatively large droplets are created first, and they subsequently desolvate into ions that are electrically guided into the MS analyzer. The exact mechanism of ion production from an electrically generated liquid mist is uncertain, and a number of theories can be found in the literature on this topic.[8–15] Figure 1.2 provides another view of the ESI phenomenon.[16]

As mentioned earlier, the ionization portion of ESI is usually aided by a number of methods. The electrosprayed ions can be found as large clusters that consist of an analyte with many solvent and/or water molecules. Protons are the charge carriers for the analyte–solvent clusters in the positive-ion mode. In addition to other criteria, solvents are chosen such that the analyte has the

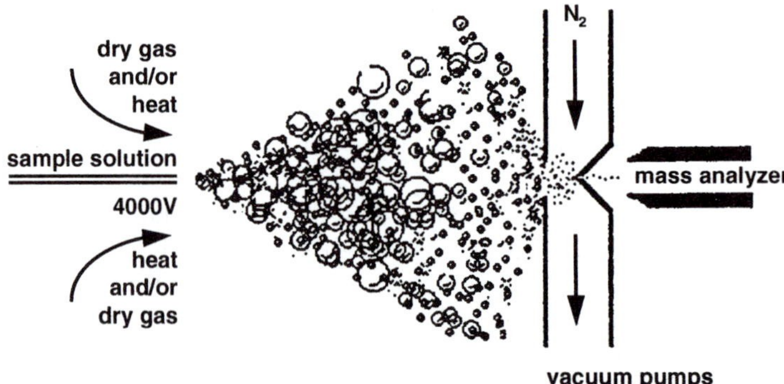

Figure 1.1 Diagram showing an atmospheric pressure interface of an ESI source and a quadrupole mass spectrometer. A conceptual picture of the ESI liquid mist plume is presented. (© 1994, National Academy of Sciences, U.S.A. Reprinted from reference 7 with permission.)

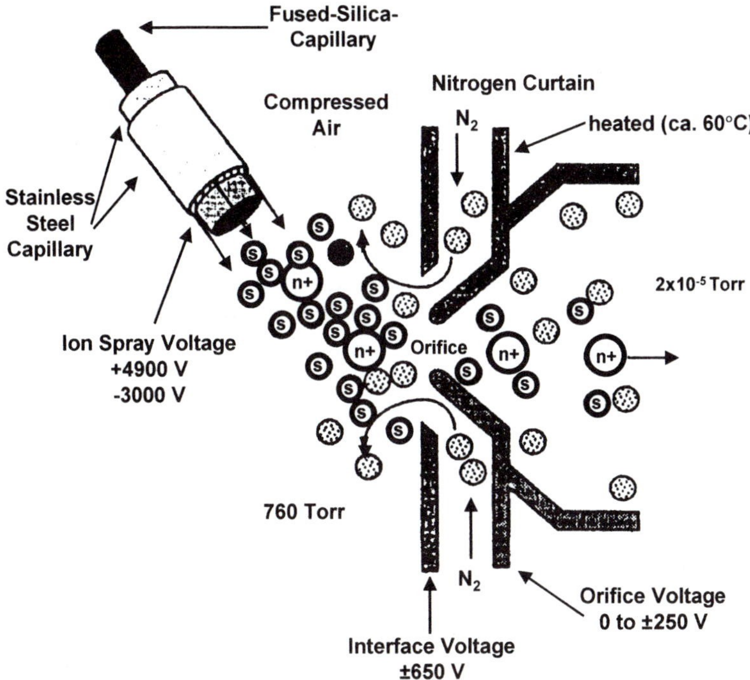

Figure 1.2 Diagram showing an atmospheric pressure interface of an ESI source and a quadrupole mass spectrometer similar to that shown in Figure 1.1. A conceptual presentation shows the solvent molecules and ions emanating from the nebulizer (air sheath)-assisted ESI source and entering the vacuum analyzer. (© 1994, Reproduced from reference 16 by permission of Wiley-VCH Publishers, Inc., Weinheim, Germany.)

higher proton affinity, and this results in the charging of the analyte by the transfer of proton(s). The clustering of a protonated analyte with neutral water and solvent molecules is a typical phenomenon for ionization methods that occur at atmospheric pressure, and ESI belongs to this class of ionization. The cluster array around an ion detracts from the direct measurement of the analyte's mass-to-charge ratio in a mass spectrometer. Figures 1.1 and 1.2 show one method to alleviate this problem, where a countercurrent of nitrogen gas is directed from the high-pressure side of the mass spectrometer sample introduction area to the electrospray needle. The countercurrent gas serves as a drying stream or curtain in order to strip off the bound cluster solvent/water molecules from the charged analyte.

Returning to the electrospray probe, there are certain procedures that can facilitate a more uniform, stable and reproducible dispersion of electrosprayed droplets. The simplest version of an ESI process[17] is that of an unassisted spray of a liquid stream of analyte (Figure 1.3). Figure 1.4 presents an assisted version[18] where an outer sheath of liquid, usually an organic solvent, is used to entrain the analyte liquid stream from the central capillary. The resulting ions

Figure 1.3 Schematic of the basic process to produce charged droplets in ESI. E, electric field. (© 1991, *Mass Spectrometry Reviews*. Reprinted from reference 17 by permission of John Wiley & Sons, Inc.)

enter the sample introduction system of the mass spectrometer via electrostatic ion-guide elements.

Figures 1.2 and 1.5 present a pneumatically assisted (compressed-air-assisted) electrospray, or IonSpray™, where a gas sheath[16,19] is used to entrain and disperse the liquid stream. Note that in Figures 1.2, 1.4 and 1.5, a counter-current stream of nitrogen gas is also used from the sample introduction side of the mass spectrometer in order to desolvate the analyte ions. Figure 1.6 shows a system where both liquid and heated gas streams[20–22] are combined in order to facilitate the rapid dispersion of the sample liquid into relatively small drops from the capillary column so as to minimize the occurrence of analyte loss in otherwise relatively large drops. Figure 1.7 shows a schematic of an ultrasoni-

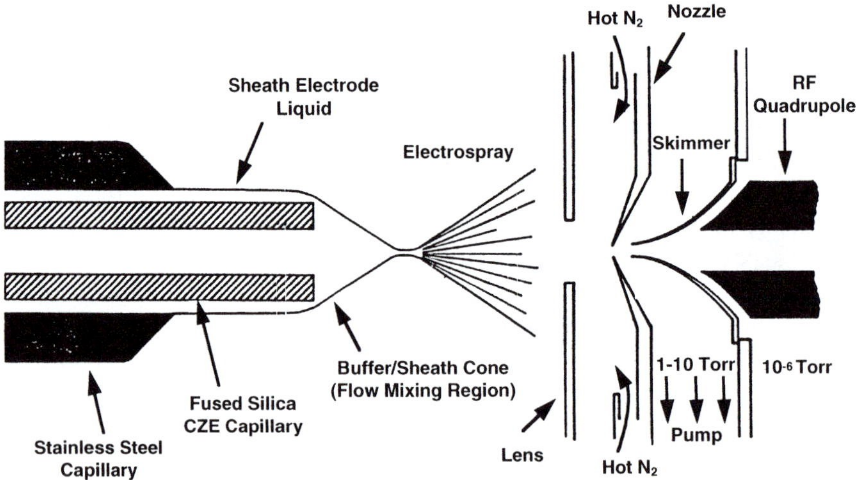

Figure 1.4 Diagram showing an atmospheric pressure interface of an ESI source and a quadrupole mass spectrometer similar to that shown in Figures 1.1 and 1.2. The mass spectrometer inlet is shown in detail. (© 1989. Reprinted from reference 18 by permission of Academic Press, Inc., Orlando, Florida.)

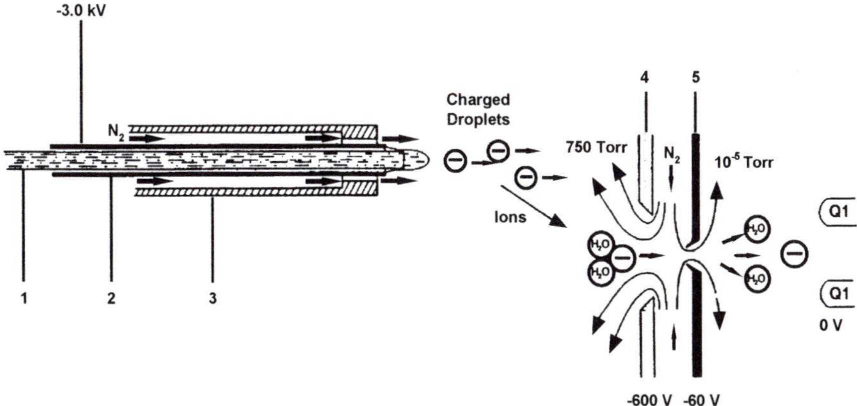

Figure 1.5 Diagram showing an atmospheric pressure interface of a nebulizer-assisted ESI source and a quadrupole mass spectrometer similar to that shown in Figure 1.2. A nitrogen curtain gas is used. (1) 50 μm i.d. silica capillary, (2) 0.2 mm i.d. stainless steel capillary, (3) 0.8 mm i.d. Teflon tube with a narrow-bore insert, (4) ion focusing lens, and (5) orifice plate with a 100 μm orifice. (© 1987, American Chemical Society. Reprinted from reference 19 with permission.)

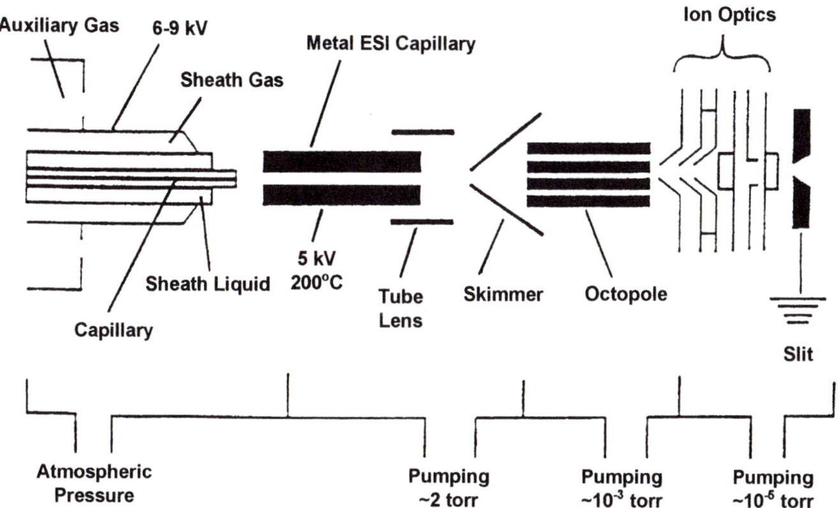

Figure 1.6 Schematic showing the interface of both a gas and liquid sheath-assisted ESI source with a mass spectrometer via a heated metal capillary tube and ion-guide optics. Note the differentially-pumped regions. (© 1995. Reprinted from reference 20 by kind permission of Elsevier Science-NL, Sara Burgerhartstraat 25, 1055 KV Amsterdam, The Netherlands.)

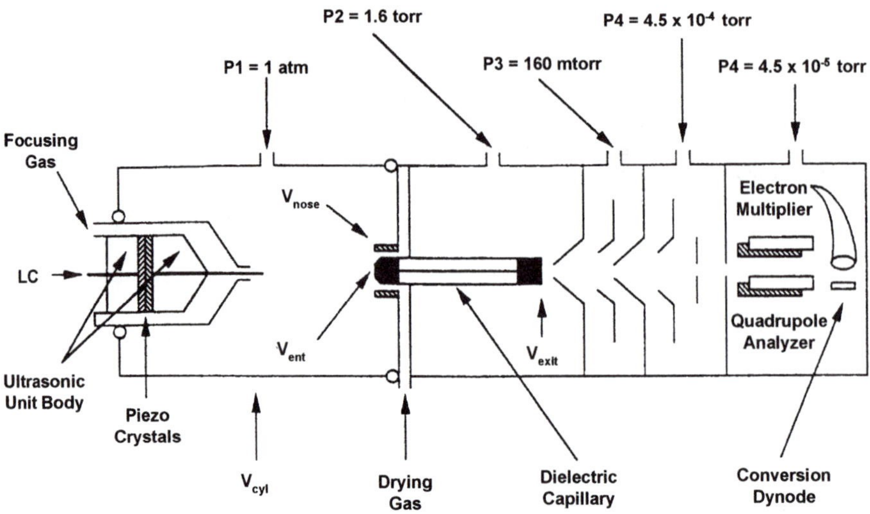

Figure 1.7 Schematic of an ESI–quadrupole mass spectrometry source. The ESI is an Ultraspray™ device that uses an ultrasonically assisted mechanical means to aid in the dispersal of the solution. The drying gas was at 180°C. (© 1994, American Chemical Society. Reprinted from reference 23 with permission.)

cally assisted electrospray method, or Ultraspray™, interfaced with a single quadrupole mass spectrometer.[23] Large drops lead to the possibility of a decrease in the amount of analyte that reaches the ionized state in the gas phase, and all four methods shown in (Figures 1.4–1.7) attempt to minimize the loss of analyte by reducing the size of the drops.

The initial drops formed by electrosprayed solutions can vary in diameter from the low micrometer size [9, 10, 13, 15, 19, 24, 25] to sizes in the range of 60 μm in diameter.[25] The diameter of an initial droplet is directly proportional to the 2/3 power of the electrospray flow rate (i.e. change in volume per time).[2] These relatively large drops then successively produce much smaller diameter droplets. The drops shrink by evaporation, and the surface experiences an increase in charge density. The electrostatic repulsion from the increase in charge density eventually produces a force greater than the surface tension, such that the drop explodes (Coulomb explosion) into smaller droplets. This process continues until the droplets are small enough to be able to desorb analyte ions into the gas phase.[12,13,25] For microelectrospray, initial droplet diameters are on the order of 0.18 μm.[2] The quality, consistency, stability and reproducibility of producing droplets, and hence mass spectral analyte signals, depend not only on the flow rate, but also on the liquid conductivity and surface tension. The addition of organic solvent to the aqueous analyte solution facilitates a lower conductivity and lower surface tension than that of a pure aqueous solution.[21,26] Additional benefits are also realized by utilizing organic solvents in the aqueous analyte milieu (Chapter 7).

The desolvation of the analyte ions was shown to occur by a countercurrent stream of nitrogen gas. Figures 1.6 and 1.7 present another method whereby the electrosprayed ionic clusters are entrained in either a metal or glass capillary, respectively. These capillaries have an electrical potential applied to them in order that they can act as an ion-focusing electrode for the ions, and the capillary is heated (Figure 1.6) so as to cause ion desolvation. The front end (atmospheric pressure side) of the glass capillary is metallized (Figure 1.8)[17] in order for it to act as an electrode for the electrosprayed ions (Figure 1.7). Figures 1.7 and 1.8 also show a countercurrent stream of nitrogen gas to facilitate desolvation of the analyte ions at the entrance of the heated capillary tube. However, a bath gas of nitrogen for a heated capillary is not necessarily required.[27]

Figure 1.9 shows that it is possible to interface an ESI source with a sector mass spectrometer,[28] and Figure 1.10 shows an ESI source as the sample introduction system for a triple quadrupole (tandem) mass spectrometer.[16] Electrospray ionization has also been used to deliver samples to Fourier transform (FT) (Figure 1.11)[29] , time-of-flight (TOF) (Figure 1.12),[30–32] and ion-trap (Figure 1.13) mass spectrometers.[31,33,34] Indeed, ESI has been shown to be versatile in its ability to be interfaced with a wide variety of mass spectrometer analyzers.

Mass spectrometers can be found as either single or tandem analyzers. Tandem MS analyzers have the capability of selecting an ion in the conventional mass spectrum such that only that particular ion (precursor ion) is allowed to pass through the first mass analyzer. The first mass analyzer is shown as quadrupole 1 in Figure 1.10. The precursor ion can be dissociated in the collision or second quadrupole (Figure 1.10) by the introduction of argon gas. Depending on the charge of the ion (see Chapter 4), lower as well as higher mass signals can

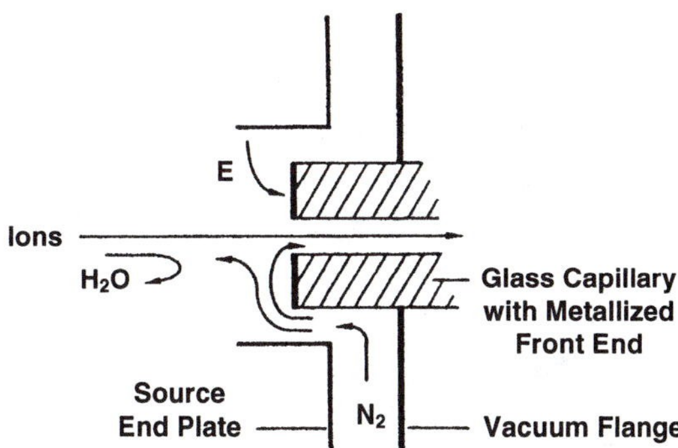

Figure 1.8 Electrosprayed ions entering the inlet of the glass capillary tube "ion channel." The inlet of the capillary is positioned in the orifice on the vacuum flange (plate separating atmospheric pressure and vacuum—e.g., in Figures 1.2 and 1.5). The nitrogen drying gas emanates from the plenum into the capillary and the ambient surroundings. (© 1991, *Mass Spectrometry Reviews*. Reprinted from reference 17 by permission of John Wiley & Sons, Inc.)

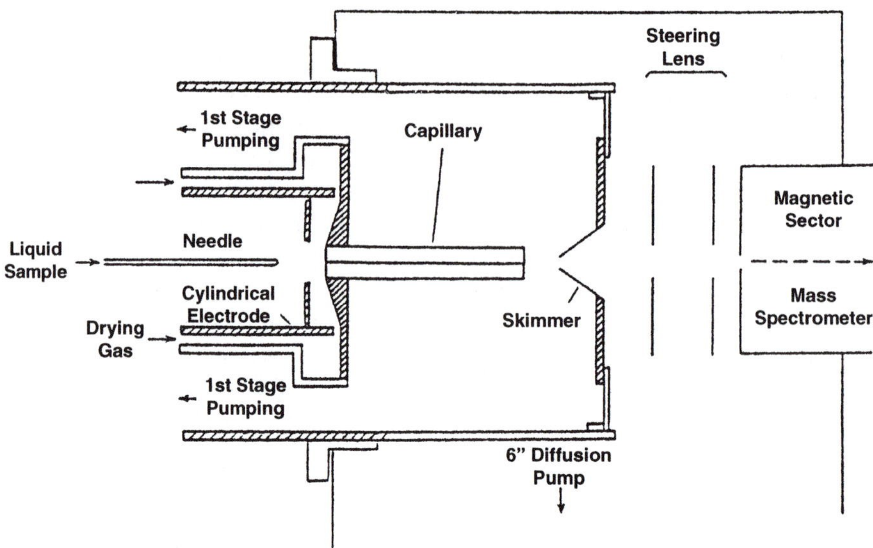

Figure 1.9 Diagram of an ESI source interfaced to a capillary-skimmer ion transport and focus arrangement for a magnetic sector mass spectrometer. Three pumping stages are used. (© 1991, American Society for Mass Spectrometry. Reprinted from reference 28 by permission of Elsevier Science, Inc.)

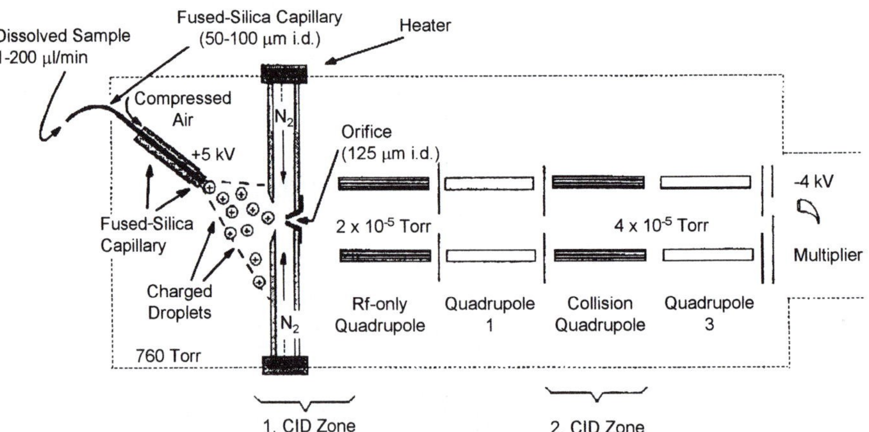

Figure 1.10 Diagram of an ESI interface similar to that shown in Figure 1.2; the mass spectrometer is of the triple quadrupole design. Note the two regions where collision-induced dissociation (CID) can occur. The first CID zone (up-front CID) allows for the dissociation of all entering ions, and the second CID zone allows for a selected ion to be fragmented. (© 1994. Reproduced from reference 16 by permission of Wiley-VCH Publishers, Inc., Weinheim, Germany.)

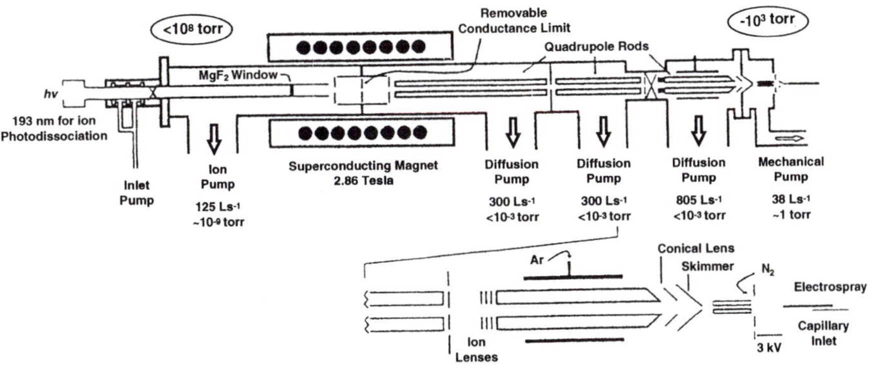

Figure 1.11 Schematic of an FT-MS system with an ESI source situated on the far right of the diagram. Note the four pump systems in the ion path prior to the front trap of the dual FT-MS cell. Cell pressure was 3×10^{-8} torr. (© 1990. Reproduced from reference 29 by permission of John Wiley & Sons, Limited.)

be observed with respect to the precursor ion. The dissociation of the precursor ion into product ions in the second quadrupole is labeled either collision-activated dissociation (CAD) or collision-induced dissociation (CID). Quadrupole 3 (Figure 1.10) then performs a mass scan of these product ions.

Another important parameter is the first or "up-front" CID zone (Figure 1.10). This region occurs in single and tandem mass spectrometers and can perform an optional capability for ion fragmentation. The drawback with respect to the up-front CID zone or CID-MS zone in mass spectrometers is

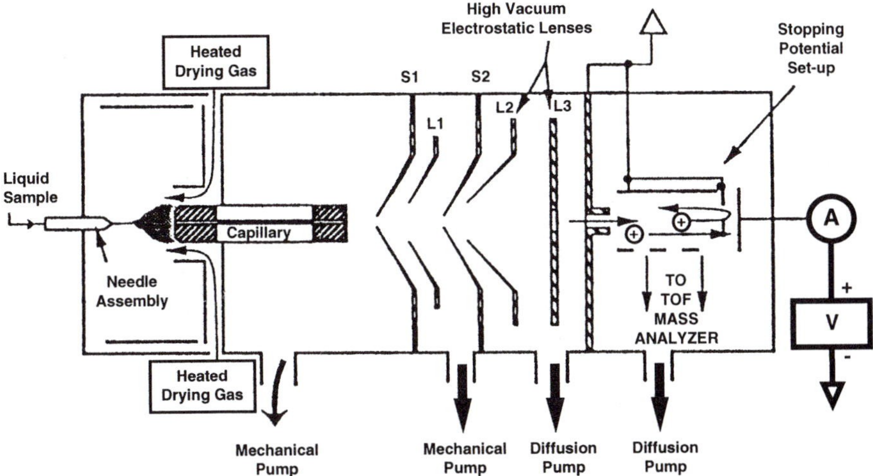

Figure 1.12 An ESI interface to a capillary-skimmer arrangement for a time-of-flight mass spectrometer.

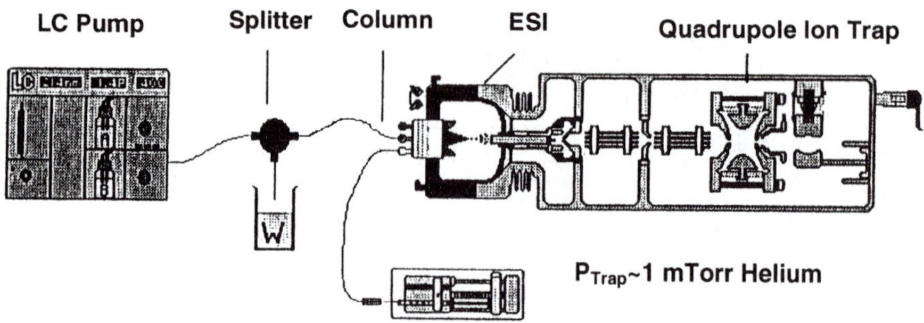

Figure 1.13 Diagram of a high-pressure liquid chromatography and direct-infusion sample introduction system to an ESI ion source; the latter is interfaced to an ion-trap mass spectrometer.

that selection of an ion cannot occur. All ions entering the mass spectrometer orifice region will experience the CID effect. This is caused by an increase in the potential between the two elements, rather than the introduction of a separate source of gas. These elements are usually located in an intermediate reduced-pressure region (Figures 1.6, 1.7, 1.9, 1.11, 1.12) or at the entrance to the high-vacuum region (Figures 1.1, 1.4, 1.5, 1.10), where a pressure differential occurs between the two electrical elements. The elements can be found as the orifice plate and the rf-only quadrupole ion guide in Figures 1.1, 1.4, 1.5, and 1.10, and the exit of the capillary tube and skimmer electrode in Figures 1.6, 1.7, 1.9, 1.11, and 1.12. This up-front CID zone has important applications in the characterization and optimization of the mass spectra of biological molecules and in the elucidation of structural features in these high-mass entities.

REFERENCES

1. Grace, J.M.; Marijnissen, J.C.M. *J. Aerosol Sci.* **1994,** *25,* pp 1005–1019.
2. Wilm, M.S.; Mann, M. *Int. J. Mass Spectrom. Ion Processes,* **1994,** *136,* pp 167–180.
3. Körner, R.; Wilm, M.; Morand, K.; Schubert, M.; Mann, M. *J. Am. Soc. Mass Spectrom.* **1996,** *7,* pp 150–156.
4. Wilm, M.; Mann, M. *Anal. Chem.*, **1996,** *68,* pp 1–8.
5. Shevchenko, A.; Wilm, M.; Vorm, O.; Mann, M. *Anal. Chem.* **1996,** *68,* pp 850–858.
6. Wilm, M.; Shevchenko, A.; Houthaeve, T.; Breit, S.; Schweigerer, L.; Fotsis, T.; Mann, M. *Nature* **1996,** *379,* pp 466–469.
7. Siuzdak, G. *Proc. Natl. Acad. Sci. U.S.A.* **1994,** *91,* pp 11290–11297.
8. Iribarne, J.V.; Thomson, B.A. *J. Chem. Phys.* **1976,** *64,* pp 2287–2294.
9. Thomson, B.A.; Iribarne, J.V. *J. Chem. Phys.* **1979,** *71,* pp 4451–4463.
10. Iribarne, J.V.; Dziedzic, P.J.; Thomson, B. A. *Int. J. Mass Spectrom. Ion Phys.* **1983,** *50,* pp 331–347.
11. Fenn, J.B.; Mann, M.; Meng, C.K.; Wong, S.F.; Whitehouse, C.M. *Mass Spectrom. Rev.* **1990,** *9,* pp 37–70.

12. Ikonomou, M.G.; Blades, A.T.; Kebarle, P. *Anal. Chem.* **1991,** *63,* pp 1989–1998.
13. Kebarle, P.; Tang, L. *Anal. Chem.* **1993,** *65,* pp 972A–986A.
14. Fenn, J.B.; Rosell, J.; Nohmi, T.; Shen, S.; Banks, F.J., Jr. In *Biochemical and Biotechnological Applications of Electrospray Ionization Mass Spectrometry;* Snyder, A.P., Ed.; ACS Symposium Series 619; American Chemical Society: Washington, DC, 1995; Chapter 3, pp 60–80.
15. Smith, R.D.; Loo, J.A.; Loo, R.R.O.; Busman, M.; Udseth, H.R. *Mass Spectrom. Rev.* **1991,** *10,* pp 359–451.
16. Metzger, J.W.; Eckerskorn, C. In *Microcharacterization of Proteins;* Kellner, R.; Lottspeich, F.; Meyer, H.E., Eds.; VCH Publishers: New York, 1994; Volume 2, pp 167–187.
17. Bruins, A.P. *Mass Spectrom. Rev.* **1991,** *10,* pp 53–77.
18. Loo, J.A.; Udseth, H.R.; Smith, R.D. *Anal. Biochem.* **1989,** *179,* pp 404–412.
19. Bruins, A.P.; Covey, T.R.; Henion, J.D. *Anal. Chem.* **1987,** *59,* pp 2642–2646.
20. Dobberstein, P.; Muenster, H. *J. Chromatog. A* **1995,** *712,* pp 3–15.
21. Smith, R.D.; Loo, J.A.; Edmonds, C.G.; Barinaga, C.J.; Udseth, H.R. *Anal. Chem.* **1990,** *62,* pp 882–899.
22. Griffin, P.R.; Coffman, J.A.; Hood, L.E.; Yates, J.R. III, *Int. J. Mass Spectrom. Ion Processes,* **1991,** *111,* pp 131–149.
23. Banks, J.F. Jr.; Shen, S.; Whitehouse, C.M.; Fenn, J.B. *Anal. Chem.* **1994,** *66,* pp 406–414.
24. Smith, R.D.; Loo, J.A.; Edmonds, C.G. In *Mass Spectrometry: Clinical and Biomedical Applications;* Desiderio, D.M., Ed.; Plenum Press: New York, 1992; Volume 1, Chapter 2, pp 37–98.
25. Mann, M.; Fenn, J.B. In *Mass Spectrometry: Clinical and Biomedical Applications*; Desiderio, D.M., Ed., Plenum Press: New York, 1992; Volume 1, Chapter 1, pp 1–35.
26. Chowdbury, S.K.; Chait, B.T. *Anal. Chem.* **1991,** *63,* pp 1660–1664.
27. Chowdhury, S.K.; Katta, V.; Chait, B.T. *Rapid Commun. Mass Spectrom.* **1990,** *4,* pp 81–87.
28. Larsen, B.S.; McEwen, C.N. *J. Am. Soc. Mass Spectrom.* **1991,** *2,* pp 205–211.
29. Henry, K.D.; McLafferty, F.W. *Org. Mass Spectrom.* **1990,** *25,* pp 490–492.
30. Gulcicek, E.E.; Whitehouse, C.M.; Boyle, J.G. *Proc. 41st ASMS Conference on Mass Spectrometry and Allied Topics,* San Francisco, CA; May 31–June 4, **1993,** pp 746a–b.
31. Michael, S.M.; Chien, B.M.; Lubman, D.M. *Anal. Chem.* **1993,** *65,* pp 2614–2620.
32. Chernushevich, I.V.; Verentchikov, A.N.; Ens, W.; Standing, K.G. *J. Am. Soc. Mass Spectrom.* **1996,** *7,* pp 342–349.
33. McLuckey, S.A.; Van Berkel, G.J.; Glish, G.L.; Huang, E.C.; Henion, J.D. *Anal. Chem.* **1991,** *63,* pp 375–383.
34. Bier, M.E.; Schwartz, J.C.; Zhou, J.; Syka, J.E.P.; Taylor, D.; Land, A.P.; James, M.; Fies, B. *Proc. 43rd ASMS Conference on Mass Spectrometry and Allied Topics,* Atlanta, GA, May 21–26, 1995, p 988.

2

Analyte Separation Methods

High-Pressure Liquid Chromatography and Capillary Electrophoresis

Water is a necessity in the majority of electrospray ionization (ESI) investigations, and it has advantages most suitable for biological molecules. However, as in any physical process, there are disadvantages.

The behavior of an electrospray ionization system is such that by using only an aqueous water medium, a physically poor and unstable electrospray plume results. This is caused by the properties of surface tension and polarity. Organic solvents have a lower surface tension than water, and, when they are mixed in certain proportions with water, the electrospray process can provide a plume mist in a more steady and reproducible fashion. The polarity of the liquid is also a factor, because the more polar the solvent, the more energy is required to dissociate the mist into finer drops. Thus, more energy is required to dissociate the analyte from the aqueous medium/solvent molecules than with a nonpolar solvent. A rough analogy to this situation is that a drop of a 50:50 methanol: water solution evaporates more quickly than a drop of pure water.

The following discussion concentrates on the major experimental parameters that affect the ESI process in the liquid state. Table 2.1 distills the important figures of merit that an experimentalist should keep in mind when performing liquid chromatography and capillary electrophoresis experiments so as to produce an optimal signal for the analyte. A comprehensive treatment of the parameters listed in Table 2.1 is beyond the scope of this book. Adequate treatments of the topics in Table 2.1 can be found in the literature, and it is believed that in order to provide for an initial experimental protocol, and taking into account the various parameters in Table 2.1, a fair amount of literature may need to be consulted. This investment in time by the experimentalist can provide for a robust setup of conditions to provide good sensitivity and an acceptable mass spectral integrity of the analyte signal.

Table 2.1 Electrospray ionization sample parameters

Parameter	References
Sample introduction system:	
liquid chromatography	1–9
capillary electrophoresis	1, 10–24
Postcolumn split	4, 5, 25
Liquid flow rate	3–5, 10, 13, 26–31
Liquid flow conductivity, surface tension	13, 15, 27–29, 31–33
Liquid dielectric constant	10, 16, 28
Solvent, mobile phase, sheath flows	1–3, 6–8, 10, 13–17, 22, 23, 28, 29, 32, 34
Electrolytes, buffers, salts, pH	3, 23, 28, 29, 31, 34, 35
Sample concentration	3, 10, 28, 30, 32
Sample sensitivity	1, 4, 5, 10, 16, 22, 26, 29, 31, 36, 37
Surfactants	38, 39

In the presentation of Table 2.1 in this chapter, separate narratives are given for the major components, elements, and combinations of each of the parameters. This procedure allows an adequate amount of overview to be presented in a relatively straightforward fashion. Appendix 1 provides a comprehensive table that lists the major operating conditions for all the analytes (i.e., figures) presented in this book. This list comprises the actual experimental protocols of small to large biological analytes reported in this book and can serve as a guideline for initial experimental protocol conditions.

Reversed-Phase, High-Pressure Liquid Chromatography

There are many liquid-based methods of separating a mixture of biological compounds or enzyme digests of a large protein. High-pressure liquid chromatography (HPLC) is one of the main techniques used for biological molecule separations. Two important categories of this technique are normal-phase and reversed-phase (RP-HPLC) chromatography. Normal-phase HPLC consists of a polar stationary phase and a nonpolar mobile liquid phase. This type of arrangement is used infrequently for moderate- to large-size biological molecules. Instead, RP-HPLC has found considerable use for biological molecule separations. The RP-HPLC technique is the opposite of the normal-phase technique in that a nonpolar stationary phase is employed with a polar liquid phase. An important class of RP-HPLC stationary-phase supports are made of silica columns derivatized with octadecylsilane, and this class is usually abbreviated as C_{18}. The stationary–mobile phase system of RP-HPLC can handle neutral polar and nonpolar analytes, as well as acidic, basic, and amphoteric compounds.[40,41] An amphoteric compound can exist in a net positive, negative, or neutral state, depending on the pH. Proteins and peptides belong to this class of analyte and are well suited for the RP-HPLC separation technique.

An analyte should be in a neutral charge state in order to provide for an optimal separation in RP-HPLC. When a compound ionizes in solution, or if it possesses a degree of polarizability as it is passing through a liquid chromatography (LC) column, the ionized/polarized analyte will be attracted to a greater degree by the stationary phase while the neutral and/or no net charge analyte will pass through the column at a faster rate. The former condition causes band-broadening or increased peak width. Thus, the pH of the solvent must also be considered. The pI of a molecule (i.e., the pH where the molecule has a net charge of zero) is the ideal pH for a biological species. The popular silica-based stationary phases perform properly only in the range of pH 2.0–8.0,[6,7] and the column can degrade to a significant extent outside of that pH range. Another method to effectively neutralize an analyte is to add a compound that interacts with the sample and causes the latter to be neutral.[7,42,43] This association of a sample and additive compound is called ion pairing, and the pairing event can electrostatically neutralize the analyte.

Solvents have a property that can cause analytes of various polarities to dislodge from the silica-based stationary phase. As analytes enter the column, they are differentially attracted to the stationary phase. A competition or equilibrium occurs because different solvents also have different affinities for the stationary phase. Thus, an equilibrium arises from the differential affinities between an analyte and the support species, and the solvent and the support species. This is a major parameter in effecting a separation.[44]

A solvent that has a low affinity for the stationary phase is said to have a weak elution power, because the analyte spends more time in the stationary phase due to its greater affinity for the stationary phase than the solvent. A solvent that has a high affinity for the stationary phase has a strong elution power, because analyte spends a relatively short time in the stationary phase than in the liquid phase. Thus, the analyte will go through the column in a relatively shorter time.[44] Mobile liquid phases are generally composed of water or a buffer solution and at least one nonpolar solvent. Common solvents listed in increasing elution power are methanol < acetonitrile < ethanol < isopropanol < dimethylformamide < *n*-propanol < dioxane < tetrahydrofuran.[45–47]

Usually, a third additive or modifier, in a relatively small amount (5–25%), is added to a binary water–nonpolar solvent system to confer even more selection for certain analytes in the elution process. An example that shows the effect of an organic modifier solvent added to an aqueous trifluoroacetic acid (TFA) binary elution system is shown in Figure 2.1.[6] The analyte mixtures comprise five synthetic decapeptides (S1–S5), and their structures are given in Table 2.2. Isopropanol was stated to have greater elution power than methanol, and Figure 2.1 shows this to be the case in that isopropanol 'chases' the analytes off the column and into the flowing mobile phase faster than methanol. A similar effect for the three organic modifier solvents is observed for an RP-HPLC analysis of the 23 tryptic peptides of porcine growth hormone[48] (data not shown). Other common organic modifiers for RP-HPLC are formic acid and acetic acid.[49]

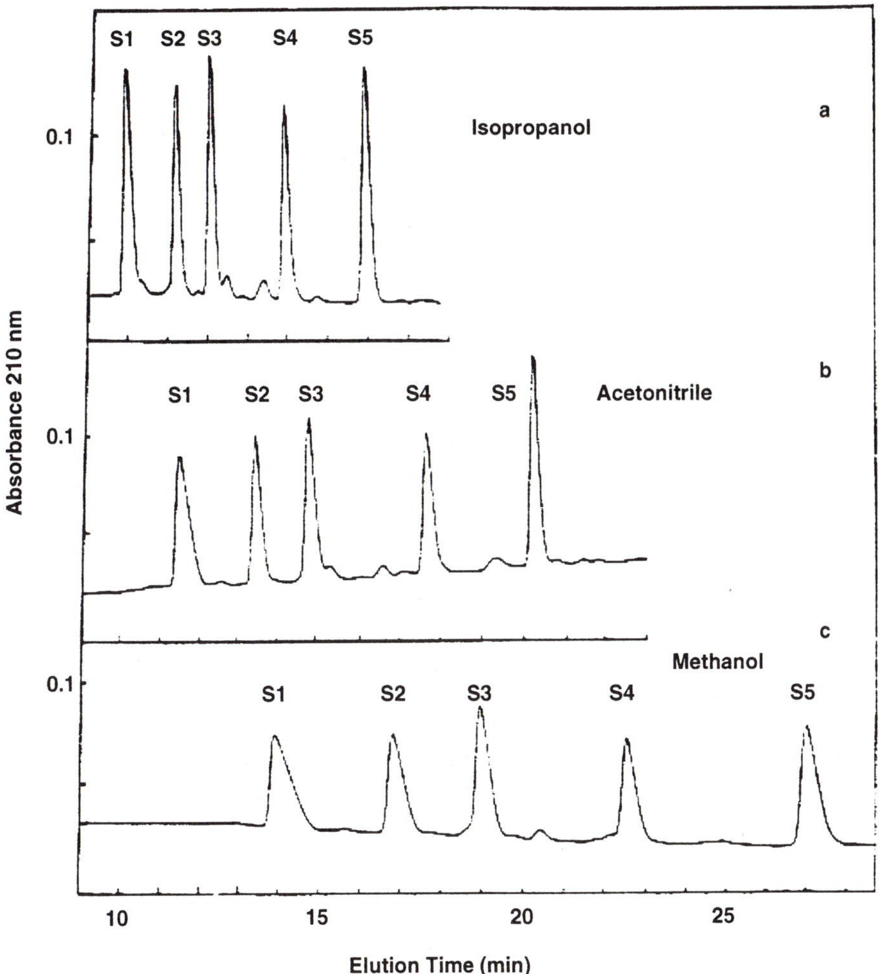

Figure 2.1 The RP-HPLC chromatograms of a mixture of five synthetic peptides. The sequence for the peptides is given in Table 2.2. An A–B gradient was used to elute the peptides where A = 0.1% aqueous TFA and B = 0.1% TFA in (a) isopropanol, (b) acetonitrile, and (c) methanol. The column was a SynChropak RP-P C_{18}, 250 × 4.1 mm i.d., 6.5 μm particle size, 300 Å pore size with a 10% carbon loading, flow rate at 1 ml/min, 26°C and pH 2.0. (© 1991, CRC Press, Boca Raton, FL. Reprinted from reference 6 with permission.)

Most investigations with RP-HPLC and protein/peptide analytes usually take place at a pH of 2.0–3.0. The silica-based columns are more stable at low pH, and low pH suppresses the proton ionization of the silanol oxygen groups, preventing excess negative charge. A low net charge on the column suppresses the unwanted electrostatic interaction of the sample with the silica stationary support species. The presence of this undesirable electrostatic interaction tends to spread the analyte band as noted above.

Table 2.2 Sequences of RP-HPLC peptide standards

Peptide standard	Sequence
S1	Arg-Gly-Ala-Gly-Gly-Leu-Gly-Leu-Gly-Lys-NH_2
S2	$CH_3C(O)$-Arg-Gly-Gly-Gly-Gly-Leu-Gly-Leu-Gly-Lys-NH_2
S3	$CH_3C(O)$-Arg-Gly-Ala-Gly-Gly-Leu-Gly-Leu-Gly-Lys-NH_2
S4	$CH_3C(O)$-Arg-Gly-Val-Gly-Gly-Leu-Gly-Leu-Gly-Lys-NH_2
S5	$CH_3C(O)$-Arg-Gly-Val-Val-Gly-Leu-Gly-Leu-Gly-Lys-NH_2

Figure 2.2 presents RP-HPLC spectra of the five peptides listed in Table 2.2. Conditions were silica-based RP-P C_{18} columns of different length, diameter and age, with identical packings (stationary phase), pH 2.0, and mobile phases of aqueous TFA with an acetonitrile modifier. All four column conditions provide a separation of the five peptides. However, note that for Figures 2a and c, little difference in elution time is observed even though the column length decreases from 250 mm to 50 mm, respectively. The TFA acts as an effective ion-pairing reagent and produces sharp peptide peaks in Figure 2.2. By changing the ion-pairing reagent from TFA to pentafluoropropionic acid and heptafluorobutyric acid, the retention times of a mixture of seven peptides increase to a significant extent, and the spacing between each peptide signal also varies.[50] Figures 2a and b show the effect of column aging, where over 4 months a column (Figure 2b) was used at least 50 times.[6,51] The new column in Figure 2a retained the peptides for a longer time than the aged column. Figures 2a and d compare column diameter. The greater volume of column packing in Figure 2d allows for a longer retention time; however, broader peaks are observed.

It should be noted that nonvolatile constituents in the liquid phase should not be used for RP-HPLC–ESI–mass spectrometry procedures. Precipitation, crystallization, and a plugged electrospray orifice can occur with nonvolatile buffer, additive, and solvent species.[52] Meyer,[45] Melander et al.,[53] and Guo et al.[51,54] provide useful examples of the relative mobility of peptides and proteins in RP-HPLC systems and modifications to various parameters for optimization of the interactions of the mobile and stationary phases with the biological molecule.

Capillary Electrophoresis

Another important method of liquid-based separation is capillary electrophoresis (CE). There are many variants of the technique that provide different levels of separation efficiency, depending on the analyte;[24,42,55–59] however, there are three commonly used versions of CE that can be interfaced with ESI.

1. *Capillary zone electrophoresis* (CZE) is the most popular and simplest form of CE.[57] A capillary column is filled with one type of buffer electrolyte, and an electric field gradient is applied through the length of the capillary column.[20,55,57,58,60,61] The electrical gradient is applied such that the analyte ions move toward the detector end of the column, and small-diameter columns (< 75

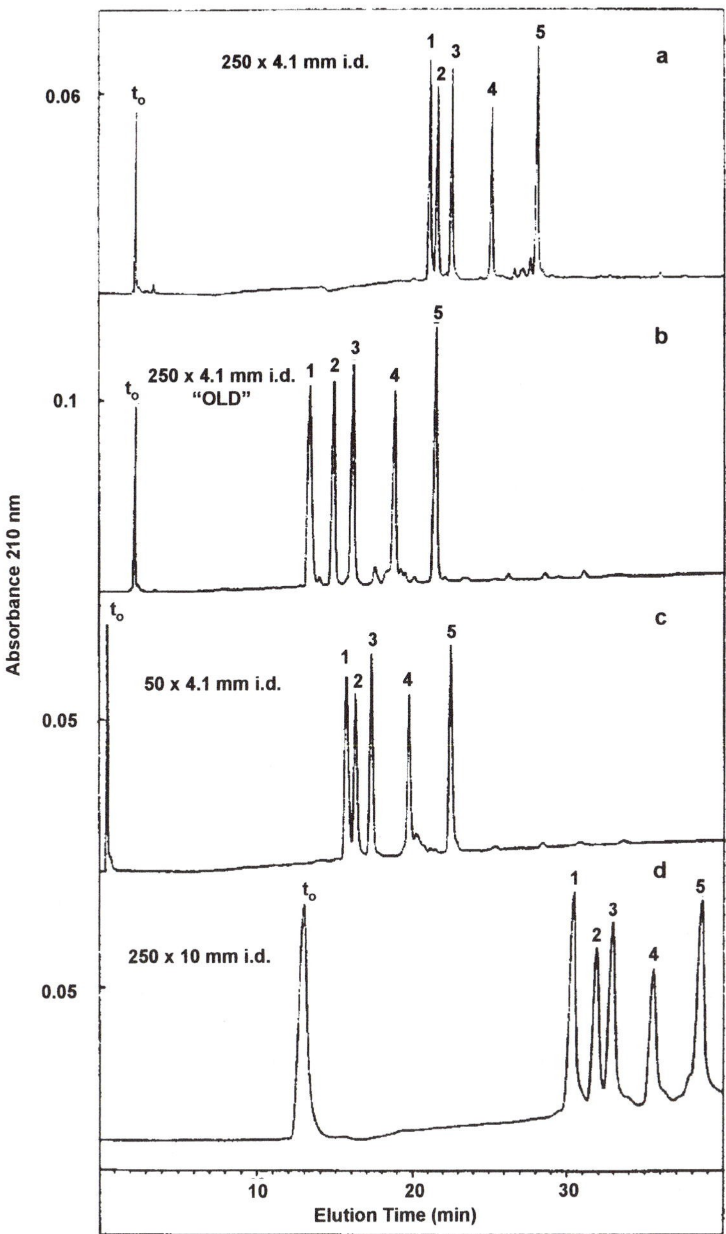

Figure 2.2 The RP-HPLC chromatograms of a mixture of five synthetic peptides (Table 2.2). A separate SynChropak RP-P C_{18} column was used for each analysis, with 6.5 μm particle size, 300 Å pore size, and 10% carbon loading; t_0 = elution time of unretained compounds. Column dimensions are noted on each chromatogram. (a, c, d) relatively new columns, (b) aged column; 4 months at > 50 experiments. Mobile phase conditions: a linear A–B gradient (1% B/min). A = 0.1% aqueous TFA and B = 0.1% TFA in acetonitrile with a flow rate of 1 ml/min at 26°C, pH 2.0. Sample volume for (d) was twice that of (a–c). (© 1991. Reprinted from reference 6 by kind permission of Elsevier Science-NL, Sara Burgerhartstraat 25, 1055 KV Amsterdam, The Netherlands.)

μm) are usually used. The ion movement is proportional to the voltage, and the ion front essentially has a flat profile compared with the usual parabolic (curved) profile of analytes in gas or liquid chromatography systems.[62]

2. *Capillary isotachophoresis* (CITP) uses the same instrumentation as CZE and can tolerate relatively larger sample loadings. For separation of a mixture of cationic (positively charged) analytes, an aliquot of a highly mobile cationic compound (leading electrolyte) is applied to the column along with buffering anions.[19,24,58,63,64] The sample is then placed on top of the first loading, and then a slow-moving buffered cationic electrolyte (trailing electrolyte) is placed behind the sample. The analytes are developed (separated) between the two electrolytes. The analytes separate into bands and eventually all move at the same rate toward an electrode. The concentration and intensity of each species in a mass spectrum is proportional to the width of the respective band of ions. Another interesting phenomenon is that all developed CITP bands tend to have similar ion concentrations; thus, the signal response of each analyte band will be fairly similar in magnitude. This technique is especially suitable for high-sensitivity and trace-analysis investigations, because the analyte ions tend to concentrate into bands as opposed to the inherent dilution or band-broadening phenomenon in CZE.

3. *Capillary isoelectric focusing* (CIEF) employs a mixture of charge-carrying compounds that have different pI values. Amphoteric compounds can exist as either an acid or a base, and an ampholyte is an amphoteric compound that can exist as either an anion or a cation, depending on the solution pH. The isoelectric point, or pI, is where the ampholyte is electrically neutral. Thus, at pH values below and above the pI, an ampholyte exists as a cation and an anion, respectively.[59,65] In an electrical gradient through the liquid in a capillary, an ampholyte mixture of polyamine and polycarboxyl compounds separate from themselves based on their pI values, and, in effect, cause zones of successively different pH values in the liquid milieu.[7,55,57,66,67] Thus, the pH can increase or decrease over the length of the column, depending on the polarity of the applied voltage, and the respective ends of the capillary are placed in low and high pH solutions to set up the static pH gradient.[67] In practice, and unlike CZE and CITP, a solution of analyte along with the different compounds fill the entire column, and the compounds and analyte migrate and separate according to their pI values[58] under an applied electric field. The separation is complete when the compounds and analyte become stationary in the electrical gradient at respective pI values. The compounds that define the pH gradient are prevented from entering the front-end (acid anolyte) and back-end (base catholyte) liquid reservoirs because the reservoirs act as pH barrier extremes to the ampholytes and analyte(s).[68] One end is then connected to an ESI source, and the compounds/analyte are eluted into the ESI–MS system with the aid of a salt buffer and positive pressure.[69]

There are a number of drawbacks to the CIEF approach, and, with respect to ESI, the relatively high amount of salt compounds in the mixture (1–5%), which are needed to set up the pI gradient, can create a problem. In order to produce acceptable sensitivity values, ESI is best operated at much lower compound/buffer concentrations. Experiments have shown that protein analysis performed under pure buffer conditions was 300 times more sensitive than when the ampholytes were added to the system for a CIEF separation. However, the CIEF sensitivity level was 1.5–2 orders of magnitude better with respect to CZE-MS of the proteins.[56,68]

Table 2.3 provides a useful guide for a number of commonly used CE buffer salts,[21,57,61,63,70–77] and includes buffers that have multiple dissociation constants, with their respective pH buffering range. Thus, for a given analyte, Table 2.3 can provide suggested buffer conditions for initial analyses. Other references can be found[78–80] that provide additional buffer compound pK_a information. Reference 63 provides an extensive list of optimal pH buffering ranges for combinations of different cations and anions. A comprehensive series of tables can be found[19,81] that contain information on buffers, their optimal pH range, optimal performance with solvent/salt additives, and analyte applications for capillary electrophoretic methods. Reference 34 provides salt–buffer combinations and concentration values suitable for ESI-MS analyte detection purposes. Depending on the amount of analyte, different buffer concentrations, and hence ionic strength, may have to be interrogated for maximal sensitivity. An increase in the salt concentration or buffer ionic strength will cause a decrease in the electroosmotic flow (EOF); that is, the ion flow in the liquid under the influence of an applied voltage.[62] Even though a low buffer concentration increases the EOF, broad or unsymmetrical ion profiles can result.

For CE methods in general, coated or noncoated silica capillary columns can be used, and solvent conditions usually rely on salt electrolytes. As in RP-HPLC mass spectrometry, involatile buffer and salt additives must be avoided due to physical and mass spectral interference phenomena. For silica capillaries in general, positively charged proteins and peptides tend to adhere on the uncoated silica walls. This adherence to the capillary walls slows the biological molecule migration and produces tailing and broad bands.[19,82–84] The pK_a of the SiOH to SiO^- negatively charged silica silanoate moiety is $2 \sim 4$.[85] Thus, this net negative charge on the bare capillary walls will retain positively charged proteins and hence broaden their elution profile.[82] For example, CE of many proteins, such as that of lysozyme with the use of bare silica capillary at pH 7.0, causes severe band-broadening because of differential adsorption of the positively charged protein on the negatively charged silica surface.[86] Coated capillaries are usually used for CE of proteins. Positively charged polymers are used so as to electrostatically provide a coat or layer over the negatively charged silica walls.[87] This layer therefore tends to repel positively charged analyte proteins; however, these coatings can attract net negatively charged proteins.[82,83,86] Therefore, uncoated, bare capillary columns can be used for net negatively charged proteins at physiological (pH 7.0) or higher pH conditions. However, under high pH conditions (pH 8–11), proteins can be denatured, and in certain situations this needs to be avoided.

Physiological pH values of the mobile phase (~pH 7.4) can be used with coated capillaries if the buffer contents are at a high enough concentration (100–200 mM).[18] High buffer concentrations cause electrical neutrality of the entire

Table 2.3 Capillary electrophoresis buffer systems and pK_a values in aqueous solution

Buffer		pK_a (20°C)	pH Range
Piperazine	a	5.33	
	b	9.78	
Sodium phosphate	a	2.12	1.6–3.2
	b	7.21	5.9–7.8
	c	12.32	10.8–13.0
Sodium citrate	a	3.06	2.1–6.5
	b	4.74	2.1–6.5
	c	5.40	2.1–6.5
Sodium formate		3.75	2.6–4.8
Sodium succinate	a	4.19	3.2–6.6
	b	5.57	3.2–6.6
Sodium acetate		4.75	3.4–5.8
BES, *N*, *N*-bis(2-hydroxyethyl)-2-aminoethane sulfuric acid		7.15	5.9–7.9
Ammonium carbonate	a	6.35	6.2–10.8
	b	10.33	6.2–10.8
Triethanolamine		7.76	7.3–8.3
Trichloroacetic acid (TCA)		0.52	
Oxalic acid	a	1.27	
	b	4.27	
Trifluoroacetic acid (TFA)		0.50	
MES, 2-(*N*-morpholino)–ethanesulfonic acid		6.15	4.9–6.9
Imidazole		7.00	6.2–7.8
Morpholine		8.49	
HEPPSO, *N*-(2-hydroxyethyl)piperazine-*N′*-(2-hydroxypropylenesulfonic acid)		8.00	7.1–8.6
CHAPSO, 3-[(3-cholamidopropyl)-dimethyl-ammonio]-2-hydroxy-propane sulfonate		9.60	8.9–10.4
Glycine		2.4	1.4–3.4
		9.8	8.8–10.8
PIPES, piperazine-*N*,*N*-bis(2-ethanesulfonic acid)		6.8	6.1–7.5
HEPES, *N*-(2-hydroxyethyl)piperazine-*N′*-(2-ethanesulfonic acid)		7.5	6.8–8.2
Cacodylate, dimethylarsinic acid		6.27	5.0–7.4
TAPS, 3-{[tris(hydroxymethyl)methyl]amino}propane sulfuric acid		8.40	7.7–9.1
Piperidine		11.12	
APS, 3-aminopropane sulfuric acid		9.89	
PIBS, 1,4-bis(4-sulfobutyl) piperazine		8.60	
Tricine, *N*-tris(hydroxymethyl)methylglycine		8.1	7.4–8.8
Triethylamine acetate		4.75	3.4–5.8
		10.72	10.0–11.5
MOPS, 3-(*N*-morpholino)propanesulfonic acid		7.20	5.9–7.9
Tris, tris(hydroxymethyl)aminomethane		8.3	6.6–8.8
Bicine, *N*, *N*-bis(2-hydroxyethyl)glycine		8.35	7.4–9.2
Sodium borate	a	9.24	8.0–10.5
	b	12.74	
	c	13.80	
CHES, 2-(*N*-cyclohexylamino)ethanesulfonic acid	a	10.4	9.7–11.1
	b	9.50	8.6–10.0
CAPS, 3-(cyclohexylamino)-1-propanesulfonic acid		10.4	9.1–11.1
Triethylamine		10.72	10.0–11.5

(*continued*)

Table 2.3 (*continued*)

Buffer		pK_a (20°C)	pH Range
Sodium malonate		3.40	2.5–4.0
ADA, *N*-(2-acetamido)iminodiacetic acid	a	<2	
	b	2–3	
	c	6.62	6.0–7.3
ACES, *N*-(2-acetamido)-2-aminoethane sulfuric acid		6.88	6.1–7.6
TES, *N*-tris(hydroxymethyl)-methyl-2-aminoethane sulfuric acid		7.50	6.8–8.3
MOPSO, 3-(*N*-morpholino)-2-hydroxypropane sulfuric acid		6.95	6.2–7.7
POPSO, piperazine-*N*,*N'*-bis(2-hydroxypropane sulfuric acid) dihydrate		7.85	7.2–8.6
TAPSO, 3-[*N*-tris(hydroxymethyl)methylamino]-2-hydroxypropane sulfuric acid		7.7	7.0–8.3
DIPSO, 3-[*N*-bis(hydroxyethyl)amino]-2- hydroxypropane sulfuric acid		7.6	7.0–8.3

liquid milieu, allow the applied electrical gradient to govern the movement of analyte molecules, and prevent local charge imbalance conditions.

Coated capillaries do impart some deficiencies in analyses in that their lifetime is usually short and their cost is not insignificant.[82] Some buffer additives and noncovalent, electrostatic coatings adsorb/desorb reversibly with the capillary silica, and this is known as dynamic coating.[62] This is desirable in regeneration of the capillary coating, and usually it does not affect spectral detection methods. However, the reversible desorption process can have deleterious effects in the ESI process for mass spectral detection.[64,82,88] In addition, it has been found that relatively large proteins tend to adsorb onto coated or uncoated silica columns, presumably due to a larger surface area of contact between the silica wall and protein.[82]

Another method to treat silica columns was discovered in 1988 by McCormick.[89] Instead of coating the silica walls, it was found that phosphate buffers with low pH values (i.e., somewhat lower than the pI of the protein analytes) significantly decreased the adsorption of positively charged proteins. This could occur by the phosphate forming a complex with the silica and allowing for a more facile protonation of the complex. This protonated complex would reduce the negative charge on the capillary wall.[89,90] Typical pH values used were in the range 1.5–4.5.[89] Good examples of the use of different types of deactivated columns are shown in Figure 2.3. A phosphate-deactivated silica column was able to separate recombinant human growth hormone (rhGH) and all its variants, and the protein was obtained as a mixture from its expression in *Escherichia coli* (Figure 2.3a). A bare unprocessed column (Figure 2.3b) and a polyvinylalcohol (PVA)-coated capillary (Figure 2.3c) could not separate all the protein derivatives of rhGH.[83,91] Another example of the effect of simple phosphate-deactivation is shown in Figure 2.4. Erythropoietin (EPO) is a glycoprotein (Chapter 12) and exists as multiple copies of the same protein with a

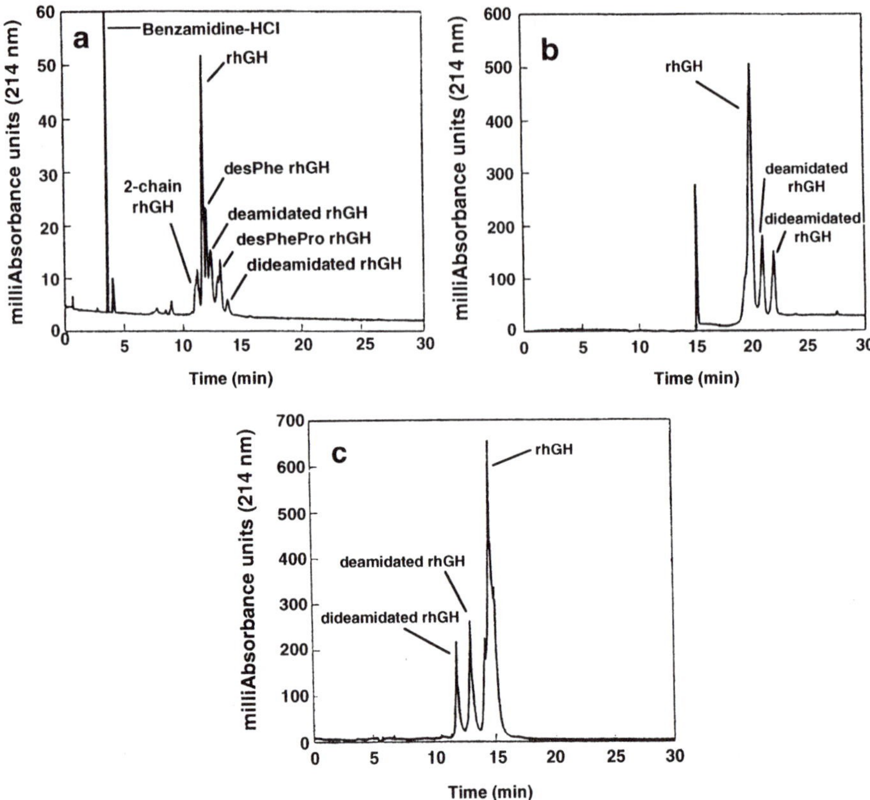

Figure 2.3 The CZE separation of the indicated rhGH native and variant forms using (a) phosphate-deactivated, fused-silica capillary; (b) bare silica; and (c) polyvinylalcohol-coated column. Field strength = 600 V/cm, capillary i.d. = 25 μm. (© 1996. Reprinted from reference 91 with kind permission of Elsevier Science-NL, Sara Burgerhartstraat 25, 1055 KV Amsterdam, The Netherlands.)

different number and type of sugar unit arrangements attached to the same amino acid residue. This multiple sugar arrangement on a particular amino acid occurs on four different residues. The recombinant human glycoprotein mixture was subjected to CZE under acetate buffer conditions, and Figure 2.4 shows the separation where the glycoprotein was neutralized to pH 4 using acetic acid, phosphoric acid, or sulfuric acid.[83,90] The phosphate additive (Figure 2.4b) clearly shows the glycoprotein separating into four major bands, where each band is a mixture of various arrangements of sugar molecules (glycoforms—cf. Chapter 12) on a particular amino acid residue.

Buffer salt composition and concentration can affect sensitivity in CE-mass spectrometry detection methods. As mentioned above, volatile buffers are superior to involatile salts for ESI-mass spectrometry. The lower the concentration of buffer, the better the sensitivity in ESI-MS. Thus, a balance must be sought

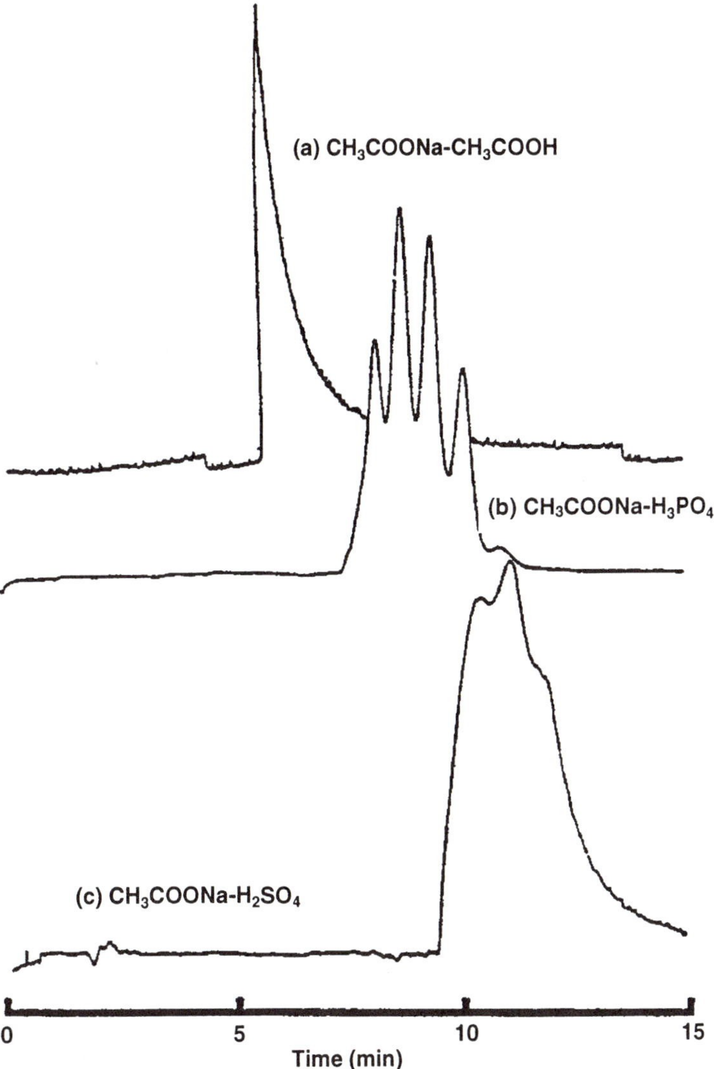

Figure 2.4 The CE of recombinant human EPO. Each experiment used a 100 mM acetate buffer neutralized to pH 4.0 with (a) acetic acid, 10 kV, 30 μA; (b) phosphoric acid, 10 kV, 120 μA; and (c) sulfuric acid, 10 kV, 200 μA. (© 1991. Reprinted from reference 90 by kind permission of Elsevier Science-NL, Sara Burgerhartstraat 25, 1055 KV Amsterdam, The Netherlands.)

between sensitivity, band shape of the analyte, and buffering capacity, and these are dictated to a large degree by buffer concentration, type, and electrical conductivity.[21,92] Buffer type includes relative volatility and organic moiety.

Buffering capacity is a very important issue. A buffer (i.e., a buffer salt solution) functions best (i.e., maximal buffering capacity) when the pH of the solution is equal to the pK_a of the buffer salt ±1.5 pH units. If organic solvent is added to the solution, then this will shift the pK_a of the buffer salt, and if the buffer salt concentration is dilute (less than 5 mM), then a reduced pH window must be considered about the pK_a value.[70] Table 2.3 provides pK_a values where various buffers exert maximal buffering capacity. A solution of sodium phosphate at pH 5.0 provides no buffering capacity, and therefore it is just a salt solution. The same solution at pH 7.0 will provide a good buffering capacity, and it is then a buffer salt solution. The sodium acetate/acetic acid pair would provide a much better buffering capacity for a solution at pH 5.0 (Table 2.3).

Ammonium acetate is a good salt to use, especially for ESI-MS, because both cation and anion are volatile in nature. Also, ammonium acetate is a good source of protons for the biological analyte. However, keep in mind that a solution of this salt at pH 7.0–7.4 will provide no buffering capacity, because the anion and cation have maximal buffering capacities at significantly different pH values (Table 2.3).

An analogy can be made between CE and RP-HPLC. The ionic strength of the liquid medium in CE provides the separation power for various analytes in a similar fashion to the organic solvent strength in RP-HPLC.[20] These properties that cause analyte separation in CE and RP-HPLC address the analyte parameters of ion charge and hydrophobicity, respectively.[20,93] For CE, when the conductivity of the analyte is greater than that of the buffer-solvent, a decrease in efficiency and resolution results.[20,94] When the conductivity of the analyte solution plug is lower than the main buffer-solvent in the capillary (i.e., higher buffer concentrations than that of the analyte), the analyte 'focuses on itself' and thus concentrates itself (sample-stacking) into a narrow band.[64,95] Figure 2.5 shows that as the sodium phosphate buffer concentration (in the capillary) increases with a concomitant increase in ionic strength, analyte concentration increases and peak resolution and sensitivity increase[20, 94, 95] using UV absorbance detection. This phenomenon causes peak narrowing and increased peak separation of analytes. At a low ionic strength (0.025 M Na_3PO_4) of the buffer in the capillary, the ionic strength of the analyte in its solution plug injection is more similar to that of the buffer. This causes a slightly larger analyte elution plug and hence a slightly lower analyte concentration at the detector. This phenomenon of sample-stacking characterizes the CZE, CITP, and CIEF processes. Figure 2.6 presents another example of this phenomenon for a mixture of peptides at 100 μg/ml.[20,96] Increasing the sample loading from a 10-s injection of analyte (Figure 2.6a) to a 20-s loading (Figure 2.6b) causes an observable decrease in band resolution. This is also noticed between Figures 2.6c and d. The greater amount of analyte causes an increase of its conductivity in the buffer in the capillary, and thus the electric field is much lower in the analyte region than in the buffer. This field strength difference causes poor peak shape and separation.[20,94,96] In addition, an increase in the concentration of sodium chloride salt has a deleterious effect on peak shape and separation (compare Figures 2.6a and b with Figures 2.6c and d).[20,66] This is in contrast to the effect shown in Figure 2.5, where an increase in concentration of the buffer salt provided a more

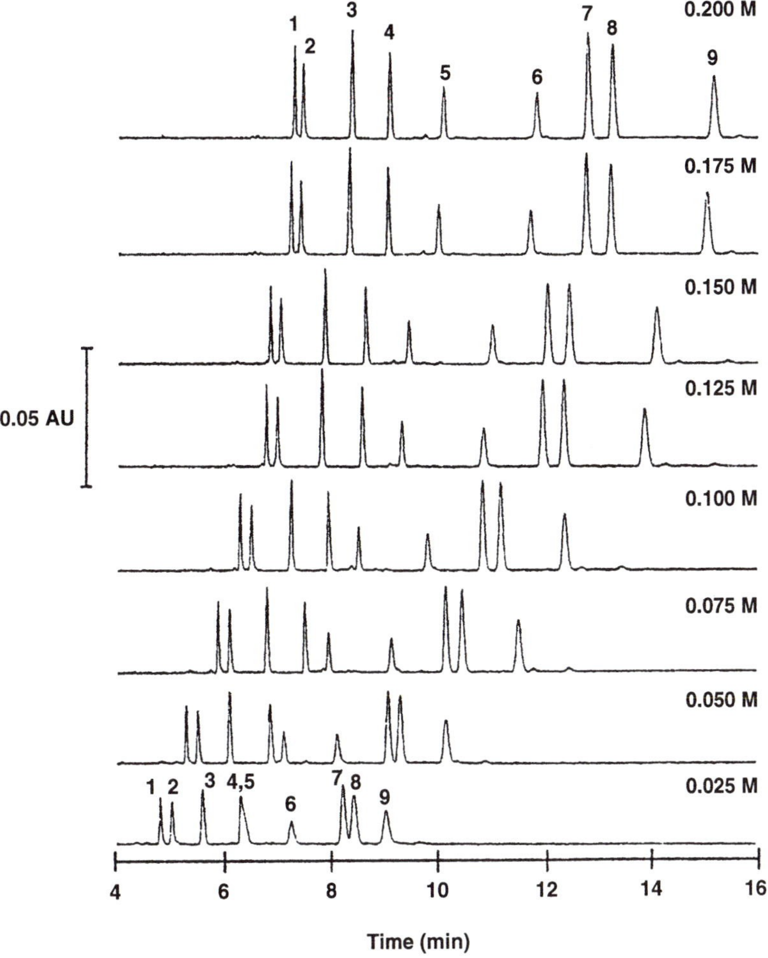

Figure 2.5 Migration of nine peptides using a 50 μm i.d. capillary in different ionic strength buffer solutions of sodium phosphate at pH 2.44. The capillary was 56.9 cm, applied voltage was 30 kV at 20°C, and detection was at 220 nm absorbance. Peptides: 1, dynorphin; 2, bradykinin; 3, angiotensin II; 4, TRH; 5, LHRH; 6, bombesin; 7, Leu-enkephalin; 8, Met-enkephalin; 9, oxytocin. AU = absorbance unit. (Permission granted by Beckman Instruments, Irvine, California.)

beneficial performance of peptide resolution and separation in the capillary electropherogram.

Another important parameter to keep in mind concerning the liquid phase is the presence of surfactants/denaturants, which are sometimes used in protein and peptide solutions. By their very nature, surfactants interact with the sample and this tends to reduce sensitivity.[17] Sodium dodecyl sulfate (SDS) is a universally used surfactant and, even though it is regularly used in gel electrophoresis,[97] it may not be a good choice to use in CE-based methods.[17] Degradation of

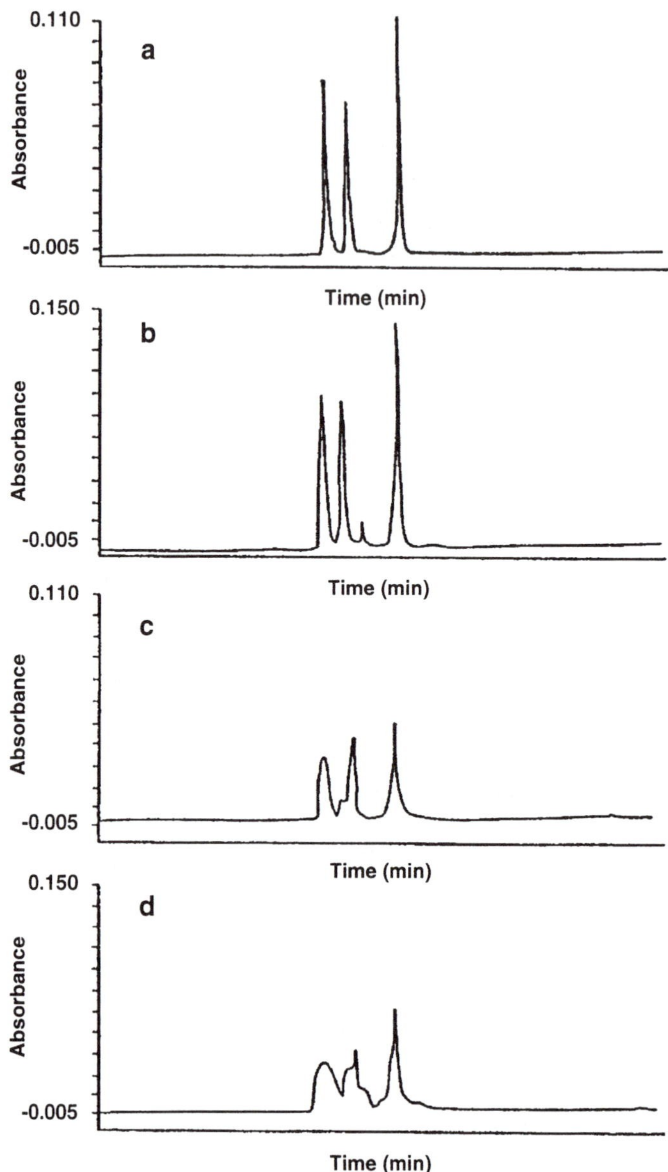

Figure 2.6 The CZE analysis of a mixture of four peptides (at 100 μg/ml) under different sodium chloride salt concentrations and time of injection of the peptide mixture: (a) 30 mM, 10 s; (b) 30 mM, 20 s; (c) 100 mM, 10 s; (d) 100 mM, 20 s. Capillary: 75 μm i.d. silica capillary, 57 cm length, separation voltage of 17 kV, detection at 200 nm absorbance, 25°C, pH 9.2. Peptides: dynorphin A, bradykinin, neurotensin, and angiotensin I. (© 1996, CRC Press, Boca Raton, Florida. Reprinted from reference 20 with permission.)

sensitivity by surfactants can take place, because many intense signals can occur in both the positive- and negative-ion modes of mass spectral analysis. Vissers et al.[39,98] removed the SDS from a tryptic protein digest with an automatic, on-line method interfaced with an RP-HPLC-MS system. The method includes a pre-concentration column and an ionic detergent trapping column. Nevertheless, under certain conditions, surfactants produce desirable properties of analyte molecules in that the association of the two species produces micelle structure. These micelles can be thought of as pseudo-stationary phases with respect to the analyte, because the micelles (a) migrate in the liquid medium under an electric field and (b) provide a partitioning phase for the analyte. This experimental setup is usually referred to as micellar electrokinetic capillary chromatography.[57,63]

The addition of small amounts of organic solvent to the CE aqueous buffer carrier can affect the analyte retention time in the capillary column. Methanol and acetonitrile are commonly used for this purpose.[19]

There are a number of reports that compare the differences of an RP-HPLC chromatogram and a CZE chromatogram with UV absorbance detection. Examples include tryptic digests (see Chapter 9 for details) of recombinant (r) human interleukin-6 (IL-6) (Figure 4 in reference 20), r-human growth hormone (Figure 2 in reference 61), human β-globin (Figure 4 in reference 61), and ESI-MS detection of a mixture of protein standards.[56]

An interesting application of CIEF-ESI-MS consists of measuring the two major proteins of human red blood cells (erythrocytes). By using a microscope to retrieve only 5–10 intact erythrocytes, both the α- and β-chains of hemoglobin could be observed with CE-ESI-MS.[22,99–101] The CIEF-ESI-MS technique was used to visualize the separate, intact hemoglobin and carbonic anhydrase

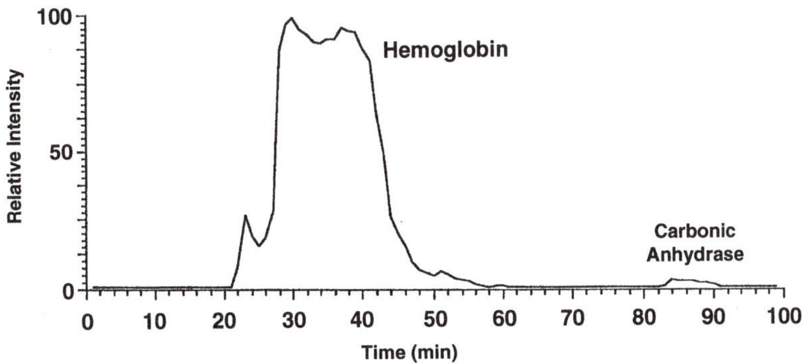

Figure 2.7 The CIEF-ESI total ion current, as detected by mass spectrometry, of a 700-nl injection of 2% human whole blood lysate. The CIEF gradient was from a mixture of a 1% aqueous solution of pH 5–7 ampholytes. Conditions: 1.1-m polyacrylamide-coated capillary, analyte = 20 mM HOAc, catholyte = 20 mM NH_4OH, ESI sheath flow = 2 µl/min with 75% MeOH/25% 10 mM HOAc. (© 1997. Reproduced from reference 22 by permission of Wiley-VCH Publishers, Inc., Weinheim, Germany.)

proteins[101,102] with a 1% ampholyte load in the liquid. Figure 2.7 shows the response of two proteins from only one red blood cell. The separation was effected by taking advantage of the different isoelectric points (pI) of 6.9 and 6.63 for hemoglobin and carbonic anhydrase, respectively. These two proteins can be separated and observed by ESI-MS, despite (a) the approximately 70-fold greater amount of hemoglobin (450 attomoles) with respect to carbonic anhydrase (7 attomoles) and (b) the more than 100 proteins normally found in a typical red blood cell. The ESI-mass spectra could be obtained for both proteins.[99–101]

REFERENCES

1. *The API Book*; Perkin-Elmer SCIEX Instruments: Canada, 1994.
2. Metzger, J.W.; Eckerskorn, C. In *Microcharacterization of Proteins,* Kellner, R.; Lottspeich, F.; Meyer, H.E., Eds.; VCH Publishers: New York, 1994; Volume 2, pp 167–187.
3. Griffin, P.R.; Coffman, J.A.; Hood, L.E.; Yates, J.R., III *Int. J. Mass Spectrom. Ion Processes* **1991,** *111*, pp 131–149.
4. Covey, T.R. *The Realities and Misconceptions of Electrospray Ionization and HPLC Flow Rates;* PE-SCIEX Ionspray Application Note; Perkin-Elmer SCIEX Instruments: Canada, **1993**.
5. Covey, T.R. In *Biological and Biotechnological Applications of Electrospray Ionization Mass Spectrometry;* Snyder, A.P., Ed.; American Chemical Society: Washington, DC, 1996; Chapter 2, pp 21–59.
6. Mant, C.T.; Hodges, R.S. In *High Performance Liquid Chromatography of Peptides and Proteins: Separation, Analysis, and Conformation;* Mant, C.T.; Hodges, R.S., Eds.; CRC Press: Boca Raton, FL, 1991, pp 289–295.
7. Smith, R.M., Ed. *Gas and Liquid Chromatography in Analytical Chemistry;* Smith, R.M., Ed; John Wiley & Sons: Chichester, U.K., 1988; Chapter 12, pp 263–310.
8. Tomer, K.B.; Moseley, M.A.; Deterding, L.J.; Parker, C.E. *Mass Spectrom. Rev.* **1994,** *13,* pp 431–457.
9. Bruins, A.P. *Trends Anal. Chem.* **1994,** *13,* pp 81–90.
10. Edmonds, C. G.; Loo, J. A.; Barinaga, C. J.; Udseth, H. R.; Smith, R. D. *J. Chromatogr.* **1989,** *474,* pp 21–37.
11. Smith, R.D.; Loo, J.A.; Barinaga, C.J.; Edmonds, C.G.; Udseth, H.R. *J. Chromatogr.* **1989,** *480,* pp 211–232.
12. Loo, J.A.; Udseth, H.R.; Smith, R.D. *Anal. Biochem.* **1989,** *179*, pp 404–412.
13. Smith, R.D.; Loo, J.A.; Edmonds, C.G. In *Mass Spectrometry: Clinical and Biomedical Applications;* Desiderio, D.M., Ed.; Plenum Press: New York, 1992; Volume Chapter 2, pp 37–98.
14. Hsieh, F.Y.L.; Cai, J.; Henion, J. *J. Chromatogr.* 1994, *679,* pp 206–211.
15. Johansson, I.M.; Huang, E.C.; Henion, J.D.; Zweigenbaum, J. *J. Chromatogr.* **1991,** *554,* pp 311–327.
16. Smith, R.D.; Loo, J.A.; Loo, R.R.O.; Busman, M.; Udseth, H.R. *Mass Spectrom. Rev.* **1991,** *10*, pp 359–451.
17. Smith, R.D.; Udseth, H.R. In *Capillary Electrophoresis Technology;* Guzman, N.A., Ed; Marcel Dekker: New York, 1993; Chapter 16, pp 525–567.
18. Camilleri, P. In *Capillary Electrophoresis, Theory and Practice;* Camilleri, P., Ed; CRC Press: Boca Raton, FL, 1993; Chapter 1, pp 1–23.

19. Pleasance, S.; Thibault, P. In *Capillary Electrophoresis, Theory and Practice;* Camilleri, P., Ed; CRC Press: Boca Raton, FL, 1993; Chapter 8, pp 311–369.
20. Rickard, E.C.; Towns, J.K. In *New Methods in Peptide Mapping for the Characterization of Proteins;* Hancock, W.S., Ed; CRC Press: Boca Raton, FL, 1996, Chapter 4, pp 97–117.
21. Moring, S.E. In *Capillary Electrophoresis in Analytical Biotechnology;* Righetti, P.G., Ed; CRC Press: Boca Raton, FL, 1996; Chapter 2, pp 37–58.
22. Severs, J.C.; Smith, R.D. In *Handbook of Capillary Electrophoresis,* 2nd Edition; Landers, J.P., Ed; CRC Press: Boca Raton, FL, 1997; Chapter 28, pp 791–826.
23. Banks, J.F. *J. Chromatogr., A* 1995, *712,* pp 245–252.
24. Krivankova, L.; Bocek, P. *J. Chromatogr., B* **1997,** *689,* pp 13–34.
25. Gelpi, E. *J. Chromatogr., A* **1995,** *703,* pp 59–80.
26. Allen, M.H.; Vestal, M.L. *J. Am. Soc. Mass Spectrom.* **1992,** *3,* pp 18–26.
27. Meng, C.K.; Mann, M.; Fenn, J.B. *Z. Phys. D: At., Mol. Clusters* **1988,** *10,* pp 361–368.
28. Smith, R. D.; Loo, J. A.; Edmonds, C. G.; Barinaga, C. J.; Udseth, H.R. *Anal. Chem.* **1990,** *62,* pp 882–899.
29. Eshraghi, J.; Chowdhury, S.K. *Anal. Chem.* **1993,** *65,* pp 3528–3533.
30. Mann, M. *Org. Mass Spectrom.* 1990, *25,* pp 575–587.
31. Banks, J.F. Jr.; Whitehouse, C.M. In *Methods in Enzymology;* Karger, B.L.; Hancock, W.S., Eds; Academic Press: San Diego, CA, 1996; Volume 270, Part A, Chapter 21, pp 486–519.
32. Edmonds, C.G.; Smith, R.D. In *Methods in Enzymology: Mass Spectrometry;* McCloskey, J. A., Ed.; Academic Press: San Diego, CA, 1990; Volume 193, Chapter 22, pp 412–431.
33. Smith, R.D.; Loo, J.A.; Edmonds, C.G.; Barinaga, C.J.; Udseth, H.R. *J. Chromatogr.* **1990**, *516,* pp 157–165.
34. Fairwell, T.; Fales, H.M.; Dutky, R.C.; Sheeley, D.M. *Proc. 42nd ASMS Conference on Mass Spectrometry and Allied Topics,* Chicago, IL, **1994,** p 640.
35. Mirza, U.A.; Chait, B.T. *Anal. Chem.* 1994, *66,* pp 2898–2904.
36. Emmett, M.R.; Caprioli, R.M. *J. Am. Soc. Mass Spectrom.* **1994,** *5,* pp 605–613.
37. Chowdhury, S.K.; Katta, V.; Chait, B.T. *Rapid Commun. Mass Spectrom.* **1990,** *4,* pp 81–87.
38. Ogorzalek Loo, R.R.; Dales, N.; Andrews, P.C. *Protein Sci.,* **1994,** *3,* pp 1975–1983.
39. Vissers, J.P.C.; Chervet, J.P.; Salzmann, J.P. *J. Mass Spectrom.* 1996, *31,* pp 1021–1027.
40. Scott, R.P.W. In *High Performance Liquid Chromatography: Principles and Methods in Biotechnology;* Katz, E.D., Ed; John Wiley & Sons: Chichester, U.K., 1996; Chapter 2, pp 25–94.
41. Eksteen, R. In *High Performance Liquid Chromatography: Principles and Methods in Biotechnology;* Katz, E.D., Ed.; John Wiley & Sons: Chichester, U.K., 1996; Chapter 3, pp 95–162.
42. Larson, J.R.; Tingstad, J.E.; Swadesh, J.K. In *HPLC: Practical and Industrial Applications;* Swadesh, J., Ed.; CRC Press: Boca Raton, FL, 1996; Chapter 1, pp 1–55.
43. Rush, R.S.; Derby, P.L.; Strickland, T.W.; Rohde, M.F. *Anal. Chem.* **1993,** *65,* pp 1834–1842.
44. Meyer, V.R., Ed. *Practical High–Performance Liquid Chromatography;,* 2nd Edition; John Wiley & Sons, Chichester, U.K., 1994, Chapter 9, pp 123–143.

45. Meyer, V.R., Ed. *Practical High–Performance Liquid Chromatography,* 2nd Edition; John Wiley & Sons, Chichester, U.K., 1994; Chapter 10, pp 144–157.
46. Mahoney, W.C.; Hermodson, M. A. *J. Biol. Chem.* **1980,** *255,* pp 11199–11203.
47. Hermodson, M.; Mahoney, W.C. In *Methods in Enzymology;* Hirs, C.H.W., Ed.; Academic Press: New York, 1983; Volume 91, Part I, Chapter 30, pp 352–359.
48. Aguilar, M.I.; Hearn, M.T.W. In *Methods in Enzymology;* Karger, B.L.; Hancock, W.S., Eds.; Academic Press: San Diego, CA, 1996, Volume 270, Part A, Chapter 1, pp 3–26.
49. Guzzetta, A.W.; Hancock, W.S. In *New Methods in Peptide Mapping for the Characterization of Proteins;* Hancock, W.S., Ed.; CRC Press: Boca Raton, FL, 1996, Chapter 7, pp 181–217.
50. Bennett, H.P.J. In *High Performance Liquid Chromatography of Peptides and Proteins: Separation, Analysis, and Conformation;* Mant, C.T.; Hodges, R.S., Eds.; CRC Press: Boca Raton, FL, 1991; pp 319–325.
51. Guo, D.; Mant, C.T.; Taneja, A.K.; Parker, J.M.R.; Hodges, R.S. *J. Chromatogr.* **1986,** *359,* pp 499–517.
52. Niessen, W.M.A.; Tinke, A.P. *J. Chromatogr.* **1995,** *703,* pp 37–57.
53. Melander, W.R.; Corradini, D.; Horvath, Cs. *J. Chromatogr.* **1984,** *317,* pp 67–85.
54. Guo, D.; Mant, C.T.; Taneja, A.K.; Hodges, R.S. *J. Chromatogr.* **1986,** *359,* pp 519–532.
55. Garfin, D.E. In *Introduction to Biophysical Methods for Protein and Nucleic Acid Research;* Glasel, J.A.; Deutscher, M.P., Eds.; Academic Press: San Diego, CA, 1995; Chapter 2, pp 53–109.
56. Tang, Q.; Harrata, K.; Lee, C.S. *J. Mass Spectrom.* **1996,** *31,* pp 1284–1290.
57. Pritchett, T.; Robey, F.A. In *Handbook of Capillary Electrophoresis,* 2nd Edition; Landers, J.P., Ed.; CRC Press: Boca Raton, FL, 1997; Chapter 9, pp 259–295.
58. Benedek, K.; Guttman, A. In *HPLC: Practical and Industrial Applications;* Swadesh, J., Ed.; CRC Press: Boca Raton, FL, 1996; Chapter 7, pp 305–346.
59. Baker, D.R., Ed. *Capillary Electrophoresis;* John Wiley & Sons: New York, NY, 1995; Chapter 3, pp 53–93.
60. Wahl, J.H.; Udseth, H.R.; Smith, R.D. In *New Methods in Peptide Mapping for the Characterization of Proteins;* Hancock, W.S., Ed.; CRC Press: Boca Raton, FL, 1996; Chapter 6, pp 143–179.
61. Rickard, E.C.; Towns, J.K. In *Methods in Enzymology;* Karger, B.L.; Hancock, W.S., Eds.; Academic Press: San Diego, CA, 1996, Volume 271, Part B; Chapter 11, pp 237–264.
62. Baker, D.R., Ed. *Capillary Electrophoresis;* John Wiley & Sons: New York, 1995; Chapter 2, pp 19–52.
63. van de Goor, T.; Apffel, A.; Chakel, J.; Hancock, W. In *Handbook of Capillary Electrophoresis,* 2nd Edition; Landers, J.P., Ed.; CRC Press: Boca Raton, FL, 1997; Chapter 8, pp 214–258.
64. Thompson, T.J.; Foret, F.; Vouros, P.; Karger, B.L. *Anal. Chem.* **1993,** *65,* pp 900–906.
65. Zumdahl, S.S., Ed. *Chemistry,* 3rd Edition; D.C. Heath and Co.: Lexington, MA, 1993; Chapter 14, pp 637–695.
66. Rodriguez–Diaz, R.; Wehr, T.; Zhu, M.; Levi, V. In *Handbook of Capillary Electrophoresis;* 2nd Edition; Landers, J.P., Ed.; CRC Press: Boca Raton, FL, 1997; Chapter 4, pp 101–138.
67. Weinberger, R. *Am. Lab.* **1996,** *28,* pp 28T–28U.
68. Tang, W.; Kamel–Harrata, A.; Lee, L.S. *Anal. Chem.* **1995,** *67,* pp 3515–3519.

69. Mazzeo, J.R.; Krull, I.S. *Anal. Chem.* **1991,** *63,* pp 2852–2857.
70. Neue, U.D. *Am. Lab.* **1997,** *29,* pp 33H–33J.
71. Good, N.E.; Winget, G.D.; Winter, W.; Connolly, T.N.; Izawa, S.; Singh, R.M.M. *Biochemistry* **1966**, *5,* pp 467–477.
72. Ferguson, W.J.; Braunschweiger, K.I.; Braunschweiger, W.R.; Smith, J.R.; McCormick, J.J.; Wasmann, C.C.; Jarvis, N.P.; Bell, D.H.; Good, N.E. *Anal. Biochem.* **1980,** *104,* pp 300–310.
73. Stoll, V.S.; Blanchard, J.S. In *Methods in Enzymology;* Deutscher, M.P., Ed.; Academic Press: New York, 1990, Volume 182; Chapter 4, pp 24–38.
74. Blanchard, J.S. In *Methods in Enzymology;* Jakoby, W.B., Ed.; Academic Press: New York, 1984; Volume 104, Part C; Chapter 26, pp 404–414.
75. Good, N.E.; Izawa, S. In *Methods in Enzymology;* San Pietro, A., Ed.; Academic Press: New York, 1972; Volume 29, Part B; Chapter 3, pp 53–68.
76. Weinberger, R. *Am. Lab.* **1997,** *29,* pp 60–62.
77. Messana, I.; Rossetti, D.V.; Cassiano, L.; Misiti, F.; Giardina, B.; Castagnola, M. *J. Chromatogr., B* **1997,** *699,* pp 149–171.
78. Rappoport, Z., Ed.; *Handbook of Tables for Organic Compound Identification;* CRC Press: Boca Raton, FL, 1987.
79. Serjeant, E.P.; Dempsey, B., Eds.; *Ionisation Constants of Organic Acids in Aqueous Solution;* Pergamon Press: Oxford, U.K., 1979.
80. Lide, D.R., Ed.; *CRC Handbook of Chemistry and Physics,* 74th Edition; CRC Press: Boca Raton, FL, 1993–1994.
81. Camilleri, P., Ed; *Capillary Electrophoresis, Theory and Practice;* CRC Press: Boca Raton, FL, 1993; Appendix II, pp 413–443.
82. Cordova, E.; Gao, J.; Whitesides, G.M. *Anal. Chem.* **1997,** *69,* pp 1370–1379.
83. Weinberger, R. *Am. Lab.* **1997,** *29,* p 220.
84. Baker, D.R., Ed.; *Capillary Electrophoresis;* John Wiley & Sons: New York, 1995; Chapter 7, pp 211–236.
85. Parks, G.A. *Chem. Rev.* 1965, *65,* pp 177–198.
86. Towns, J.K.; Regnier, F.E. *J. Chromatogr.* **1990,** *516,* pp 69–78.
87. Morand, M.; Blaas, D.; Kenndler, E. *J. Chromatogr., B* **1997,** 691, pp 192–196.
88. Niessen, W.M.A.; Tjaden, U.R.; van der Greef, J. *J. Chromatogr.* **1993,** *636,* pp 3–19.
89. McCormick, R.M. *Anal. Chem.* **1988,** *60,* pp 2322–2328.
90. Tran, A.D.; Park, S.; Lisi, P.J.; Huynh, O.T.; Ryall, R.R.; Lane, P.A. *J. Chromatogr.* **1991,** *542,* pp 459–471.
91. McNerney, T.M.; Watson, S.K.; Sim, J.-H.; Bridenbaugh, R.L. *J. Chromatogr., A* **1996,** *744,* pp 223–229.
92. Lee, H.G.; Desiderio, D.M. *J. Chromatogr., B* **1997,** *691,* pp 67–75.
93. Apffel, A.; Chakel, J.; Udiavar, S.; Hancock, W.S.; Souders, C.; Pungor, E., Jr. In *Biochemical and Biotechnological Applications of Electrospray Ionization Mass Spectrometry;* Snyder, A.P., Ed.; American Chemical Society: Washington, DC, 1996; Chapter 23, pp 432–471.
94. McLaughlin, G.; Biehler, R.; Anderson, K.; Schwartz, H. E. *Capillary Dimensions with P/ACE 2000 Series Instruments: 50 μm vs. 75-μm–i.d. Capillaries;* Beckman Technical Information Bulletin TIBC-106; Beckman Instruments; Palo Alto, CA, 1991.
95. Baker, D.R., Ed. *Capillary Electrophoresis;* John Wiley & Sons: New York, 1995; Chapter 4, pp 94–158.

96. Satow, T.; Machida, A.; Funakushi, K.; Palmieri, R. *J. High Resolut. Chromatogr.* **1991,** *14,* pp 276–279.

97. Walker, J. M., Ed.; *Methods in Molecular Biology: Basic Protein and Peptide Protocols;* Humana Press: Totowa, NJ, 1994; Vol. 32.

98. Vissers, J.P.C.; Hulst, W.P.; Chervet, J.P.; Snijders, H.M.J.; Cramers, C.A. *J. Chromatogr., B* **1996,** *686,* pp 119–128.

99. Hofstadler, S.A.; Swanek, F.D.; Gale, D.C.; Ewing, A.G.; Smith, R.D. *Anal. Chem.* 1995, *67,* pp 1477–1480.

100. Hofstadler, S.A.; Severs, J.C.; Smith, R.D.; Swanek, F.D.; Ewing, A.G. *Rapid. Commun. Mass Spectrom.* 1996, *10,* pp 919–922.

101. Hofstadler, S.A.; Severs, J.; Swanek, F.D.; Ewing, A.G.; Smith, R.D. *Proc. 44th ASMS Conference on Mass Spectrometry and Allied Topics,* Portland, OR, **1996,** p 1287.

102. Severs, J.C.; Hofstadler, S.A.; Zhao, Z.; Senh, R.T.; Smith, R.D. *Electrophoresis* **1996,** *17,* pp 1–10.

3

The Isotope Effect at High Mass

When is the mass of a peak not a real mass? This question can be considered to be misleading,[1] because most atomic elements can be found as more than one mass in nature. As elements are combined into compounds, the compounds themselves can be found at different masses. Generally, as the number of elements increase in the formation of a compound, more discrete masses of the same compound are created. Is it possible that these different masses of one compound are of negligible intensity except for one intense mass? Except for very-low-mass compounds, this is not the case.[2] In a series of compounds spanning a few hundred mass units (daltons, Da) to 1000–1500 Da, a second and even a third mass peak are observed at intensities well above 10% of the most intense mass (base peak). These additional masses are found at higher mass values with respect to the base peak of a particular compound. In this same series of compounds, when the mass approaches and exceeds 1000 Da, these additional masses become significant in intensity to the point that they approach and exceed the intensity of the lowest mass value.

In a series of molecules that span the mass range of 1000 Da to as high as 100,000,000 Da[3] and beyond, a number of mass spectral phenomena take place. As one proceeds from a lower mass compound to a higher mass compound, the nominal mass and lower mass isotopes successively decrease in relative abundance or intensity, and the higher mass isotopes increase in relative intensity. These higher mass isotopes are many tens to hundreds of daltons removed with respect to the lower mass isotopes of the particular compound, and these lower mass isotopes are of negligible intensity.

All mass peaks, including the lowest (monoisotopic) mass peak of a compound or biological molecule, are called isotopes. In contrast to the many elements that have more than one naturally occurring mass, there are 20 elements that exist as non-man-made (i.e., found naturally on Earth), and each can be characterized as having a discrete mass. For example, every atom of cesium, iodine, or fluorine has a discrete mass of 132.905, 126.9044, and 18.9984 Da, respectively.

However, to further complicate matters, each observed isotope peak (except for the nominal and monoisotopic masses) consists of many additional isotope peaks that span an even smaller mass range. This chapter presents a qualitative and quantitative treatment of the dissection of this molecular envelope of masses. This chapter can also be considered as material to provide an appreciation of the complicated nature of a high-mass ESI-generated mass spectral peak. A mass spectral treatment of atomic isotopes can be found in *Interpretation of Mass Spectra*,[2] where compounds of relatively low mass (under 1000 Da) are presented.

Biological molecules are composed of combinations of the following elements: carbon (C), hydrogen (H), nitrogen (N), oxygen (O), sulfur (S), and phosphorus (P). A number of biological molecules also bind metal ions, and these include proteins and DNA. If the atomic masses of all the isotopes of C, H, N, O, S, and P were integers with only zeroes to the right of the decimal point, mass spectral interpretation would be simplified to a limited extent. However, if these elements were found only as respective, single-mass species (i.e., only one exact mass per element, characterized as either an integer or rational number), mass spectral interpretation would be greatly simplified. Reality dictates a more comprehensive investigation of the mass constitution of an ionized molecule in a mass spectrum. The operator of a mass spectrometer can take some comfort in the fact that a particular mass or isotope of an atom usually dominates to a significant extent in relative abundance. That is, only a certain exact mass (a particular isotope) of an element is distributed in nature in greater amounts or percentages than its other isotopes. Thus, the probability of observing two or more of the same high-abundance isotope of an atom in a molecule is very high. The following provides a quantitative framework for the concepts of isotope distribution in biological molecules.

Table 3.1 presents a tabulation of the isotopes of the six most common atoms found in biological molecules. Note that in the "*A*" or first section of the table, the mass of highest percentage (abundance) of each atom is listed. Therefore, from a randomly chosen set of 100 atoms of carbon, 98.9% (or 98.9 atoms) will have a mass of 12.0000 Da. Similarly, for 100 atoms of hydrogen, 99.985% (or 99.985 atoms) will have a mass of 1.007825 Da, etc.

Since fractions of atoms do not exist, this concept can be restated in a more practical manner. For every 100,000 atoms of carbon, 98.9% (or 98,900 atoms) will have a mass of 12.0000 Da, and for 100,000 atoms of hydrogen, 99.985% (or 99,985 atoms) will have a mass of 1.007825 Da, etc.

There is only one isotope of phosphorus, and every phosphorus atom has the mass 30.973762 Da. Note, in particular, that the nominal mass of phosphorus (31 Da) is higher than its monoisotopic mass value, and a similar situation is found with oxygen (nominal mass of 16 Da). Each mass in the first section of Table 3.1 is very close to an integer value, and, by convention, the most

Table 3.1 Table of isotopes[a]

		Mass	Mass defect	Abundance (%)
A (monoisotopic)[b]	^{12}C	12.0000	0.0	98.90
	^{1}H	1.007825	0.007825	99.986
	^{14}N	14.003074	0.003074	99.63
	^{16}O	15.994915	–0.005085	99.762
	^{32}S	31.972071	–0.027929	95.02
	^{31}P	30.973762	–0.026238	100.00
$A+1$	^{13}C	13.003354	0.003354	1.10
	^{2}H	2.014102	0.014102	0.015
	^{15}N	15.000109	0.000109	0.37
	^{17}O	16.999131	–0.000869	0.038
	^{33}S	32.971457	–0.028543	0.75
$A+2$	^{18}O	17.999160	–0.00084	0.200
	^{34}S	33.967867	–0.032133	4.21
$A+4$	^{36}S	35.967081	–0.032919	0.02
Weighted average	C	12.011	—	—
	H	1.00797	—	—
	N	14.0067	—	—
	O	15.9994	—	—
	S	32.066	—	—
	P	30.973762	—	100.00

[a]From references 4–9.
[b]Nomenclature from McLafferty and Tureček.[2] A, mass number of an atom.

abundant isotope of carbon was chosen to have an integer value. Hydrogen has a mass that is +0.007825 Da above 1.0 Da; nitrogen has a mass that is 0.003074 Da above 14.0 Da. Oxygen has a mass that is 0.005085 Da below 16.0 Da; sulfur has a mass that is 0.027929 Da below 32.0 Da. Phosphorus has a mass that is 0.0262378 Da below 31.0 Da. These differences with respect to the nominal or integer mass value are called mass defects, and are characterized as positive or negative mass defects.

In the ($A+1$) section of Table 3.1, select isotopes of the C, H, N, O, and S atoms are grouped, because they are approximately 1 Da in mass above their respective most abundant "A" atom. They also have positive and negative mass defect values in comparison with the respective whole number value of the ($A+1$) mass. However, they have relatively low abundance, such that for every 1000 carbon atoms, only 1.1% (11 atoms) have a mass of 13.003354 Da, etc. In the ($A+2$) section, only oxygen and sulfur have a (small) percentage of atoms that are approximately 2 Da above the respective most abundant "A" atom. Thus, C, H, and N can be found in two different masses of unequal proportions, while oxygen can be found in three different masses of unequal proportions. Sulfur atoms can be found as four different masses with different abundances (unequal proportions).

The last section of Table 3.1 provides a weighted average of each atom. A weighted average is calculated by multiplying each isotope mass of an atom by its respective abundance and adding the resulting values. For carbon, we have 12.0000(0.989) + 13.003354(0.0110) = 11.868 + 0.143037 = 12.011 Da; for hydrogen, the average mass is 1.007825(0.99985) + 2.014102(0.00015) = 1.00767 + 0.000302 = 1.00797 Da, etc.

Among all of the $(A + 1)$, $(A + 2)$, and $(A + 4)$ isotope species, ^{34}S is the most abundant isotope; however, sulfur is not an atom that occurs to a significant extent for a given biological molecule. Carbon is found in every organic molecule, and is repeated to a significant extent in biological molecules. Carbon has the second highest occurrence (abundance), of 1.1%, among the $(A + 1)$, $(A + 2)$, and $(A + 4)$ isotopes in Table 3.1. For a biological molecule, hydrogen can usually be found in greater number than any other element. The ^{1}H isotope is very high in abundance, and the ^{2}H isotope occurs only 15 times for every 100,000 atoms of hydrogen. To equal the abundance of one ^{13}C atom, 73.3 atoms of ^{2}H are required; that is, $73.3 \times 0.015 = 1.10$, or 733 atoms of ^{2}H are required to equal the abundance of 10 ^{13}C atoms. Seventy-three ^{2}H atoms are not found in biological molecules, because the probability of having that many ^{2}H atoms in a single large biological molecule is statistically negligible (Appendix 2). Because of its frequency and relatively high probability of occurrence, carbon has the largest statistical contribution with respect to intensity or relative abundance of a mass spectral peak within the $(A + 1)$ to $(A + 4)$ isotopes in biological molecules. Detailed descriptions of this concept are provided below.

The closest integer value of the most abundant mass of an atom is the nominal or integer atomic mass: C, 12; H, 1; N, 14; O, 16; S, 32. The A grouping of atomic masses in Table 3.1 are called monoisotopic masses, and the weighted averages are the average atomic mass values. Note that the nominal mass, 12, and monoisotopic mass, 12.00, of carbon are the same. This concept is also similar for a molecule, because its mass can be expressed as a nominal, monoisotopic, and average mass. The literature sometimes refers to the monoisotopic peak as the exact mass, the molecular mass, or the principal peak of a compound.[10, 11] The average (molecular or chemical) mass of a compound is sometimes referred to as the chemical or relative molecular mass[11,12] and is denoted by the symbol M_r.[13,14]

Despite the above, the reader may declare that there is only one mass for one molecule. As Table 3.1 shows, elements, which are the very substance of molecules, can exist as multiple masses or isotopes. Thus, molecules can also exist as multiple masses. Textbooks on mass spectrometry clearly outline the utility of the isotope species, since that is an important piece of information in the elucidation of the structure of an organic molecule. Isotope species are especially important in this regard for molecules in the low to mid-hundreds of mass units. Textbooks[2,15,16] underscore this point for electron ionization mass spectra, because the isotope species are very useful in ascertaining the type and frequency of select elements that constitute the molecule of interest.

Tables 3.2–3.6 provide the contribution by mass of the constituent elements in selected biological molecules. The mass is represented by three categories of masses that can be used to characterize molecules, and the elemental masses used

Table 3.2 Element contributions to the mass values of the tyrosyl tyrosyl tyrosine tripeptide

Element	Number of atoms	Mass (Da) Nominal	Mono-isotopic	Mass defect	%	Average	Isotope shift Da	%
C	27	324	324.000	0.0	0.0	324.297	0.297	86.34
H	29	29	29.227	0.227	83.5	29.231	0.004	1.16
N	3	42	42.009	0.009	3.3	42.020	0.011	3.20
O	7	112	111.964	0.036	13.2	111.996	0.032	9.30
S	0	—	—	—	—	—	—	—
Total	—	507	507.20	0.272	100.0	507.544	0.344	100.00

H/C number ratio = 1.07.

Table 3.3 Element contributions to the mass of the gramicidin S molecule

Element	Number of atoms	Mass (Da) Nominal	Mono-isotopic	Mass defect	%	Average	Isotope shift Da	%
C	60	720	720.000	0.0	0.0	720.66	0.660	86.73
H	92	92	92.720	0.720	89.11	92.733	0.013	1.71
N	12	168	168.037	0.037	4.6	168.080	0.043	5.65
O	10	160	159.949	0.051	6.31	159.994	0.045	5.91
S	0	—		—	—	—	—	—
Total	—	1140	1140.706	0.808	100.02	1141.467	0.761	100.00

H/C number ratio = 1.53.

Table 3.4 Element contributions to the mass of the porcine insulin molecule

Element	Number of atoms	Mass (Da) Nominal	Mono-isotopic	Mass defect	%	Average	Isotope shift Da	%
C	256	3072	3072.000	0.0	0.0	3074.816	2.816	70.21
H	381	381	383.981	2.981	79.81	384.037	0.056	1.40
N	65	910	910.200	0.20	5.35	910.435	0.235	5.86
O	76	1216	1215.614	0.386	10.33	1215.954	0.340	8.48
S	6	192	191.832	0.168	4.5	192.396	0.564	14.06
Total	—	5771	5773.627	3.735	99.99	5777.638	4.011	100.01

H/C number ratio = 1.49.

Table 3.5 Element contributions to the mass of the glucagon trimer molecule

		Mass (Da)					Isotope shift	
Element	Number of atoms	Nominal	Mono-isotopic	Mass defect	%	Average	Da	%
C	459	5,508	5,508.000	0.0	0.0	5,513.049	5.049	76.98
H	672	672	677.258	5.258	81.0	677.356	0.098	1.49
N	126	1,764	1,764.387	0.387	5.96	1,764.844	0.457	6.97
O	150	2,400	2,399.237	0.763	11.75	2399.91	0.673	10.26
S	3	96	95.916	0.084	1.29	96.188	0.282	4.30
Total	—	10,440	10,444.80	6.492	100.00	10,451.357	6.559	100.00

H/C number ratio = 1.46.

Table 3.6 Element contributions to the mass values of the porcine proinsulin molecule

		Mass (Da)					Isotope shift	
Element	Number of atoms	Nominal	Mono-isotopic	Mass defect	%	Average	Da	%
C	769	9,228	9,228.000	0.0	0.0	9,236.459	8.459	80.07
H	1,212	1,212	1,221.484	9.484	83.97	1,221.660	0.176	1.67
N	210	2,940	2,940.645	0.645	5.71	2,941.407	0.762	7.21
O	218	3,488	3,486.891	1.109	9.82	3,487.870	0.979	9.27
S	2	64	63.944	0.056	0.49	64.132	0.188	1.78
Total	—	16,932	16,940.964	11.294	99.99	16,951.528	10.564	100.00

H/C number ratio = 1.58.

are listed in Table 3.1. The mass defect in the tables refers to the difference between the nominal and monoisotopic mass of the biological compound. In general, the mass defect can refer to the difference in mass of the nominal and any (higher mass) isotope of the compound. Note that, in the tables, ^{1}H provides the dominant contribution (80–90%) to the mass defect.

The isotope shift refers to the difference between the average and monoisotopic masses. Masses of molecules in Tables 3.2–3.6 span the range of 500–16,951 Da, and carbon provides the dominant contribution in mass as well as in the isotope shift (70–86%) in the molecules. Keep in mind that all the $(A+1)$ and higher mass isotopes contribute to the "mass displacement" between the monoisotopic and average mass of a molecule. At this point, the reader should practice this concept further with the biological molecules listed in Problems 3.1–3.5.

PROBLEM 3.1 Construct a table similar to Tables 3.2–3.6 for the hemoglobin β-chain 41–59 amino acid subunit, which has a molecular formula of $C_{93}H_{135}N_{21}O_{30}S$.

PROBLEM 3.2 Construct a table similar to Tables 3.2–3.6 for the glucagon (monomer) molecule, which has a molecular formula of $C_{153}H_{224}N_{42}O_{50}S$.

PROBLEM 3.3 Construct a table similar to Tables 3.2–3.6 for the bovine insulin molecule (α- and β-chains), which has a molecular formula of $C_{254}H_{377}N_{65}O_{75}S_6$.

PROBLEM 3.4 Construct a table similar to Tables 3.2–3.6 for the porcine proinsulin molecule, which has a molecular formula of $C_{398}H_{617}N_{109}O_{123}S_6$.

PROBLEM 3.5 Construct a table similar to Tables 3.2–3.6 for the porcine proinsulin dimer molecule, which has a molecular formula of $C_{796}H_{1234}N_{218}O_{246}S_{12}$.

It is instructive to provide a comprehensive description of the role that the elemental isotopes have in the mass spectral interpretation of biological molecules with respect to mass and relative abundance (intensity). The initial literature studies which investigated this phenomenon utilized fast atom bombardment (FAB) and secondary-ion mass spectrometry (SIMS) techniques; therefore, the introduction to individual isotope peaks and isotope envelopes will make use of these studies. The elemental atomic mass designations of C, H, and N follow the order nominal < monoisotopic < average (Table 3.1). However, for oxygen, the atomic mass designations follow the order monoisotopic < average < nominal, and this is caused by the negative mass defect. For sulfur, the mass designations follow the order monoisotopic < nominal < average; this is explained by the negative mass defect of monoisotopic ^{32}S and the positive mass defect displayed by its weighted average mass (Table 3.1). For biological molecules that contain the C, H, N, O, and S elements, the mass designations for a given molecule generally follow the order nominal < monoisotopic < average.

In the series of compounds in Tables 3.2–3.6, as the mass increases, the difference between the nominal and monoisotopic masses increases. One must keep in mind that for any organic compound, the nominal mass does not really exist. There is always a mass defect value associated with the nominal mass for every atom (except carbon) and molecule (except C_n). Thus, when the nominal mass is used as a label for the mass spectral peak of a biological compound, the actual mass is the monoisotopic mass. As Tables 3.2–3.6 show, the monoisotopic mass is higher than its nominal mass counterpart. The H/C number ratio (Tables 3.2–3.6) indicates that for biological molecules greater than 1000 Da, the ratio of hydrogen to carbon atoms is approximately 1.5:1.0. The results of Tables 3.2–3.6 and Problems 3.1–3.5 are used in Figure 3.1 to provide insights into the relationship of the isotope shift with the mass of biomolecules. The nominal, monoisotopic, and average masses of a number of peptides and proteins are presented in Appendix 3.

Figure 3.1 displays a fairly close relationship between the isotope shift and the mass of a biological molecule, where, for approximately every 1500 Da in mass, the average chemical mass increases by an additional 1 Da over that of the monoisotopic mass.[17] Note that with respect to Tables 3.2–3.6 and the answers to Problems 3.1–3.5, a significant percentage (>70%) of the isotope shift increment is due to the ^{13}C isotope. This is in contrast to that of the mass defect, where the ^{2}H isotope dominates to a significant extent (>80%). Tables 3.2–3.6

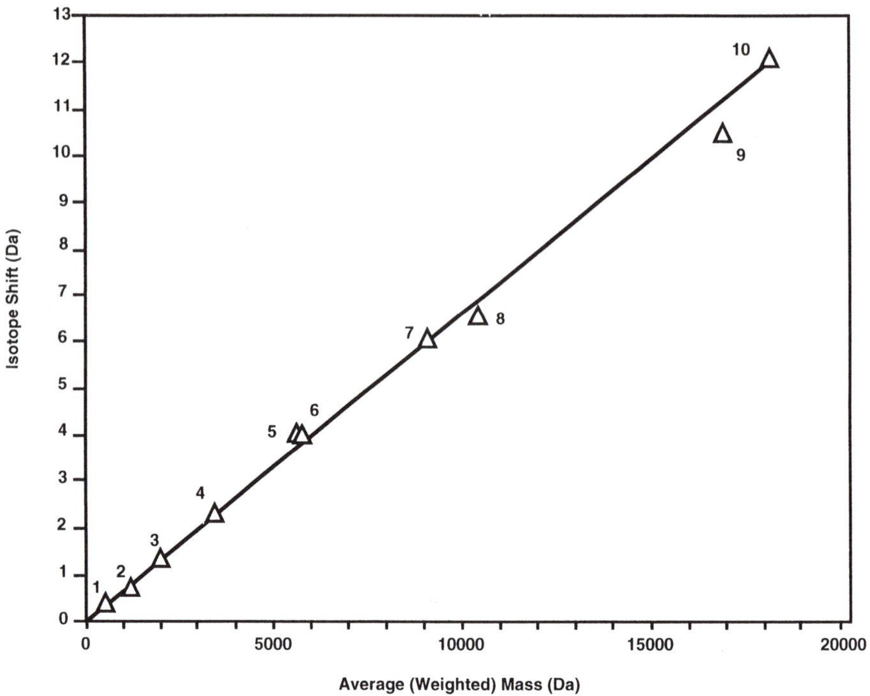

Figure 3.1 Plot of the average weighted mass of selected peptides and proteins vs the isotope shift (Da). (1) Tyr-Tyr-Tyr, (2) gramicidin S, (3) hemoglobin β–chain residues 41–59, (4) glucagon, (5) bovine insulin, (6) porcine insulin, (7) porcine proinsulin, (8) glucagon trimer, (9) equine myoglobin, and (10) porcine proinsulin dimer. (© 1991. Reprinted from reference 17 by kind permission of Elsevier Science–NL, Sara Burgerhartstraat 25, 1055 KV Amsterdam, The Netherlands.)

and Problems 3.1–3.5 provide only a partial understanding of the molecular ion region of a high-mass biological molecule, because essentially only three masses were calculated for a given molecule. No provision for intensity or magnitude was provided, and usually many more masses are present in the molecular ion region (envelope of isotope masses) of an ESI-mass spectrum.

Equations can be used to generate a theoretical or calculated mass spectrum of the molecular mass region of a biological molecule.[5,18–23] These calculations reveal important observations about the role of isotopes (the multiplicity of different mass peaks from the same molecule) and the significance of the separate isotope peaks with regard to their calculated relative abundances. The following provides a model which simulates the ESI-mass spectrum for a chosen biological molecule. In practice, minor deviations in relative intensity are sometimes found in a comparison of an experimental and a calculated mass spectrum; however, this was the subject of recent isotope distribution calculations.[18]

Electron ionization mass spectra are usually characterized by the presence of an intense ion (base peak) that represents the molecular formula of the compound of interest, and this ion is called the molecular ion[2,15,16] Invariably, smaller signals are found in a narrow mass range at higher mass from the base peak, and these lower intensity ions are almost always the isotope species of the molecular ion. In ESI-mass spectra that contain different high-mass ions, and as was stated in general terms, the higher mass isotope ions of each compound can become significant in intensity as the nominal/monoisotopic mass features become insignificant in intensity. The best way to present this concept is to calculate the relative abundances of the isotopes of the molecular region of a relatively high-mass biological molecule and to compare the results with an experimental spectrum. The calculations are purely statistical in nature, and the results track an experimentally derived spectrum fairly closely.

A comprehensive set of calculations for the relative intensities of selected isotope species for the amino acid residues 41–59 of the hemoglobin β-chain can be found in Appendix 2. After careful inspection of Appendix 2, the reader should perform Problem A2.1. Appendix 2 and Problem A2.1 present information on neutral, zero-charge mass spectral distributions of the $^{13}C_n$ isotope species of two biological molecules, where each ^{13}C designation (isotope) is used as a surrogate for a number of very-close-in-mass isotope species.

In contrast to the ^{13}C isotopes, the monoisotopic mass is unique with respect to a given compound. This mass comprises that mass alone, and it is not a collection of closely spaced isotopes. Thus, the monoisotopic mass is the most accurate experimentally available mass of a compound.[4,24,25] A comparison of theoretical and experimentally derived abundance distributions that span a relatively wide analyte mass range about the molecular species are shown in Figures 3.2–3.7. Proceeding from Figure 3.2 to Figure 3.7, the average molecular mass increases from 3483.7 Da (glucagon monomer) to 16,951.5 Da (equine myoglobin), respectively. For all five proteins, the nominal mass is not observed in either the experimental or theoretical mass spectra.[5, 26–29] As shown in the calculated and observed information in Tables 3.2–3.6 and Figures 3.2–3.6, respectively, no mass features exist between the nominal and monoisotopic mass values of a compound. The monoisotopic mass of bovine ubiquitin (m/z 8559.616) (Figure 3.5) is observed at a 20% abundance level in the theoretical, zero-charge spectrum, but is barely visible in the experimental mass spectrum at 8559.6 Da. As a rule of thumb for biological molecules, the most abundant (base) peak in an isotope envelope shifts to higher mass from the monoisotopic mass by approximately 1 Da for every 1500-Da increase in molecular mass[25,30] (Figure 3.1). Appendix 2 provides the rationale and calculations for the isotope species of a portion of the molecular ion region (envelope) of the amino acid residues 41–59 fragment of the hemoglobin β-chain.

Figure 3.6 shows the very low abundance for the monoisotopic mass in a calculated mass spectrum for the glucagon trimer, and Figures 3.7a and b present evidence that for both theoretical and experimental spectra of equine myoglobin, the monoisotopic mass has an abundance that is negligible. The magnitude of the abundance of the monoisotopic species for the glucagon trimer (Figure 3.6) and equine myoglobin (Figure 3.7) can be calculated based on the procedures detailed in Appendix 2:

Glucagon trimer

$^{12}C_{459}$	$^{1}H_{672}$	$^{14}N_{126}$	$^{16}O_{150}$	$^{32}S_{3}$
$(0.989)^{459}$	$(0.99985)^{672}$	$(0.9963)^{126}$	$(0.99762)^{150}$	$(0.9502)^{3}$
0.006238	0.9041	0.6268	0.6995	0.8579

Equine myoglobin

$^{12}C_{769}$	$^{1}H_{1212}$	$^{14}N_{210}$	$^{16}O_{218}$	$^{32}S_{2}$
$(0.989)^{769}$	$(0.99985)^{1212}$	$(0.9963)^{210}$	$(0.99762)^{218}$	$(0.9502)^{2}$
0.0002023	0.8337	0.4591	0.5948	0.9029

The product of each of the terms listed above yields probabilities or absolute abundances for the monoisotopic mass of 0.0011 and 4.16×10^{-5} for the glucagon trimer and equine myoglobin, respectively. These probabilities must be adjusted with respect to the base isotope absolute peak intensity in order to yield

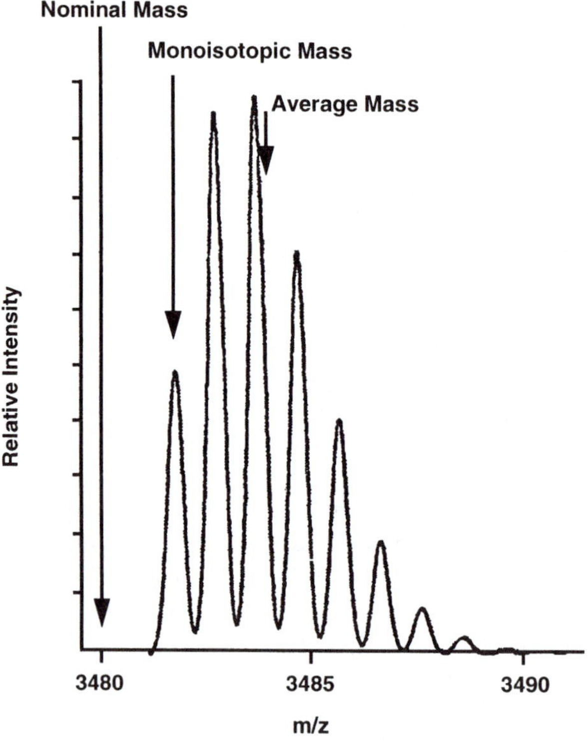

Figure 3.2 Theoretically generated mass spectral isotope distribution of the molecular ion region of glucagon ($M_r = 3483.8$ Da). (© 1983. Reprinted from reference 26 by kind permission of Elsevier Science-NL, Sara Burgerhartstraat 25, 1055 KV Amsterdam, The Netherlands.)

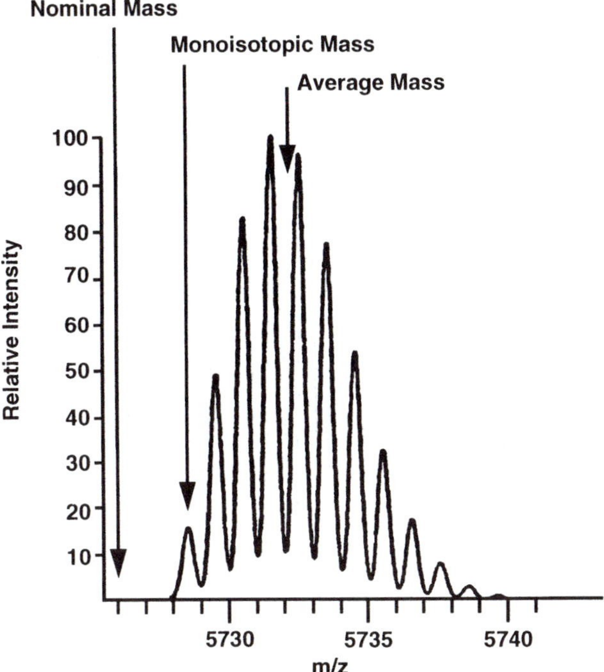

Figure 3.3 Theoretically generated mass spectral isotope distribution of the molecular ion region of bovine insulin (α- and β-chains) ($M_r = 5733.6$ Da. (© 1983. Reprinted from reference 5 by kind permission of Elsevier Science–NL, Sara Burgerhartstraat 25, 1055 KV Amsterdam, The Netherlands.)

a zero-charge mass spectrum with 0–100% relative intensities (Appendix 2). With the above data and the information in Appendix 2, the ^{13}C-containing isotopes can be viewed as defining the abundance and isotope mass distribution of the molecular region to a significant extent. Figures 3.2–3.4 and 3.6 also show that the average mass is not equal to the most abundant peak in a ^{13}C-isotope profile. The average mass is usually within 0.5–1.0 Da on the high-mass side of the most abundant peak. As an example, Figure 3.5 delineates this phenomenon in that the $^{13}C_5$ isotope of bovine ubiquitin is at 8564.666 Da while the average mass is calculated as 8565.01 Da. Here, the average mass lies 0.344 Da above the $^{13}C_5$-isotope peak. The exact numerical value that characterizes the average mass in Figure 5b—which, remember, is a calculated value—is of negligible abundance as obtained on a Fourier-transform mass spectrometer. Figure 3.8 provides a more detailed diagram of the relative position of the average mass with respect to an isotope envelope of mass spectral peaks[31,32] of a peptide. The average mass is the statistical average value of an isotope envelope, and it is usually not a discrete, real analyte peak. Note that the most abundant isotope peak is slightly smaller in mass than the average mass value. The concept of mass spectrometer resolution shown in Figure 3.8 is presented in Chapter 5.

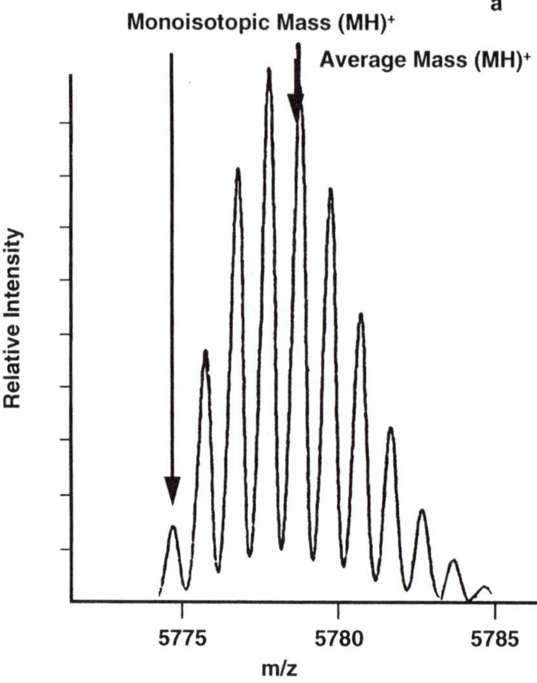

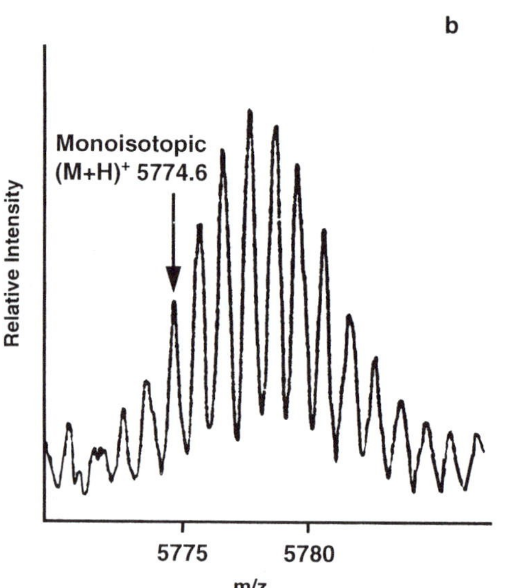

Figure 3.4 (a) Theoretically generated mass spectral isotope distribution of the $(M + H)^+$ region of porcine insulin and (b) the fast atom bombardment (FAB) mass spectral region of the $(M + H)^+$ ion of porcine insulin ($M_r = 5777.6$ Da). (© 1983. Reprinted from reference 26 by kind permission of Elsevier Science-NL, Sara Burgerhartstraat 25, 1055 KV Amsterdam, The Netherlands.)

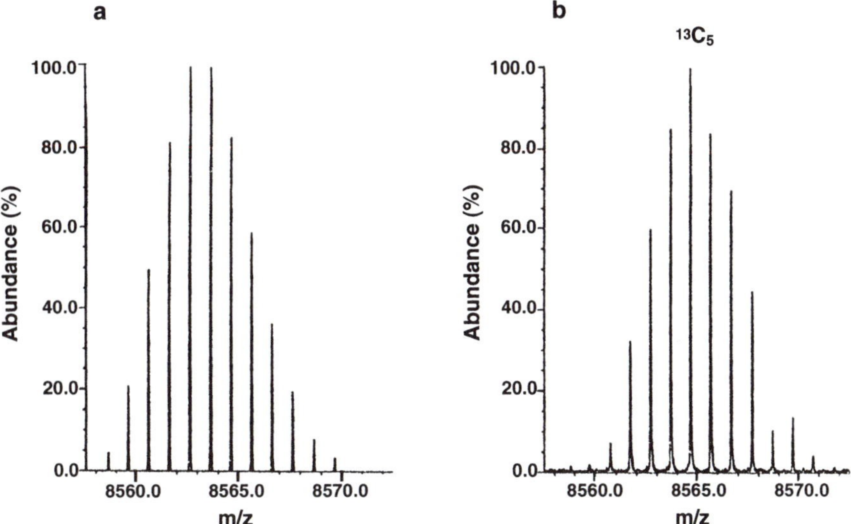

Figure 3.5 Mass spectral isotope distribution of the molecular ion region of bovine ubiquitin ($M_r = 8565.01$): (a) theoretical, (b) experimental. (© 1994. Reproduced from reference 29 by permission of John Wiley & Sons, Limited.)

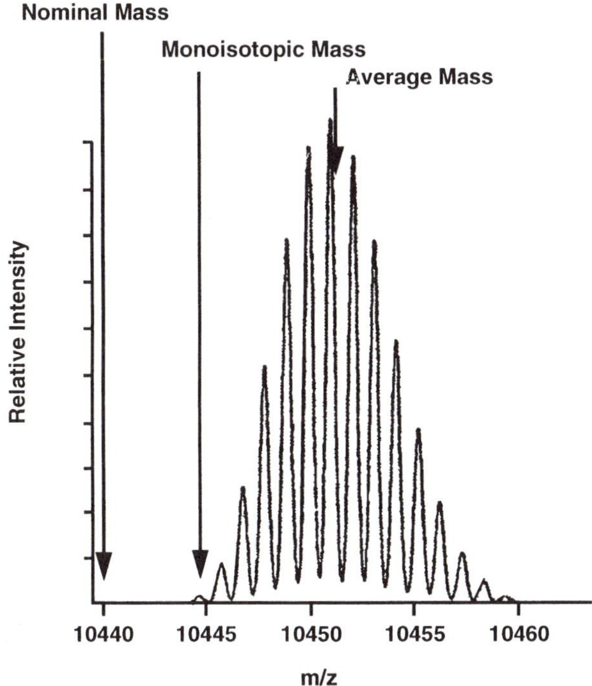

Figure 3.6 Theoretically generated mass spectral isotope distribution of the molecular ion region of the trimer of glucagon ($M_r = 10,451.3$ Da). (© 1983. Reprinted from reference 26 by kind permission of Elsevier Science-NL, Sara Burgerhartstraat 25, 1055 KV Amsterdam, The Netherlands.)

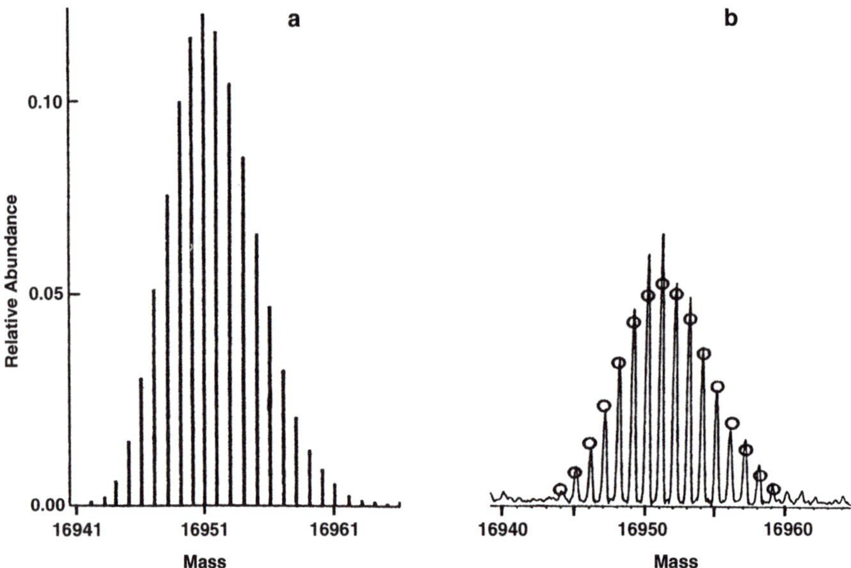

Figure 3.7 Mass spectra of horse heart myoglobin ($M_r = 16,951.5$ Da). (a) Theoretically generated isotope distribution of the molecular ion region. The monoisotopic mass of the molecular species is 16,940.964 Da (16,941 Da). (b) Deconvolution mass spectrum (cf. Chapter 4) of the ESI experimentally observed, multiply charged peaks in the conventional mass spectrum. Circles represent the calculated mass spectral distribution as shown in part a. [(a) © 1993. Reproduced from reference 28 by permission of John Wiley & Sons, Limited. (b) © 1995, American Society for Mass Spectrometry. Reprinted from reference 27 by permission of Elsevier Science, Inc.].

Appendix 2 and Problem A2.1 present examples of the calculations and resultant diversity of the individual "hidden" isotope species that constitute each $^{13}C_n$ isotope peak observed in the isotope envelope of a biological molecule. Table 3.7 presents a partial tabulation[4] of the individual isotope species that constitute the most abundant isotope peak ($^{13}C_6$) for the glucagon trimer (Figure 3.6). Note that the most abundant isotope can be characterized as the $^{13}C_6$ species. Only the most abundant individual isotopes, approximately 6 Da higher in mass than the monoisotopic peak, are listed. For example, the $^{13}C_3{}^{15}N_1{}^{34}S_1$ nomenclature in Table 3.7 represents the $^{12}C_{456}{}^{13}C_3{}^{1}H_{672}{}^{14}N_{125}{}^{15}N^{16}O_{150}{}^{32}S_2{}^{34}S_1$ isotope molecule, and note the narrow mass range of molecular isotopes that constitute the $^{13}C_6$ most abundant individual isotope peak.

McLafferty et al. have conducted ESI-FT-MS experiments on the chondroitinase I enzyme and these provide further insights on the phenomenon of isotope species, especially at relatively high mass.[33] The molecular formula of the enzyme is $^{12}C_{5039}{}^{1}H_{7770}{}^{14}N_{1360}{}^{16}O_{1525}{}^{32}S_{22}$ and the ^{13}C formula that represents the most abundant isotope peak is $^{12}C_{4970}{}^{13}C_{69}{}^{14}N_{1360}{}^{16}O_{1525}{}^{32}S_{22}$. The most abundant mass of the enzyme exists at approximately 69 Da above the mono-

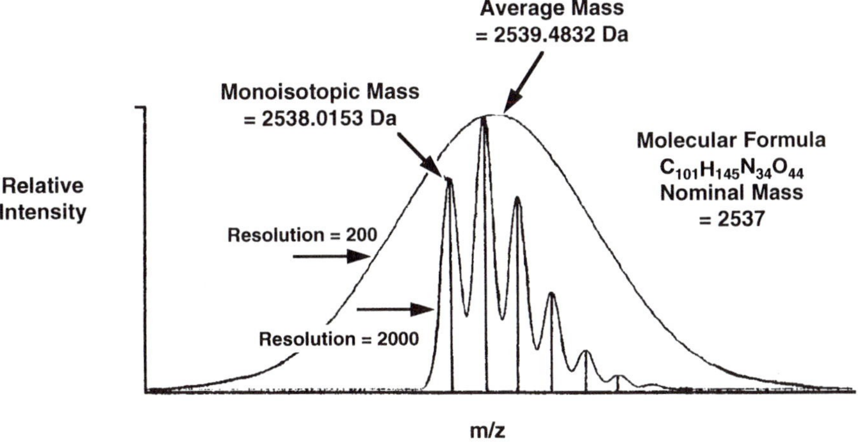

Figure 3.8 Theoretical mass spectra of the $(M + H)^+$ region of a peptide calculated at three different mass spectrometer resolution values. The calculated mass spectra at resolution values of 200, 2000, and 200,000 were assumed for a time-of-flight, quadrupole, and Fourier transform mass analyzer, respectively. (© 1996. Reprinted from reference 31 by permission of Academic Press, Inc., Orlando, Florida.)

Table 3.7 Partial composition of the most abundant isotope peak ($^{13}C_6$) of the glucagon trimer[4]

Isotope	Mass (Da)	% Abundance
$^{13}C_3{}^{15}N_1{}^{34}S_1$	10,450.80	5.61
$^{13}C_2{}^{18}O_1{}^{34}S_1$	10,450.80	2.18
$^{13}C_3{}^{15}N_3$	10,450.80	1.47
$^{13}C_4{}^{34}S_1$	10,450.80	15.45
$^{13}C_4{}^{15}N_1{}^{33}S_1$	10,450.80	1.28
$^{13}C_2{}^{15}N_2{}^{18}O_1$	10,450.81	1.74
$^{13}C_4{}^{15}N_2$	10,450.81	12.34
$^{13}C_3{}^{15}N_1{}^{18}O_1$	10,450.81	12.90
$^{13}C_3{}^{2}H_1{}^{34}S_1$	10,450.81	1.21
$^{13}C_4{}^{15}N_1{}^{17}O_1$	10,450.81	3.02
$^{13}C_5{}^{33}S_1$	10,450.81	2.83
$^{13}C_5{}^{15}N_1$	10,450.81	54.69
$^{13}C_2{}^{18}O_2$	10,450.82	2.49
$^{13}C_4{}^{18}O_1$	10,450.82	35.53
$^{13}C_4{}^{2}H_1{}^{15}N_1$	10,450.82	5.38
$^{13}C_3{}^{17}O_1{}^{18}O_1$	10,450.82	1.56
$^{13}C_5{}^{17}O_1$	10,450.82	6.64
$^{13}C_6$	10,450.82	100.00
$^{13}C_3{}^{2}H_1{}^{18}O_1$	10,450.82	2.79
$^{13}C_5{}^{2}H_1$	10,450.82	11.84

isotopic peak, and the latter peak is essentially absent in the mass spectrum.[33] The following masses can be calculated from Table 3.1:

	Mass (Da)
Nominal	112,382
Monoisotopic	112,438.6119
Most abundant	112,507.8433
Weighted average	112,509.0049

Note the considerable difference in mass between the nominal/monoisotopic and the monoisotopic/average mass pairs. The most abundant mass ($^{13}C_{69}$) is 1.16 Da lower than the average mass, and the most abundant mass actually consists of 509 separate isobaric peaks of different isotope compositions that have an intensity $\geqslant$14% in the isotope cluster that contains the $^{13}C_{69}$ base peak.

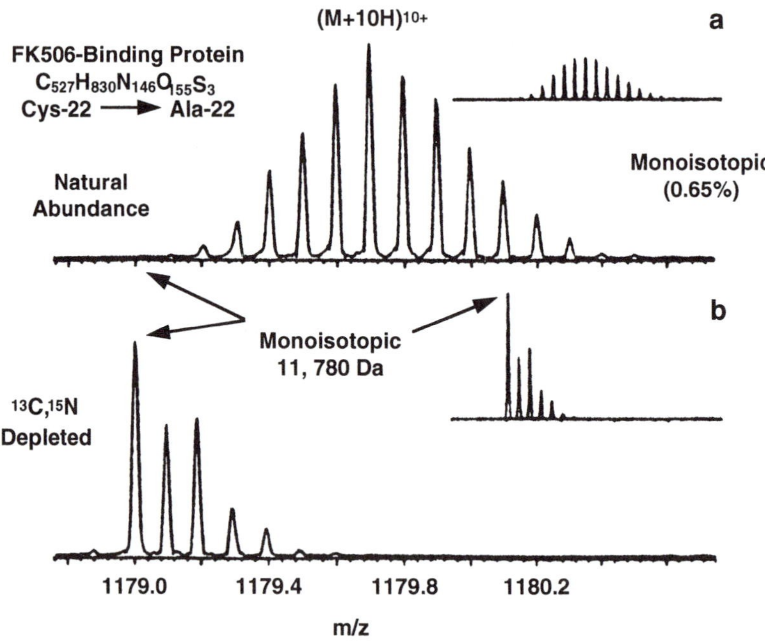

Figure 3.9 The ESI-FT-MS mass spectra of a mutant of the FK506–binding protein. The mutation causes a Cys-22 to Ala residue conversion. (a) Normal mass spectral isotope distribution of the $(M + 10H)^{10+}$ ion (cf. Chapter 4) with the normal 98.89% (^{12}C) and 99.63% (^{14}N) weighted distribution. Inset represents the theoretically generated isotope distribution. (b) Mass spectral isotope distribution of the $(M + 10H)^{10+}$ ion for the same protein produced and isolated from *E. coli* grown on a medium of nutrients that contained 99.95% ^{12}C and 99.99% ^{14}N. Inset represents the theoretically generated isotope distribution for the ^{13}C,^{15}N-depleted protein. (© 1997, American Chemical Society. Reprinted from reference 25 with permission.)

The monoisotopic mass of a compound has a unique role with respect to its exact mass benchmark status. However, its presence in an experimentally derived mass spectrum diminishes to a negligible amount in compounds over 10–15 kDa in mass.[24,30] Insightful experiments were performed[25] where the *E. coli* organism was grown in a medium enriched with the monoisotopic form of carbon and nitrogen (i.e., with glucose that had 99.95% ^{12}C in the glucose sugar and 99.99% ^{14}N in the ammonium sulfate salt). The FK506-binding protein ($^{12}C_{527}$ $^{1}H_{830}{}^{14}N_{146}{}^{16}O_{155}{}^{32}S_{3}$) was isolated and subjected to ESI-FT-MS analysis. Monoisotopic enrichment of a biological molecule should lead to a skewed mass spectrum such that the monoisotopic species should appear with a higher intensity than without the enrichment process. Figure 3.9 displays this concept in that the native protein (unenriched) in Figure 3.9a shows a normal isotope intensity distribution of the $(M + 10H)^{10+}$ charged ion (cf. Chapter 4). Note the negligible intensity of the monoisotopic species; the inset provides the theoretically calculated spectrum. Figure 3.9b shows the ^{13}C- and ^{15}N-depleted spectrum of the protein, and the monoisotopic peak is the base peak in the spectrum of a protein with M_r = 11,780.7 Da! Thus, the monoisotopic peak is directly observed with significant intensity in a protein with a relatively high mass. The inset of Figure 3.9b shows the theoretical distribution of the ^{13}C,^{15}N-depleted protein, and both experimental spectra in Figure 3.9 closely track their respective theoretical spectra. Furthermore, calculations showed that similar enrichment procedures would allow the monoisotopic mass to be observed at 5% relative intensity for a protein of 60 kDa.[25]

REFERENCES

1. Biemann, K. In *Methods in Enzymology;* McCloskey, J.A., Ed.; Academic Press: New York, **1990**; Volume 193, pp 295–305.
2. McLafferty, F.W.; Tureček, F. *Interpretation of Mass Spectra,* 4th Edition; University Science Books: Mill Valley, CA, 1993.
3. Chen, R.; Cheng, X., Mitchell, D.W.; Hofstadler, S.A.; Wu, Q.; Rockwood, A.L.; Sherman, M.G.; Smith, R.D. *Anal. Chem.* **1995,** *67,* pp 1159–1163.
4. Yergey, J.; Heller, D.; Hansen, G.; Cotter, R.J.; Fenselau, C. *Anal. Chem.* **1983,** *55,* pp 353–356.
5. Yergey, J.A. *Int. J. Mass Spectrom. Ion Phys.* **1983,** *52,* pp 337–349.
6. Wada, Y.; Tamura, J.; Musselman, B.D.; Kassel, D.B.; Sakurai, T.; Matsuo, T. *Rapid Commun. Mass Spectrom.* **1992,** *6,* pp. 9–13.
7. Weast, R.C.; Astle, M.J.; Beyer, W.H. *The CRC Handbook,* 67th Edition; CRC Press: Boca Raton, FL, 1986–1987; pp B220–B224.
8. IUPAC Atomic Weights of the Elements 1989. *Pure Appl. Chem.* **1991,** *63,* p 994.
9. McCloskey, J.A. In *Methods in Enzymology*; McCloskey, J.A., Ed.; Academic Press: New York, 1990; Volume 193, Appendix I, pp 869–870.
10. Crawford, L.R.; *Int. J. Mass Spectrom. Ion Phys.* **1972,** *10,* pp 279–292.
11. Matsuo, T.; Sakurai, T.; Matsuda, H.; Wollnik, H.; Katakuse, I. *Biomed. Mass Spectrom.* **1983,** *10,* pp 57–60.
12. Biemann, K.; Martin, S.A. *Mass Spectrom. Rev.* **1987,** *6,* pp 1–76.
13. Mann, M.; Wilm, M. *Trends Biochem. Sci.* **1995,** *20,* pp 219–224.

14. Smith, R.D.; Light-Wahl, K.J.; Winger, B.E.; Goodlett, D.R., *Biological Mass Spectrometry Present and Future;* Matsuo, T.; Caprioli, R.M.; Gross, M.L.; Seyama, Y., Eds.; John Wiley and Sons: Chichester, U.K., 1994; Chapter 2.2, pp 41–74.
15. Creswell, C.J.; Runquist, O.A.; Campbell, M.M. *Spectral Analysis of Organic Compounds 2nd Edition*; Burgess Publishing: Minneapolis, MN, 1972.
16. *Mass Spectrometry: Analytical Chemistry by Open Learning,* Davis, R.; Frearson, M.; Prichard, F.E., Eds.; John Wiley & Sons: Chichester, U.K., 1987.
17. Zubarev, R.A. *Int. J. Mass Spectrom. Ion Processes* **1991,** *107,* pp 17–27.
18. Rockwood, A.L.; Orden, S.L.V. *Anal. Chem.* **1996,** *68,* pp 2027–2030.
19. Yamamoto, H.; McCloskey, J.A. *Anal. Chem.* **1977,** *49,* pp 281–283.
20. Hibbert, D.B. *Chemom. Intell. Lab. Syst.* **1989,** 6, pp 203–212.
21. Grange, A.H.; Brumley, W.C. *J. Am. Soc. Mass Spectrom.* **1997,** *8,* pp 170–182.
22. McCloskey, J.A. In *Methods in Enzymology;* McCloskey, J.A., Ed; Academic Press: New York, 1990; Volume 193, pp 882–886.
23. *Mass and Abundance Tables for Use in Mass Spectrometry*; Beynon, J.H.; Williams, A.E., Eds., Elsevier Publishing Co.: Amsterdam, The Netherlands, 1963.
24. Winger, B.E.; Hofstadler, S.A.; Bruce, J.A.; Udseth, H.R.; Smith, R.D.; *J. Am. Soc. Mass Spectrom.* **1993,** *4,* pp 566–577.
25. Marshall, A.G.; Senko, M.W.; Li, W.; Li, M.; Dillon, S.; Guan, S.; Logan, T.M., *J. Am. Chem. Soc.* **1997,** *119*, pp 433–434.
26. Fenselau, C.; Yergey, J.; Heller, D. *Int J. Mass Spectrom. Ion Phys.* **1983,** *53,* pp 5–20.
27. Senko, M.W.; Beu, S.C.; McLafferty, F.W. *J. Am. Soc. Mass Spectrom.* **1995,** *6,* pp 229–233.
28. Ashton, D.S.; Beddell, C.R.; Cooper, D.J.; Green, B.N.; Oliver, R.W.A. *Org. Mass Spectrom.* **1993,** *28,* pp 721–728.
29. Winger, B.E.; Hein, R.E.; Becker, B.L.; Campana, J.E. *Rapid Commun. Mass Spectrom.,* **1994,** *8,* pp 495–497.
30. McLafferty, F.W., *Acc. Chem. Res.* **1994,** *27,* pp 379–386.
31. Siuzdak, G. *Mass Spectrometry for Biotechnology;* Academic Press: San Diego, CA, 1996; p 67.
32. Siuzdak, G., *Proc. Natl. Acad. Sci. U.S.A.* **1994,** *91,* 11290–11297.
33. Kelleher, N.L.; Senko, M.W.; Siegel, M.M.; McLafferty, F.W. *J. Am. Soc. Mass Spectrom.* **1997,** *8,* pp 380–383.

ANSWERS

PROBLEM 3.1 Element contributions to the mass values of the hemoglobin β-chain 41–59 amino acid subunit

		Mass (Da)					Isotope shift	
Element	Number of atoms	Nominal	Mono-isotopic	Mass defect	%	Average	Da	%
C	93	1116	1116.000	0.0	0.0	1117.023	1.023	75.83
H	135	135	136.056	1.056	81.17	135.076	0.020	1.48
N	21	294	294.064	0.064	4.92	294.141	0.077	5.71
O	30	480	479.847	0.153	11.76	479.982	0.135	10.01
S	1	32	31.972	0.028	2.15	32.066	0.094	6.97
Total	—	2057	2057.939	1.301	100.00	2059.288	1.349	100.00

H/C number ratio = 1.45.

PROBLEM 3.2 Element contributions to the mass values of the glucagon (monomer) molecule

		Mass (Da)					Isotope shift	
Element	Number of atoms	Nominal	Mono-isotopic	Mass defect	%	Average	Da	%
C	153	1836	1836.000	0.0	0.0	1837.683	1.683	76.85
H	224	224	225.753	1.753	80.82	225.785	0.032	1.46
N	42	588	588.129	0.129	5.95	588.281	0.152	6.94
O	50	800	799.741	0.259	11.94	799.970	0.229	10.46
S	1	32	31.972	0.028	1.29	32.066	0.094	4.29
Total	—	3480	3481.596	2.169	100.00	3483.785	2.19	100.00

H/C number ratio = 1.46.

PROBLEM 3.3 The tabular format of the element contributions to the mass of the bovine insulin (α- and β-chains) molecule can be found in reference 17.

PROBLEM 3.4 Element contributions to the mass values of the porcine proinsulin molecule

		Mass (Da)					Isotope shift	
Element	Number of atoms	Nominal	Mono-isotopic	Mass defect	%	Average	Da	%
C	398	4776	4776.000	0.0	0.0	4780.378	4.378	73.25
H	617	617	621.828	4.828	81.06	621.917	0.089	1.49
N	109	1526	1526.335	0.335	5.92	1526.730	0.395	6.61
O	123	1968	1967.375	0.625	10.49	1967.926	0.551	9.22
S	6	192	191.832	0.168	2.82	192.396	0.564	9.44
Total	—	9079	9083.370	5.956	99.99	9089.347	5.977	100.01

PROBLEM 3.5 Element contributions to the mass values of the porcine proinsulin molecule

		Mass (Da)					Isotope shift	
Element	Number of atoms	Nominal	Mono-isotopic	Mass defect	%	Average	Da	%
C	796	9,552	9,552.00	0.0	0.0	9,560.756	8.756	73.24
H	1,234	1,234	1,243.656	9.656	81.06	1,243.835	0.179	1.50
N	218	3,052	3,052.670	0.67	5.62	3,053.461	0.791	6.62
O	246	3,936	3,934.749	1.251	10.50	3,935.852	1.103	9.22
S	12	384	383.665	0.335	2.81	384.792	1.127	9.43
Total	—	18,158	18,166.74	11.912	99.99	18,178.696	11.956	100.01

H/C number ratio = 1.55.

4

Calculation of the Analyte Mass

Electrospray ionization (ESI) is a gentle technique for charging a substance, and this process predominantly occurs by transferring one or more protons onto the analyte from proton-bound (protonated) reagent/solvent molecules in the liquid phase. Chemical ionization (CI), on the other hand,[1] relies on the presence of desorbed or directly introduced neutral gaseous analyte, and it is ionized by attachment or adduction of a charged (ionized) reagent gas. Mechanistically, ESI and CI have different origins; however, ESI can technically be considered as a form of CI. Peptides, enzymes, proteins, and their derivatives play an important role and occupy a significant percentage of the biological mass spectrometry literature. This class of biological molecule is the analyte of interest in this book. This chapter provides an initial impression of the interpretation of protein/peptide ESI-mass spectra.

The multiplicity of organic functional groups in a biological molecule allows for the presence of multiple charges, because these functional groups can attract protons. This characteristic allows ESI-mass spectra to contain more than one mass spectral peak for a biological analyte. In most cases, multiple peaks of the same biological molecule will be evident in a mass spectrum, and these peaks will differ only by the number of protons attached to the biological species. For peptides and proteins in the positive-ion mode, the terminal amine (NH_2), the amine functionality in the side chains of lysine (Lys) and arginine (Arg), and the nitrogen atoms in the histidine amino acid (His) residue are the preferred sites of protonation because of their relatively high proton affinities.[2] The methionine enkephalin peptide is composed of the amino acid sequence Tyr-Gly-Gly-Phe-Met (Tyr, tyrosine; Gly, glycine; Phe, phenylalanine; Met, methionine) and has a neutral average mass of 573.2 Da. This peptide has a terminal amine group and does not contain the Lys, Arg, and His amino acid residues. Thus, one would expect to observe only one peak in an ESI-mass spectrum that represents a one-proton ionized molecular species of methionine enkephalin at a mass-to-charge ratio of $(573 + 1H^+)/1H^+ = m/z$ 574. Figure 4.1a shows this to be the case.[3]

Table 4.1 presents standard three-letter and one-letter abbreviations of the commonly occurring amino acids, and these abbreviations will be used in place of the complete name of the amino acid where appropriate. Bradykinin, average mass (M_r) of 1059.56 Da, is presented in Figure 4.1b, and the amino acid sequence of this peptide is Arg-Pro-Pro-Gly-Phe-Ser-Pro-Phe-Arg.[3] Bradykinin contains a free terminal amine and two arginine amino acid residues, and one of the arginine residues is also the N-terminal residue. Three separate peaks in an ESI-mass spectrum are predicted, and this is observed in Figure 4.1b. The average mass of bradykinin is 1059.56 Da, and a mass spectral peak is

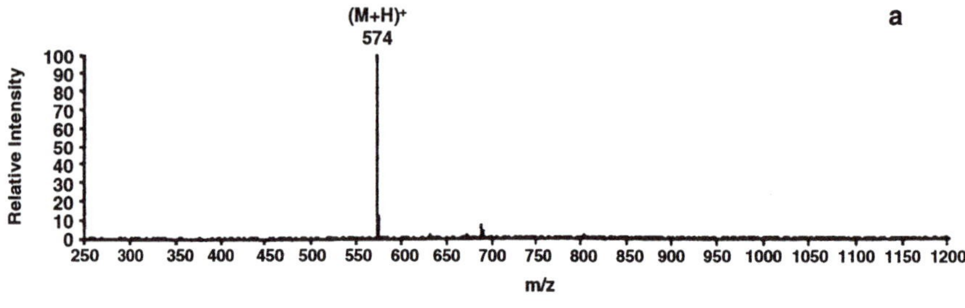

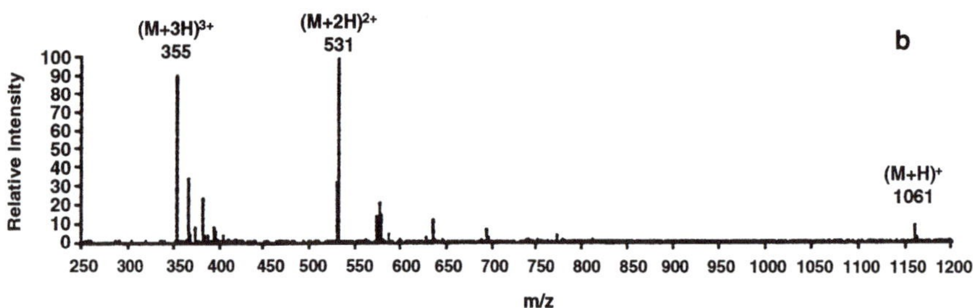

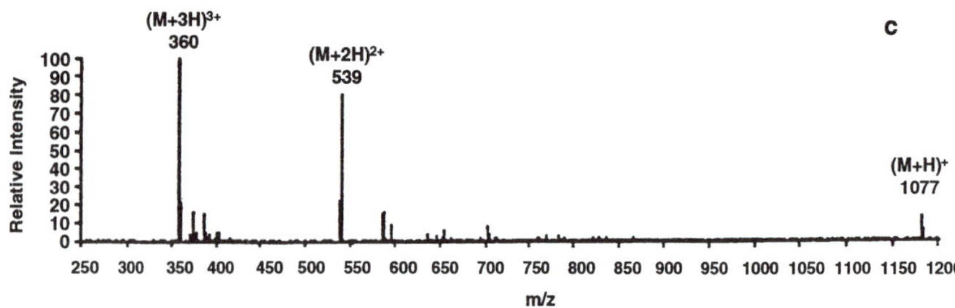

Figure 4.1 The ESI-mass spectra of (a) methionine enkephalin, (b) bradykinin, and (c) Tyr-8-bradykinin. (Permission granted by PE-SCIEX, Concord, Ontario, Canada.)

Table 4.1 Abbreviations of the 20 common amino acids

Amino acid	Three-letter	One-letter
Alanine	Ala	A
Arginine	Arg	R
Asparagine	Asn	N
Aspartic acid	Asp	D
Cysteine	Cys	C
Glutamine	Gln	Q
Glutamic acid	Glu	E
Glycine	Gly	G
Histidine	His	H
Isoleucine	Ile	I
Leucine	Leu	L
Lysine	Lys	K
Methionine	Met	M
Phenylalanine	Phe	F
Proline	Pro	P
Serine	Ser	S
Threonine	Thr	T
Tryptophan	Trp	W
Tyrosine	Tyr	Y
Valine	Val	V

observed at m/z 1061 that is the $(M+H)^+$ species, where M denotes the neutral molecular species of interest. The two intense peaks in Figure 4.1b appear to be the $(M+2H)^{2+}$ and $(M+3H)^{3+}$ species as the following calculations show:

		m/z
$(M+2H)^{2+}$	$(1060+2H)^+/2H^+ =$	531
$(M+3H)^{3+}$	$(1060 + 3H)^+/3H^+ =$	354.3

Three peaks are still observed upon changing the phenylalanine residue at position 8 to tyrosine (Figure 4.1c), because the end terminal amine and the two arginine amino acid residues are still present in the peptide.[3] The mass would have to shift 16 Da higher from bradykinin to Tyr-8-bradykinin, because the mass difference of a Phe and Tyr residue is 16 Da. This mass shift is observed from m/z 1061 to m/z 1077 of the one proton-ionized molecular species. The mass calculations and peak identities of Tyr-8-bradykinin follow in a similar fashion to that of bradykinin:

		m/z
$(M+H)^+$	$(1076+1H)/1H^+ =$	1077
$(M+2H)^{2+}$	$(1076+2)/2 =$	539
$(M+3H)^{3+}$	$(1076+3)/3 =$	359.7

In these mass calculations, two assumptions were made. The first is that a continuous, integral change in charge occurs (Figures 4.1b and c)—that is, the $(M+2H)^{2+}$ must be present in order to avoid a discontinuity in the charge-state

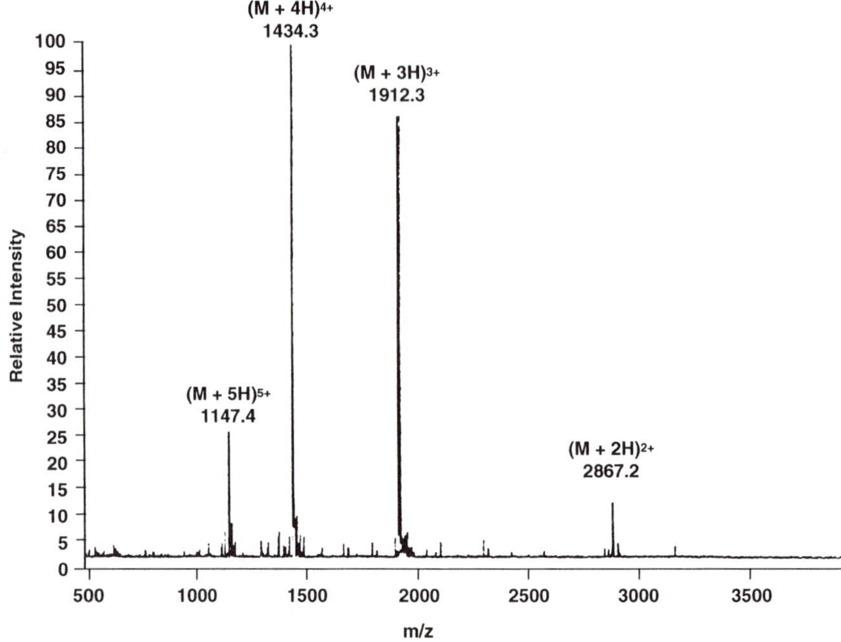

Figure 4.2 The ESI-mass spectrum of bovine pancreas insulin. (© 1990. Reproduced from reference 4 by permission of John Wiley & Sons, Limited.)

distribution—and the second assumption is that the ionizing species is a proton with a nominal mass of 1 Da. These assumptions are operative in Figure 4.1, and they are even more important in the next series of spectra.

Suppose the $(M+H)^+$ is absent in the ESI-mass spectrum of a biological molecule. A number of reasons can contribute to the absence of the benchmark $(M+H)^+$ species: for example, (a) the mass of the $(M+H)^+$ species is beyond the range of the mass spectrometry analyzer; (b) as the analyte becomes higher in mass, the $(M+H)^+$ species becomes negligible in abundance due to the low probability of occurrence. The latter thus refers to the occurrence of the analyte in the multiple-proton ionized states. Relatively low-charged species, such as $(M+2H)^{2+}$, $(M+3H)^{3+}$, etc., also can be higher in mass than the mass range of most mass spectrometry instruments, and/or have a statistically low probability of occurrence. In this case, a single algorithm can be used to arrive at the neutral molecular mass of the analyte. Figure 4.2 presents an ESI-mass spectrum of bovine insulin,[4] which has an average mass of 5733.6 Da, and a complete accounting of the spectrum for mass information is presented in Table 4.2. For two adjacent peaks in the mass spectrum that have masses of m_1 and m_2 with, respectively, n_1 and n_2 number of integer (integral) charges, the following formulae are used:

$$m_1 = (M + n_1)/n_1 \qquad m_2 = (M + n_2)/n_2$$

Table 4.2 Experimental mass calculation for bovine insulin

m_2	m_1	Δm	Approx. n_2	n_2	M
2867.2	1912.3	954.9	2.00	2	5732.4
1912.3	1434.3	478.0	3.00	3	5733.9
1434.3	1147.4	286.9	3.99	4	5733.2

A fourth data point can be obtained by having $m_2 = m/z$ 1147.4; n_2 must be 5 by following the trend of the 2, 3, and 4 charges: $M = 5(1147.4 - 1) = 5732.0$ Da

M_{expt} = Average experimental mass = 5732.9 ± 0.73 Da

M_r = Weighted (theoretical) average mass = 5733.58 Da

where M is a single experimental determination of the mass of bovine insulin and $m_2 > m_1$, and $n_2 < n_1$. Both m_1 and m_2 can be either nominal, monoisotopic, or average mass values. Assuming continuity of charge states and proton ionization:[5,6]

(1) $n_2 = n_1 - 1$ The 1 value indicates one charged species, not 1 Da.

(2) $n_2 = \dfrac{m_1 - 1}{m_2 - m_1}$ The 1 value indicates the mass of a proton, or 1.0079 Da for a more accurate value.

(3) $M = n_2(m_2 - 1)$ The 1 value indicates the mass of a proton, or 1.0079 Da for a more accurate value.

The calculation in equation (2) yields an approximate n_2 value, and is converted to the closest integer value when used in equation (3). If the approximate n_2 value is not close to an integer, then either a miscalculation may have occurred, or a modified form of the analyte or a contaminant may be present. The n_2 value can provide a check on the validity and integrity of the calculations and in ascribing the correct series of peaks, especially if there are more than one series of peaks or impurities present in the spectrum. This deconvolution procedure provides independent M values for each peak in the mass spectrum, and these experimentally derived M values are averaged to provide an arithmetic average mass (M_{expt}). The value is close to the average weighted mass (M_r) (Chapter 3). However, the m/z values used from the mass spectra are the most abundant values for each charge state. The chapter on isotopes (Chapter 3) shows that the mass of the most abundant mass spectral peak is slightly less than the (weighted) average mass (M_r). The arithmetic mean-derived and calculated average mass values are not identical, yet, with a satisfactorily tuned mass spectrometer, the experimentally derived value is usually between the most abundant and average mass values.

Note the separation in mass between the peaks in Figure 4.2. Proceeding from high to low mass, the charge state of each peak increases by one unit, and the separation in mass between adjacent peaks (Δm, Table 4.2) progressively decreases in a nonlinear fashion. These items are important, as observed in Figure 4.3 and detailed in Table 4.3 for the *E. coli* thioredoxin molecule.[7] A

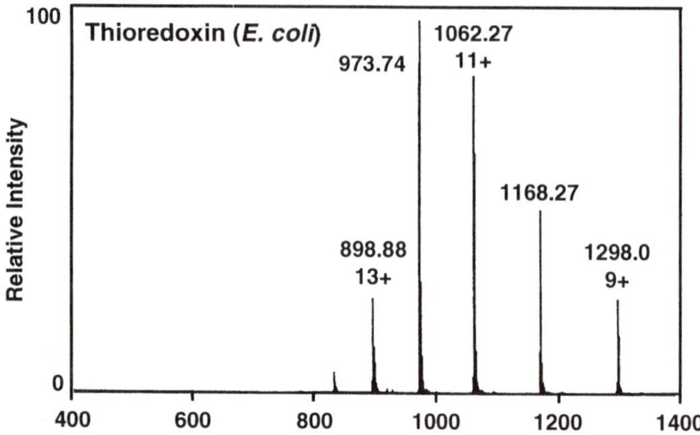

Figure 4.3 The ESI-mass spectrum of *E. coli* thioredoxin. (© 1990, American Chemical Society. Reprinted from reference 7 with permission.)

Table 4.3 Experimental mass calculation for *E. coli* thioredoxin

m_2	m_1	Δm	Approx. n_2	n_2	M
1,298.0	1,168.27	129.73	9.0	9	11,673.0
1,168.27	1,062.27	106.0	10.01	10	11,672.7
1,062.27	973.74	88.53	10.99	11	11,674.0
973.74	898.88	74.86	11.99	12	11,672.88

Another data point can be obtained by having $m_2 = m/z$ 898.88; n_2 must be 13 by following the trend of the charges: $M = 13(898.88 - 1) = 11,672.4$ Da

$M_{expt} = 11,673.0$ Da

$M_r = 11,673.4$ Da

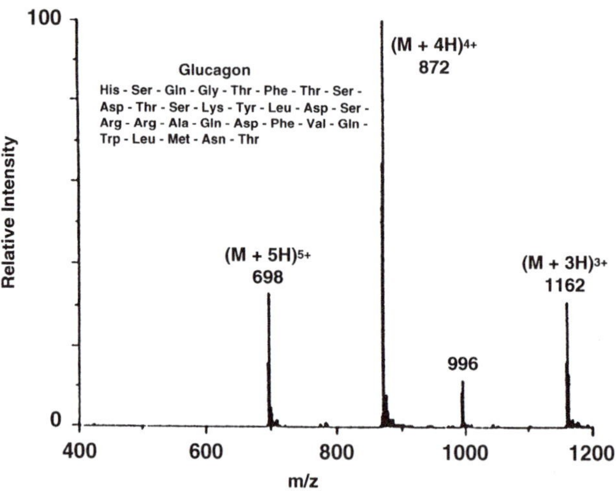

Figure 4.4 The ESI-mass spectrum of glucagon. The ion at 996 Da is due to a contaminant peptide. (© 1990. Reprinted from reference 8 by permission of Academic Press, Inc., Orlando, Florida.)

series of peaks are observed that generally follow the trend in Figure 4.2, and the experimentally determined arithmetic average mass is very close to the calculated average mass value (Table 4.3).

The next example has a number of 'anomalies' in the spectrum and should provide an appreciation of the mathematical and decision-making processes on the assumptions made in an experimental mass calculation. Figure 4.4 provides an ESI-mass spectrum of glucagon (M_r = 3483.8 Da), and Table 4.4 provides the mass spectral interpretation.[8] The spectrum appears to be easily interpretable because of the presence of only a few sharp peaks; however, the Table 4.4 analysis appears to be more problematic. Calculations yield n_2 values, from high to low mass, of 6, 7, 4, 5, and 3. Clearly, something is wrong in the interpretation; however, all derived n_2 values are integral units. The first three entries in Table 4.4 follow from considering the high to low mass values, and $m_1 = m/z$ 698 can be used to generate a fourth mass value. Which values are correct? One of the four masses must be an impurity or contaminant peptide.[8] The fifth entry in Table 4.4 considers the lowest intensity, m/z 996, to be an impurity, and thus the mass calculation using m/z 1162 and 872 skips over the m/z 996 value. One more consideration could be that m/z 872 is an impurity and would be skipped in the mass analysis of m/z 996 and 698. This is presented as the sixth entry in Table 4.4. The m/z 996–698 pair produces a nonintegral n_2 value (2.33) and the m/z 872–698 pair produces an integral value. Those pairs with m/z 996 should be deleted (first two entries), and this mass is revealed as the fugitive peak. The m/z 996 peak is indeed a contaminant, and a very good match of M_{expt} and M_r occurs with consideration of the other three peaks. Figure 4.5 and Table 4.5 present an experimental mass analysis of β-lactoglobulin.[5] Note that the M_{expt} is lower in mass than the average mass.

Table 4.4 Experimental mass calculation for glucagon

m_2	m_1	Δm	Approx. n_2	n_2	M
1162	996	166	5.99[a]	6	6966
996	872	124	7.01[a]	7	6965
872	698	174	4.01	4	3484

Another data point can be obtained by having $m_2 = m/z$ 698; n_2 must be 5 in order to follow 4: $M = 5(698 - 1) = 3485$ Da

m_2	m_1	Δm	Approx. n_2	n_2	M
1162	872	290	8.00	3	3483
996	698	298	2.33[a]	—	—

$M_{expt} = 3484 \pm 0.82$ Da
$M_r = 3483.8$ Da

[a]Deleted from mass calculation.

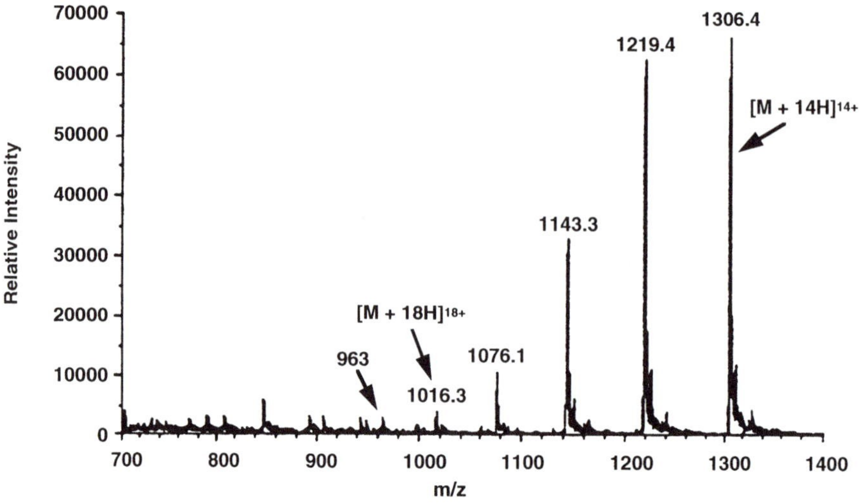

Figure 4.5 The ESI-mass spectrum of β-lactoglobulin. (© 1988. Reproduced from reference 5 by permission of John Wiley & Sons, Limited.)

The next series of ESI-mass spectra are relatively more complicated in nature and reflect the greater complexity of the analytes. Figure 4.6 is an ESI-mass spectrum[9] of horseradish peroxidase (HRP). The HRP does not exist as a pure substance. It consists of a peptide backbone with sugar units attached at certain amino acid residues. However, the same peptide backbone can have different amounts of sugar units in the same protein preparation. The protein is actually a heterogeneous ensemble of related glycoproteins, and thus differs in mass (cf. Chapter 12). A major series of broad peaks are observed with charge states in the range 19–22 added protons. Each major peak consists of a number of closely spaced peaks. The most intense peak of this series has an M_{expt} of 43,164.8 ± 3.2 Da and has two measurable smaller peaks at 43,248 and 43,330 Da. A minor series of peaks that spans the same charge distribution as the major series consists of an intense peak at 42,343.2 Da and two minor peaks at 42,427.0 and 42,509.0 Da. Figure 4.7 shows an ESI-mass spectrum of another glycoprotein,

Table 4.5 Experimental mass calculation for β-lactoglobulin

m_2	m_1	Δm	Approx. n_2	n_2	M
1,306.4	1,219.4	87.0	14.00	14	18,275.6
1,219.4	1,143.3	76.1	15.01	15	18,276.0
1,143.3	1,076.1	67.2	16.00	16	18,276.8
1,076.1	1,016.3	59.8	16.98	17	18,276.7
1,016.3	963	53.3	18.05	18	18,275.4

$M_{expt} = 18276.1 \pm 0.62$ Da
$M_r = 18277.1$ Da

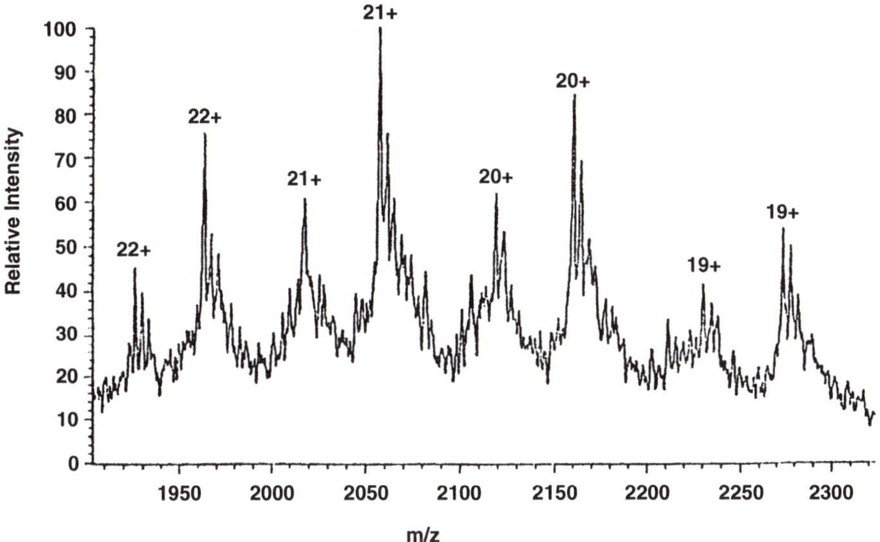

Figure 4.6 The ESI-magnetic sector mass spectrum of the horseradish peroxidase glycoprotein. (© 1992. Reproduced from reference 9 by permission of John Wiley & Sons, Limited.)

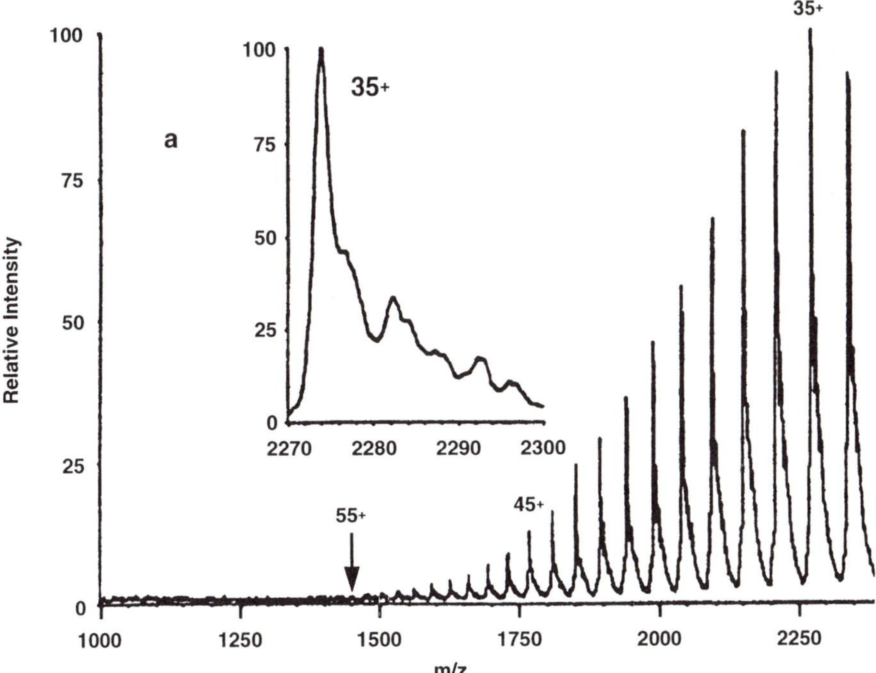

Figure 4.7 The ESI-mass spectrum of human serum apo-transferrin. Inset shows the detail of the $(M + 35H)^{35+}$ peak. (© 1991, American Society for Mass Spectrometry. Reprinted from reference 10 by permission of Elsevier Science, Inc.)

human serum apo-transferrin; note the relative width of the peaks.[10] The apo prefix means that the iron-heme moiety has been removed from the glycoprotein. Multiple glycoproteins of relatively similar mass are indicated in the expanded mass spectrum of the 35+ charge-state peak [i.e., $(M + 35H)^{35+}$], as shown in the inset of Figure 4.7. The major intense peak in each charge state has an M_{expt} of 79556.8±1.7 Da.

Figure 4.8 shows a spectrum of bovine serum albumin (BSA)[10] with an M_{expt} of 66,431.5 ± 1.2 Da. The peaks are well resolved despite the relatively high mass. The inset of the 41+ peak shows that each peak actually has two other components of lower intensity, labeled M_1 and M_2, and they are 108 Da and 149 Da higher in mass than the most intense peak (M_0) of each charge state. Note that the low intensity series of peaks is the BSA dimer molecule. Under different experimental conditions, the BSA dimer (Figure 4.9) dominates.[9] Because of its relatively high mass, approximately 132,900 Da, each charge state is poorly resolved, because the peaks are very close together.

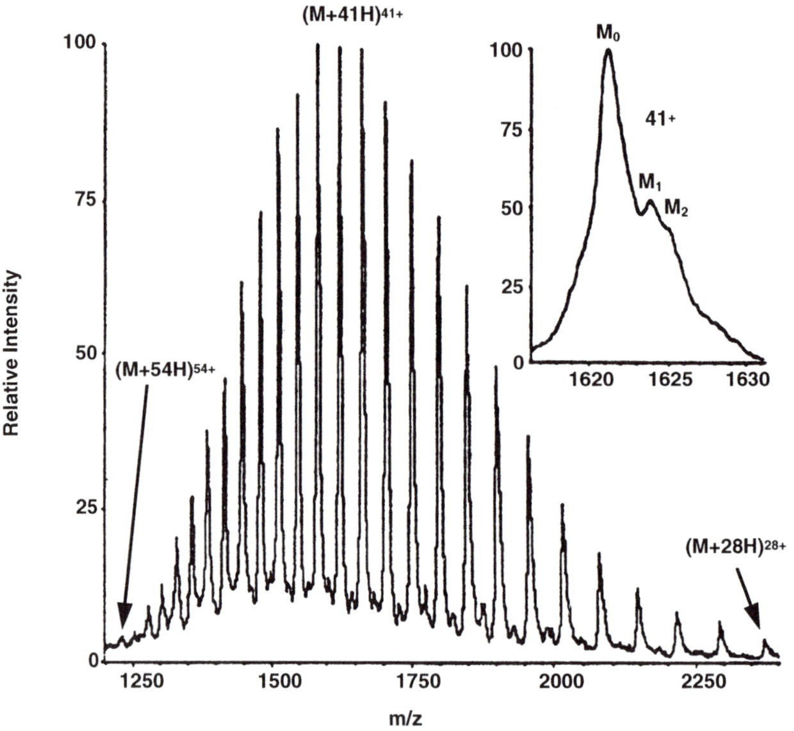

Figure 4.8 The ESI-mass spectrum of bovine serum albumin (BSA). The very-low-intensity peaks between the intense signals are due to the BSA dimer. Inset shows the detail of the $(M + 41H)^{41+}$ peak where the BSA and two unknown proteins (M_1 and M_2) are observed. Solvent was 9:2 H_2O:MeOH with 0.2% TFA, and the disulfide bonds remained intact. (© 1991, American Society for Mass Spectrometry. Reprinted from reference 10 by permission of Elsevier Science, Inc.)

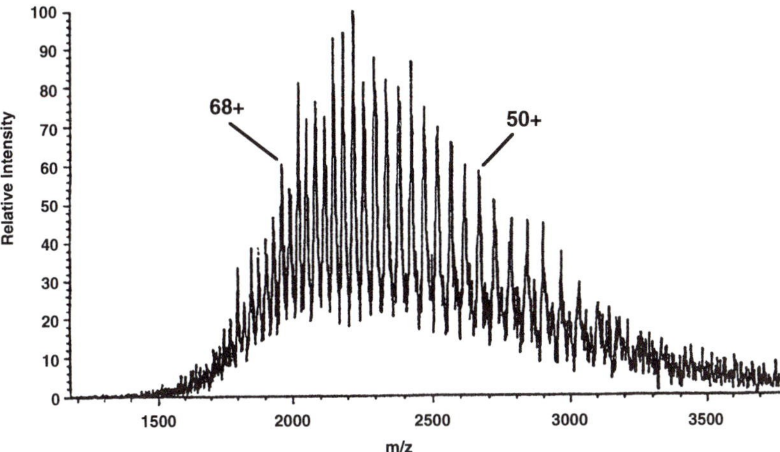

Figure 4.9 The ESI-mass spectrum of bovine serum albumin (BSA) dimer. Direct infusion was used with a 50% aq MeOH/1% HOAc solvent, and the disulfide bonds remained intact. The ion acceleration voltage, V_{acc}, was 4 kV. (© 1992. Reproduced from reference 9 by permission of John Wiley & Sons, Limited.)

Figure 4.10 provides an ESI-mass spectrum of bovine serum apo-transferrin, and the expansion of the 52+ ion region (inset) shows the multiple peaks (components) of this glycoprotein.[10] The two major peaks labeled I and II have average masses of 78,630.5 ± 1.8 and 78,326.8 ± 3.3 Da, respectively. Above 2000 Da, a very small series of peaks is observed between the relatively more intense peaks, M_{expt} of 77,170 Da. They coalesce and are hidden by their m/z superposition on the intense lower mass peaks.

Problems 4.1–4.4 provide further practice for the reader on the subject of calculating the experimental mass values of proteins. For the following problems, calculate the M_{expt} by arranging the m_2, m_1, Δm, approximate n_2, n_2 and individual experimental mass measurements (M) in a tabular format.

PROBLEM 4.1 Figure 4.11, chicken egg lysozyme.[11]

PROBLEM 4.2 Figure 4.12, human hemoglobin β-chain variant A.[12]

PROBLEM 4.3 Figure 4.13, human hemoglobin β-chain variant Willamette.[12]

PROBLEM 4.4 Figure 4.14, equine myoglobin.[8]

In practice, an experimental mass calculation does not need to be done in a manual fashion. Mass spectrometry computer algorithms exist to deconvolute the various m/z charge states in a mass spectrum into not only an M_{expt} value, but also into a spectrum where the x-axis is in units of mass, not m/z. This latter (deconvoluted) spectrum is essentially the neutral mass spectrum of the ESI-multiply-charged mass spectrum. Algorithmic procedures can provide information on the presence of closely spaced peaks originating from other compounds, including single amino acid changes, carbohydrate heterogeneity, and alkali

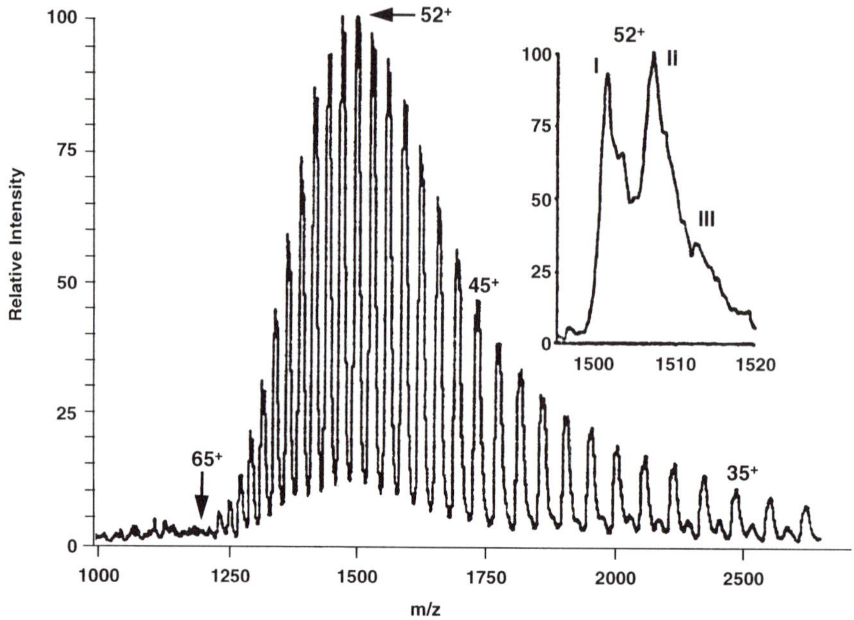

Figure 4.10 The ESI-mass spectrum of bovine serum apo-transferrin. Inset shows the detail of the $(M + 52H)^{52+}$ peak where two extra protein components are observed. (© 1991, American Society for Mass Spectrometry. Reprinted from reference 10 by permission of Elsevier Science, Inc.)

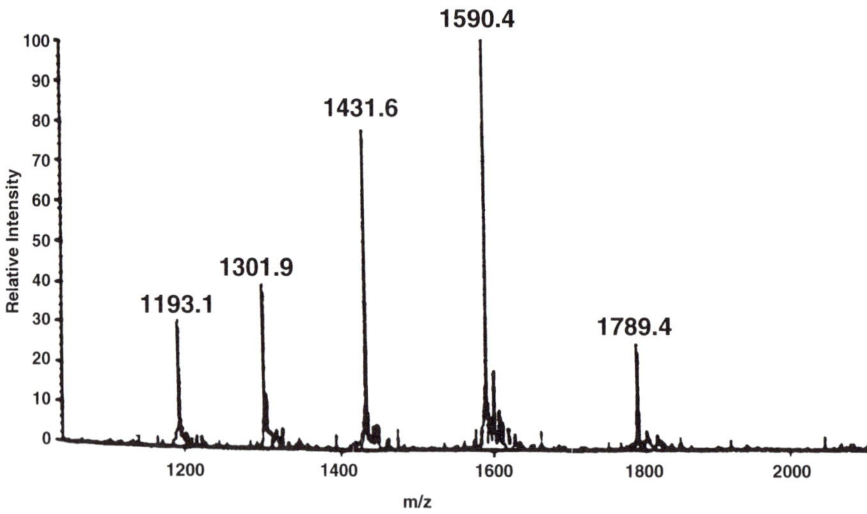

Figure 4.11 The ESI-mass spectrum of chicken egg lysozyme. Solvent was 49.5% MeOH and 1% HOAc, and the disulfide bonds remained intact. $V_{acc} = 2$ kV. (© 1996. Reproduced from reference 11 by permission of John Wiley & Sons, Limited.)

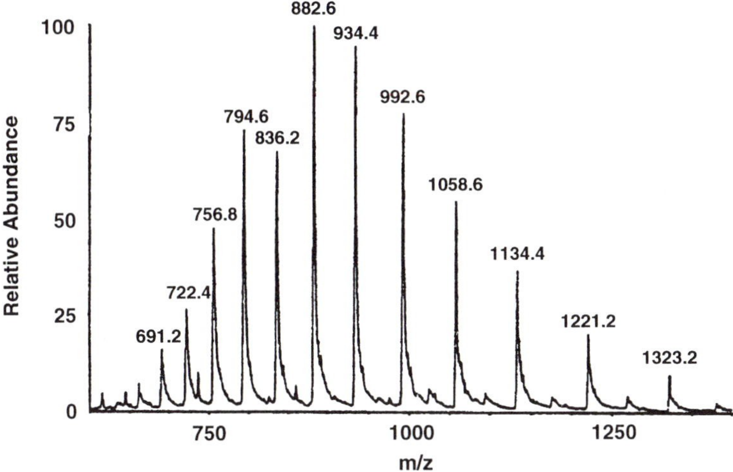

Figure 4.12 The ESI-mass spectrum of human hemoglobin β-chain variant A. (© 1993. Reproduced from reference 12 by permission of John Wiley & Sons, Limited.)

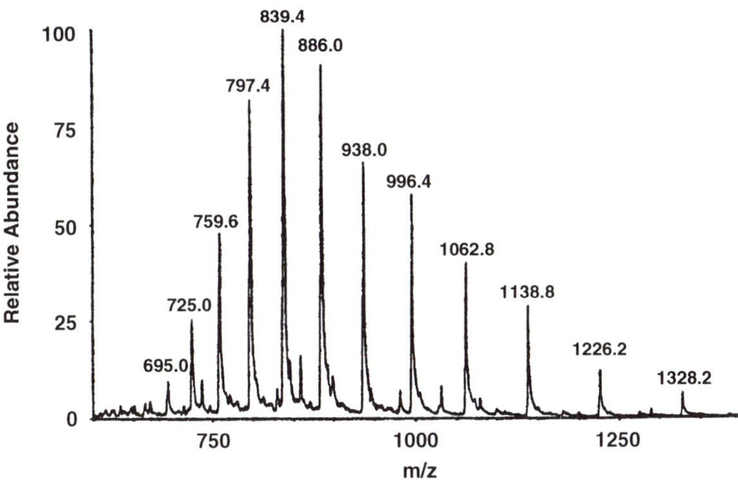

Figure 4.13 The ESI-mass spectrum of human hemoglobin β-chain variant Willamette. (© 1993. Reproduced from reference 12 by permission of John Wiley & Sons, Limited.)

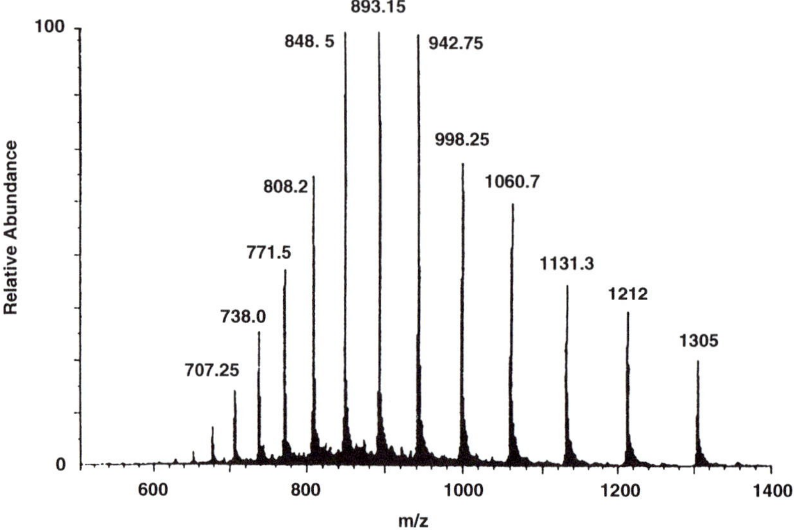

Figure 4.14 The ESI-mass spectrum of equine myoglobin. (© 1990. Reprinted from reference 8 by permission of Academic Press, Inc., Orlando, Florida.)

adducts (*vide infra*) on the analyte. Figures 4.6, 4.8, and 4.10 provide examples of the presence of significant amounts of several species in a multiply charged mass spectrum. Each substance produces a sequential pattern of charged peaks, and it would be very difficult to manually separate the m/z species from each peak, especially in Figures 4.8 and 4.10.

Figure 4.15a provides a raw ESI-mass spectrum of carbonic anhydrase, and note that the deconvoluted spectrum in Figure 4.15b shows essentially an intense single peak with an experimental mass of 29,014 Da.[13] A number of mathematical protocols can be used in a deconvolution algorithm, and examples of these procedures can be found in the literature by Fenn et al.,[14–16] the maximum entropy (MaxEnt) algorithm,[17] and an entropy-based algorithm by Reinhold and Reinhold.[18] Figure 4.16a provides a raw ESI-mass spectrum of β_H bovine lens crystallin protein. Figure 4.16b shows that each peak cluster in the pattern of peaks in Figure 4.16a is composed of three distinct proteins.[19] Figure 4.17 shows two clear separate series of charge states for the human growth hormone (hGH). Note the distinction of the two series; namely, m/z 2329.9 and 2108.4 are elements of the lower intensity series and m/z 2213.4 and 2012.4 belong to the dominant series of peaks.[20]

PROBLEM 4.5 Calculate the M_{expt} of the species that represent each series of peaks in Figure 4.17. Based on the M_{expt} values, what is the relationship between the two proteins? In the calculation of the M_{expt} of the protein in the minor intensity series, the n_2 value should be rounded to the nearest 0.1 Da of the calculated approximate n_2 value.

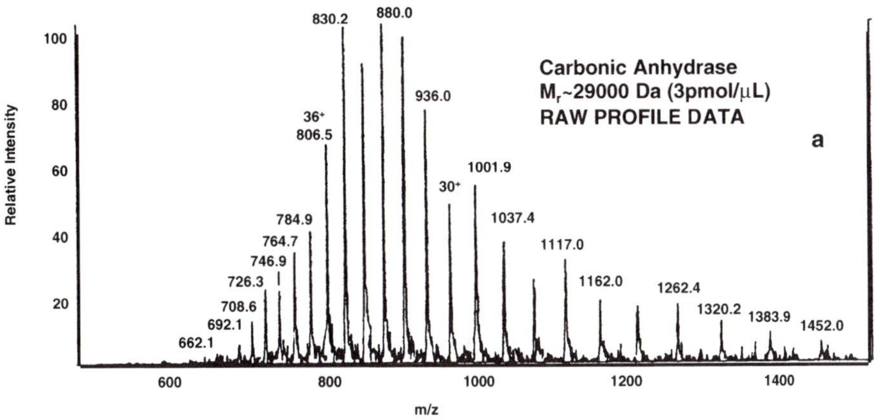

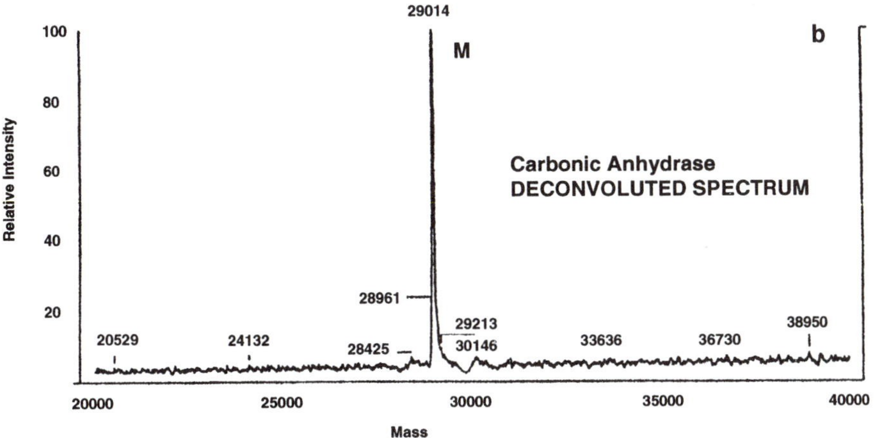

Figure 4.15 (a) The ESI-experimental mass spectrum and (b) deconvoluted, neutral-scale mass spectrum of carbonic anhydrase. (© 1990. Permission granted by Finnigan Corporation, San Jose, California.)

Figure 4.18a provides an ESI-mass spectrum of a mixture of two proteins and a protein impurity, and Figure 4.18b shows the deconvoluted neutral-scale mass spectrum.[7,21] This spectrum comprises a mixture of separate proteins, and each component displays its own series of multiply charged peaks. The three series of peaks are superimposed on each other in Figure 4.18a because they were infused into the mass spectrometer as a mixture with no preseparation (Chapter 2). Table 4.6 provides the manual mass analysis for the protein mixture. The major peaks are listed in Table 4.6, and an analysis clearly yields two sets of masses. The M_{expt} values for thioredoxin and ubiquitin are very close to their calculated average mass values. This analysis of thioredoxin is similar to that of the analysis in Table 4.3 for the pure protein solution.[7]

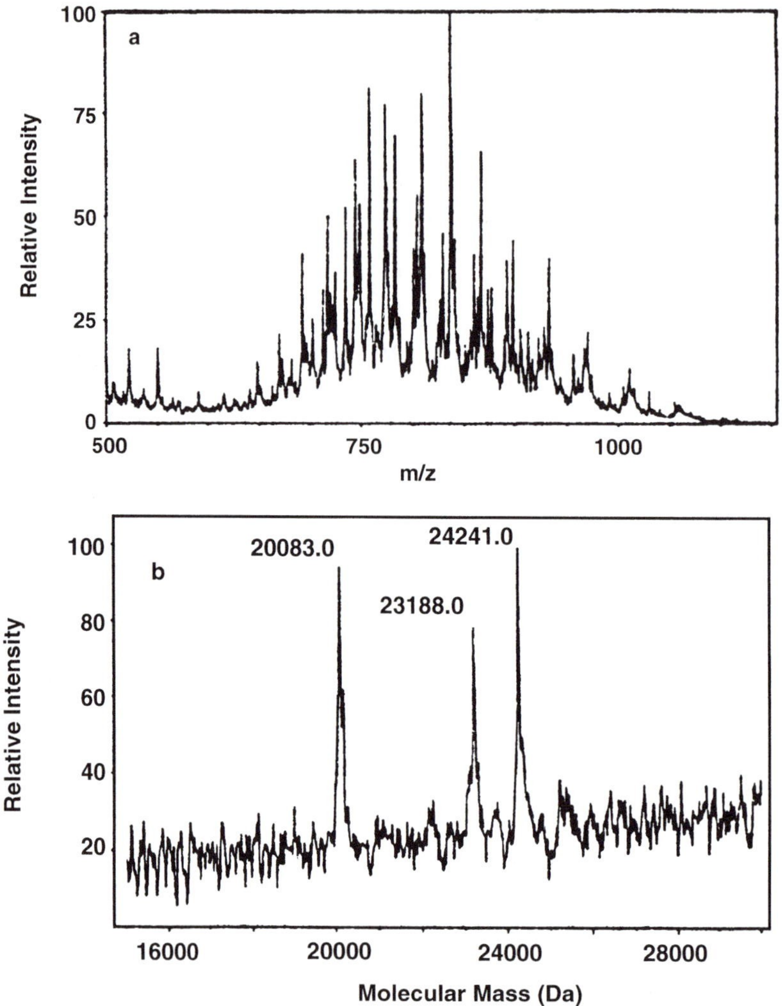

Figure 4.16 (a) The ESI-experimental mass spectrum and (b) deconvoluted, neutral-scale mass spectrum of a mixture of β_H bovine lens crystallin proteins. (© 1991. Reprinted from reference 19 by permission of Portland Press Limited, Essex, U.K.)

In this analysis, the manual procedure reveals an anomaly at the 898.8–857.4-Da ion pair and thus was not considered in the average mass calculations. These ions are shown to originate from the two different proteins.

Figure 4.19 presents mass-labeled peaks of an ESI-mass spectrum of the ovalbumin glycoprotein.[22] The inset to Figure 4.19 shows two dominant glycoproteins in the sample, and they are relatively close in mass. However, these peaks are very broad, with a peak width at half-height of at least several hundred daltons for both proteins. This indicates that many individual species constitute each major peak. Table 4.7 presents an M_{expt} analysis of Figure 4.19, and,

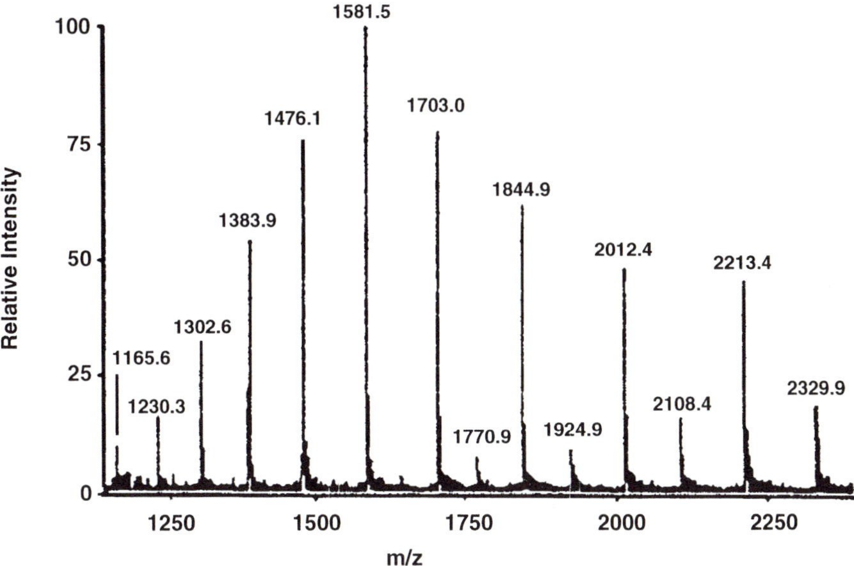

Figure 4.17 The ESI-mass spectrum of human growth hormone. (© 1995. Reprinted from reference 20 by kind permission of Elsevier Science-NL, Sara Burgerhartstraat 25, 1055 KV Amsterdam, The Netherlands.)

indeed, very little information can be obtained from a manual analysis. The Δm and n_2 values provide no coherent trends. Thus, this spectrum must be analyzed with a computer algorithm. The ESI-mass spectrum of recombinant plasminogen-activating inhibitor (PAI) glycoprotein[22] is presented in Figure 4.20, and a similar analysis to that of ovalbumin (Figure 4.19) is observed in Table 4.8. Despite the relatively sharp multiply charged peaks in Figure 4.20, the deconvoluted spectrum in the inset shows a broad range of impurities with possible cation adducts. It is these other materials that most likely prevent a satisfactory manual analysis (Table 4.8). Thus, computer deconvolution techniques are also required for this spectrum.

Figure 4.21 provides a theoretical analysis of the distribution of peaks in an ESI-mass spectrum.[9] Five different biological molecules of average mass 10,000; 20,000; 50,000; 100,000; and 300,000 Da are presented. Note that as the M_r increases, more charge states appear in a given mass region. For example, six peaks can be found between m/z 1000 and 2000 for a 10,000-Da protein. However, for a protein that has a mass of 100,000 Da, 51 peaks are found between m/z 1000 and 2000, and the spacing between the peaks is considerably smaller than that of a 10,000-Da protein. As was observed in examples such as Figures 4.19 and 4.20, multiply charged peaks can be broadened due to the presence of more than a few compounds. An ideal situation is shown in Figure 4.21. The majority of quadrupole and sector mass spectrometry instruments have mass scanning limits at about 2000–2500 Da and 5000–6000 Da,

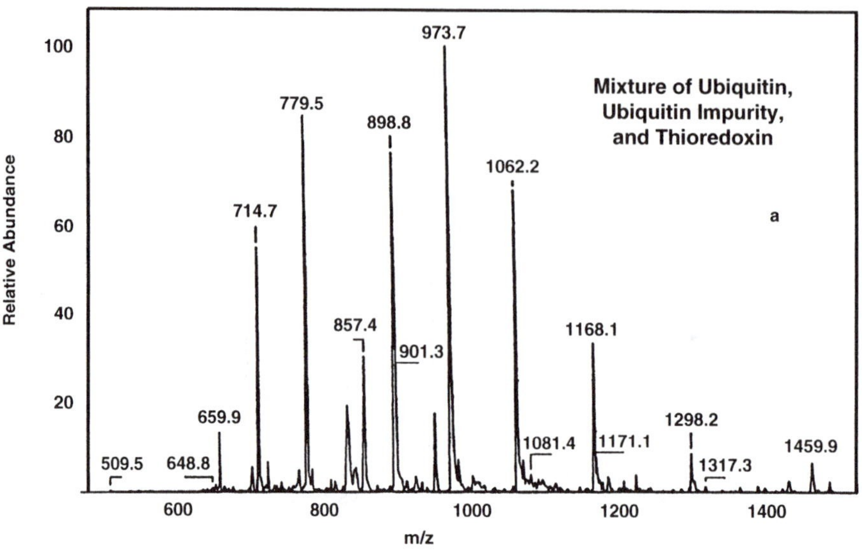

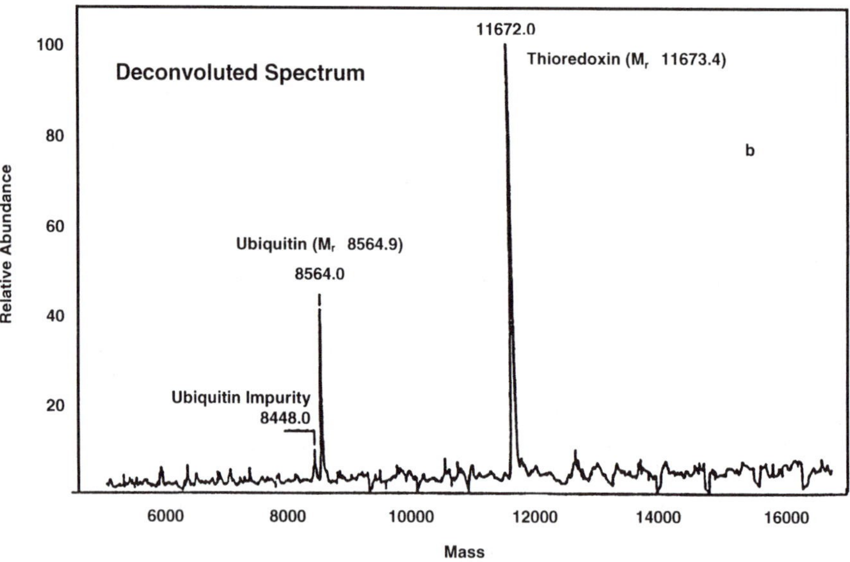

Figure 4.18 (a) The ESI-experimental mass spectrum and (b) deconvoluted, neutral-scale mass spectrum of a mixture of ubiquitin, ubiquitin variant protein, and thioredoxin. (© 1990, American Chemical Society. Reprinted from reference 7 with permission.)

Table 4.6 Experimental mass calculation for a mixture of three proteins

m_2	m_1	Δm	Approx. n_2	n_2	M
1,459.9	1,298.2	161.7	8.02	8	11,671.2
1,298.2	1,168.1	130.1	8.97	9	11,674.8
1,168.1	1,062.2	105.9	10.02	10	11,671.0
1,062.2	973.7	89.2	10.90	11	11,673.2
973.7	898.8	74.9	11.99	12	11,672.4
898.8	857.4	41.4	20.69	—	—
857.4	779.5	77.9	9.99	10	8,564.0
779.5	714.7	64.8	11.01	11	8,563.5
714.7	659.9	54.8	12.02	12	8,564.4

Another data point can be obtained by having $m_2 = m/z$ 659.9; n_2 must be 13 by following the trend of the charges: $M = 13(659.9 - 1) = 8{,}565.7$ Da

$M_{expt} = 11,672.52$ Da For thioredoxin
$M_r = 11,673.4$ Da

$M_{expt} = 8{,}564.4$ Da For ubiquitin
$M_r = 8{,}565.0$ Da

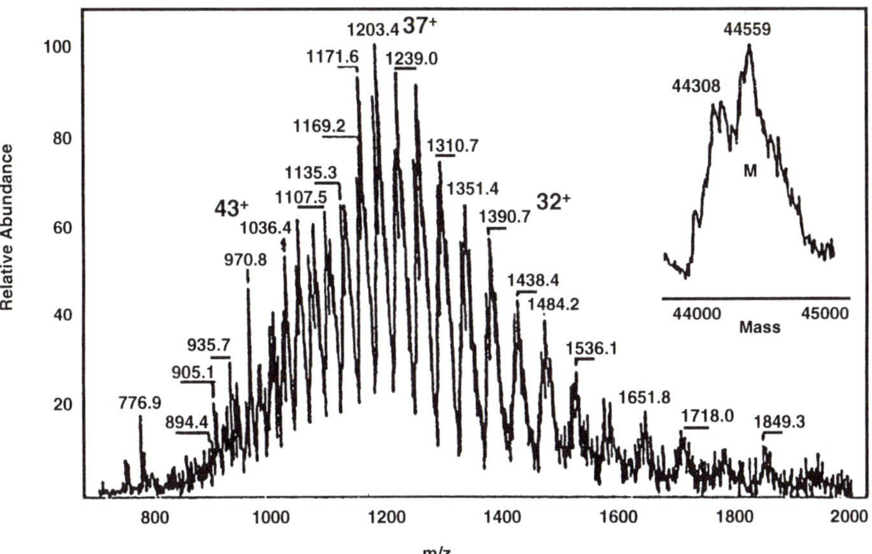

Figure 4.19 The ESI-mass spectrum of the chicken egg ovalbumin glycoprotein. Inset shows the deconvoluted, neutral-scale mass spectrum. Solvent was 50% MeOH/5% HOAc and the disulfide bonds remained intact. M = molecular mass. (© 1990. Reprinted from reference 22 by permission of Academic Press, Inc., Orlando, Florida.)

Table 4.7 Experimental mass calculation of the glycoprotein ovalbumin

m_2	m_1	Δm	Approx. n_2	n_2	M
1,484.2	1,438.4	45.8	30.34	30	44,496.0
1,438.4	1,390.7	47.7	29.13	29	41,684.6
1,390.7	1,351.4	39.3	34.36	34	47,249.8
1,351.4	1,310.7	40.7	32.18	32	43,212.8
Unlabeled peak					
1,239.0	1,203.4	35.6	33.77	38	47,044
1,203.4	1,171.6	31.8	36.81	37	44,525.8
1,171.6	1,135.3	36.3	31.25	31	36,288.6

Table 4.8 Experimental mass calculation of recombinant PAI protein

m_2	m_1	Δm	Approx. n_2	n_2	M
1,157.9	1,127.4	30.5	36.93	37	42,805.3
1,127.4	1,097.9	29.5	37.18	37	41,676.8
1,097.9	1,069.9	28	38.17	38	41,682.2
1,069.9	1,045.8	24.1	43.35	43	45,962.7
1,045.8	1,019.3	26.5	38.43	38	39,702.4
930.7	911.2	19.5	46.68	47	43,695.9

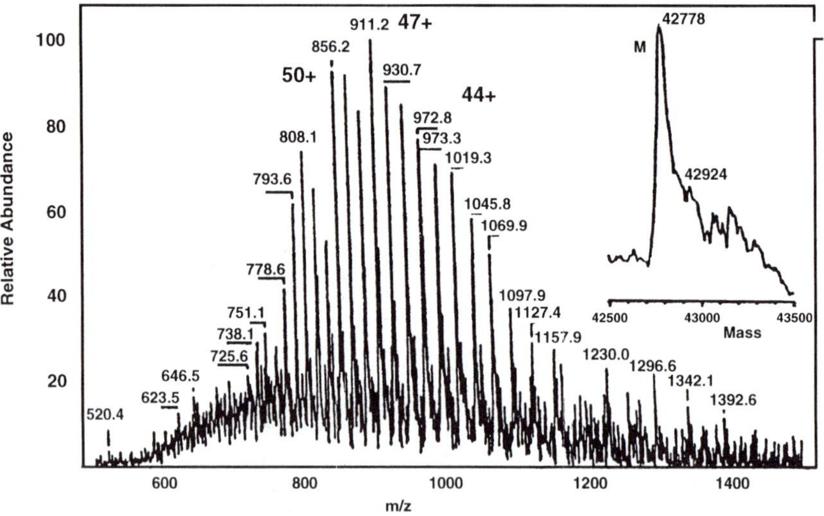

Figure 4.20 The ESI-mass spectrum of recombinant plasminogen-activating inhibitor. Inset shows the deconvoluted, neutral-scale mass spectrum. (© 1990. Reprinted from reference 22 by permission of Academic Press, Inc., Orlando, Florida.)

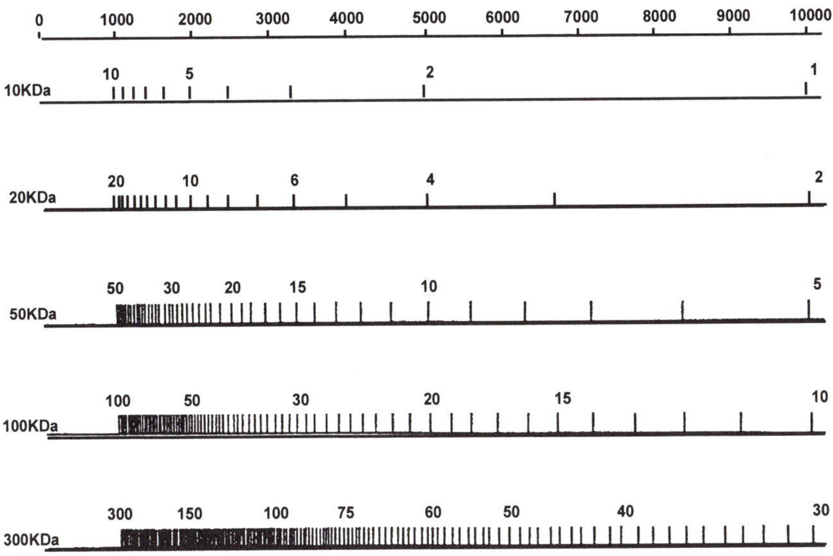

Figure 4.21 Theoretically calculated m/z peaks for the labeled charge states $(M + nH)^{n+}$ for molecules of mass: 10,000; 20,000; 50,000; 100,000; and 300,000 Da. The 10-kDa example shows the m/z for n = 1–10. (© 1992. Reproduced from reference 9 by permission of John Wiley & Sons, Limited.)

respectively. These instruments access the relatively low end of the m/z scale in Figure 4.21, and as analyte mass increases, the tendency for spectral congestion increases. Fourier-transform and time-of-flight instruments can access electrosprayed ions that are much higher in mass than can quadrupole or sector instruments. The higher the analyte's mass, the lower the probability of the appearance of the low-charge-state, high-mass peaks in the mass spectrum. A perusal of the figures in this chapter attests to this phenomenon.

PROBLEM 4.6 In the ESI-mass spectrum of the 10,000-Da molecule in Figure 4.21, calculate the theoretical m/z values for all ten peaks.

For pure analytes that yield relatively narrow ESI-mass spectral peaks, the width of the multiply charged peaks can be calculated. Consider the peaks of the glucagon monomer with an average mass of 3483.8 Da (Figure 3.2). The 3+, or $(M+3H)^{3+}$, and 4+, or $(M + 4H)^{4+}$, proton-ionized mass spectral peaks (data not shown) of glucagon would occur at m/z 1162.3 and 872.0, respectively. Note that in Figure 3.2, there is a total of nine isotope peaks, including the monoisotopic peak for the glucagon monomer. Therefore, the glucagon monomer can be labeled as a protein of average mass of 3483.8 Da with an absolute mass spread of approximately 9 Da. The $(M + H)^{+}$ peak would occur at (3483.8 + 1)/1 = 3484.8 Da in an ESI-mass spectrum and have a mass spread of approximately 9 Da. Thus, the 3+ and 4+ peaks at m/z 1162.3 and 872.0, respectively, would have theoretical widths of 9 Da/3+

= 3 Da and 9/4+ = 2.25 Da, respectively, for the entire isotope envelope of peaks (data not shown).

Consider the 9+ and 10+ peaks of the glucagon trimer (data not shown), which has an average mass of 10451 Da (Figure 3.6). The $(M+9H)^{9+}$ and $(M+10H)^{10+}$ peaks at m/z 1162.2 and 1046.1, respectively, would have theoretical widths of 15 Da/9+ = 1.7 Da and 15/10+ = 1.5 Da, respectively, for the envelope of isotope peaks. For a constant m/z value (m/z 1162.2), as the mass of the analyte increases [e.g., from 3483.3 (glucagon monomer) to 10451 Da (glucagon trimer)], the theoretical (calculated) peak width actually decreases (i.e., from 3 to 1.7 Da, respectively).

The calculated peak width for the multiply charged isotope envelope is usually not observed because of peak-broadening phenomena (*vide infra*). Consider the equine myoglobin molecule with an average mass of 16,951.5 Da. Figure 3.7a shows that there are approximately 24 ^{13}C-based isotopes for the molecular ion. For the $(M+13H)^{13+}$ and $(M+19)^{19+}$ peaks[23] at m/z 1304 and 893 (data not shown), respectively, peak widths of the isotope envelopes are calculated as 24/13 = 1.8 and 24/19 = 1.3 Da, respectively. Another example of increasing M_r with decreasing peak width, from the discussion above, follows where relatively similar m/z values are considered:

	M_r	m/z(nH+)	number of observed isotopes	peak width (Da)
Glucagon monomer	3483.8	872.0 (4+)	9	2.25
Myoglobin	16951	893.0 (19+)	24	13

For the 13+ peak in myoglobin, instead of 1.8 Da, the observed mass spectral peak width was 3.2 Da,[23] and this was the peak width at half-height, not the baseline peak width value. Bovine serum albumin, average mass of 66,443 Da, would be expected to produce a $(M+50H)^{50+}$ peak at m/z 1329, with a peak width at half-height of approximately 2 Da.[23] The observed value was significantly higher at 4.3 Da. The observed deviation from calculated values for the mass width of the isotope envelope of multiply charged ions is to be expected, and there are a number of reasons for this: (a) impure sample, (b) mixture of closely-related sample species, (c) carbohydrate heterogeneity or microheterogeneity (Chapter 12), (d) alkali adducts, and (e) instrumental performance.[23]

An observation in the ESI-mass spectra of proteins is the highest, experimentally observed charge state versus the theoretical maximum charge state. This was briefly introduced for simple peptides in Figure 4.1. As stated, the amine side chains of the Lys, Arg, and His residues, along with the terminal amine moiety, are the predominant sites that attract the proton charge in peptides and proteins. Smith et al.[7] plotted the average mass of over 50 proteins and peptides versus the highest observed charge state in ESI-mass spectra. Figure 4.22 presents this plot, and a roughly linear correlation is observed. Many of the proteins plotted in Figure 4.22 are found in Appendix 3.

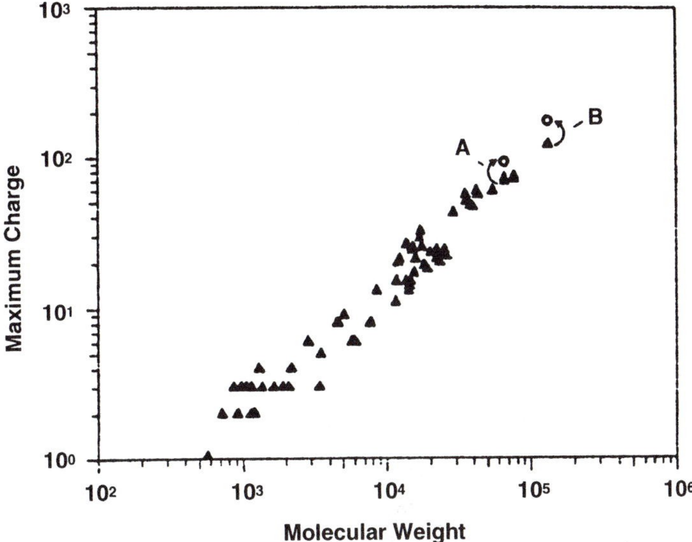

Figure 4.22 Plot of the maximum (M + nH)$^{n+}$ charge state vs M_r in ESI-mass spectra of proteins and peptides given in reference 7. A and B represent an increase in observed charge state upon disulfide linkage reduction (○) for BSA monomer (≃ 66,300 Da) and dimer (≃ 133,000 Da), respectively. (© 1990, American Chemical Society. Reprinted with permission from reference 7.)

There are many factors that influence the observed charge-state distribution of a protein. The literature provides evidence that, for the same protein, different charge-state distributions can be obtained depending on the instrument, instrumental voltage parameters, solvent/buffer conditions, and the physical and biochemical state of the protein (Chapter 7). Examples of the latter include the degree of tertiary structure and the presence of disulfide bridges, where the latter can restrict the freedom of movement of the polypeptide backbone. The greater the number of disulfide bridges in a protein, the more likely that some amino acids will not be able to protonate because of solvent inaccessibility. For example, compare Figures 4.11 and 4.23[9] for lysozyme, M_r of 14306.6 Da; Figures 4.19 and 4.24[24] for ovalbumin, M_r approximately 44,400 Da; Figures 4.8 and 4.25[25] for BSA, M_r 66,431.5 Da; and Figures 4.9 and 4.26 for the BSA dimer,[24] M_r approximately 133,000 Da.

Figures 4.11 and 4.23 present the ESI-mass spectra of lysozyme, and similar solvent conditions were used. Both spectra were acquired on magnetic sector instruments; however, the acceleration voltage of the ions was considerably higher in Figure 4.23 (≃4 kV) than that of Figure 4.11 (2 kV). The relatively more gentle energy-deposition condition for Figure 4.11 can most likely allow for the significant intensities of the $(M + n\text{H})^{n+}$, where $n = 10$–12. In a comparison of Figures 4.19 and 4.24, the relatively gentle V_{acc} voltage and different solvent compositions can account for the higher charge states for ovalbumin in

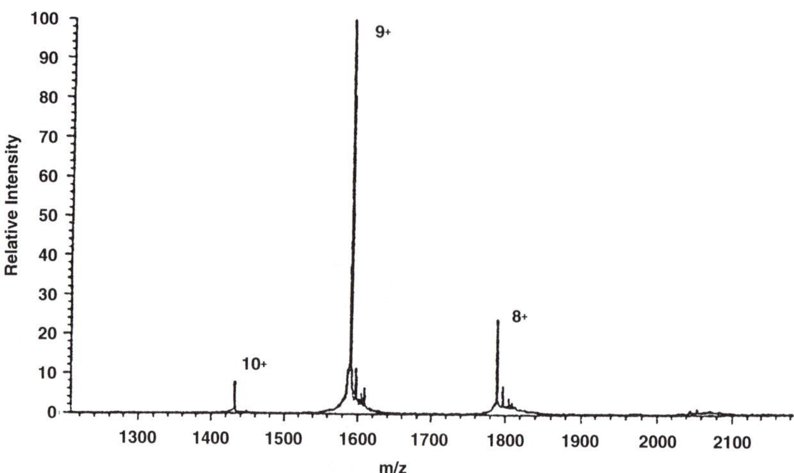

Figure 4.23 The ESI-mass spectrum of chicken egg lysozyme. Direct infusion was used with a 50% aq MeOH/1% HOAc solvent, and the disulfide bonds remained intact. V_{acc} = 4 kV. (© 1992. Reproduced from reference 9 by permission of John Wiley & Sons, Limited.)

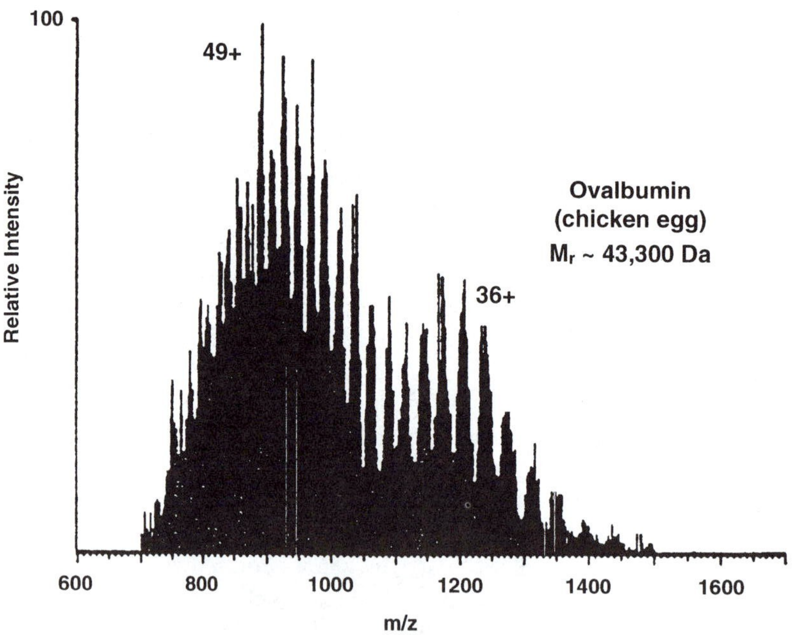

Figure 4.24 The ESI-mass spectrum of chicken egg ovalbumin glycoprotein. Solvent was 1–5% HOAc in water, with a MeOH/AcCN liquid sheath, and the disulfide bonds remained intact. V_{acc} = 200 V. (© 1989. Reprinted from reference 24 by permission of Academic Press, Inc., Orlando, Florida.)

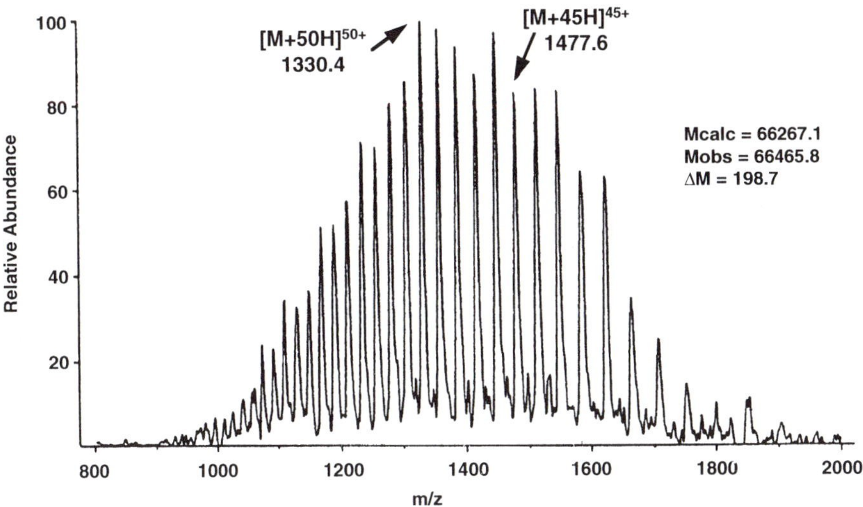

Figure 4.25 The ESI-mass spectrum of BSA. The disulfide bonds were reduced to thiol groups. (© 1990. Reprinted from reference 25 by permission of Academic Press, Inc., Orlando, Florida.)

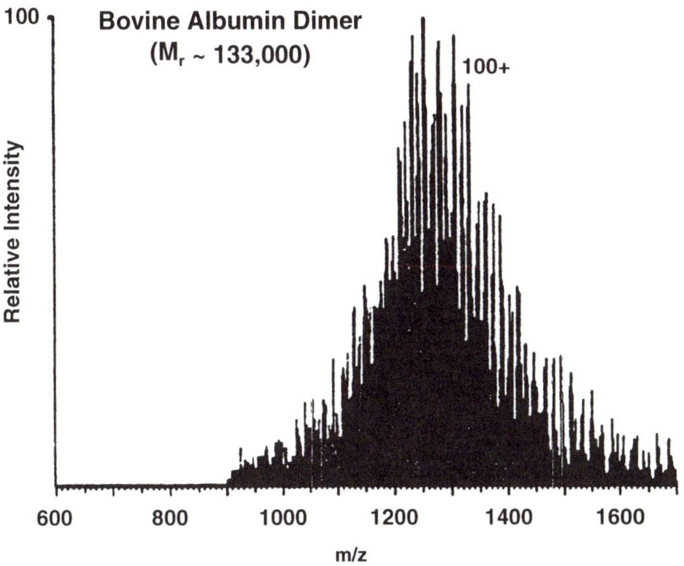

Figure 4.26 The ESI-mass spectrum of the BSA dimer. Solvent was an aqueous 1% HOAc solution, and the disulfide bonds remained intact. V_{acc} = 200 V. (© 1989. Reprinted from reference 24 by permission of Academic Press, Inc., Orlando, Florida.)

Figure 4.24 than in Figure 4.19. The shift to higher amounts of protonation in the disulfide-bond-reduced (to thiol groups) BSA protein (Figure 4.25) compared with the oxidized sulfur atoms in the disulfide bonds (native protein, Figure 4.8) clearly can be rationalized by a greater accessibility of the protein residues to the protons in the liquid media. Figure 4.26 shows significantly higher charge states of $(M + nH)^{n+}$ where $n > 100$, than that of Figure 4.9 for the BSA dimer. A gentler V_{acc} for Figure 4.26 than Figure 4.9 and different solvent conditions can account for these observations.

Interestingly, some proteins can produce peaks whose experimentally derived maximum charge state is higher than the respective calculated value.[7,26] The reason for this observation (Table 4.9) is uncertain, and explanations for this phenomenon may require attributes of the basic mechanisms that govern the formation of ions from an electrosprayed sample.

McLuckey and coworkers[27,28] present an interesting set of experiments that provided strong evidence of a straightforward method for arriving at the total number of basic residues in a peptide. The basic residues consist of Lys, Arg, His, and the N-terminal amine moiety. In order to accomplish this, an ion-trap mass spectrometer was used, but it appears that an FT-MS could also be used. An electrosprayed peptide under denaturing conditions (unfolded peptide) is allowed to react with gaseous HI in the helium atmosphere of the ion trap. It

Table 4.9 Proteins that have the highest multiple positive charge state in their mass spectra, *n* (Expt.), greater than the calculated *n* value[7,26]

Protein	M_r	*n* (Expt.)	*n* (Calc.)[a]
Vasotocin	1,050	3	2
Bombesin	1,620	3	2
Gramicidin	1,882	3	2
Thioredoxin (*E. coli*)	11,673	15	13
α-Lactalbumin (bovine) (carboxymethylated)	14,647	20	17
Interleukin-1 α (for recombinant/human)	17,646	25	18
Casein (dephosphorylated)	22,974	29	26
Trypsinogen (bovine)	23,981	22	21
Chymotrypsin (bovine)	23,234	24	22
Chymotrypsinogen (bovine)	25,656	22	21
Subtilisin (Carlsberg)	27,288	27	19
Carbonic anhydrase (bovine)	29,021	42	39
Thioesterase II	29,230	41	37
Thermolysin	34,334	35	30
Lactate dehydrogenase	35,500	55	51
Alcohol dehydrogenase (yeast *N*-acetyl)	36,748	52	44
Recombinase A (*E. coli*)	37,842	53	44
Actin (rabbit muscle)	42,000	59	46
α_1-Antitrypsin (recombinant truncated variant)	43,911	62	54
Ovalbumin	44,300	53	42
Protein A (recombinant)	43,531	70	64
Outer membrane protein (*B. pertussis*) 1-575	53,133	63	54

[a]*n* (Calc.) = The sum of the basic amino acid residues (K, H, R) plus 1 for the N-terminal amine moiety.

$H_3CC(O)$ - GDEV**K**G**KK**IFVQ**K**CAQC**H**TVE**K**GG**K**H**K**TGPNL**H**GLFG**RK**TGQAPGFSYTDA
N**K**N**K**GITWGEETLMEYLENP**KK**YIPGT**K**MIFAGI**KKK**GE**R**EDLIAYL**KK**ATNE - COOH

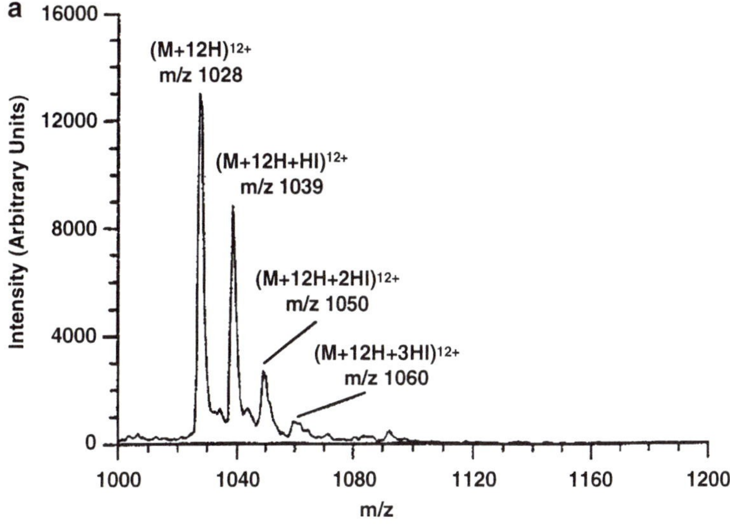

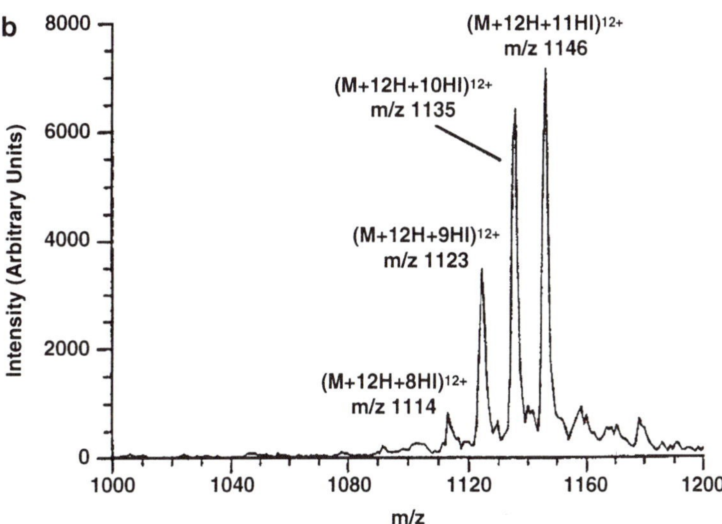

Figure 4.27 The ESI-ion-trap mass spectra of the $(M + 12H)^{12+}$ charge state of bovine cytochrome c. The protein was equilibrated with HI for (a) 25 ms and (b) 300 ms. The amino acid sequence is given in (a) and the basic residues are shown in boldface larger type. (© 1997, American Chemical Society. Reprinted from reference 27 with permission.)

was observed that by allowing sufficient time for the multiply charged peptide ions to adduct to HI, the sum of the number of protons and HI molecules attached to the analyte yielded the sum of the basic amino acid residues. Ion storage times of up to 3 s were used in order to effect saturation with HI of the nonprotonated basic residues. These relatively long equilibrium times could not occur in linear quadrupole or sector mass spectrometers since they do not store ions.

Figure 4.27 provides an example where bovine cytochrome c was allowed to react with gaseous HI for a short (25 ms) and a long (300 ms) time prior to mass analysis. Note that both Figures 4.27a and b provide the $(M + 12H)^{12+}$; however, many more HI molecules attach to the protein under longer equilibrium conditions. Addition of 12 hydrogen ions and a maximum of 11 HI molecules yields 23, and this also represents the total number of basic sites in the protein. The 23 basic sites are shown in boldface larger type in the cytochrome c sequence above Figure 4.27a, and note that the N-terminus is blocked as an acetylated derivative. Out of 21 peptides and proteins, 20 were found to obey this observation.[27]

REFERENCES

1. Harrison, A.G. *Chemical Ionization Mass Spectrometry*; CRC Press: Boca Raton, FL, 1983.
2. Stryer, L., Ed. *Biochemistry*, 4th Edition; W.H. Freeman: New York, 1995, Chapter 2, pp 17–44.
3. *The API Book*, Perkin-Elmer SCIEX Instruments: Canada, 1994; p 80.
4. Meng, C.; McEwen, C.N.; Larsen, B.S. *Rapid Commun. Mass Spectrom.* **1990,** *4,* pp 147–150.
5. Covey, T.R.; Bonner, R.F.; Shushan, B.I.; Henion, J., *Rapid Commun. Mass Spectrom.* **1988,** *2,* pp 249–256.
6. Watson, J.T. In *Introduction to Mass Spectrometry;* Third Edition; Lippincott-Raven Publishers: Philadelphia, PA, 1997; Chapter 11, pp 303–319.
7. Smith, R.D.; Loo, J.A.; Edmonds, C.G.; Barinaga, C.J.; Udseth, H.R. *Anal. Chem.* **1990,** *62,* pp 882–899.
8. Edmonds, C.G.; Smith, R.D. In *Methods in Enzymology: Mass Spectrometry;* McCloskey, J.A., Ed; Academic Press: San Diego, CA, 1990; Volume 193, Chapter 22, pp 412–431.
9. Chapman, J.R.; Gallagher, R.T.; Barton, E.C.; Curtis, J.M.; Derrick, P.J. *Org. Mass Spectrom.* **1992,** *27,* pp 195–203.
10. Feng, R.; Konishi, Y.; Bell, A.W. *J. Am. Soc. Mass Spectrom.* **1991,** 2, pp 387–401.
11. Gallagher, R.T.; Chapman, J.R.; Mann, M. *Rapid Commun. Mass Spectrom.* **1990,** 4, pp 369–372.
12. Light–Wahl, K.J.; Loo, J.A.; Edmonds, C.G.; Smith, R.D.; Witkowska, H.E.; Shackleton, C.H.L.; Wu, C.C., *Biol. Mass Spectrom.* **1993,** *22,* pp 112–120.
13. *Anal. News,* January/February 1990, p 5.
14. Mann, M.; Meng, C.K.; Fenn, J.B. *Anal. Chem.* **1989,** *61,* pp 1702–1708.
15. Fenn, J.B.; Mann, M.; Meng, C.K.; Wong, S.F.; Whitehouse, C.M. *Mass Spectrom. Rev.* **1990,** *9,* pp 37–70.

16. Labowsky, M.; Whitehouse, C.; Fenn, J.B. *Rapid Commun. Mass Spectrom.* **1993,** 7, pp 71–84.
17. Ferrige, A.G.; Seddon, M.J.; Green, B.N.; Jarvis, S.A.; Skilling, J. *Rapid Commun. Mass Spectrom.* **1992,** *6,* pp 707–711.
18. Reinhold, B.B.; Reinhold, V.N. *J. Am. Soc. Mass Spectrom.* **1992,** 3, pp 207–215.
19. Edmonds, C.G.; Loo, J.A.; Ogorzalek-Loo, R.R.; Udseth, H.R.; Barinaga, C.J.; Smith, R.D. *Biochem. Soc. Trans.* **1991,** *19,* pp 943–947.
20. Nguyen, D.N.; Becker, G.W.; Riggin, R.M. *J. Chromatogr.* **1995,** *705,* pp 21–45.
21. Hail, M.; Lewis, S.; Zhou, J.; Schwartz, J.; Jardine, I. In *Biological Mass Spectrometry;* Burlingame, A.L.; McCloskey, J.A., Eds.; Elsevier Science B.V.: Amsterdam, The Netherlands, 1990: pp 101–117.
22. Jardine, I. In *Methods in Enzymology: Mass Spectrometry,* McCloskey, J.A., Ed; Academic Press: San Diego, CA, 1990; Volume 193, Chapter 24, pp 441–455.
23. Loo, J.A.; Edmonds, C.G.; Smith, R.D. *Anal. Chem.* **1991,** *63,* pp 2488–2499.
24. Loo, J.A.; Udseth, H.R.; Smith, R.D. *Anal. Biochem.* **1989,** *179,* pp 404–412.
25. Hirayama, K.; Akashi, S.; Furuya, M.; Fukuhara, K. *Biochem. Biophys. Res. Commun.* **1990,** *173,* pp 639–646.
26. Ashton, D.S.; Beddell, C.R.; Cooper, D.J.; Green, B.N.; Oliver, R.W.A. *Org. Mass Spectrom.* **1993,** *28,* pp 721–728.
27. Stephenson, J.L., Jr.; McLuckey, S.A. *Anal. Chem.* **1997,** *69,* pp 281–285.
28. Stephenson, J.L., Jr.; McLuckey, S.A. *J. Am. Chem. Soc.* **1997,** *119,* pp 1688–1696.

ANSWERS

PROBLEM 4.1 Experimental mass calculations for chicken egg lysozyme

m_2	m_1	Δm	Approx. n_2	n_2	M
1,789.4	1590.4	199.0	7.99	8	14,307.2
1,590.4	1431.6	158.8	9.01	9	14,304.6
1,431.6	1301.9	129.7	10.03	10	14,306.0
1,301.9	1193.1	108.8	10.96	11	14,309.9
1,193.1				12	14,305.2

$M_{expt} = 14,306.6 \pm 1.87$ Da
$M_r = 14,305.2$ Da

PROBLEM 4.2 Experimental mass calculations for the A-designated variant of the human hemoglobin β-chain

m_2	m_1	Δm	Approx. n_2	n_2	M
1,323.2	1,221.2	102	11.96	12	15,866.4
1,221.2	1,134.4	86.8	13.06	13	15,862.6
1,134.4	1,058.6	75.8	13.95	14	1,5867.6
1,058.6	992.6	66.0	15.02	15	15,864.0
992.6	934.4	58.2	16.04	16	15,865.6
934.4	882.6	51.8	17.02	17	15,867.8
882.6	836.2	46.4	18.00	18	15,868.8
836.2	794.6	41.6	19.08	19	15,868.8
794.6	756.8	37.8	19.99	20	15,872.0
756.8	722.4	34.4	20.97	21	15,871.8

$M_{expt} = 15,867.1 \pm 2.6$ Da
$M_r = 15,867.2$ Da

PROBLEM 4.3 Experimental mass calculations for the Willamette variant of the human hemoglobin β-chain

m_2	m$_1$	Δm	Approx. n_2	n_2	M
1,328.2	1,226.2	102	12.01	12	15,926.4
1.226.2	1,138.8	87.4	13.02	13	15,927.6
1,138.8	1,062.8	76.0	13.97	14	15,929.2
1,062.8	996.4	66.4	14.998	15	15,927.0
996.4	938.0	58.4	16.04	16	15,926.4
938.0	886.0	52.0	17.02	17	15,929.0
886.0	839.4	46.6	17.99	18	15,930.0
839.4	797.4	42.0	18.96	19	15,929.6
797.4	759.6	37.8	20.07	20	15,928.0
759.6	725.0	34.6	20.92	21	15,930.6

$M_{expt} = 15,928.4 \pm 1.4$ Da
$M_r = 15,926.2$ Da

PROBLEM 4.4 Experimental mass calculations for equine myoglobin

m_2	m_1	Δm	Approx. n_2	n_2	M
1305	1212	93	13.02	13	16,952.0
1212	1,131.3	80.7	14.01	14	16,954.0
1,131.3	1,060.7	70.6	15.01	15	16,954.5
1,060.7	998.25	62.45	15.97	16	16,955.2
998.25	942.75	55.5	16.97	17	16,953.25
942.75	893.15	49.6	17.99	18	16,951.5
893.15	848.5	44.65	18.98	19	16,950.85
848.5	808.2	40.3	20.03	20	16,950.0
808.2	771.5	36.7	20.99	21	16,951.2
771.5	738.0	33.5	22.0	22	16,951.0
738.0	707.25	30.75	22.97	23	16,951.0

$M_{expt} = 15{,}952.2 \pm 1.6$ Da
$M_r = 16{,}951.5$ Da

PROBLEM 4.5 Calculation of the experimental mass of the dominant series of peaks of hGh in Figure 4.17

m_2	m_1	Δm	Approx. n_2	n_2	M
2,213.4	2,012.4	201	10.01	10	22,124.0
2,012.4	1,844.9	167.5	11.01	11	22,125.4
1,844.9	1,703.0	141.9	11.99	12	22,126.8
1,703.0	1,581.5	121.5	13.01	13	22,126.0
1,581.5	1,476.1	105.4	14.00	14	22,127.0
1,476.1	1,383.9	92.2	15.00	15	22,126.5
1,383.9	1,302.6	81.3	16.01	16	22,126.4
1,302.6	1,230.3	72.3	17.00	17	22,127.2
1,230.3				18	22,127.4

$M_{expt} = 22{,}126.2 \pm 0.95$ Da

Calculation of the experimental mass of the minor series of peaks of hGH in Figure 4.17

m_2	m_1	Δm	Approx. n_2	n_2	Corrected n_2	M
2,329.9	2,108.4	221.5	9.51	9.50	19	44,249.1
2,108.4	1,924.9	183.5	10.48	10.50	21	44,255.4
1,924.9	1,770.9	154.0	11.49	11.50	23	44,249.7
1,770.9				12.50	25	44,247.5

$M_{expt} = 44{,}250.4 \pm 2.97$ Da

The two series of charge states behave in a similar fashion to BSA in Figures 4.8 and 4.9. The BSA can exist as a monomer and dimer depending on the solvent composition of the buffer, and this phenomenon also occurs for hGH. The minor series results in numbers that lie between integer values. This observation provides evidence that the relationship of the two series of peaks constitute multiple numbers of the analyte; namely, monomer and dimer. For hGH, an aqueous solvent yields primarily the dimer form, but, with an addition of ≥30%

of acetonitrile, dissociation takes place and the monomer form predominates as shown in Figure 4.17. Thus, the series of half-charge states (n_2) must be multiplied by a factor of 2 (corrected n_2 term) prior to multiplication of the $m_2 - 1$ term.

PROBLEM 4.6 For a single-proton ionized 10,000-Da species that has a charge of 1+, the m/z value is (10,000 + 1)/1 = 10,001 Da. For a doubly charged species, (10,000 + 2)/2 = m/z 5001; and for

3+,	(10,000 + 3)/3 = m/z 3334.3;
4+,	(10,000 + 4)/4 = m/z 2501;
5+,	(10,000 + 5)/5 = m/z 2001;
6+,	(10,000 + 6)/6 = m/z 1667.7;
8+,	(10,000 + 8)/8 = m/z 1251.0;
9+,	(10,000 + 9)/9 = m/z 1112.1;
10+,	(10,000 + 10)/10 = m/z 1001.0.

5

Analytical Figures of Merit for a High-Mass Isotope Envelope

Compounds of greater than a few thousand daltons have an instrumental/alkali/mixture contribution and an inherent (isotope spread) contribution in the broadening of their mass spectral peaks, as was noted in Chapters 3 and 4. Therefore, regardless of the biological molecule that is analyzed, its high-mass nature requires a number of additional analytical figures of merit which help to determine the precision, relative accuracy, and assignment confidence of the mass spectral peaks.

These analytical figures of merit take into account the peak-broadening issues with respect to closely spaced and overlapping peaks, and, in a less rigorous sense, isolated single peaks. The basic parameter is resolution, and this is used to characterize the spectral quality of select mass spectral peaks.[1–5] The resolution or resolving power parameter defines the relative separation of two closely spaced mass spectral peaks. Figure 5.1a shows a situation where two peaks overlap to a limited extent.[6] Note that the solid tracing is observed in an experimental mass spectrum. Theoretically, the peaks intersect each other at 5% of their respective maximal intensity, yet they coalesce to form an observed merged minimal intensity at 10% of the peak height. This is usually referred to as a 10% valley. For overlapping peaks, the Δm value is the width in daltons at a select relative intensity for each peak, as well as the mass separation between the maxima of both peaks. It is the 5% Δm value that serves as the link between a true resolution measurement of two overlapping peaks (Δm at 5% of maximal intensity) and that of one of the isolated peaks (Δm at 50% of maximal intensity) (Figure 5.1b).[7]

Another situation that can be encountered is when two peaks are very close together and display a significant amount of overlap,[7] as in Figure 5.1c. The mass separation is very small indeed, and the resolution of both peaks at half-

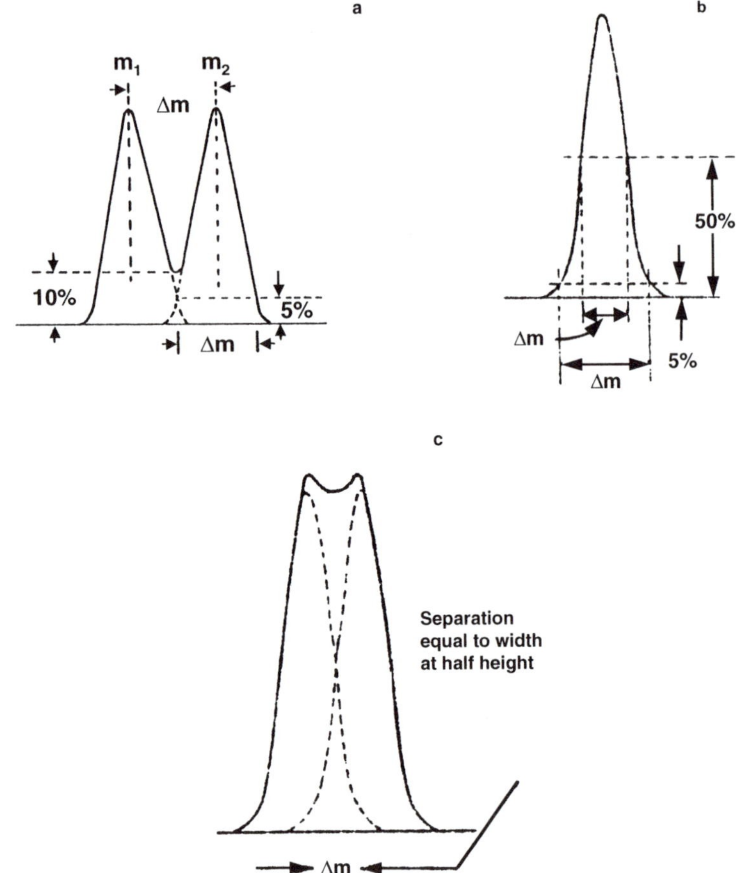

Figure 5.1 The concept of resolution between two adjacent mass spectral peaks is presented in contour plots: two peaks that overlap at the (a) 10% and (c) 50% relative intensity level; (b) resolution parameters for the contour plot of an isolated mass spectral peak. [(a) © 1994. Reprinted from reference 6 by permission of Plenum Publishing Company, New York. (b and c) © 1974. Reproduced from reference 7 by permission of Thomson Science & Professional, New York.]

height (50%), or peak width (in daltons) at half-height, is usually used. The difference in daltons between the peak maxima also define the Δm (Figure 5.1c). Thus, the resolution of two overlapping peaks is usually taken as 10% of the height of the peaks (10% valley) and that of a single peak is usually calculated at the 5% peak height. Resolution is also used to provide an accuracy figure of merit in the comparison of a theoretical or calculated mass value with that of the experimentally observed value. Thus, resolution defines the peak width, and it is used in an equation to yield a parts-per-million value (ppm) that is an estimate of the accuracy of an experimental mass assignment.[8]

The following equations are used:

$$R = \frac{M}{\Delta m}$$

$$\text{ppm} = \frac{10^6}{R} = \frac{10^6 \Delta m}{M}$$

where R = resolution, experimentally measured M = mass of the peak of interest, and Δm = width in daltons of the peak or width in daltons between overlapping peaks. For example, it is desired to measure the mass of a compound of m/z 1000 to within ± 0.5 Da. The ppm value in the estimate of the mass accuracy is $10^6(0.5)/1000 = 500$ ppm. Therefore, at m/z 1000, 0.5 Da represents an accuracy of 500 ppm and corresponds to the m/z 999.5–1000.5 mass range. Table 5.1 provides further examples; note that as the mass tolerance decreases (i.e., decreasing Δm value) for a given mass, the resolution increases. Achieving smaller Δm values yields narrower mass spectral peaks (smaller Δm). The accuracy of mass measurement increases and is mirrored in the decreasing ppm values. In the understanding of Table 5.1, note that the figure 50 ppm does not mean 50-millionths of a dalton or m/z value. Fifty-millionths of a dalton is 0.000050 Da. The ppm value is derived from the measured width of a peak *in conjunction with* the m/z value of the compound. The m/z value can originate from a singly or multiply charged mass assignment.

The mass range in Table 5.1 can also be used as a means to compare a calculated or theoretical mass value with that of the experimentally observed value. Table 5.2 provides data from D'Agostino et al. for five synthetic peptides.[9] For the first entry, the calculations are as follows:

$$\frac{(\text{experimental} - \text{calculated})10^6}{\text{calculated}} = \text{ppm}$$

$$\frac{3708.99 - 3708.9494}{3708.9494} = \frac{0.0406(10^6)}{3708.9494} = 11.0 \text{ ppm}$$

where M is the calculated mass.

Table 5.1 The resolution analytical parameter for mass spectral peaks

m/z	±Mass tolerance (Δm)[a]	Resolution	±ppm	Mass range (Da)
1,000	0.5	2,000	500	999.5–1,000.5
1,000	0.05	20,000	50	999.95–1000.05
1,000	0.005	200,00	5	999.995–1,000.005
2,000	0.5	4,000	250	1,999.5–2,000.5
2,000	0.05	40,000	25	1,999.95–2,000.05
2,000	0.005	400,000	2.5	1,999.995–2,000.005
10,000	0.5	20,000	50	9,999.5–10,000.5
10,000	0.05	200,000	5	9,999.95–10,000.05
10,000	0.005	2,000,000	0.5	9,999.995–10,000.005

[a]Measured peak width.

Table 5.2 Calculated and experimentally observed monoisotopic masses for five synthetic peptides

Peptide	Experimental mass mean[a] ± SD	Calculated mass	Δm_{diff}	ppm	Resolution employed
Ac-KCEALEGKLGAVEEKLGAV EEKLGAVEEKLEALEG-NH_2	3708.99 ± 0.02	3708.9494	+0.0406	11.0	4500
Ac-EIAELKKEIAELKK-NH_2	1682.04 ± 0.01	1681.9978	+0.0422	25.1	2500
Ac-ELEKLLKELEKLLKEKEK-NH_2	2280.34 ± 0.04	2280.3668	−0.0268	11.8	3000
FPVKLFPVKL (ring)	1168.90 ± 0.01	1168.7372	+0.1628	139.3	2500
ATKKEVPLGVAADANKLGEIEAL KAEIEALKAGGDEQFIPKGGEIEA LKAEIEALKA-NH_2	5883.44 ± 0.12	5883.2247	+0.2153	36.6	6000

[a] Average of three separate measurements ± standard deviation (SD); masses are in daltons.

PROBLEM 5.1 Table 5.3 presents data from D'Agostino et al. for three synthetic peptides.[9] Calculate the Δm_{diff} and ppm values.

An overlap of two closely spaced peaks was not a consideration in Problem 5.1; therefore, the Δm_{diff} value was not derived from a peak width but from the difference between an experimentally observed peak and its theoretical (calculated) mass value. The standard deviation is a measure of mass precision, not accuracy, and will be dealt with later in this chapter. In Tables 5.2 and 5.3, the monoisotopic mass was calculated by the method presented in Chapter 3 that takes into consideration a 1-charge-unit difference between the m_1 and m_2 ESI-mass spectral peaks. The standard deviation was derived from three separate mass spectral determinations.

The "resolution employed" values in Table 5.2 were not derived from the Δm_{diff} values in Table 5.2. If the Δm_{diff} value of 0.0406 (first entry in Table 5.2) was actually a Δm value between two closely spaced mass spectral peaks, the calculation

$$R = \frac{M}{\Delta m_{\text{calc}}} = \frac{3708.99}{0.0406} = 91{,}354$$

is understood to provide for a separation of two distinct peaks at m/z 3708.99 and 3708.9494 to within a 10% valley. This would only be done if these two peaks were actually present in the mass spectrum. The mass separation of only 0.0406 Da would require a mass spectrometer resolution of 91,300, as calculated for two peaks at 3708.99 and 3708.94 Da. Instead, the resolution values in Table 5.2 are those needed to produce a 10% valley for adjacent ^{13}C isotope peaks in a given multiply charged envelope of ^{13}C peaks, where $\Delta m = 1.0033$ Da. The value of 1.0033 Da is the difference in mass between a carbon-12 and a carbon-13 atom. As stated in Chapter 3, the ^{13}C isotope peaks are separated from each other by approximately 1 Da, and the following discussion presents examples for derivation of the resolution; that is, the resolution employed, or required, between ^{13}C isotope peaks. Figure 5.2 presents an ESI-mass spectrum of ω-conotoxin MVIIC,[10] and the inset shows the ^{13}C isotope peak separation of the $(M + 5H)^{5+}$ species. To determine the approximate mass spectrometer resolution, one can use either the 10% or 50% valley definition. The peaks appear to be closer to the 10% valley definition of peak resolution. The most abundant peak is usually used, and extension lines that follow the contour of the peak (dotted lines in Figure 5.2 inset) are drawn to the baseline. At 10% peak height, the peak width is measured, and the value is used to calculate the R value. Using a millimeter ruler on the page itself, the distance between m/z 550 and 551 in the inset of Figure 5.2 is 14.5 mm, and the distance between the two dotted lines at 10% peak height is approximately 3 mm. Thus, the equivalent value in m/z units for the 3-mm distance is 0.207 Da. Division of m/z 550.6 by 0.207 Da yields 2661, which is the approximate R or resolution value of the instrument. D'Agostino et al. list an applied figure of 3000 resolution.[9]

Table 5.3 Calculated and experimentally observed monoisotopic masses for three synthetic peptides

	Peptide	Experimental mean[a] ± SD	Calculated
1.	Ac-EIEALKAEIEALKAGGDEQF IPKGGEIEALKAEIEALKA-NH_2	4,162.30 ± 0.02	4,162.2412
2.	(Ac-QCGALQKQVGALQKQVGA LQKQVGALQKQVGALQK-NH_2)$_2$ dimer with C-C bridge	7,368.36 ± 0.08	7,368.1718
3.	(Ac-EIEALKAEIEALKAEIEALK AGGDEQFIPKGGEIEALKAEIEAL KAEIEACKA-NH_2)$_2$ dimer with C–C bridge	11,326.33 ± 0.15	11,326.9960[b]

[a] Average of three separate measurements ± standard deviation (SD).
[b] (Weighted) average mass.

Another example will now be shown. Figure 5.3a shows the ESI-mass spectrum of the last synthetic peptide in Table 5.2, and Figure 5.3b shows the isotope distribution of the $(M + 8H)^{8+}$ ion. Treated in a similar fashion to that of the inset of Figure 5.2, a 10% valley definition is used.[10] The distance is 17 mm for a span of 0.4 Da in Figure 5.3b, and the most abundant ^{13}C isotope at 10% peak height is 5.5 mm (0.129 Da) wide. The *R* value of 736.817/0.129 = 5711 is close to the 6000 resolution value employed.

PROBLEM 5.2 Figure 5.4a presents an ESI-mass spectrum of the second peptide entry[9] in Table 5.3. Calculate the approximate mass spectrometer *R* value used for Figure 5.4b. With respect to Figure 5.1a, note, in particular, that the calculation for Δm can be obtained by using one peak or two adjacent peaks. Both methods yield the same value for Δm, and this is true for Δm in Figures 5.2–5.4.

PROBLEM 5.3 Figure 5.5 presents a partial ^{13}C isotope pattern of the $(M + 2H)^{2+}$ ion for bradykinin.[11] Calculate the mass spectrometer resolution using the most intense peak. The inset of Figure 5.5 provides the same spectrum at an increased resolution (with a concomitant loss of absolute signal strength). Calculate the resolution used.

PROBLEM 5.4 Figure 5.6 presents the ^{13}C isotopic distribution envelope of the $(M + 5H)^{5+}$ ion of bovine insulin.[11] Calculate the mass spectrometer *R* value.

PROBLEM 5.5a Figure 5.7a presents an ESI-mass spectrum of the substance P(1–9) peptide.[12] The average (weighted) mass of substance P(1–9) is 1103.6

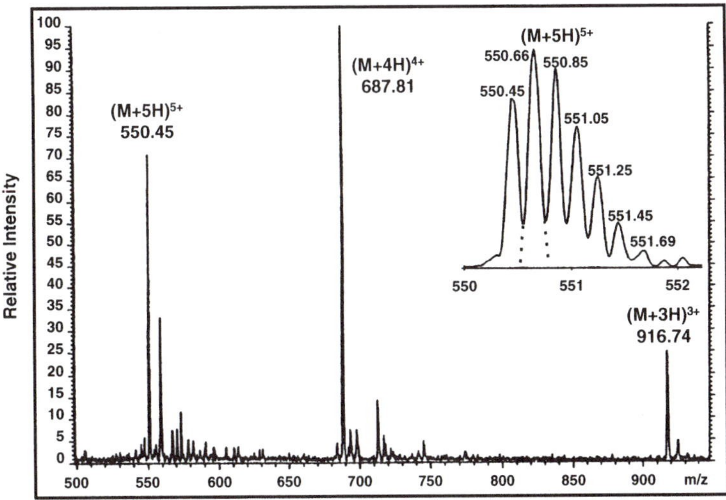

Figure 5.2 The ESI-mass spectrum of ω-conotoxin MVIIC; the inset shows the ^{13}C isotope resolution of the $(M + 5H)^{5+}$ peak. Dashed lines are extended from the *m*/*z* 550.66 peak in the inset to the baseline for resolution calculation purposes.

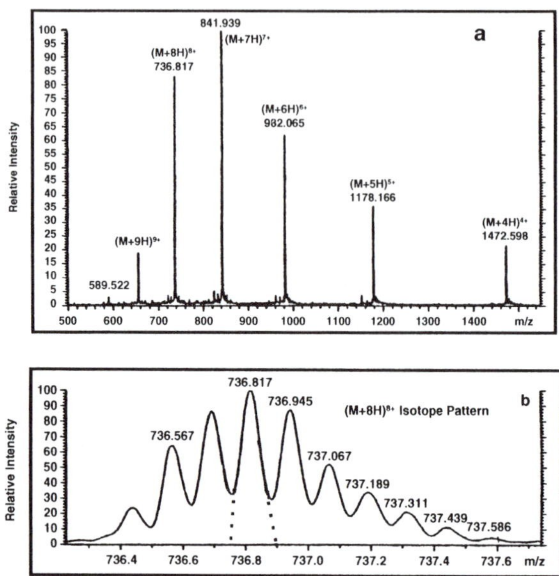

Figure 5.3 (a) The ESI-mass spectra of the last synthetic peptide in Table 5.2; (b) presents the isotope distribution of the $(M + 8H)^{8+}$ peak. The dashed lines extending from the m/z 736.817 peak to the baseline are for resolution calculation purposes.

Da. What is the ppm error for the experimentally derived mass? Hint: Arrive at the experimental zero charge mass, and use the average mass of H.

PROBLEM 5.5b Tandem mass spectrometry or collision-induced dissociation (CID) of the substance P(1–9) peptide is shown in Figure 5.7b. Figure 5.7c is the ^{13}C isotope distribution of the peak labeled b_7 (Chapter 9) in Figure 5.7b.[12] Using a 5% valley definition, calculate the mass spectrometer resolution for the b_7 CID fragment ion.

PROBLEM 5.6 Suppose you had a mixture of two peptides of approximately equal concentrations. The two peptides are identical except that for a given amino acid residue position, one peptide has an Asn residue and the other has a Leu. If the peptide with the Leu has a mass of 10,000 Da, what resolution is required to produce two peaks at a 10% valley separation?[13]

PROBLEM 5.7 Repeat Problem 5.6 but replace the Leu–Asn pair with a Gly–Ala pair.

PROBLEM 5.8 Repeat Problem 5.6 but replace the Leu–Asn pair with a Gln–Lys pair.

PROBLEM 5.9 (a) Figure 5.8 shows a high-resolution Fourier transform mass spectrum of a mixture of the following three octapeptides:

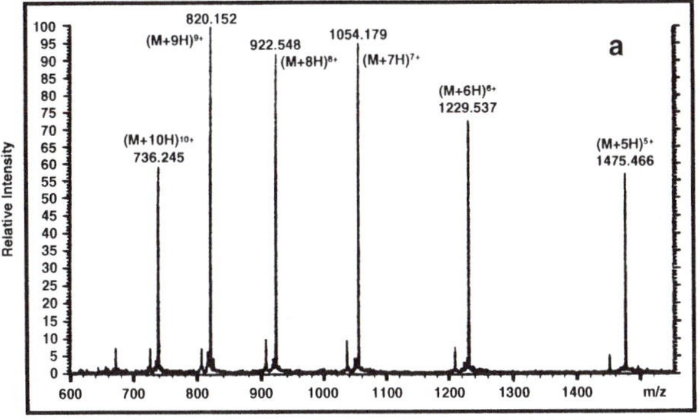

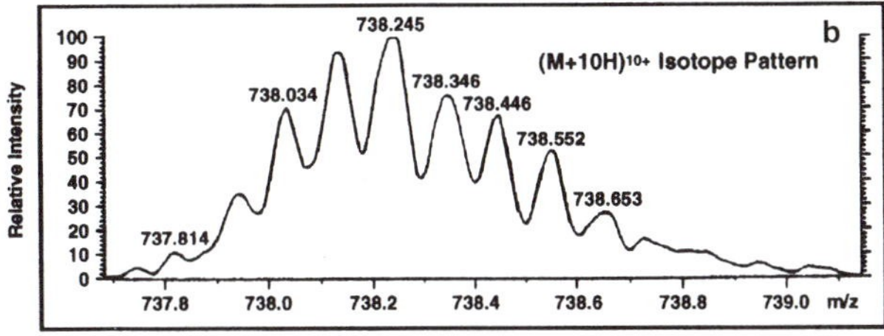

Figure 5.4 (a) The ESI-mass spectra of the second peptide entry in Table 5.3; (b) presents the isotope distribution of the $(M + 10H)^{10+}$ peak. (© 1995. Reproduced from reference 9 by permission of John Wiley & Sons, Limited.)

$$(\text{SIINKEKL} + \text{H})^+$$
$$(\text{SIINEEKL} + \text{H})^+$$
$$(\text{SIINQEKL} + \text{H})^+$$

Identify each of the six most intense peaks using the monoisotopic mass of the amino acid residues.[14]

(b) Cite the two peaks used to calculate the mass spectrometer resolution of the mass spectrum in Figure 5.8, and find the R value.

The resolutions obtained in Problems 5.6–5.9 were based on the degree of separation of two peaks (Δm). Figures 5.9a and b present multiply charged ions of two different proteins and an impurity species of approximately 25% by weight that has a mass 16 Da higher than the ion of interest.[15] This mass difference can reflect an oxidation impurity. The theoretical effect that a mass spectrometric resolution adjustment has on the overlap of the two species is shown for three different resolution values for each ion pair and is highlighted in Problem 5.10.

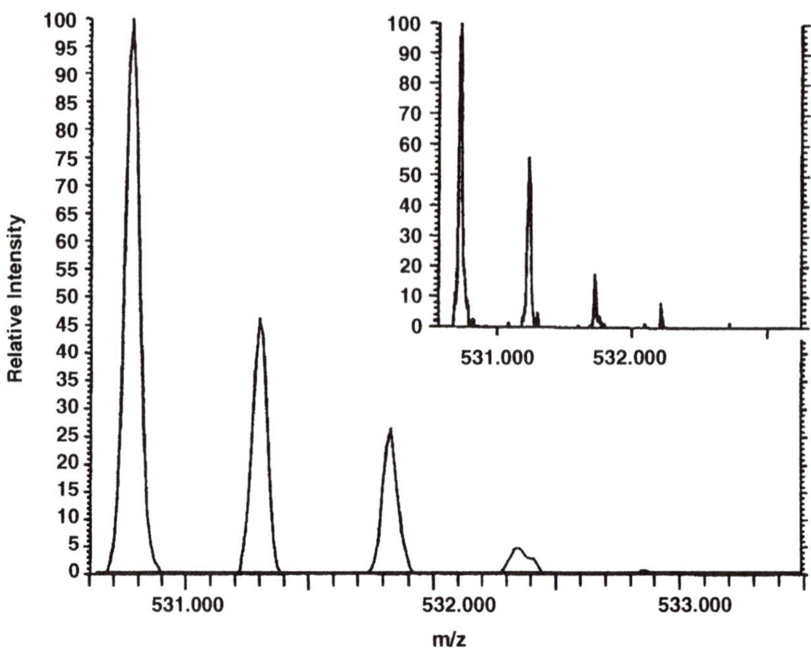

Figure 5.5 The ESI-partial mass spectral isotope distribution of bradykinin for the $(M + 2H)^{2+}$ peak. Inset shows the same spectrum at an increased resolution. © 1991. Reproduced from reference 11 by permission of John Wiley & Sons, Limited.)

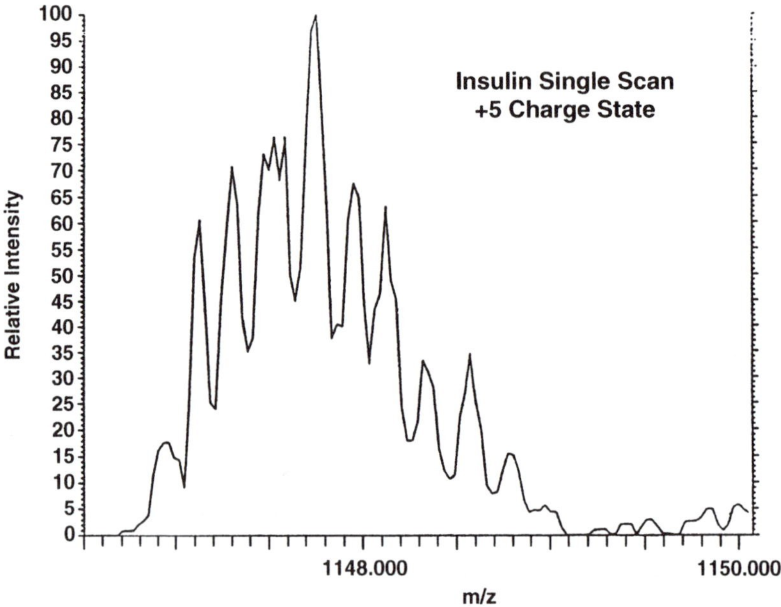

Figure 5.6 The ESI-mass spectral isotope distribution of bovine insulin for the $(M + 5H)^{5+}$ peak. (© 1991. Reproduced from reference 11 by permission of John Wiley & Sons, Limited.)

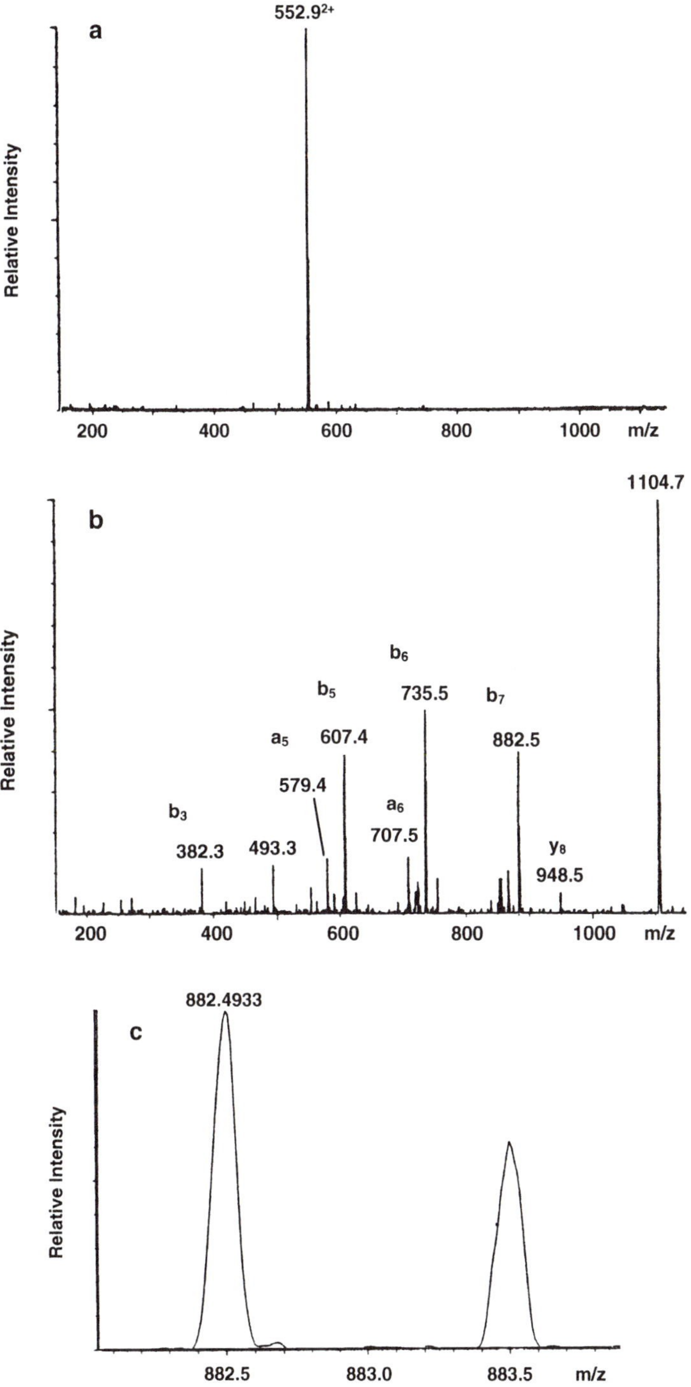

Figure 5.7 (a) The ESI-mass spectrum of substance P(1–9), (b) product-ion mass spectrum from the *m/z* 552.9^{2+} parent in (a), and (c) the ^{13}C isotope distribution of the b_7 ion in (b). (© 1993. Reproduced from reference 12 by permission of John Wiley & Sons, Limited.)

PROBLEM 5.10 For Figures 5.9a and b, each resolution can be calculated to an approximate value by the $M/\Delta m$ expression. Calculate the three resolution values that are given for the most intense analyte peaks in Figures 5.9a and b. For the 1000 resolution trace in Figure 5.9a, use the 50% peak width at half-height for the calculation of Δm.

Figure 5.10 presents a case where two peaks are resolved at about the 50% point.[16] This represents the $(M + 15H)^{15+}$ peak of bovine trypsin, and both peaks represent two distinct, but close in mass, proteins. The resolution of the spectrum can be calculated by dividing the difference of the m/z of both peaks (Δm, cf. Figure 5.1a) into m/z 1553.7. This results in an R value of 1195.

The human hemoglobin (Hb) protein is an interesting molecule from a mass spectrometry viewpoint. The intact molecule consists of two different protein subunits, and each subunit exists as a duplicate in the Hb molecule, or $\alpha_2\beta_2$. The β_2 Hb molecule also has a central iron-heme unit. The four protein subunits and the heme complex are all noncovalently bound together by electrostatic forces. The interest for mass spectrometry utility is not so much with the intact unit as it is with the individual subunits themselves. The Hb molecule exists as native (normal) α- and β-chains, and each chain can have a point (i.e., a single amino acid) mutation at many different amino acid residues.[17] Thus, a person can have both normal α- and β-chains, both variant (mutant) α- and β-chains, or

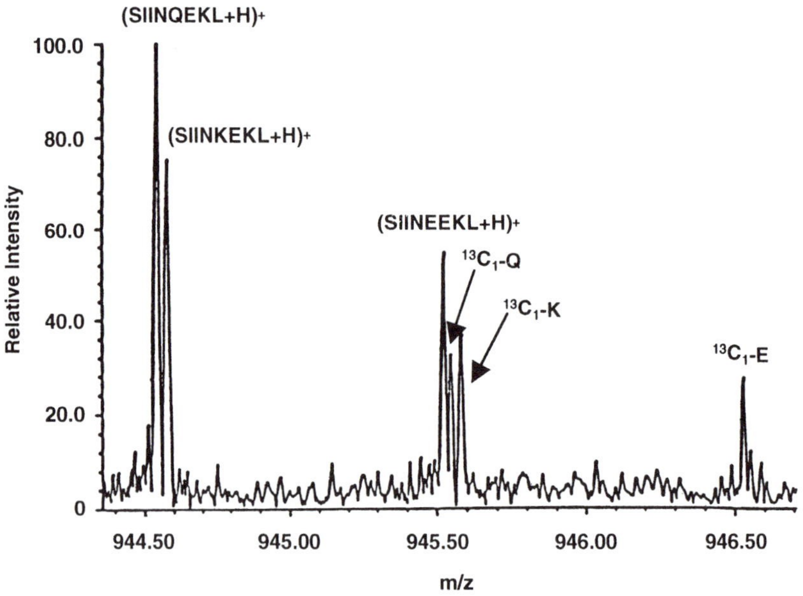

Figure 5.8 High-resolution Fourier transform mass spectrum of a mixture of three octapeptides. (© 1996. Reproduced from reference 14 by permission of John Wiley & Sons, Limited.)

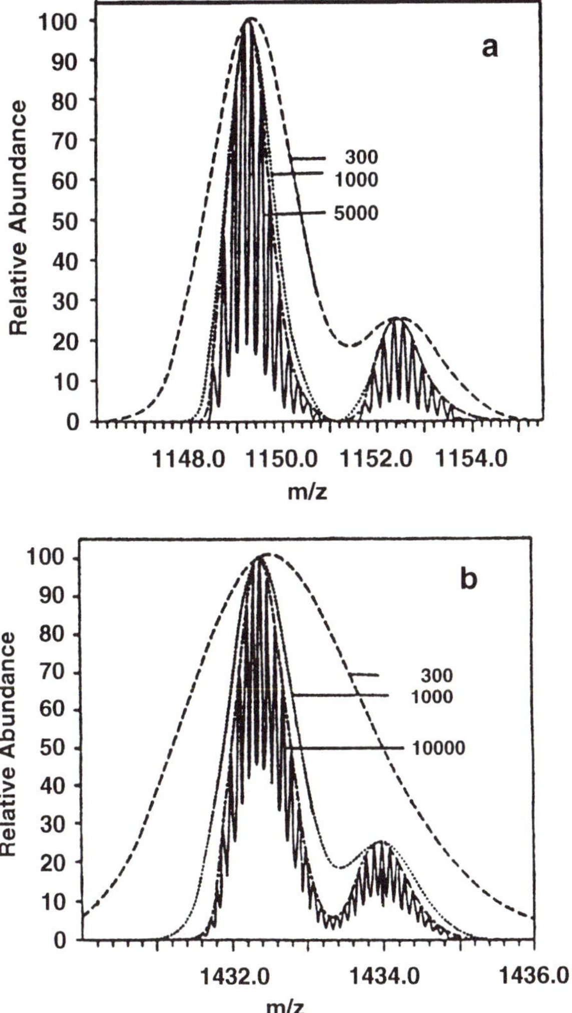

Figure 5.9 The ESI-mass spectra obtained at the indicated mass spectrometer resolution (R) values for (a) the $(M + 5H)^{5+}$ ion of bovine insulin and (b) the $(M + 10H)^{10+}$ ion of lysozyme. An oxidation impurity is observed at higher mass and 25% relative intensity in (a) and (b). (© 1991. Reproduced from reference 15 by permission of Portland Press Limited, Essex, U.K.)

a combination of the four types. The majority of variant α- and β-chains are caused by a change in only one amino acid residue for a given hemoglobin molecule. A person carrying the normal $\alpha_2\beta_2$ complement is said to be homozygous, and a person is heterozygous, or a carrier of a variant chain, with the following combinations: $\alpha\alpha'\beta_2$, $\alpha_2\beta\beta'$, and $\alpha\alpha'\beta\beta'$, where the prime indicates a variant. A person is homozygous for a variant for the following combinations: $\alpha'_2\beta_2$, $\alpha_2\beta'_2$, and $\alpha'_2\beta'_2$.

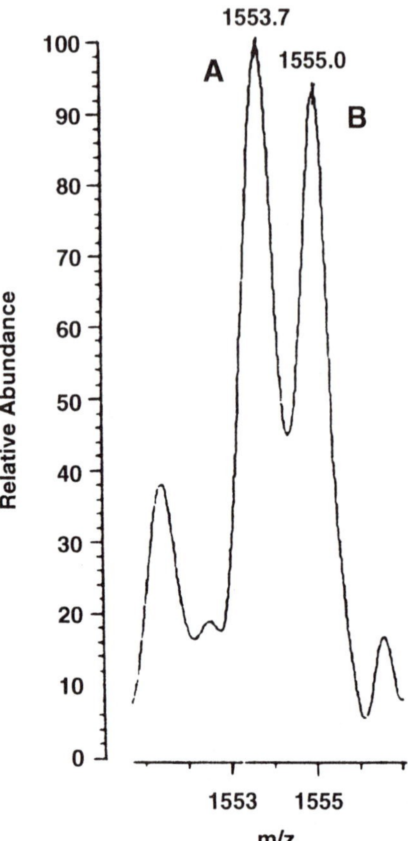

Figure 5.10 The ESI-mass spectrum of bovine trypsin for the $(M + 15H)^{15+}$ peak. The deconvoluted M_r value for peak A is 23,290.1 Da and that for peak B is 23,309.7 Da. (© 1991. Reproduced from reference 16 by permission of John Wiley & Sons, Limited.)

The normal α- and β-subunit protein average masses are 15,126.5 and 15,867.5 Da, respectively. The most widely known mutant Hb is the cause of sickle-cell anemia. This mutation occurs as a single point mutation in a β-chain. A person can be a carrier of the disease, or heterozygous (i.e., $\alpha_2\beta\beta^S$) or homozygous for the disease (i.e., $\alpha_2\beta_2{}^S$). The point mutation in the β-chain for the sickle Hb molecule occurs when the sixth amino acid (Glu, in the normal β-protein) is replaced with a Val residue. This causes a decrease in mass of 30 Da, and the β^S-molecule has a mass of 15,837.5 Da. This can easily be detected by ESI-MS.[17,18] Problem 5.11 deals with combination of a heterozygous normal and mutant β-chain that differ by less than 30 Da.

PROBLEM 5.11 In Figure 5.11a, the ESI-mass spectrum is shown for the human hemoglobin α- and β-chains, and the carrier is heterozygous for a mutant on either the α- or β-chain.[19] Determine which chain carries the mutant and calcu-

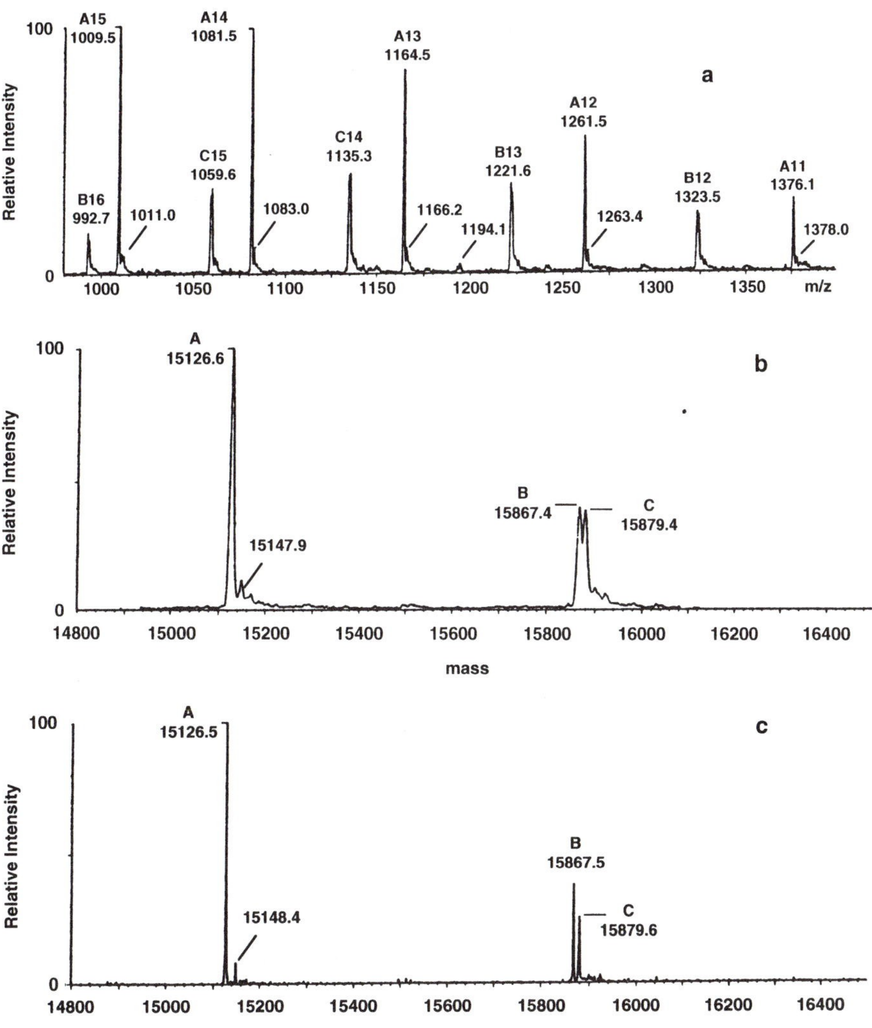

Figure 5.11 (a) The ESI-mass spectrum of human hemoglobin Quebec-Chori β-variant (the numbers after the letters indicate the charge state of the mass), (b) neutral-charge mass spectrum from a conventional deconvolution algorithm, and (c) MaxEnt algorithm-derived mass spectrum. (© 1992. Reproduced from reference 19 by permission of John Wiley & Sons, Limited.)

late the experimental mass values of all three proteins. These proteins are labeled A, B, and C in Figure 5.11a. An exact calculation can be obtained for the native subunit, and an estimate of both masses can be obtained for the two heterozygous protein species. Look carefully at Figure 5.11a and you may be able to observe the heterozygous set of peaks. Treat the B and C peaks as one series of peaks and draw conclusions based on the calculated M_{expt} values.

PROBLEM 5.12 In Figure 5.12a, the ESI-mass spectrum is shown for human Hb α- and β-chains.[19] Can the manual experimental mass deconvolution procedure identify the heterozygous chain? In Figure 5.12a, can a visual determination be made on which set of multiply charged peaks is heterozygous?

Note that in both Problems 5.11 and 5.12, measurements were made on two proteins of different mass where one pair differed by 12 Da and the other pair differed by 9 Da. One may think that their differences in mass are very easy to

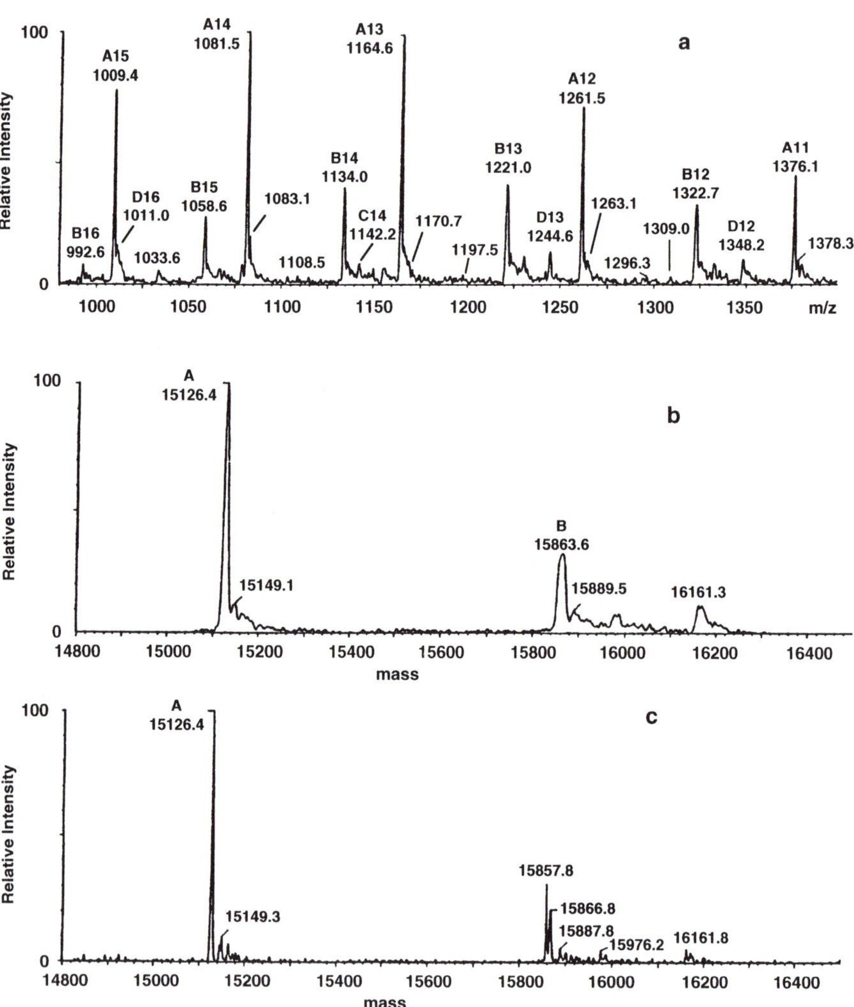

Figure 5.12 (a) The ESI-mass spectrum of human hemoglobin Hafnia β-variant (the numbers after the letters indicate the charge state of the mass), (b) neutral-charge mass spectrum from a conventional deconvolution algorithm, and (c) MaxEnt algorithm-derived mass spectrum. (© 1992. Reproduced from reference 19 by permission of John Wiley & Sons, Limited.)

detect, and they would be if the original masses were much lower than 15,860 Da. Figures 5.11b and 5.12b show significant overlap of these peaks, and that is due to the ^{13}C isotope spread of each peak. The ^{13}C peaks of the normal β-chains and variant β-chains overlap to a significant extent. To put this into perspective, Figure 5.13 presents a theoretical isotope distribution of the normal human β-hemoglobin chain.[22] Under the high resolution condition, 7 or 8 ^{13}C isotope masses are observed on either side of the most abundant peaks. For the hemoglobin β-chain variant, either side of the isotope envelope also consists of 6–8 ^{13}C isotope peaks. Thus, the high-mass side of the ^{13}C isotope envelope of the β-Hafnia chain variant and that of the low-mass side of the normal β-chain overlap to a significant extent since they differ by only 9 Da (Figure 5.12).

The MaxEnt algorithm effectively separates each of the β-chain peaks in Figures 5.11c and 5.12c such that there is baseline or near-baseline resolution. Therefore, despite their closeness in mass, the accuracy of the mass analyses can be calculated separately for each peak without overlap considerations. Table 5.4 provides this analysis. The standard deviation (SD) values for the β-Hafnia experiments deserve consideration at this point. The 15,858.4 $\pm$ 0.8 Da (β-Hafnia) value in Table 5.4 refers to the mean $\pm$ SD (95.5%) of seven individual determinations. The mean $\pm 2\sigma$ = mean SD where 2σ (or the SD) value on either side of the mean mass encompasses 95.5% of all the mean values.[8,23,24] This yields the following values for R and ppm:

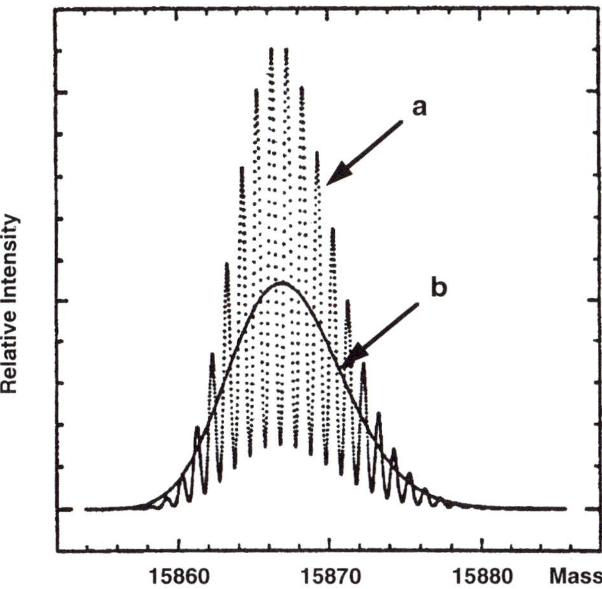

Figure 5.13 Theoretical mass spectral isotope distribution of the molecular region of the hemoglobin β-chain at a resolution R of (a) 15,000 and (b) 2000 using a 10% valley definition. (© 1994 John Wiley & Sons, Limited. Reproduced from reference 22 with permission.)

Table 5.4 Experimental and calculated masses and mass parameters for native and mutant Hb β-chains[a]

Hemoglobin chain	Theoretical mass (Da)	Experimental mass ± SD	Δm_{diff}[b]	R[c]	ppm
β	15,867.2	15,867.5	+0.3	52,890	18.9
β-Quebec-Chori	15,879.3	15,879.6	+0.3	52,930	18.9
β[d]	15,867.2	15,867.6 ± 0.7	+0.4	39,669	25.2
β-Hafnia[d]	15,858.2	15,858.4 ± 0.8	+0.2	79,290	12.61

[a]Table taken from reference 19.
[b]Experimental – theoretical mass (Da).
[c]Resolution.
[d]Results are shown for the average of seven separate mass measurement experiments.

$$R = \frac{15{,}858.4}{2(0.8)} = 9911.5 \qquad \text{and} \qquad \frac{10^6}{9911.5} = 101 \text{ ppm}$$

This average ppm value is much higher than the 12.61 ppm value derived from the β-Hafnia theoretical and average experimental mass value (Table 5.4). However, the SD value indicates that a single experimental measurement of the β-chain will yield a mass within about 100 ppm of the theoretical value, and this is observed as 12.61 ppm in the ppm column in Table 5.4. The precision of the mass measurement of the β-Hafnia chain is as follows:

$$\frac{(15{,}858.4 - 15{,}858.2)100}{15{,}858.2} = \frac{20}{15{,}858.2} = 0.0013\%$$

The difference between the experimental measurement and the theoretical value is 0.0013%, or the measurement is within 0.0015% of the theoretical value. For the normal β-chain,

$$\frac{(15{,}867.6 - 15{,}867.2)100}{15{,}867.2} = 0.0025\%$$

PROBLEM 5.13 In Table 5.3, calculate the R and the ppm equivalent of the SD values for the experimental mean mass value for each peptide.[9] Calculate the precision of the mass measurement for each peptide.

REFERENCES

1. Siuzdak, G. *Mass Spectrometry for Biotechnology;* Academic Press: San Diego, CA, 1996.
2. Burlingame, A.L.; Carr, S.A., Eds. *Mass Spectrometry in the Biological Sciences;* Humana Press: Totowa, NJ, 1996; Appendix XI, pp 546–553.
3. Jennings, K.R.; Dolnikowski, G.G. In *Methods in Enzymology: Mass Spectrometry;* McCloskey, J.A., Ed.; Academic Press: San Diego, CA, 1990; Volume 193, Chapter 2, pp 37–61.
4. Price, D.; Williams, J.E. *Dynamic Mass Spectrometry;* Heydon & Son: Philadelphia, PA, 1970; Volume 1, pp 199–200.
5. Beynon, J.H. *Mass Spectrometry and its Applications to Organic Chemistry;* Elsevier Publishing Co.: New York, 1960.
6. Dass, C. In *Mass Spectrometry: Clinical and Biomedical Applications;* Desiderio, D.M., Ed.; Plenum Press, New York, 1994; Volume 2, Chapter 1, pp 1–52.
7. Willard, H.H.; Merritt, L.L, Jr.; Dean, J.A. *Instrumental Methods of Analysis,* 5th Edition; D. Van Nostrand & Co.: New York, 1974, Chapter 16, p 467.
8. Chapman, J.R.; Gallagher, R.T.; Barton, E.C.; Curtis, J.M.; Derrick, P.J. *Org. Mass Spectrom.* **1992**, *27,* pp 195–203.
9. D'Agostino, P.A.; Hancock, J.R.; Provost, L.R. *Rapid Commun. Mass Spectrom.* 1995, *9,* pp 597–603.
10. D'Agostino, P.A.; Hancock, J.R.; Provost, L.R. *Proc. 43rd ASMS Conference on Mass Spectrometry and Allied Topics,* Atlanta, GA, **1995,** p 267.
11. Larsen, B.S.; McEwen, C.N. *J. Am. Soc. Mass Spectrom.* **1991,** *2,* pp 205–211.
12. Starrett, A.M.; DiDonato, G.C. *Rapid Commun. Mass Spectrom.* **1993,** 7, pp 12–15.

13. Carr, S.A.; Hemling, M.E.; Bean, M.F.; Roberts, G.D. *Anal. Chem.* **1991,** *63,* pp 2802– 2824.
14. Winger, B.E.; Campana, J.E. *Rapid Commun. Mass Spectrom.* **1996,** *10,* pp 1811–1813.
15. McEwen, C.N.; Larsen, B.S. *Rapid Commun. Mass Spectrom.* **1992,** *6,* pp 173–178.
16. Chapman, J.R.; Gallagher, R.T.; Mann, M. *Biochem. Soc. Trans.* **1991,** pp 940–943.
17. Shackleton, C.H.L.; Witkowska, H.E. In *Mass Spectrometry: Clinical and Biomedical Applications*; Desiderio, D.M., Ed.; Plenum Press, New York, 1994; Volume 2, Chapter 4, pp 135–199.
18. Sakairi, M. *Rapid Commun. Mass Spectrom.* **1993,** *7,* pp 1108–1112.
19. Ferrige, A.G.; Seddon, M.J.; Green, B.N.; Jarvis, S.A.; Skilling, J. *Rapid Commun. Mass Spectrom.* **1992,** *6,* pp 707–711.
20. Mann, M.; Meng, C.K.; Fenn, J.B. *Anal. Chem.* **1989,** *61,* pp 1702–1708.
21. Labowsky, M.; Whitehouse, C.; Fenn, J.B. *Rapid Commun. Mass Spectrom.* **1993,** *7,* pp 71–84.
22. Wada, Y.; Matsuo, T. In *Biological Mass Spectrometry: Present and Future;* Matsuo, T., Caprioli, R.M., Gross, M.L., Seyama, Y., Eds.; John Wiley & Sons: Chichester, U.K., 1994; Chapter 3.7, pp 369–399.
23. Hamburg, M. *Statistical Analysis for Decision Making,* 3rd Edition; Harcourt Brace Jovanovich,: New York, 1983, pp 42–44.
24. Box, G.E.P.; Hunter, W.G.; Hunter, J.S. *Statistics for Experimenters;* John Wiley & Sons: New York, 1978; Chapter 2.

ANSWERS

PROBLEM 5.1 For the three entries in Table 5.3, the Δm_{diff} and ppm values are as follows:

Peptide	$\Delta\ m_{\text{diff}}$	ppm
1	+0.0588	1.4
2	+0.1882	25.5
3	−0.665	58.7

PROBLEM 5.2 Linear extensions are drawn and the 10% and 50% intensity values of the most intense isotope (m/z 738.245) are found as shown in Figure 5.4c. In the figure, 20 mm spans 0.4 Da on the x-axis, and the distances in

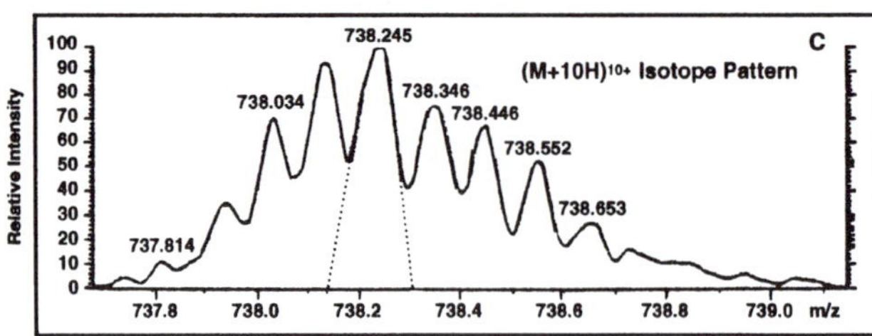

Figure 5.4 (c) The ESI-mass spectral isotope distribution of the second peptide entry in Table 5.3 for the $(M + 10H)^{10+}$ peak of part (a). Dashed lines extending from the m/z 738.245 peak to the baseline are for resolution calculation purposes. (© 1995. Reproduced from reference 9 by permission of John Wiley & Sons, Limited.)

millimeters (daltons) at intensity values of 10% and 50% are 8(0.16) and 4.75(0.95), respectively.

The R values for the 10% and 50% intensity values are 738.245/0.16 = 4614 and 738.245/0.095 = 7771, respectively. D'Agostino et al.[9] list a mass spectrometer resolution value of 8000, and this is to be expected based on the relatively poorer resolution of the ^{13}C isotope peaks. Thus, the 50% definition of resolution was used.

PROBLEM 5.3 A "10% valley" or 10% of the maximum intensity is used.

For Figure 5.5, the mass conversion of 30 mm (0.8 Da on the x-axis) and a 4.5-mm width of the m/z 530.8 peak at the 10% intensity level yield a peak width of 0.12 Da. R = 530.8/0.12 = 4423, which is close to the published experimental value[11] of 5000.

For the inset of Figure 5.5, the mass conversion of 19.5 mm (1 Da on the x-axis) and a 1.25-mm width of the m/z 530.8 peak at the 10% intensity level yield a peak width of 0.064 Da. R = 530.8/0.064 = 8294, which is relatively close to the published value[11] of 10,000.

PROBLEM 5.4 Using either method (from Figure 5.1a) to calculate Δm, Δm = 0.24 Da and R = 1147.8/0.24 = 4783, which is close to the published value of R = 5000.

PROBLEM 5.5a m/z 552.9^{2+} = $(M + 2H)^{2+}$ of substance P(1–9). Use the following rationale:

$$\frac{M + H}{1(+)} = M + 1.00797 \qquad \text{and} \qquad \frac{M + 2H}{2(+)} = 552.9$$

Therefore, M_{expt} = 2(552.9) − 2(1.00797) = 1103.784 Da = zero charge mass. The experimental error for the mass is

$$\frac{10^6 \Delta m_{diff}}{M_{calc}} = \frac{10^6(1103.784 - 1103.6)}{1103.6} = 166.7 \text{ ppm}$$

PROBLEM 5.5b The x-axis of Figure 5.7c yields 19 mm (0.5 Da) and the 6.75-mm width at 5% intensity for the m/z 882.4933 b_7 ion is equivalent to 0.178 Da. Thus, 882.4933/0.178 = 4958, where Δm = 0.178, and this value is close to the 5000 resolution value stated in the literature.[12]

PROBLEM 5.6 The average mass of the Leu residue in the peptide is 113.1590 Da and that of Asn is 114.1036 Da. The average mass of Asn is 0.9446 Da higher than that of Leu.

$$\text{Resolution} = \frac{M}{\Delta m} = \frac{10{,}000}{10{,}000.9446 - 10{,}000} = \frac{10{,}000}{0.9446} = 10{,}586$$

PROBLEM 5.7 The amino acid residue mass of Gly is 57.0518 Da and that of Ala is 71.0786 Da. Thus,

$$R = \frac{M}{\Delta m} = \frac{10{,}000}{10{,}014.0268 - 10{,}000} = 713$$

PROBLEM 5.8 The amino acid residue mass of Lys is 128.1736 Da and that of Gln is 128.1304 Da. Thus,

$$R = \frac{M}{\Delta m} = \frac{10{,}000}{10{,}000.0432 - 10{,}000} = 231{,}481$$

PROBLEM 5.9 (a) The six peaks and their masses can be identified as follows:

$(\mathrm{SIINQEKL} + \mathrm{H})^+$	944.5413
$(\mathrm{SIINKEKL} + \mathrm{H})^+$	944.5777
$(\mathrm{SIINEEKL} + \mathrm{H})^+$	945.5254
$(\mathrm{SIIN}(^{13}\mathrm{C}_1\text{–}\mathrm{Q})\mathrm{EKL} + \mathrm{H})^+$	945.5447
$(\mathrm{SIIN}(^{13}\mathrm{C}_1\text{–}\mathrm{K})\mathrm{EKL} + \mathrm{H})^+$	945.5811
$(\mathrm{SIIN}(^{13}\mathrm{C}_1\text{–}\mathrm{E})\mathrm{EKL} + \mathrm{H})^+$	946.5288

(b) The $(\mathrm{SIINEEKL} + \mathrm{H})^+$ and $(\mathrm{SIIN}(^{13}\mathrm{C}_1\text{–}\mathrm{Q})\mathrm{EKL} + \mathrm{H})^+$ peaks are used because they have the smallest difference in mass:

$$R = \frac{M}{\Delta m} = \frac{945.5254}{945.5447 - 945.5254} = 48{,}991$$

An R value of 60,000 was actually used in the experiment.[14]

PROBLEM 5.10 In Figure 5.9a, 25.5 mm spans 4 Da on the m/z axis. For the 300 resolution analysis, the distance between the centers of each peak at 10% relative abundance (10% valley)—that is, m/z 1148.7 and 1152.25—is 20 mm. Thus,

$$\frac{25.5\ \mathrm{mm}}{20\ \mathrm{mm}} = \frac{4\ \mathrm{Da}}{3.137\ \mathrm{Da}} \qquad \frac{M}{\Delta m} = \frac{1148.7}{3.137} = 366 \simeq 300\ \text{resolution}$$

For the 1000 resolution analysis and at the 50% valley level, the full width at half–height of the intense peak is 7.5 mm or 1.176 Da:

$$\frac{25.5\ \mathrm{mm}}{7.5\ \mathrm{mm}} = \frac{4\ \mathrm{Da}}{1.176\ \mathrm{Da}} \qquad \frac{M}{\Delta m} = \frac{1148.6}{1.176} - 977 \simeq 1000\ \text{resolution}$$

Note the m/z shift of the peak to 0.1 Da less than the peak for the 300 resolution analysis. For the 5000 resolution, an estimate of about 1 mm in width can be made for each of the individual ^{13}C isotopic peaks. This yields an R value of $1148.6/0.157 = 7316 \simeq 5000$.

In Figure 5.9b, 41 mm spans 4 Da on the x-axis. For the 300 resolution analysis, the width of the broad-dashed peak at 10% valley is 55 mm or 5.365 Da; $R = 1432.6/5.365 = 267 \simeq 300$. For the 1000 resolution analysis, the distance between the centers of each peak at m/z 1432.4 and 1434.0 is 1.6 Da; therefore, $R = 1432.4/1.6 = 895 \simeq 1000$. For the 10,000 resolution, a rough estimate of the width of each isotope peak is 1 mm, which is equivalent to 0.0976 Da; $R = 1432.4/0.0976 = 14{,}676 \simeq 10{,}000$.

PROBLEM 5.11 From Figure 5.11a, the following are obtained:

A peaks					
m_2	m_1	Δm	Approx. n_2	n_2	M
1,376.1	1,261.5	114.6	11.00	11	15,126.1
1,261.5	1,164.5	97	12.00	12	15,126.0
1,164.5	1,081.5	83	13.02	13	15,125.5
1,081.5	1,009.5	72	14.01	14	15,127.0
1,009.5				15	15,127.5

M_r for the A peak $= 15{,}126.4 \pm 0.76 \simeq 15{,}126.5$ Da, the MaxEnt value (Figure 5.11c)

B & C peaks					
m_2	m_1	Δm	Approx. n_2	n_2	M
1,323.5	1,221.6	101.9	11.98	12	15,870.0
1,221.6	1,135.3	86.3	13.14	13	15,867.8
1,135.3	1,059.6	75.7	13.98	14	15,880.2
1,059.6	992.7	66.9	14.82	15	1,5879.0

Note that the Δm used here (from Chapter 4) has a different meaning than Δm in Figure 5.1.

On initial inspection, the B and C peaks are treated as one series of peaks. The B and C calculations indicate that there may be two closely related peaks where the lower mass (M) is from the +12 and +13 charge states, and the higher mass peak is from the + 14 and +15 mass peaks. Also, careful inspection of the spectrum in Figure 5.11a shows that the B and C series appear to have a split at the top of these multiply charged peaks. The normal deconvolution procedure (see Chapter 4 and references 20 and 21) yields Figure 5.11b, and, in addition to the "A" peak, the hypothesis that there are two close–in–mass peaks in the less intense series in Figure 5.11a is shown as the B and C peak split in Figure 5.11b. The MaxEnt data-processing algorithm completely separates (Figure 5.11c) the split peak in Figure 5.11b. Thus, the two peaks are separated by only 12.1 Da. Note that the manual processing technique yields a close mass correspondence with Figure 5.11c. Therefore, it is the β-chain that is heterozygous for a variant. The peak at 15,126.5 Da is the normal α-chain, and the peaks at 15,867.5 Da and 15,879.6 Da are the normal β and β-Quebec-Chori variant chain masses, respectively. The mutant occurs at the 87 amino acid position in the β-chain, and the normal Thr (residue mass is 101.1048 Da) is changed to Ile (residue mass is 113.1590 Da). This results in a +12.05 Da increase.

PROBLEM 5.12 From Figure 5.12a, the following are obtained:

A peaks					
m_2	m_1	Δm	Approx. n_2	n_2	M
1,376.1	1,261.5	114.6	11.00	11	15,126.1
1,261.5	1,164.6	96.9	12.01	12	15,126.0
1,164.6	1,081.5	83.1	13.00	13	15,126.8
1,081.5	1,009.4	72.1	13.99	14	15,127.0
1,009.4				15	15,126.0

M_r for the A peak $= 15{,}126.4 \pm 0.31$ Da

B peaks					
m_2	m_1	Δm	Approx. n_2	n_2	M
1,322.7	1,221.0	101.7	12.00	12	15,860.4
1,221.0	1,134.0	87	13.02	13	15,860.0
1,134.0	1,058.6	75.4	14.03	14	15,862.0
1,058.6	992.6	66.0	15.02	15	15,864.0
992.6				16	15,865.6

M_r for the B peak = 15,862.4 $\simeq$ 15,863.6 Da (Figure 5.12b)

No clue is given in the manual and the conventional deconvolution (Figure 5.12b) analyses that the B peak is actually two closely spaced, but separate, proteins. Figure 5.12c presents the MaxEnt output, and two proteins of m/z 15,857.8 and 15,866.8 differ by only 9.0 Da. The A peak manual analysis provides an excellent match with that of the deconvolution and MaxEnt mass analyses. The peak at 13,126.4 Da in Figure 5.12c is the normal α-chain, and the 15,866.8-Da and 15,857.8-Da chains are the normal β and β-Hafnia variant chain experimental mass values, respectively. The mutation occurs at the 116 amino acid residue position in the β-chain, where the normal His (residue mass of 137.1408 Da) is changed to Gln (residue mass of 128.1304 Da), and this results in a 9.01 Da decrease in mass from the normal β-chain.

PROBLEM 5.13

Peptide 1:

$$R = \frac{4162.3}{2(0.02)} = 104{,}057.5 \quad \text{and} \quad \frac{10^6}{104{,}057.5} = 9.61 \text{ ppm}$$

$$\text{Precision} = \frac{4162.3 - 4162.24)100}{4162.24} = 0.0014\%$$

Peptide 2:

$$R = \frac{7368.36}{2(0.08)} = 46{,}052 \quad \text{and} \quad \frac{10^6}{46,052} = 21.71 \text{ ppm}$$

$$\text{Precision} = \frac{(7368.36 - 7368.17)100}{7368.17} = 0.0026\%$$

Peptide 3:

$$R = \frac{11{,}326.33}{2(0.15)} = 37{,}774 \quad \text{and} \quad \frac{10^6}{37{,}754} = 26.49 \text{ ppm}$$

$$\text{Precision} = \frac{(11,326.33 - 11,326.995)100}{11,326.995} = 0.0059\%$$

6

Isotope Distribution of a Multiply Charged Mass Spectral Peak

The fundamental backbone of an ESI-mass spectrum is the multiply charged nature of the spectrum. This consists of redundancy where the analyte appears repeatedly at different *m/z* values, and the peaks differ only in the number protons. Chapter 4 presented a procedure which can yield the direct charge state of a mass spectral peak, and, at the same time, a determination of the average mass of the compound could also be obtained. There are situations when the above procedure cannot be used, such as when tandem mass spectrometry is performed on a multiply charged ion. The question becomes: of what charge state(s) are the product ion(s)? This prompts the question of whether there is a method to determine the charge state of an ion in a direct fashion without the aid of the presence of other ions. This would be useful in conventional and tandem mass spectra. A very simple and basic observation is that by applying the tandem mass spectrometry technique to an ion, the ion has more than one charge if masses greater than the chosen precursor ion are present in the product-ion mass spectrum.[1]

A direct method exists and its usefulness depends on the resolution used to obtain a mass spectrum. If the resolution is high enough, the charge state can be determined by counting the number of ^{13}C peaks in an isotope envelope. The charge state of the ion is equal to the number of ^{13}C isotope peaks contained in a 1-mass-unit window, or the charge state is equal to the reciprocal of the mass difference between any two adjacent ^{13}C peaks in the resolved isotope envelope.[2] Figure 5.5 provides a good first example where, for the ESI-mass spectrum of bradykinin, there are two ^{13}C peaks in the *m/z* 531.0–532.0 mass window.[3] Alternatively, there is 0.5 Da between the two peaks in the 1-mass-unit window, and the reciprocal of this value yields 2. Thus, the charge state expression of the bradykinin peak at *m/z* 531.0 is $(M + 2H)^{2+}$. Figure 6.1

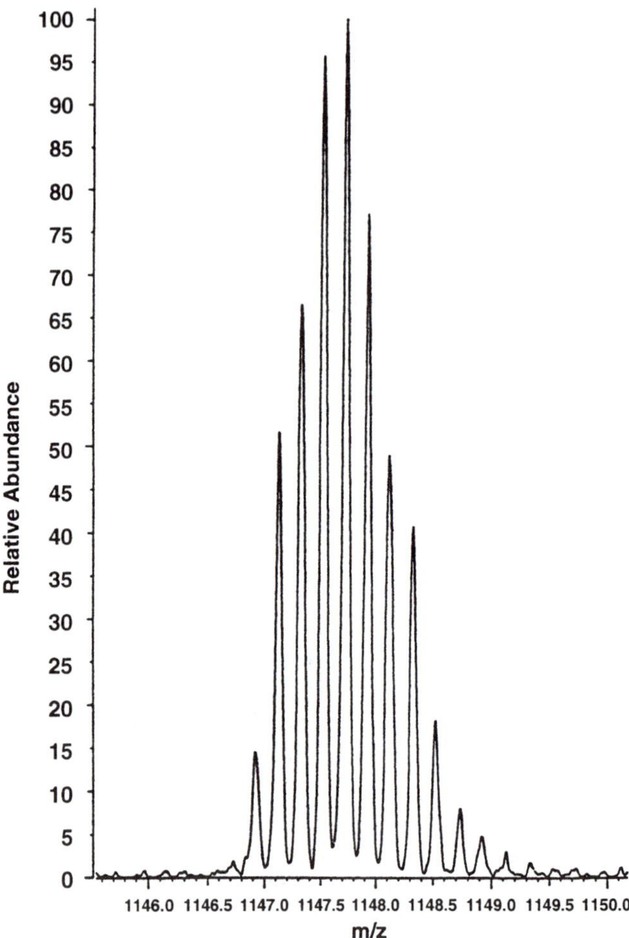

Figure 6.1 The ESI-mass spectrum of the isotope distribution of the $(M + 5H)^{5+}$ ion of bovine insulin at $R = 8100$ and a 10% valley definition (cf. Chapter 5). (© 1992, American Chemical Society. Reprinted from reference 4 with permission.)

shows the ESI-mass spectrum of one of the multiply charged ions for bovine insulin.[4] By applying a metric ruler to the page, 20 mm spans 1.3 Da, and the mass increment between two adjacent ^{13}C peaks is 3 mm or 0.195 Da. The reciprocal of 0.195 Da is 5.1, which is close to 5+ charge units. Using the peak-counting method, the second most intense peak is centered about m/z 1147.5, and this counts as one-half of a peak. Another peak is centered about m/z 1148.5, and this counts as one-half of a peak. There are four full peaks in the range m/z 1147.5 - 1148.5 Da, and the total yields five peaks. Thus, the spectrum represents the $(M + 5H)^{5+}$ peak of bovine insulin.

PROBLEM 6.1 Prove the assignment of the $(M + 5H)^{5+}$ peak in the ESI-mass spectrum[5] of ω-conotoxin MVIIC in Figure 5.2.

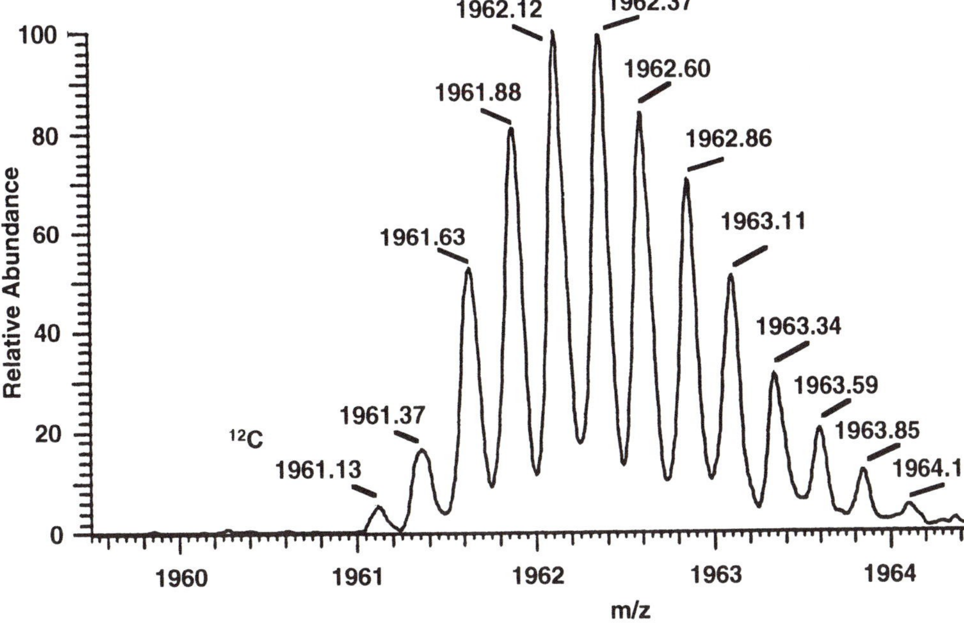

Figure 6.2 The ESI-mass spectrum of the isotope distribution of an $(M + nH)^{n+}$ ion of interleukin-8.

PROBLEM 6.2 Prove the assignment of the $(M + 8H)^{8+}$ peak in the ESI-mass spectrum of a synthetic peptide[6] in Figure 5.3b.

PROBLEM 6.3 Determine the charge state and proper multiply charged designation of the isotope envelope of interleukin-8 in Figure 6.2.[7]

This ability to determine the charge state of an ion is a powerful technique, because it is the only method to directly (unambiguously) determine the charge of an ion in tandem mass spectra. This is especially important for high-mass biological molecules. Tandem mass spectrometry of a selected precursor ion from a conventional mass spectrum can yield product ions of different charge states. This becomes evident as the ion of interest increases in mass and charge. Figure 6.3 presents an ESI-mass spectrum of horse angiotensin I, and the insets show the typical spacings of the ^{13}C peaks for the $(M + H)^+$ to $(M + 4H)^{4+}$ ions.[6,7] After a direct determination of the charge state for each multiply charged ion, the experimental mass of the peptide can be determined by the method described in Chapter 4. Tandem mass spectrometry of the $(M + 3H)^{3+}$ ion of angiotensin I, m/z 433, yields the spectrum shown in Figure 6.4.[7] This spectrum provides significant information for sequence characterization of the protein. However, the charge state must be addressed first, prior to structural determination procedures. A series of properly assigned charges for the mass spectral ions then allow for an accounting to be made for structural aspects (see Chapter 9). For example, the relatively intense peaks are labeled with the a, b,

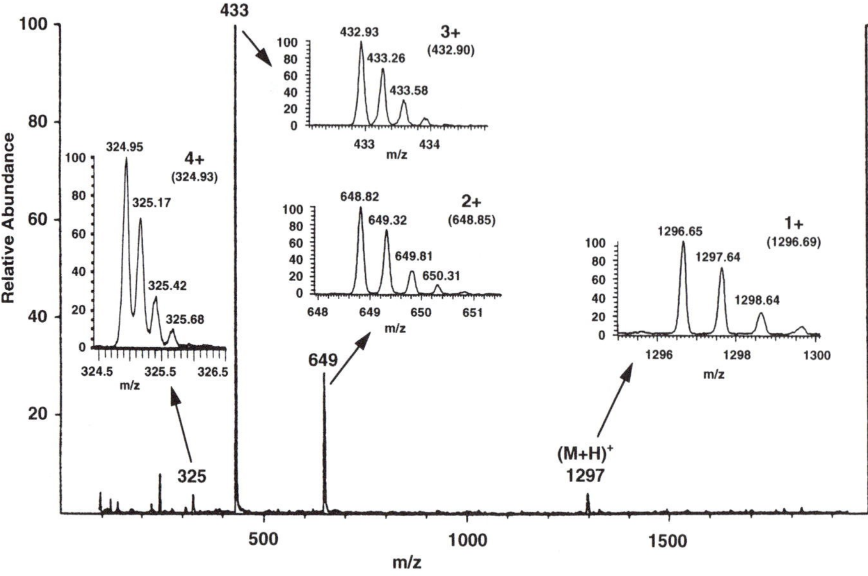

Figure 6.3 The ESI-ion-trap mass spectra of horse angiotensin I. Insets represent the isotope distribution of the $(M + nH)^{n+}$ ions where $n = 1$–4, and the average mass is 1296.4 Da. (© 1991, American Society for Mass Spectrometry. Reprinted from reference 6 by permission of Elsevier Science, Inc.)

and y designation. Note that 1+ and 2+ species are present from the original 3+ ion. The isotope method outlined above was used to assign the charge state of each ion. Look carefully at the m/z 392.7 product ion in Figure 6.4a. There appears to be a lower intensity ion on the high-mass side of the b_6^{2+} ion. Adequate mass resolution was provided so as to delineate this region, and two sets of isotope peaks are observed (Figure 6.4b). The charge state is determined, and then structural designations can be made for the ions. Note the close mass proximity, yet complete isotope separation, of the 2+ and 3+ ions. Subsequent analysis provided the b_6 and y_9 designation of the two ions by following the procedure outlined in Chapter 9.

Isotope resolution also aids in the characterization of mass differences between two peaks in a mass spectrum. Figure 6.5 shows a high resolution of the $(M + 9H)^{9+}$ peak of hen egg white lysozyme and a lower intensity, ^{13}C isotope-resolved ion envelope.[4] Mass analyses were performed on each ion envelope, and the mass difference resulted in 18.009 Da between the two envelopes. This difference could be due to a minor amount of protein that is 18 mass units less than that of the native lysozyme molecule, such as a Leu to a Met or a Glu to a Phe residue at 17.956 and 18.026 Da, respectively. However, if a loss of water occurred from the intact lysozyme molecule, the difference would be 18.0106 Da. The observed difference of 18.009 Da is much closer to 18.0106 Da than that of the amino acid substitution values. Thus, it appears that the

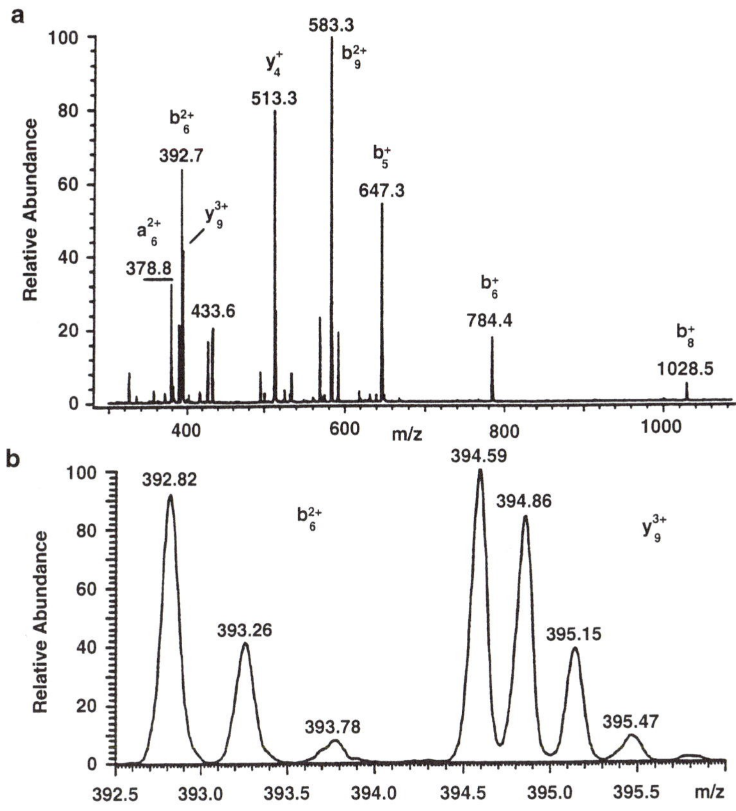

Figure 6.4 (a) The ESI-tandem mass spectrum of the $(M + 3H)^{3+}$ ion at m/z 433 of horse angiotensin I (Figure 6.3), and (b) isotope distribution of the m/z 394 region in (a).

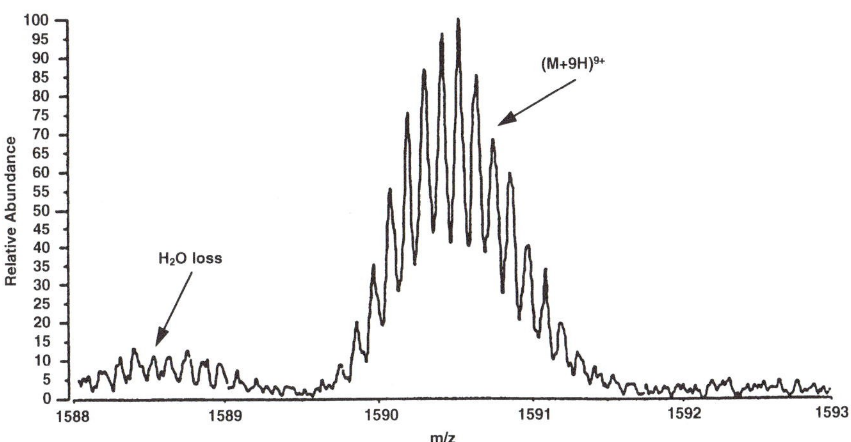

Figure 6.5 The ESI-mass spectral isotope distribution of the $(M + 9H)^{9+}$ ion for hen egg white lysozyme at a mass spectrometer resolution (R) of 10,000. (© 1992, American Chemical Society. Reprinted from reference 4 with permission.)

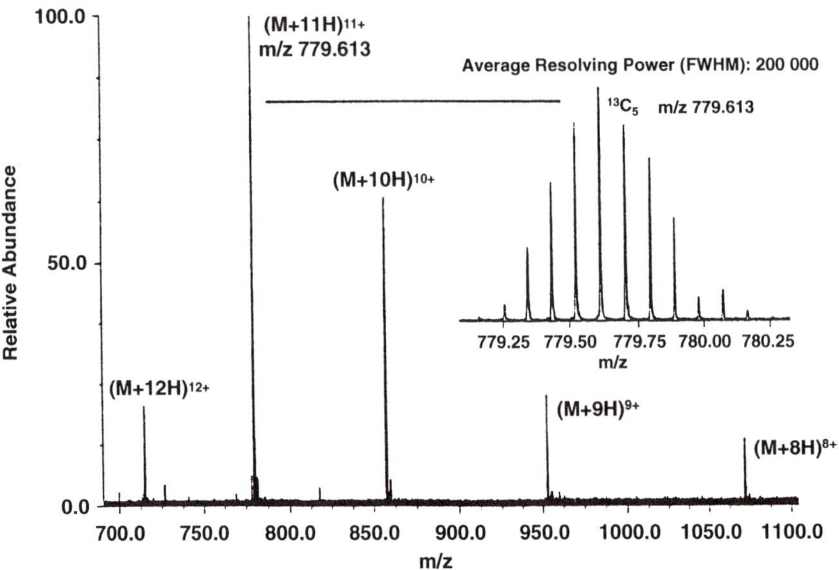

Figure 6.6 The ESI-FT mass spectrum of bovine ubiquitin; the inset provides the ^{13}C isotope distribution of the $(M + 11H)^{11+}$ ion. (© 1994. Reproduced from reference 8 by permission of John Wiley & Sons, Limited.)

low-intensity isotope envelope originates from a water loss from the intact enzyme.

Perhaps the best mass spectrometer for high-resolution and very sharp mass spectral peaks is the Fourier transform mass spectrometer (FT-MS). Figure 6.6 presents as ESI-FT mass spectrum of bovine ubiquitin, and the inset provides the ^{13}C spacings for the $(M + 11H)^{11+}$ peak.[8] Compared with Figure 6.5, it is very easy to arrive at an 11+ charge state, because 11 full peaks are clearly observed between *m/z* 779.25 and 780.25. The resolution of the peaks was 200,000 at the 50% intensity point (i.e., full width at half-maximum), while the resolution in Figure 6.5, obtained on an electromagnetic double-focusing mass spectrometer, was approximately 10,000.

Distinction of two overlapping ion species within a 1-Da mass range is possible. A doubly-charged peptide ion was subjected to collision-induced dissociation, and, among the various product ions, two ions differing by 1 Da were observed.[7] A high-resolution product-ion mass scan of the region that contained the two product ions is shown in Figure 6.7, and this was acquired on an ESI-ion-trap quadrupole mass spectrometer. Note that there is an overlap of a 1+ and a 2+ ion. The 1+ ion has masses at *m/z* 702.4 and 703.4, while the 2+ ion has masses at *m/z* 703.2, 703.7, and 704.2.

Problem A2.1 in Appendix 2 challenged the reader to calculate the absolute abundances of the monoisotopic, ^{13}C, and $^{13}C_2$ peaks of the zero-charge mass spectrum of bovine insulin. Figure 6.8 presents the $(M + 5H)^{5+}$ isotope envel-

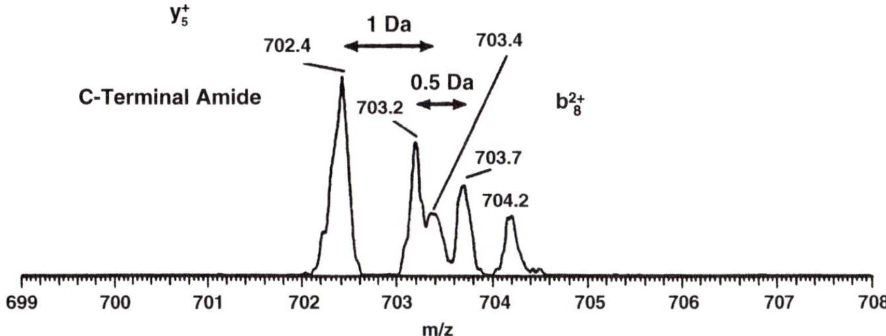

Figure 6.7 The ESI-tandem mass spectrum of the $(M + 2H)^{2+}$ ion of a peptide. The mass spectral region of the y_5^+ and b_8^{2+} product ions (see Chapter 9) is shown.

ope of the protein.[9] Note that m/z 1147.5 is the most abundant isotope peak ($^{13}C_3$) in the envelope, and the ratios of the monoisotopic (m/z 1146.9), ^{13}C, and $^{13}C_2$ isotope peaks are very similar to the ratios of the absolute abundances as given in Problem A2.1 in Appendix 2. These observations are also confirmed in Figure 6.1 and in the isotope envelope of the $(M + 4H)^{4+}$ ion of bovine insulin as shown in references 10 and 11.

PROBLEM 6.4 Figure 6.9 presents the ESI-CID tandem mass spectrum[12,13] of the $(M + 7H)^{7+}$ ion of human growth hormone releasing factor ($M_r = 5040.40$) Masses are labeled according to the conventional peptide fragment nomencla-

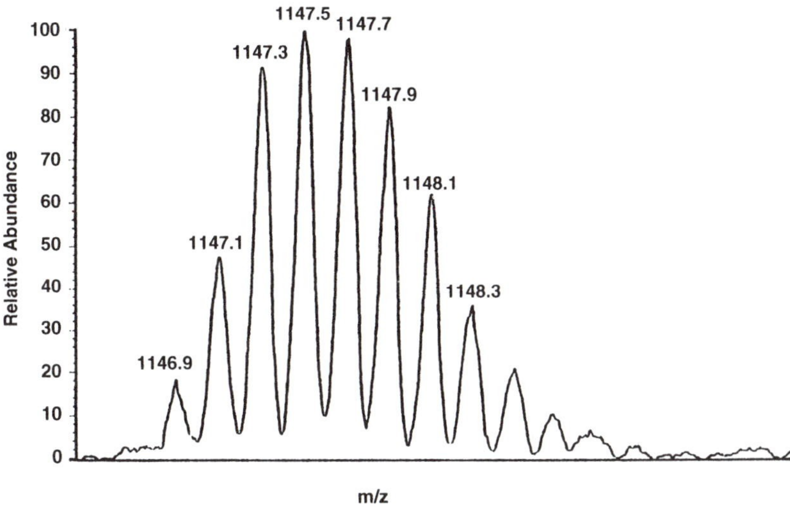

Figure 6.8 The ESI-mass spectral isotope distribution of the $(M + 5H)^{5+}$ ion of bovine insulin at $R = 7000$. (© 1993, Kratos Analytical, Manchester, U.K. Reproduced from reference 9 with permission.)

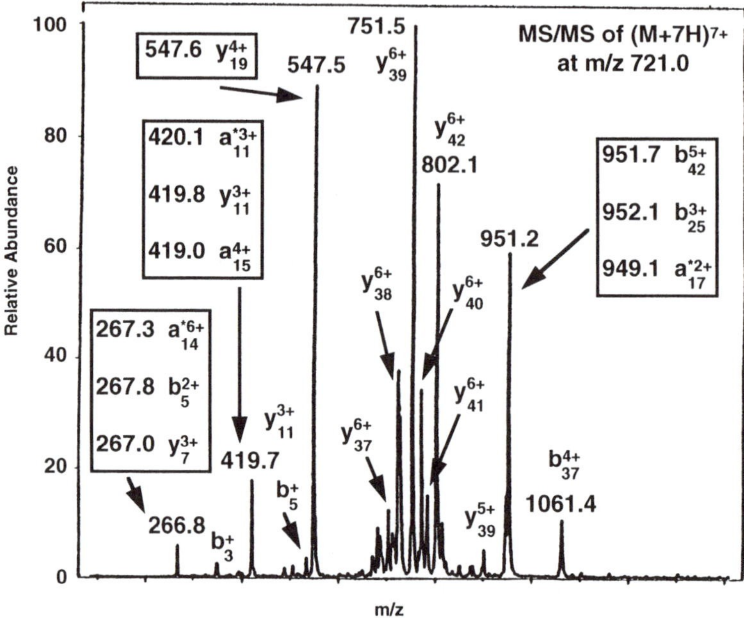

Figure 6.9 The ESI-tandem mass spectrum of the $(M + 7H)^{7+}$ ion at m/z 721.0 of recombinant human growth hormone releasing factor ($M_r = 5040.40$ Da). Note the similarity of mass-to-charge for the different sets of ions. (© 1996. Reprinted from reference 12 by permission of Academic Press, Inc., Orlando, Florida.)

ture (see Chapter 9) Three masses, in particular, are highlighted and show that three different, close-in-mass peptide fragments with different charge states could cause the appearance of the respective product ion for each of the three product ions. Detailed analyses of these three product ions[12,13] are presented in Figure 6.10. Based on the high-resolution product-ion mass spectra in Figure 6.10, match each of the three peaks with the correct alphanumeric nomenclature.

McLafferty et al. provided an example of high-resolution power[14,15] with ESI-FT-MS, as shown in Figure 6.11. An overlap of the ^{13}C isotope peaks occurs with the 18+ and 13+ ion envelopes of equine myoglobin and chicken cytochrome c, respectively. The charge states can be conveniently calculated with the given isotope peak spacings. Note that the FT-MS resolution clearly shows that a ^{13}C isotope peak of cytochrome c is bracketed by two ^{13}C isotope peaks of myoglobin at m/z 942.5, and that the peak at m/z 942.56 is actually a superposition of the m/z 942.564 and 942.562 ^{13}C isotope peaks of myoglobin and cytochrome c, respectively.

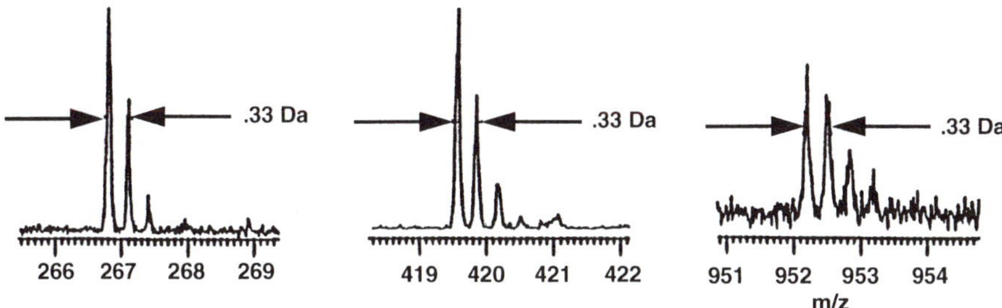

Figure 6.10 Slow scan, high-resolution ESI-tandem mass spectra of the three *m/z* regions from Figure 6.9. (© 1996. Reprinted from reference 12 by permission of Academic Press, Inc., Orlando, Florida.)

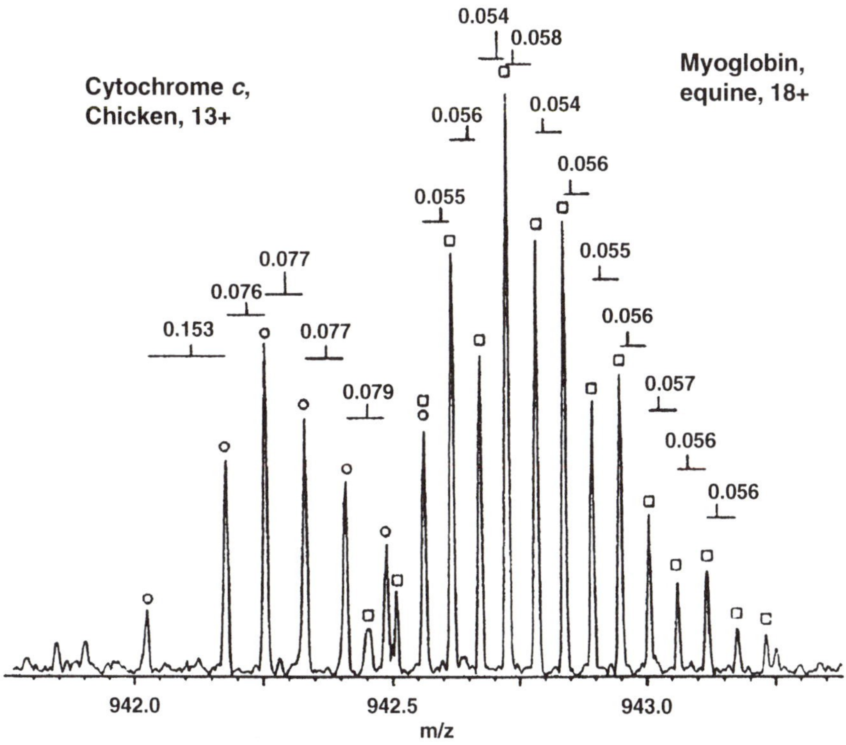

Figure 6.11 The ESI-mass spectral isotope distribution of a protein mixture in the region of the $(M + 13H)^{13+}$ ion for chicken cytochrome c and the $(M + 18H)^{18+}$ ion for equine myoglobin. (© 1993, American Society for Mass Spectrometry. Reprinted from reference 14 by permission of Elsevier Science, Inc.)

REFERENCES

1. Bruins, A.P.; Covey, T.R.; Henion, J.D. *Anal. Chem.* **1987,** *59,* pp 2642–2646.
2. Henry, K.D.; McLafferty, F.W., *Org. Mass Spectrom.* **1990,** *25,* pp 490–492.
3. Larsen, B.S.; McEwen, C.N. *J. Am. Soc. Mass Spectrom.* **1991,** *2,* pp 205–211.
4. Cody, R.B.; Tamura, J.; Musselman, B.D. *Anal. Chem.* **1992,** *64,* pp 1561–1570.
5. D'Agostino, P.A.; Hancock, W.S.; Provost, L.R. *Proc. 43rd ASMS Conference on Mass Spectrometry and Allied Topics* **1995,** p 267.
6. Schwartz, J.C.; Syka, J.E.P.; Jardine, I. *J. Am. Soc. Mass Spectrom.* **1991,** *2,* pp 198–204.
7. Land, A.; Sanders, M.; Schwartz, J.; Gale, D.; Chaudhary, T. *Proc. 44th ASMS Conference on Mass Spectrometry and Allied Topics* 1996, p 343.
8. Winger, B.E.; Hein, R.E.; Becker, B.L.; Campana, J.E. *Rapid Commun. Mass Spectrom.* **1994,** *8,* pp 495–497.
9. Chapman, J.R., Ed.; *Practical Organic Mass Spectrometry,* 2nd Edition; John Wiley & Sons: Chichester, England, U.K., 1993, Chapter 6, p 203.
10. Dobberstein, P.; Schroeder, E.; *Rapid Commun. Mass Spectrom.* **1993,** *7,* pp 861–864.
11. He, M.; Xu, J.; Chan, T.W.D.; Lau, R.L.C. *Rapid Commun. Mass Spectrom.* **1996,** *10,* pp 897–902.
12. Schwartz, J.C.; Jardine, I. In *Methods in Enzymology;* Karger, B.L.; Hancock, W.S., Eds.; Academic Press: San Diego, CA, 1996; Volume 270, Part A, Chapter 23, pp 552–586.
13. McLuckey, S.A.; VanBerkel, G.J.; Glish, G.L., Schwartz, J.C. In *Practical Aspects of Ion Trap Mass Spectrometry;* March, R.E.; Todd, J.F.J., Eds.; CRC Press: New York, 1995; Volume 2, Chapter 3, pp 89–140.
14. Beu, S.C.; Senko, M.W.; Quinn, J.P.; McLafferty, F.W. *J. Am. Soc. Mass Spectrom.* **1993,** *93,* pp 190–192.
15. McLafferty, F.W.; Beu, S.C.; Quinn, J.P.; Senko, M.W.; Shi, Y.; Suckau, D.; Wampler, F.M., III; Zhang, M.Y. In *Biological Mass Spectrometry: Present and Future;* Matsuo, T.; Caprioli, R.M.; Gross, M.L.; Seyama, Y., Eds; John Wiley & Sons: New York, 1994, Chapter 2.10, pp 199–214.

ANSWERS

PROBLEM 6.1 The reciprocal of the difference of any two adjacent isotope peaks will be sufficient in the figure inset; for example,

$$1/(551.25 - 551.05) = 1/0.2 = 5$$
$$1/(551.05 - 550.85) = 1/0.2 = 5, \text{etc.}$$

PROBLEM 6.2 There are exactly eight full ^{13}C peaks between m/z 736.4 and 737.4. Alternatively,

$$1/(737.189 - 737.067) = 8.2 \simeq 8$$
$$1/(737.067 - 736.945) = 8.2 \simeq 8$$
$$1/(736.945 - 736.817) = 7.8 \simeq 8$$

PROBLEM 6.3 There are four full peaks between m/z 1962 and 1963; therefore, the charge state is 4+ and this is designated as $(M + 4H)^{4+}$.

$$1/(1963.11 - 1962.86) = 4.0$$
$$1/(1962.86 - 1962.60) = 3.84$$
$$1/(1962.60 - 1962.37) = 4.3$$
$$1/(1962.37 - 1962.12) = 4.0$$

The charge state from the reciprocal of the mass difference between adjacent isotope peaks appears to be 4.0.

PROBLEM 6.4 (a) For the m/z 266.8 ion in Figure 6.9, Figure 6.10 shows three isotope peaks in a 1-Da region; thus, this peak must be the y_7^{3+} peptide.

(b) For the m/z 419.7 ion in Figure 6.9, Figure 6.10 shows three isotope peaks in a 1-Da region, and this could be either one of the 3+ species. Close inspection shows the two intense isotope peaks lie to the low-mass side of 420 Da; therefore, the y_{11}^{3+} ion at m/z 419.8 is the proper designation.

(c) For the m/z 951.2 ion, three isotope peaks are clearly observed in the span of 1-Da; therefore the b_{25}^{3+} designation characterizes this ion in Figure 6.10.

7

Experimental Parameters Affecting Electrospray Ionization Mass Spectra

The appearance of a protein/peptide charge distribution can vary to a significant extent, and this concept was briefly introduced in Chapter 4. This chapter investigates the many methods and solution conditions which are used to alter the shape or conformational state of a protein or peptide. The conformation of a protein significantly affects the accessibility of the residue side chains to the buffer medium. Solvent accessibility is important for the core N-terminal amine, Lys, Arg, and His locations on a peptide or protein because they determine the degree of presence of the $(M + nH)^{n+}$ mass spectral peaks and maximum charge state of the molecules.[1,2] The extent of protonating a protein analyte dictates its presence/absence in an ESI-mass spectrum. For example, if a relatively high-mass protein experiences too few proton charges, it may not be observed in the mass spectrum of conventional, mass-limited, quadrupole mass spectrometers.[3] Also, an adequate amount of charging (protonation) of a protein is desired so as to produce enough multiply charged peaks to arrive at a reliable mass value (Chapter 4).

The majority of this chapter provides information (reagents and procedures) that is at odds with the physiological status of fluids in cellular systems. The protein/peptide analyte is normally found in aqueous and narrow-pH, salt concentration, ionic strength, and temperature regimes in cellular systems. However, for ESI, the protein analytes are usually placed in a milieu that has uncharacteristic degrees of analytical parameters, with respect to that of cellular in vivo and in vitro environments, so as to allow versatility in their mass spectral expression. This mass spectral expression can be "fine-tuned", depending on the desired analyte characteristic, by varying the amount or degree of the various reagents and methods in an experimental situation. The important factors are presented in separate sections.

pH of Solution

The pH of the protein solution is an obvious candidate of importance, mainly because this is a source of the charge-carrying proton species that adduct to the protein sample of interest. A simple observation of this solution characteristic can be found in Figure 7.1. Gramicidin S has the 1+ and 2+ charge-state species displayed in the mass spectra, and as the pH increases from a low pH of 3.0 to neutral and higher pH values of 9.5–11.9, the 1+ species gradually increases in abundance with respect to the 2+ charge state.[4] This observation generally follows the trend that as the source of protons decreases (i.e., from low to high pH), the lower charge state(s) tend to dominate in the positive-ion mode.

Figure 7.2 provides a similar analysis with cytochrome c.[1,5,6] As the pH is lowered (by the addition of acetic acid) from pH 5.2 to pH 2.6, the higher charge states predominate. The pH is known to cause denaturing effects on the backbone of proteins where, as the pH is lowered, proteins generally lose their three-dimensional folded structure, the active enzymatic site is no longer present, and the backbone tends toward a roughly linear "stretched-out" conformation. This occurs because as the number of ionizing protons that attach to the protein increases (i.e., increase in charge state), more repulsion forces are introduced to the molecule. These repulsion forces combine to extend the polypeptide backbone, thus allowing for more potential protonating sites to be exposed to the proton-containing medium. However, it must be kept in mind that these are solution-phase phenomena, while the free ion form ultimately exists in the gas phase. Gas-phase ionization equilibrium can be quite different from that of the solution phase.

The effect of acid on the mass spectral distribution of equine myoglobin (Figure 7.3)[7] follows a shift analogous to that of cytochrome c in Figure 7.2. Near-physiological pH conditions at pH 6.7 and 10 mM ammonium acetate produce the 9+ and 8+ peaks for myoglobin in Figure 7.3a. The deconvoluted mass of 17,568 Da signifies that the protein remains in the intact, protein–heme complex form (open circles). Sharp mass spectral peaks are observed. The inset shows an equivalent mass spectrum except that the solution was contaminated with potassium salt; hence, cation adduction occurred, which broadened the 9+ and 8+ peaks (see Chapter 8). Some loss of the heme unit occurred (616.5 Da) from the 9+ and 8+ peaks due to the observation of low-intensity, lower mass peaks (closed circles). Upon lowering the pH to 2.2 (10% acetic acid), higher charge states are observed (Figure 7.3b) which indicate denaturation, loss of the heme group, and thus a greater exposure of the polypeptide backbone to the aqueous environment. The solution in Figure 7.3b had its pH raised to 5 by the addition of NH_4OH, and a partial renaturation to the intact protein–heme complex is observed by the presence of the 9+ and 8+ ion species (Figure 7.3c). However, none of the peaks in the pH 5 solution contained heme. Both charge-state distributions produced masses that reflected the apoprotein form of the myoglobin protein.[7]

In the negative-ion mode, the effect is just the opposite from a mass spectral viewpoint. Figure 7.4 shows two spectra of the bovine insulin α-chain where

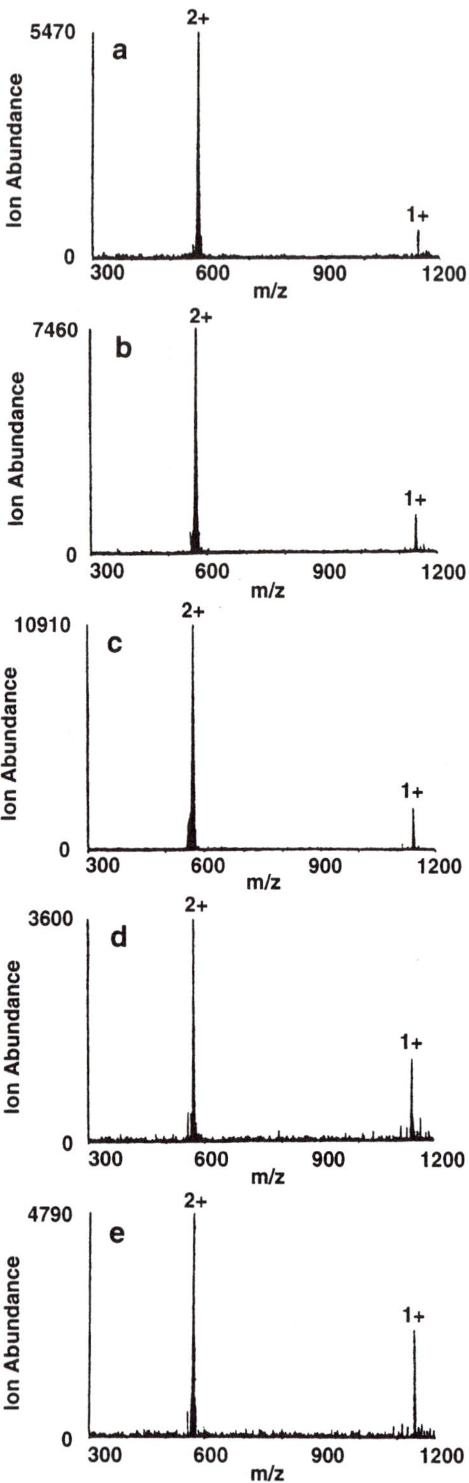

Figure 7.1 The ESI-mass spectra of gramicidin S in 50% aqueous MeOH that contains the following solution/pH conditions: (a) 5.0% HOAc, pH 2.90; (b) 1.0% HOAc, pH 3.31; (c) no added acid, pH 7.0, (d) 1.0% conc. NH_4OH, pH 10.30; and (e) 5.0% conc. NH_4OH, pH 10.54. The skimmer–cone voltage was less than 3 V in order to minimize energy deposition in the ions. (© 1994. Reproduced from reference 4 by permission of John Wiley & Sons, Limited.)

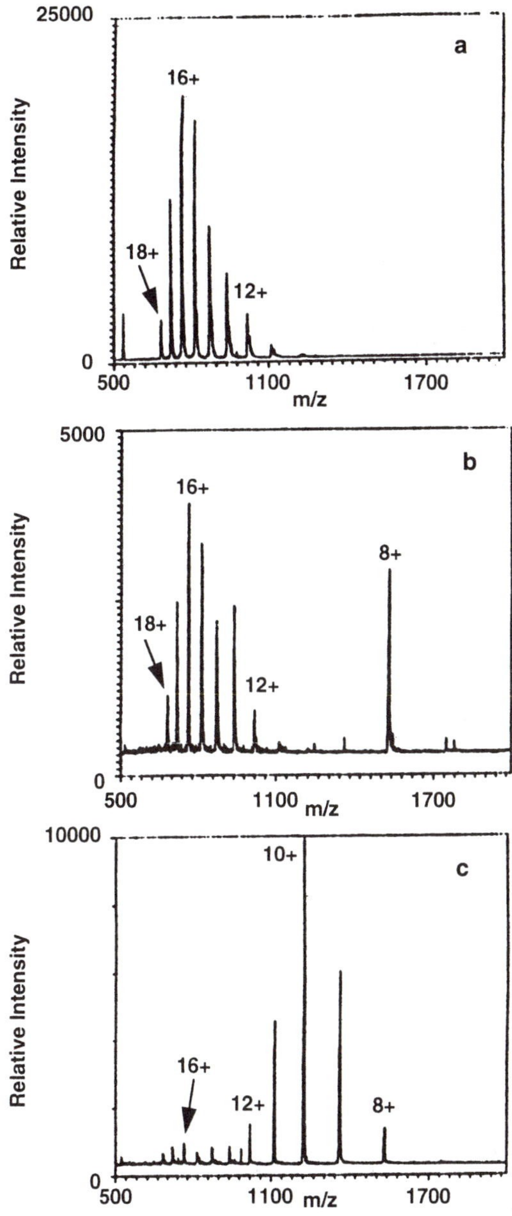

Figure 7.2 The ESI-mass spectra of bovine cytochrome c at the following aqueous acetic acid concentration and pH values: (a) 4%, pH 2.6; (b) 0.2%, pH 3.0; and (c) no acid, pH 5.2. (© 1990, American Chemical Society. Reprinted from reference 1 with permission.)

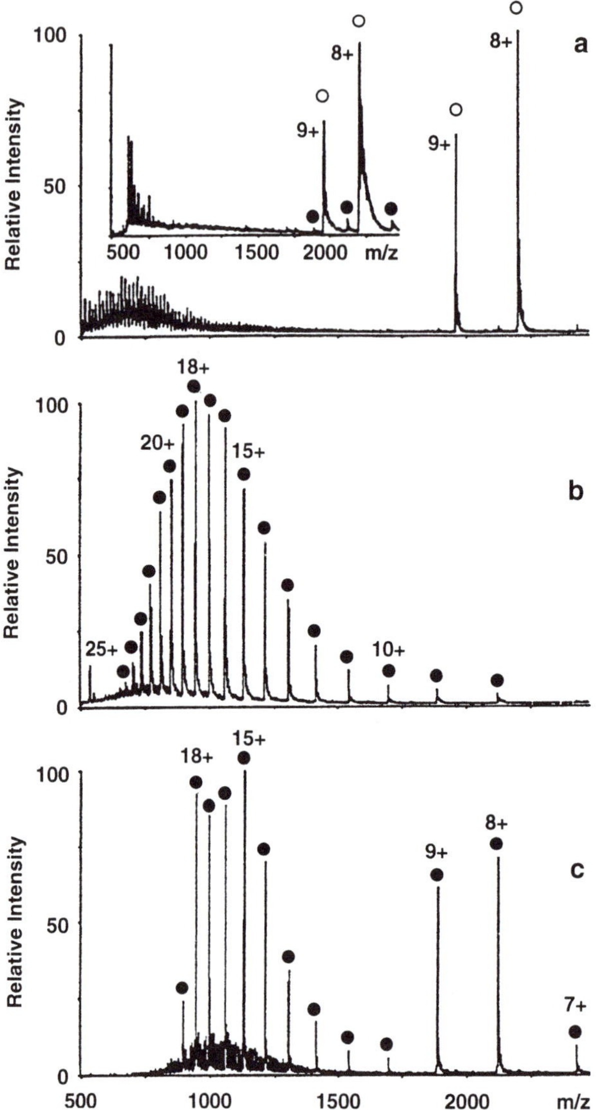

Figure 7.3 The ESI-mass spectra of equine myoglobin: (a) in a 10 mM NH_4OAc, pH 6.7 buffer, no added organic solvent, and orifice or up-front CID voltage of 80 V. Higher orifice voltages (100 V) caused some heme–polypeptide dissociation as observed in the inset. The closed circles represent the apo-myoglobin and the open circles represent the holoenzyme (heme + polypeptide). The broad protein peaks are due to potassium salt contamination. (b) As (a) except the pH was dropped to 2.2 by the addition of HOAc to a final concentration of 10% in order to cause protein denaturation. (c) As (b) except the protein was "partially renatured" by raising the pH from 2.2 to 5.0 by the addition of NH_4OH. (© 1992, American Society for Mass Spectrometry. Reprinted from reference 7 by permission of Elsevier Science, Inc.)

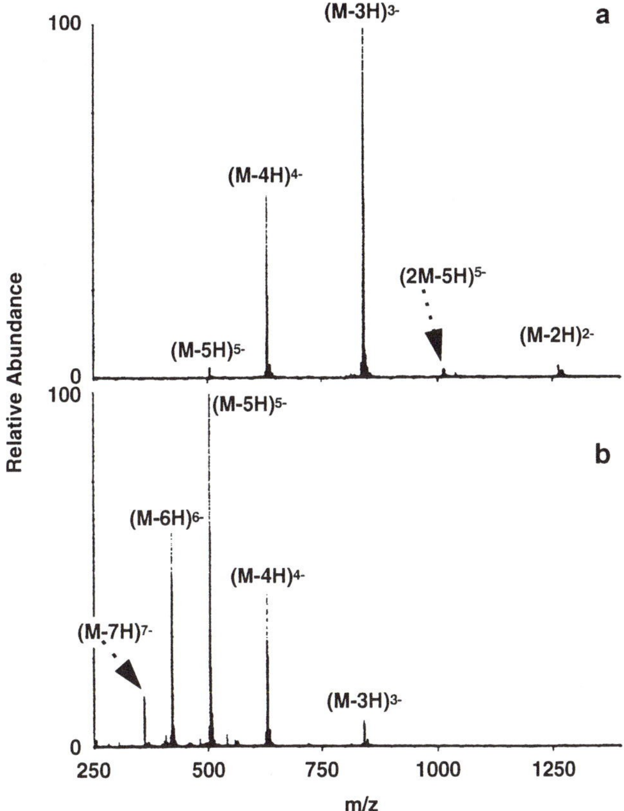

Figure 7.4 Negative-ion ESI-mass spectra of bovine α-chain insulin where the four cysteine residues were oxidized to cysteic acid: (a) aqueous solution, pH 7.0, (b) 1% NH_4OH. (© 1991, *Mass Spectrometry Reviews*. Reprinted from reference 5 by permission of John Wiley & Sons, Inc.)

both disulfide bonds from the four cysteine residues are oxidized to four cysteic acid residues.[5,8,9] Most proteins show very little if any negative-ion spectra below pH 5.0–6.0, because proteins have a net positive charge. Thus, a net positively charged protein entity would produce little to no observable peaks in the negative-ion mass spectrum. At and above pH 7.0, however, the acidic amino acid residues such as aspartic acid and glutamic acid readily deprotonate and shift the charge balance to a net negatively charged protein. Figure 7.4a presents the mass spectrum of oxidized bovine insulin α-chain at pH 7.0, and four negative charge states are observed for $(M + nH)^{n+}$ where $n = 2$–5. This can be associated with the deprotonation of the four cysteic amino acid residues plus the terminal carboxyl moiety. In a basic solution (1% NH_4OH), the two glutamic acid residues in the polypeptide deprotonate, which contributes to the seven negative charge states (Figure 7.4b).

Solvent

Various organic solvents can cause denaturation by disrupting the three-dimensional conformation of proteins, thereby exposing amino acid residue sites to the proton-laden milieu. Figure 7.5 presents spectra of bovine ubiquitin, and up to 13+ charge states are observed, which correlates well with the 12 basic amino acid residues and a free amine terminal moiety.[5,8] The amount of the acetonitrile (AcCN) organic solvent was varied with respect to water while keeping a constant concentration of the acetic acid proton source. With a relatively high amount of acetonitrile (Figures 7.5a and b), the high positive charge states are present. Even a slight change in organic solvent produces noticeable effects (Figure 7.5b) where a bimodal charge state distribution is evident. Figure 7.5c shows the ESI-mass spectrum where no acetonitrile is present in the protein solution, and the lower charge states dominate. This represents aqueous solution conditions conducive for the protein to reform into a tertiary structure, thus burying some of the basic amino acid residues within itself and reducing their accessibility to protons in the medium. Despite the disruption of the smooth intensity distribution envelope (Figure 7.5b and c), the average mass remains the same. A small amount of denatured ubiquitin remains in the spectrum in Figure 7.5c in the form of the 13+ to 9+ charge state envelope. This was postulated to originate from the acetonitrile sheath liquid (Figure 1.4) which came into contact with the central analyte spray. This hypothesis was proven by changing the acetonitrile sheath liquid to water. In an all-aqueous liquid environment,

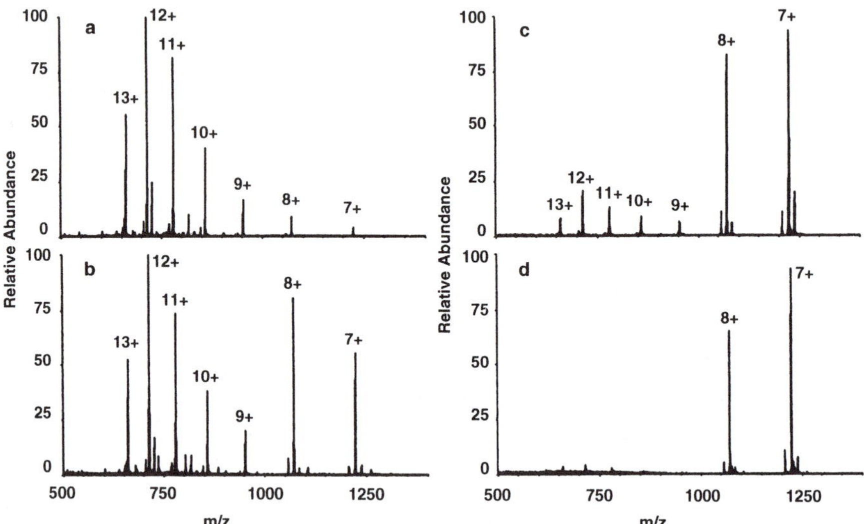

Figure 7.5 The ESI-mass spectra of bovine ubiquitin in (a) 18%:5%, (b) 12%:5%, and (c) 0%:5% aqueous AcCN:HOAc solutions with an AcCN ESI-liquid-sheath flow. (d) As (c) except water was the liquid-sheath flow. (© 1991, Mass Spectrometry Reviews. Reprinted from reference 5 by permission of John Wiley & Sons, Inc.)

Figure 7.5d is obtained, which essentially shows that the protein is in a folded, globular-type conformation (7+ and 8+ charge states), and the higher charge states are absent.

The effects of various alcoholic solvents were tested on the bradykinin and gramicidin S proteins, where MeOH is methanol; EtOH, ethanol; 2-PrOH, 2-propanol; 1-PrOH, 1-propanol; *t*-BuOH, *tert*-butanol; and 1-BuOH, 1-butanol. Figure 7.6a shows that as the solvents[4] change from MeOH to 1-BuOH, and keeping instrument and pH parameters relatively constant, the $(M + 2H)^{2+}/(M + H)^{+}$ ratio decreases for both proteins. Figures 7.6b and c show that the $(M + 2H)^{2+}$ ion for each protein decreases in intensity with respect to the $(M + H)^{+}$ ion upon changing the electrospray organic solvent component from 1-MeOH to 1-BuOH. Thus, the smaller mass alcohol imparts a relatively greater denaturation effect on the proteins than the larger alkyl alcohols.

Capillary Temperature

Temperature can have a significant effect on the stability of multiply charged ions. Figure 7.7 shows various ESI-mass spectra of bovine ubiquitin at different solution temperatures.[2] These temperatures were applied on the electrospray needle ceramic syringe, and the ions subsequently traversed a heated metal capillary tube at a constant temperature of 85°C inside the partial vacuum of the mass spectrometer.[10] Note that as the electrospray syringe temperature is increased from 25°C (Figure 7.7a) to 93°C (Figure 7.7f), the higher charge states successively become more pronounced in intensity. In Figure 7.7c, two charge-state distributions or protein populations are observed, centered about the 11+ and 7+ ions. It is thought that they represent two different conformations of the ubiquitin molecule where they differ by their overall degree of folding; that is, more compact versus less compact conformation. Upon elevated temperature conditions, the denaturation (unfolded state) of the protein allows more basic sites to be exposed to the aqueous surroundings, and an increased opportunity for protonation can occur. This allows for higher charge states to form.

Elevated temperatures are commonly used to probe the reversibility potential of protein folding. Some proteins can be heated to denaturation (unfolded state) and upon cooling they can refold to the native, active state. Other proteins cannot refold to an active, native state upon cooling of the denatured form.[6] Figure 7.8 shows ESI-mass spectra[2] of bovine cytochrome c at low and high temperatures (Figures 7.8a and b, respectively). At the elevated temperature, the higher charge states are evident. The protein solution was then heated at 90°C for 4 min and then slowly cooled to 25°C in a 25-min time period. The resulting mass spectrum (Figure 7.8c) showed that the protein did not refold in 25 min because of the close similarity to the denatured mass spectrum (Figure 7.8b).

However, for bovine ubiquitin, it was reported that allowing a denatured solution at high temperature to cool to room temperature gave an ESI-mass spectrum that mirrored the original native-state mass spectrum (data not shown).[2] Bovine ubiquitin has a conformation that is reversible upon heat denaturation and this was observed in the ESI-mass spectra.

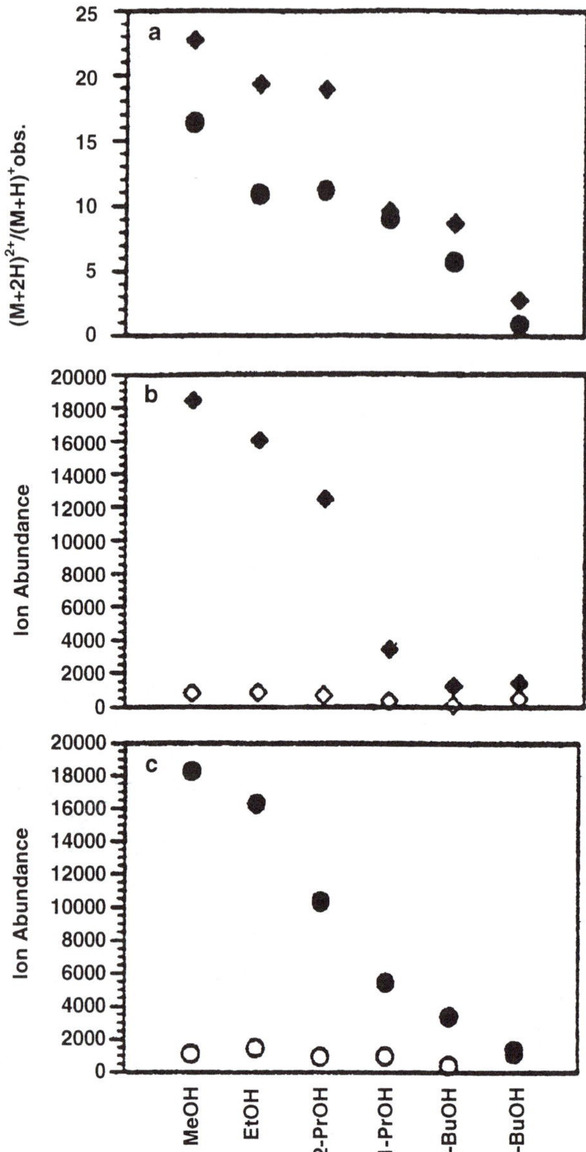

Figure 7.6 (a) Plot of the $(M + 2H)^{2+}/(M + H)^{+}$ ESI-mass spectral ion intensity ratios in the given solvents for (◆) bradykinin and (●) gramicidin S; (b) plot of the ESI-mass spectral ion intensity vs solvent for bradykinin, (◆) = $(M + 2H)^{2-}$ and (◇) = $(M + H)^{+}$; (c) plot of the ESI-mass spectral ion intensity vs solvent for gramicidin S, (●) = $(M + 2H)^{2+}$ and (○) = $(M + H)^{+}$. (© 1994. Reproduced from reference 4 by permission of John Wiley & Sons, Limited.)

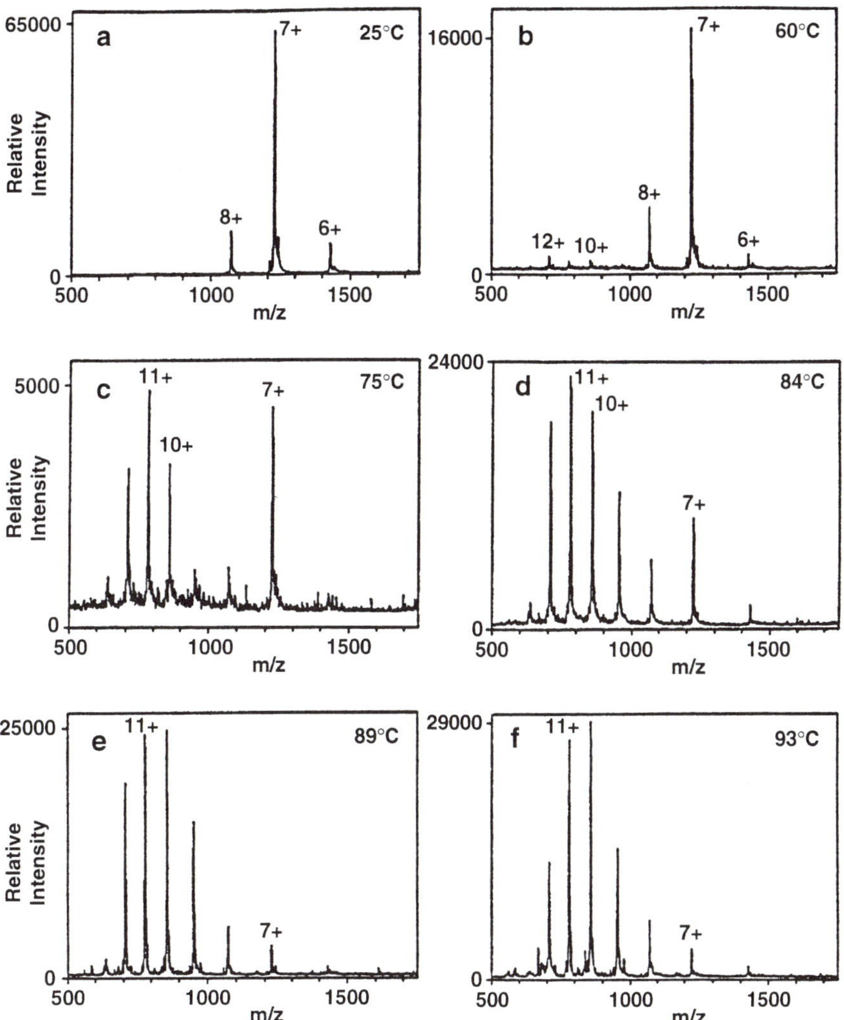

Figure 7.7 The ESI-mass spectra of bovine ubiquitin in aqueous HOAc solution at pH 2.8. The temperature is that of the analyte solution passing through a heated ceramic rod, which terminated in the electrospray needle. Ceramic rod temperatures are (a) 25°C, (b) 60°C, (c) 75°C, (d) 84°C, (e) 89°C, (f) 93°C. (© 1993, American Chemical Society. Reprinted from reference 2 with permission.)

In another series of experiments, an ambient temperature electrospray source produced melittin ions, which were admitted into a differentially heated metal capillary tube[11] prior to mass analysis. This heated capillary was used to effect thermally induced dissociation (TID) (fragmentation of the peptide backbone) of the charge states of melittin. Figure 7.9 shows a plot of temperature versus intensity of the 3+ to 6+ charge states of melittin.[11] As the temperature of the capillary was raised, the TID effected a decrease in all charge states,

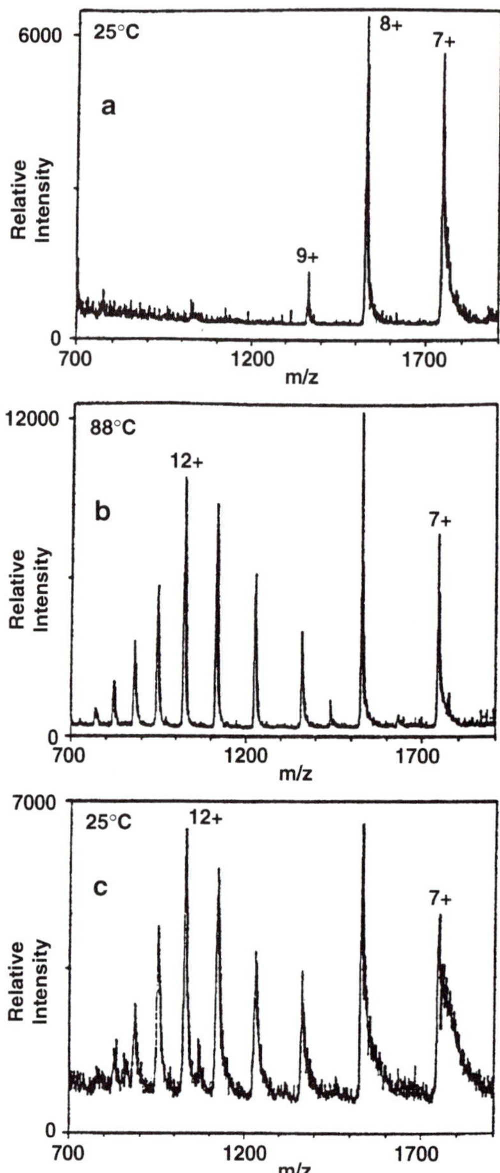

Figure 7.8 The ESI-mass spectra for probing the conformation reversibility of bovine cytochrome c. Solution was 5 mM NH_4OAc, pH 6.6, and the ceramic rod heater (cf. Figure 7.7) was at (a) 25°C and (b) 88°C. (c) The ESI-mass spectrum obtained after heating the cytochrome c solution at 90°C for 4 min with a subsequent cooling to 25°C. (© 1993, American Chemical Society. Reprinted from reference 2 with permission.)

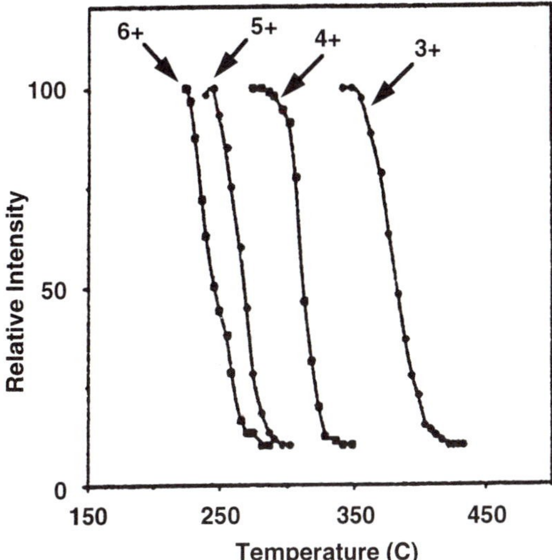

Figure 7.9 The ESI-mass spectral intensity plot vs temperature of the $(M + nH)^{n+}$, where $n = +3$ to $+6$, of melittin. The temperature was applied to a heated metal capillary between the ESI ion source and skimmer-ion focusing quadrupole rods. (© 1992, American Chemical Society. Reprinted from reference 11 with permission.)

although not in a simultaneous fashion. Higher charge states were more susceptible to heating effects, and the dissociation effect of TID produced product ions from the charge states—namely, higher charge-state parent ions fragmented to produce product ions (data presented in reference 12)—rather than producing a greater abundance of lower charge-state protonated molecular (parent) species. Relatively higher temperatures were needed to fragment a molecular species as its charge state decreased. In this case, higher temperatures on the capillary ion conduit provided enough energy to cause fragmentation of the various protein charge states as opposed to redistributing the intact protein molecule among charge states.

Disulfide Bond Oxidation

Proteins have covalent bonds that directly connect one part of the molecule to a relatively distant part of itself. These bonds are called disulfide bonds and occur between two cysteine residues. The two sulfhydryl groups are reduced, which eliminates the hydrogen ions and forms a disulfide –S–S– covalent bond (see Chapter 9). This bond can restrict the relative denaturation of the protein such that not all basic sites may be exposed upon denaturing the protein. 1,4-dithiothreitol (DTT) reduces the disulfide bond to form a sulfhydryl group on each of the affected cysteine residues. Figures 7.10a and b show ESI-mass spectra of bovine α-lactalbumin before and after treatment with DTT, respectively.[3,13] The protein has a total of 17 basic sites for protonation, and only 13 sites are

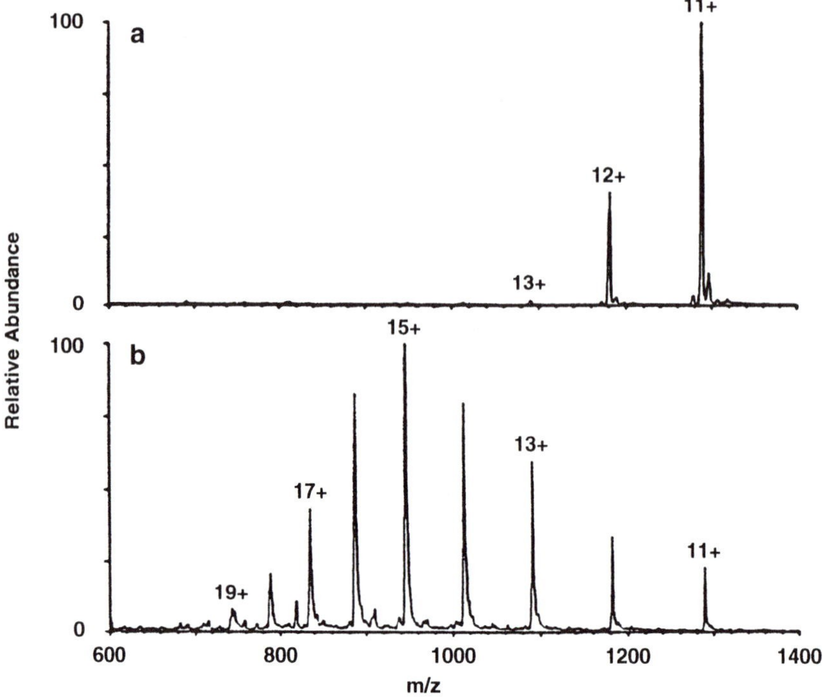

Figure 7.10 The ESI-mass spectra of bovine α-lactalbumin: (a) native protein and (b) reduction of the four disulfide bonds with dithiothreitol (DTT). (© 1990, Mass Spectrometry Reviews. Reprinted from reference 13 by permission of John Wiley & Sons, Inc.)

observed in Figure 7.10a. Upon reduction of the four disulfide bonds, solvent and proton accessibility increase such that not only are 17+ charge states observed, but also the 18+ and 19+ charge states are present. These extra two charge states are postulated to arise from two of the five glutamine residues in the protein.[13] This phenomenon is called "over-charging," and Figure 7.11 presents another case of this phenomenon with bovine proinsulin. The protein has a total of nine basic protonation sites (two lysines, two histidines, four arginines, and the N-terminal amine) and upon DTT reduction of the three disulfide bonds, up to 12+ charges are observed. Obviously, other sites on the molecule are being protonated, and glutamine residues can be the cause of this.[3]

Figure 7.12 shows the effect of disulfide group reduction[5,14] with bovine serum albumin (BSA). After treatment with DTT, BSA provides many more charge states, as shown in Figure 7.12b.

Skimmer Voltage

The skimmer voltage is the electric field that the ions experience in a region of reduced pressure near the sample introduction source of the mass spectrometer.

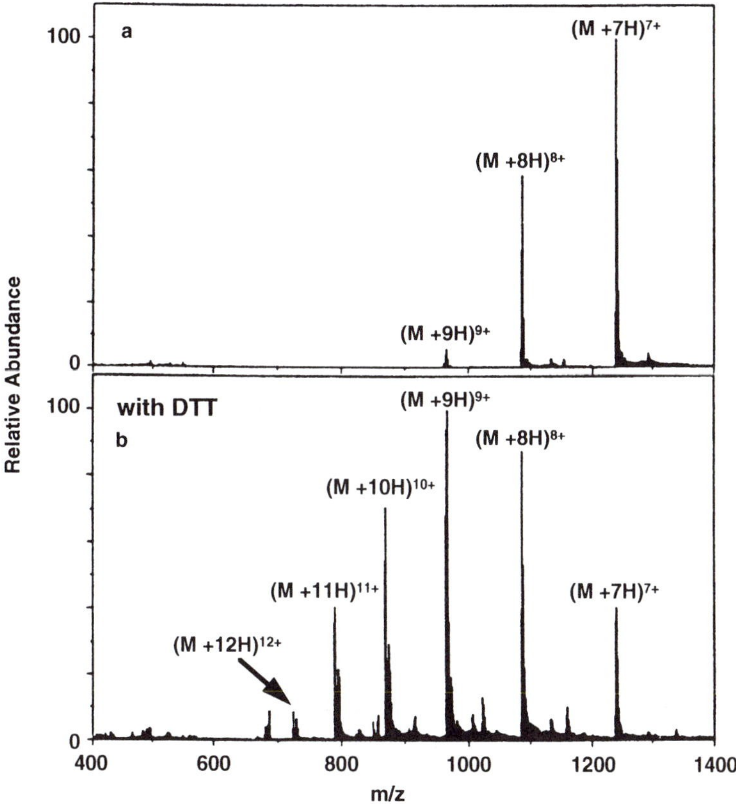

Figure 7.11 The ESI-mass spectra of bovine proinsulin: (a) native protein and (b) reduction of the three disulfide bonds with DTT. (© 1992, *Mass Spectrometry Reviews*. Reprinted from reference 3 by permission of John Wiley & Sons, Inc.)

This voltage is also labeled "up-front" CID or CID-MS and has been presented in Chapter 1. Multiply charged ions are known to adduct with a few to many water molecules. The CID-MS region aids in the reduction of these clusters so as to yield the unclustered, multiply charged analyte ion. Figure 7.13 presents ESI-mass spectra of bradykinin from a constant-temperature (85°C) metal capillary-skimmer arrangement with no countercurrent nitrogen gas.[10] Note the diversity of clustered ions (Figure 7.13a) along with the $(M + 2H)^{2+}$ ion of interest. As the potential difference of the CID-MS region is increased by increasing the capillary voltage, the spectra shown in Figures 7.13b and c are produced, where the myriad of clusters disappear to leave a spectrum essentially containing only the $(M + 2H)^{2+}$ bradykinin ion.

Clustering of water about the analyte ions is also a common phenomenon in the atmospheric pressure region of the ESI source (Figure 7.14), and the clustering can easily be eliminated as shown in Figure 7.13. A mass spectrometry system similar to that shown in Figure 1.10 was used, and the ESI

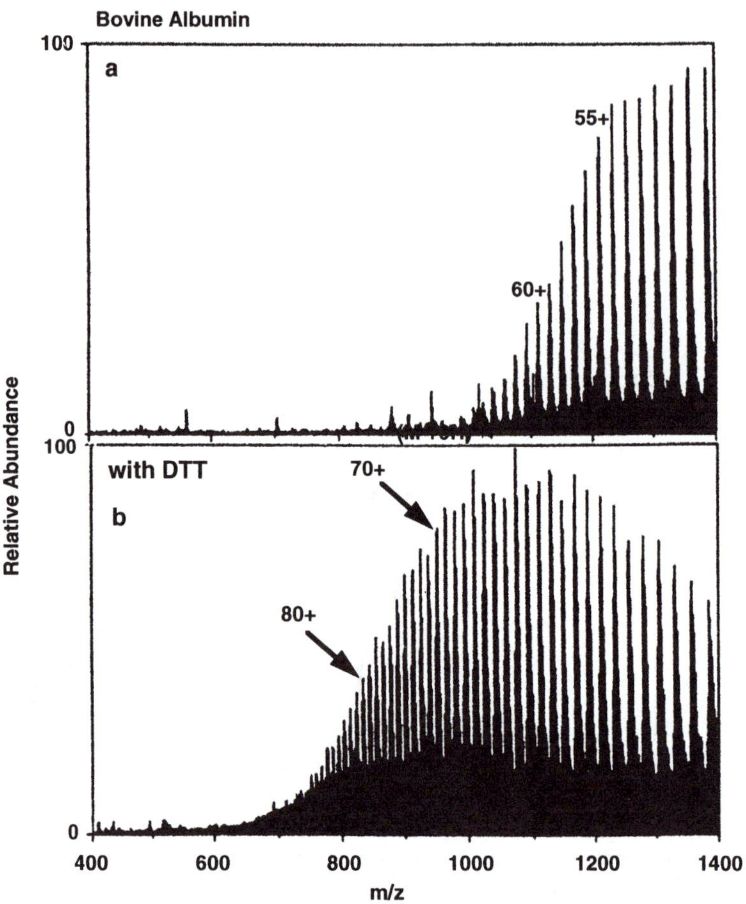

Figure 7.12 The ESI-mass spectra of bovine serum albumin: (a) native and (b) reduction of disulfide bonds with DTT. (© 1990, *Mass Spectrometry Reviews*. Reprinted from reference 14 by permission of John Wiley & Sons, Inc).

Figure 7.13 (facing page) The ESI-mass spectra of bradykinin. Collision-induced desolvation of bradykinin was accomplished by varying the voltage on a metal capillary tube. A capillary tube-skimmer arrangement was used where skimmer voltage = 17.5 V and temperature of the tube = 85°C. Capillary tube voltages (V_c) were (a) 100 V, (b) 120 V, and (c) 180 V. (© 1990. Reproduced from reference 10 by permission of John Wiley & Sons, Limited.)

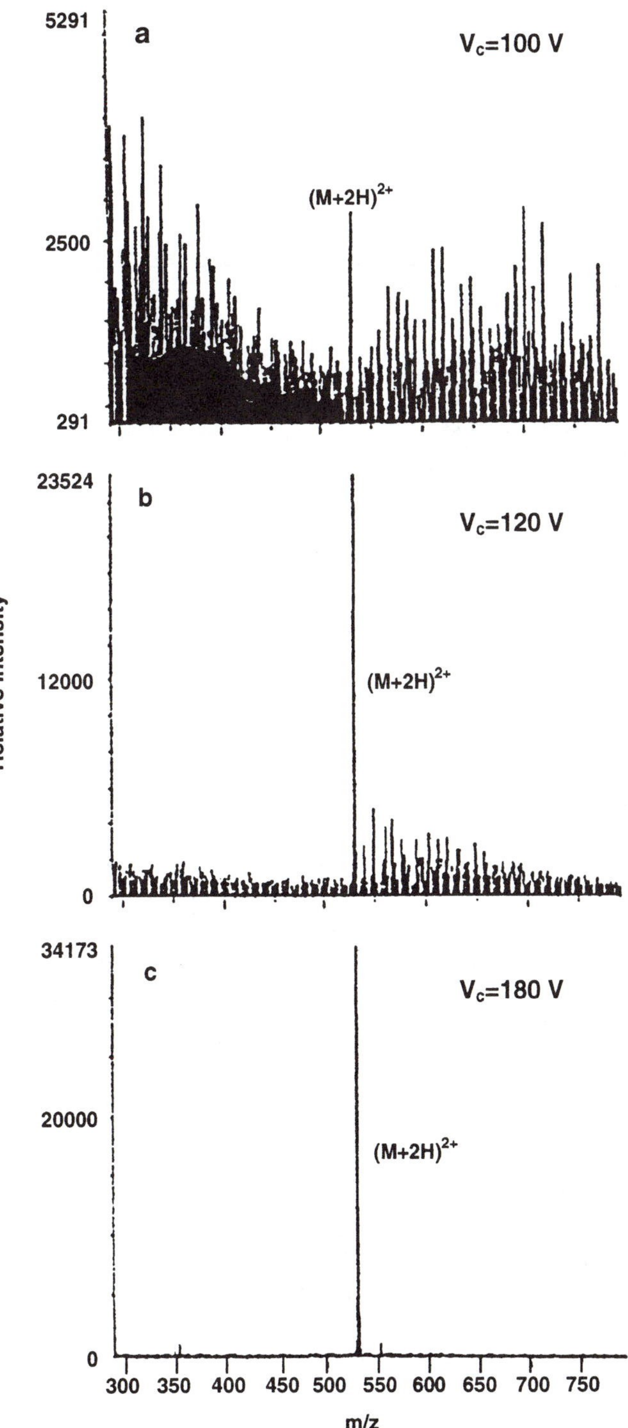
5291
a
V_c=100 V
(M+2H)2+
2500
291
23524
b
V_c=120 V
Relative Intensity
12000
(M+2H)2+
0
34173
c
V_c=180 V
20000
(M+2H)2+
0
300 350 400 450 500 550 600 650 700 750
m/z

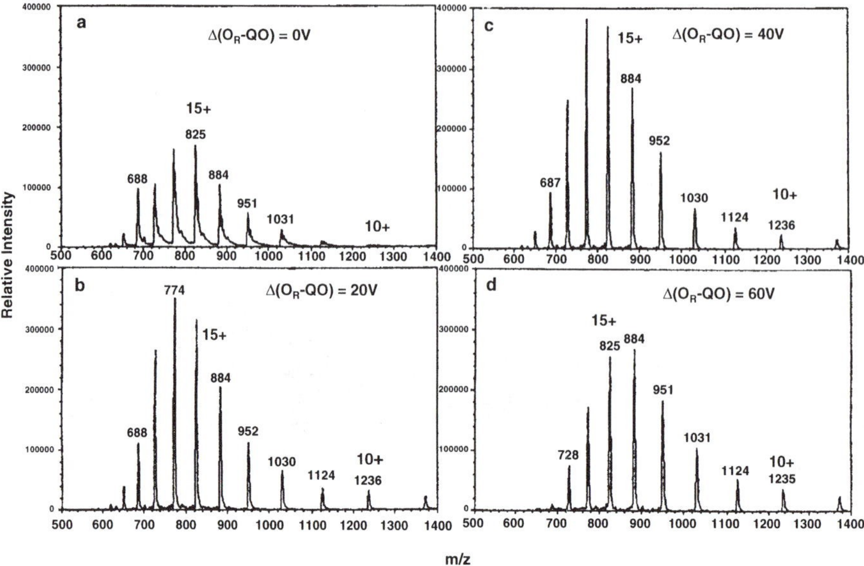

Figure 7.14 The ESI-mass spectra of horse heart cytochrome c. The up-front CID voltage was varied by changing the voltage between the orifice and ion-focusing quadrupole rod lens, $\Delta V = \Delta(O_R - Q0)$, at the atmosphere–vacuum interface (Figure 1.10). ΔV = (a) 0 V, (b) 20 V, (c) 40 V, (d) 60 V. (© 1989, American Society for Mass Spectrometry. Reprinted from reference 15 by permission of Elsevier Science, Inc.)

sample introduction device was that shown in Figure 1.4. Figure 7.14 shows a series of multiply charged peaks for equine cytochrome c.[15] As the CID-MS voltage is increased between the heated plenum and rf-only quadrupole region, the dominant effect is a noticeable sharpening of the peaks, including an increase in intensity of the peaks. Just as in Figure 7.13, the myriad of cluster ions associated with each charge state, which cause band-broadening, disappear to leave the more stable $(M + nH)^{n+}$ ion. Using relatively higher voltages,[16] as shown in Figure 7.15, equine cytochrome c displayed the typical intensity increase in the low charge-state ions and an intensity decrease in the high charge-state ions when the voltage of the skimmer was increased from 250 V to 370 V.

Using the ESI-MS system shown in Figure 1.4, the melittin protein was investigated for charge-distribution equilibria.[16] Figure 7.16 shows that as the skimmer voltage is increased from 170 V to 370 V, lower charge states successively appear, and the higher charge states disappear. Below 170 V in the CID-MS region, sensitivity was reduced. This is most likely because the relatively more gentle nature of the CID-MS region allows extensive water adduct formation to disperse the signal, as shown in Figure 7.13a.

This phenomenon also is operative for analyte adducts where the adduct is other than water. Figure 7.17 shows that for melittin, as the potential difference

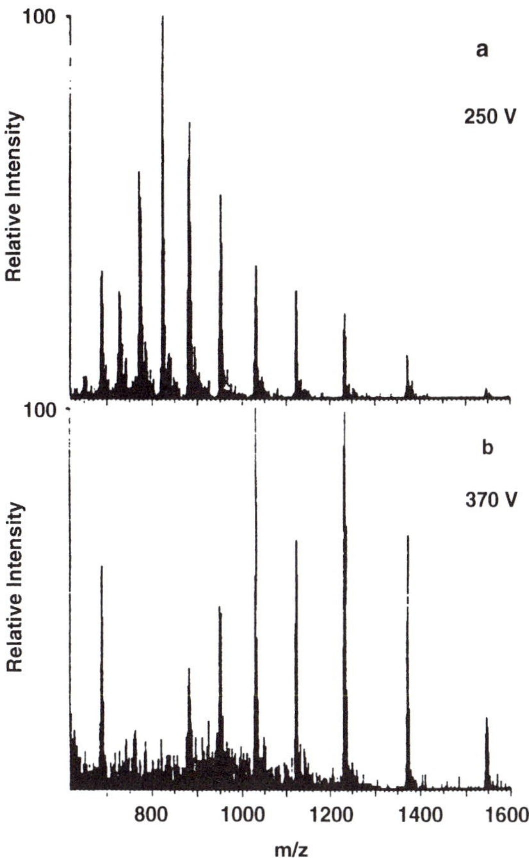

Figure 7.15 The ESI-mass spectra of horse heart cytochrome c in phosphate/NaOH solution at pH 11.0. ΔV = (a) 250 V, (b) 370 V. (© 1988. Reproduced from reference 16 by permission of John Wiley & Sons, Limited.)

between the capillary tube and skimmer arrangement is increased, trifluoroacetic acid adducts of melittin disappear, and the 3+ ion increases significantly with respect to the 4+ multiply charged ion.[17]

An increase in skimmer voltage can also yield information from relatively high-mass proteins. Figure 7.18 shows ESI-mass spectra[18] of the intact $\alpha_2\beta_2$ bovine hemoglobin (Hb) protein using an ESI-mass spectrometry system similar to that shown in Figure 1.4. As the skimmer voltage is increased, the intact Hb protein in Figure 7.18a shifts to lower charge states in Figure 7.18b. A "loosening" or dissociation of the four chains is evident because of the *m/z* 617 peak observed in Figure 7.18b. This ion corresponds to the central iron-heme moiety responsible for the electron processing of molecular oxygen.

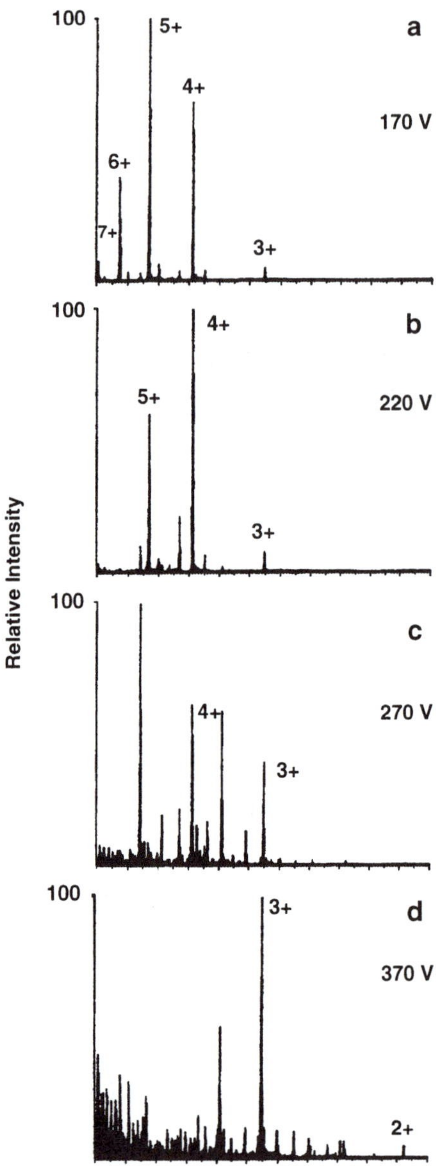

Figure 7.16 The ESI-mass spectra of melittin at ΔV voltages of (a) 170 V, (b) 220 V, (c) 270 V, and (d) 370 V. (© 1988. Reproduced from reference 16 by permission of John Wiley & Sons, Limited.)

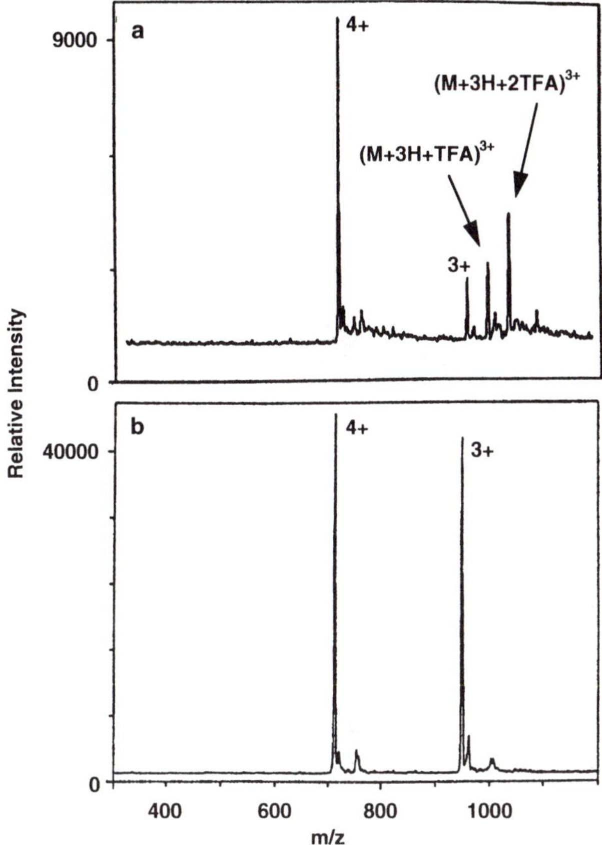

Figure 7.17 The ESI-mass spectra of bee venom melittin. Ion-desolvation voltages between the metal capillary transfer column and the skimmer were ΔV = (a) 30 V and (b) 90 V. (© 1994, American Chemical Society. Reprinted with permission from reference 17.)

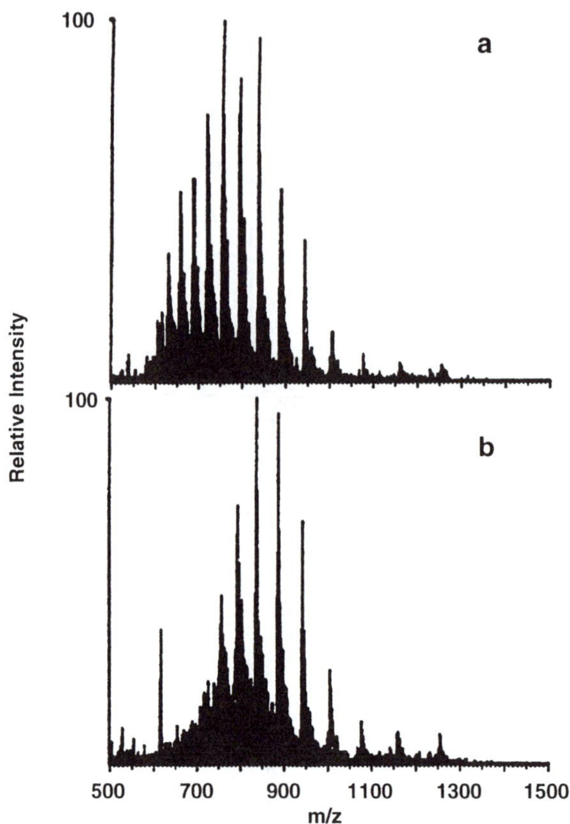

Figure 7.18 The ESI-mass spectra of bovine hemoglobin with a ΔV potential of (a) 200 V and (b) 300 V. (© 1989. Reprinted from reference 18 by kind permission of Elsevier Science-NL, Sara Burgerhartstraat 25, 1055 KV Amsterdam, The Netherlands.)

REFERENCES

1. Chowdhury, S.K.; Katta, V.; Chait, B.T. *J. Am. Soc. Mass Spectrom.* **1990,** 112, pp 9012– 9013.
2. Mirza, U.A.; Cohen, S.L.; Chait, B.T. *Anal. Chem.* 1993, *65,* pp 1–6.
3. Smith, R.D.; Loo, J.A.; Edmonds, C.G. In *Mass Spectrometry; Clinical and Biomedical Applications*, Desiderio, D.M., Ed.; Plenum Press: New York, 1992; Volume 1, Chapter 2, pp 37–98.
4. Wang, G.; Cole, R.B. *Org. Mass Spectrom.* **1994,** *29,* pp 419–427.
5. Smith, R.D.; Loo, J.A.; Ogorzalek Loo, R.R.; Busman, M.; Udseth, H.R. *Mass Spectrom. Rev.* 1991, *10,* pp 359–451.
6. Ghelis, C.; Yon, J., Eds.; *Protein Folding;* Academic Press: New York, 1982.
7. Feng, R.; Konishi, Y. *J. Am. Soc. Mass Spectrom.* **1993,** *4,* pp 638–645.
8. Loo, J.A.; Ogorzalek Loo, R.R.; Udseth, H.R.; Edmonds, C.G.; Smith, R.D. *Rapid Commun. Mass Spectrom.* **1991,** *5,* pp 101–105.
9. Loo, J.A.; Ogorzalek, R.R.; Light, K.J.; Edmonds, C.G.; Smith, R.D. *Anal. Chem.* **1992,** *64,* pp 81–88.
10. Chowdhury, S.K.; Katta, V.; Chait, B.T. *Rapid Commun. Mass Spectrom.* **1990,** *4,* pp 81–87.
11. Busman, M.; Rockwood, A.L.; Smith, R.D. *J. Phys. Chem.* **1992,** *96,* pp 2397–2400.
12. Rockwood, A.L.; Busman, M.; Udseth, H.R.; Smith, R.D. *Rapid Commun. Mass Spectrom.* **1991,** *5,* pp 582–585.
13. Loo, J.A.; Edmonds, C.G.; Udseth, H.R.; Smith, R.D. *Anal. Chem.* 1990, *62,* pp 693–698.
14. Smith, R.D.; Loo, J.A.; Edmonds, C.G.; Barinaga, C.J.; Udseth, H.R. *Anal. Chem.* 1990, 62, pp 882–899.
15. Smith, R.D.; Loo, J.A.; Barinaga, C.J.; Edmonds, C.G.; Udseth, H.R. *J. Am. Soc. Mass Spectrom.* **1989,** *90,* pp 53–65.
16. Loo, J.A.; Udseth, H.R.; Smith, R.D. *Rapid Commun. Mass Spectrom.* 1988, *2,* pp 207– 210.
17. Mirza, U.A.; Chait, B.T. *Anal. Chem.* 1994, *66,* pp 2898–2904.
18. Edmonds, C.G.; Loo, J.A.; Barinaga, C.J.; Udseth, H.R.; Smith, R.D. *J. Chromatogr.* **1989,** *474,* pp 21–37.

8

The Cation Effect on Electrospray Ionization Mass Spectra

This chapter analyzes the effects that impurity and buffer-related cations have on the mass spectral response of a protein. These species produce additional features in an ESI-mass spectrum and usually manifest themselves as peaks near singly and multiply charged analyte ion signals. These modifications inherently cause a reduction in analyte signal, because the signal is spread over multiple *m/z* peaks. On the other hand, cation adduction by metalloproteins and metalloenzymes, with respect to the presence of the holoenzyme (full complement of metal ions) and/or apoenzyme (metals removed) in an ESI-mass spectrum, can be very useful for determining metal-binding constants. Sodium, potassium, and ammonium adducts of proteins are sometimes considered as nuisances because of their lower information content (with respect to the divalent cations usually found in proteins) and competition with proton-charging. Usually, the presence of alkali ion in ESI-mass spectra indicates that clean-up or sample processing techniques are required.[1–7]

An easy way of deducing the identity of extraneous peaks is to add deliberately small amounts of the suspect cation in the protein solution and then observe the relative intensity of the peak(s) in question.[8] Equations such as those used for the calculation of the average mass of a multiply charged protein can be used to deduce the identity of metal ion–protein adduct peaks (Chapter 4). One assumes a cation and uses the appropriate mass in place of that of the hydrogen ion in the equations. However, mixed cation and proton adducts are usually observed as well as small amounts of cation-adducted protein analyte, and these can be difficult to solve by the standard average mass equations (Chapter 4). A three-dimensional macrosurface algorithm and graphical data reduction has been proposed by Fenn et al. to tackle the problem of mixed proton–cation–protein adducts.[8]

Figure 8.1 presents an ESI-mass spectrum of the $(M + 14H)^{14+}$ ion of bovine β-lactoglobulin.[9] This protein is composed of two identical subunits, and the calculated average mass of each subunit is 18,277.1. The dominant peak at m/z 1306.4, when multiplied by 14+ charges and subtracting 14 hydrogen ions from the result, yields an experimental mass of 18,275.6 Da. For this particular measurement, the difference is [(18,275.6 – 18,277.1)100]/18,277.1 = –0.00821%. Note also that despite the many ^{13}C isotope peaks that make up a protein of a mass in the 18-kDa range, the peak width for m/z at half-height is approximately 1 Da. The peak at m/z 1308.0 is assumed to be a cation adduct. A straightforward accounting of the mass difference between m/z 1306.4 and 1308.0 is to convert the former to its zero mass value and the latter m/z value to its ion-adduct value. The difference of the two values yields the mass contribution of the protons and cations for m/z 1308. Thus we have the following:

1306.4		1308.0
× 14	H+	×14
18,289.6		18,312.0
−14	H+	−18,275.6
18,275.6		36.4

There is an ionic mass of 36.4 Da adducted to the intact β-lactoglobulin subunit. In addition, 14 charged entities must contribute to the 36.4 Da in order to yield a 14+ species. An arrangement of H^+ and Na^+ can be 13 hydrogen ions and one sodium ion or $(M + 13H + Na)^{14+}$. Therefore, 22.9 + 13(1.00797) = 36.0 Da ≃ 36.4 Da. Alternatively, the calculations can be made easier as follows:

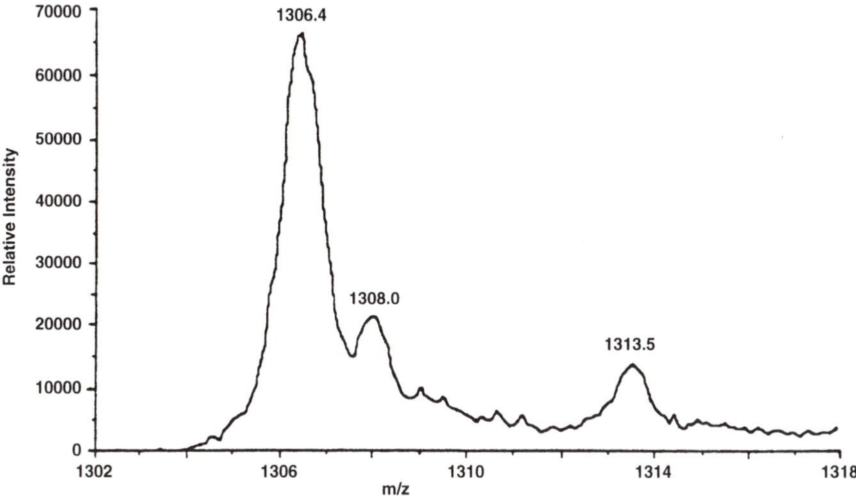

Figure 8.1 The ESI-mass spectrum of the $(M + 14H)^{14+}$ and sodium adduct ions from β-lactoglobulin. (© 1988. Reproduced from reference 9 by permission of John Wiley & Sons, Limited.)

$$
\begin{array}{r} 1306.4 \\ \times\ 14 \\ \hline 18{,}289.6 \end{array} = \text{mass of } (M+14H)^{14+} \qquad \begin{array}{r} 1308.0 \\ \times\ 14 \\ \hline 18{,}312.0 \\ -18289.6 \\ \hline 22.4 \end{array} \begin{array}{l} \\ \\ \\ = \text{mass of } M+14H+\text{alkali} \\ = \text{mass of } M+14H \\ \simeq \text{mass of } Na^+ \end{array}
$$

The next example of cation attachment to a protein is shown in Figure 8.2, where a bradykinin solution ($M_r = 1060.0$) was adjusted to a basic pH with potassium hydroxide.[10] The envelope represents doubly charged ions of the formula $(M + nH + mK)^{2+}$. The constitution of m/z 550 can be calculated as follows:

$$
\begin{array}{r} 550 \\ \times 2 \\ \hline 1100.0 \\ -1060.0 \\ \hline 40.0 \end{array}
$$

Two charges of a total mass of 40.0 Da need to be considered, and they can represent hydrogen and potassium ions, or $(M + H + K)^{2+}$. The ion at m/z 569 can be analyzed as $569 \times 2 = 1138$; $1138 - 1060 = 78$ Da. A mass of 78 Da that contains two charged ions can comprise two K^+ ions ($2 \times 39 = 78$), or $(M + 2K)^{2+}$. The ion at m/z 588 must be treated differently:

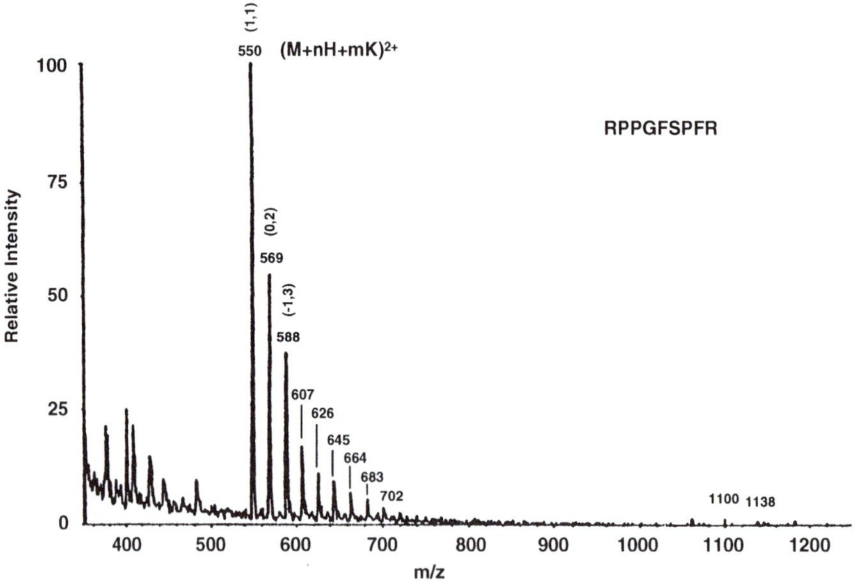

Figure 8.2 The ESI-mass spectrum of bradykinin in 50:50 MeOH:H_2O adjusted to pH 12.0 with KOH. Figures in parentheses are the (n, m) values.

$$\begin{array}{r} 588 \\ \times 2 \\ \hline 1176 \\ -1060 \\ \hline 116 \end{array}$$

Three potassium ions yield a mass of 117 Da, and, by removing hydrogen from the protein, a negative site appears that can be neutralized by one of the three potassium ions. Thus, the formula of the ion at m/z 588 can be $(M - H + 3K)^{2+}$ where $n = -1$ and $m = 3$ (Figure 8.2). The origin of the protons that are removed from the protein is speculated to be amide protons on the polypeptide backbone.

PROBLEM 8.1 Calculate the formulae for m/z 607 to 702 in Figure 8.2 in the form of $(M + nH + mK)^{(n+m)^+}$.

PROBLEM 8.2 Figure 8.3 shows a partial ESI-mass spectrum of a synthetic calcitonin protein ($M_r = 3630.3$) and represents the 3+ charged cluster of peaks.[11] The masses of the five peaks are at m/z 1211.1, 1216.8, 1218.4, 1224.0, and 1225.7. Calculate the $(M + nH + mX)^{(n+m)^+}$ adduct formulae for each peak. At least one of the peaks contains an adduct species $(M + nH + mX + yY)^{(n+m+y)^+}$, where X and Y are two different cations.

The next ESI-mass spectral interpretation is taken from an investigation of the calcium-binding metalloprotein parvalbumin from the *Rana tegrinka* frog.[12,13] Figure 8.4 provides a mass spectrum of the 9− ion envelope of parvalbumin protein in the presence of calcium acetate, and three peaks are labeled at m/z 1328.5, 1332.75, and 1337.0. An analysis of these peaks follows; note that hydrogen is lost from the intact parvalbumin in order to produce the negative

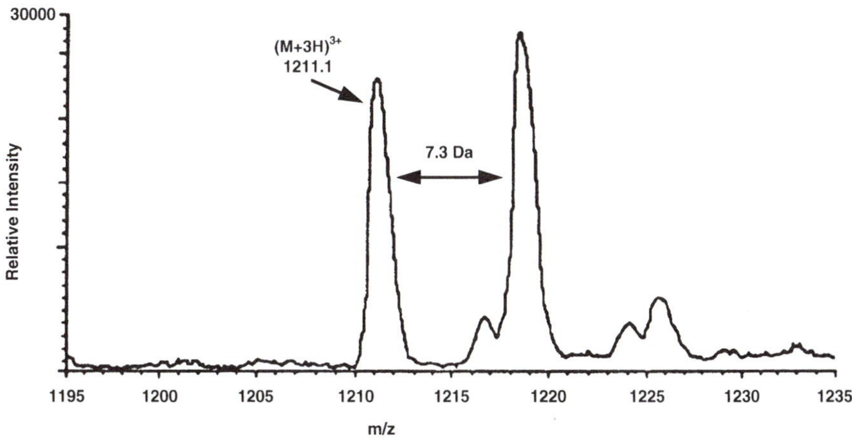

Figure 8.3 The ESI-mass spectrum of the $(M + 3H)^{3+}$ region from the synthetic human calcitonin. (Permission granted by PE-SCIEX, Concord, Ontario, Canada.)

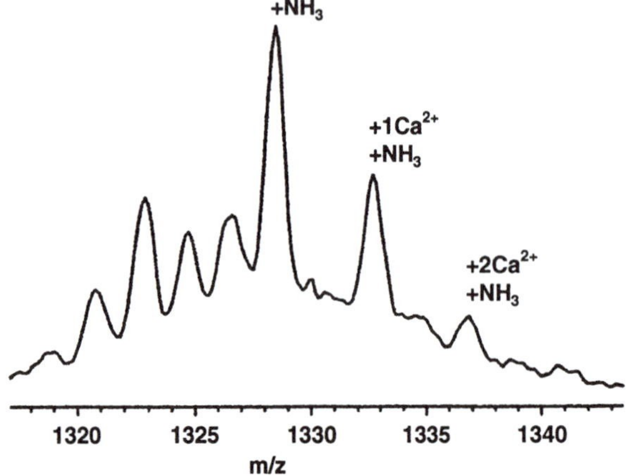

Figure 8.4. The ESI-mass spectrum of the $(M - 9H)^{9-}$ region from *Rana tegrinka* frog parvalbumin with 0.4 mM $Ca(OAc)_2$ for calcium ion-binding studies.

charge. Keep in mind that when Ca^{2+} chelates to the amino acid ligands in the protein, two hydrogen ions are lost in order to maintain the 9− state. Also, the mass analysis of the m/z 1328.5 shows that ammonia binds (adds 17 Da) to the mass of the apoprotein parvalbumin. The ammonia species is a constant in the three peaks.

$$\begin{array}{r} m/z\ 1328.5 \\ \times 9 \\ \hline 11{,}956.5 \end{array} \quad \text{and} \quad \begin{array}{r} m/z\ 1332.75 \\ \times 9 \\ \hline 11{,}994.75 \end{array}$$

Both mass values represent the 9− state, not the zero charge mass.

$$\begin{array}{r} 11{,}994.75 \\ -11{,}956.50 \\ \hline 38.25 \end{array}$$

This value is close to the mass of Ca^{2+} (40 Da), and a further loss of two hydrogen ions yields 38 Da $\simeq$ 38.25 Da. Thus, the m/z 1332.75 ion represents the $(M - 11H + Ca)^{9-}$ ion, where M is the mass of parvalbumin and an ammonia molecule.

$$\begin{array}{r} m/z\ 1337.0 \\ \times 9 \\ \hline 12{,}033.0 \\ -11{,}956.5 \\ \hline 76.5\ \text{Da} \end{array}$$

Here, 76.5 Da represents the chelation of two calcium ions (80 Da) by parvalbumin, with the concomitant loss of four hydrogen ions, and this peak is the $(M - 13H + 2Ca)^{9-}$ ion, where M includes an ammonia molecule.

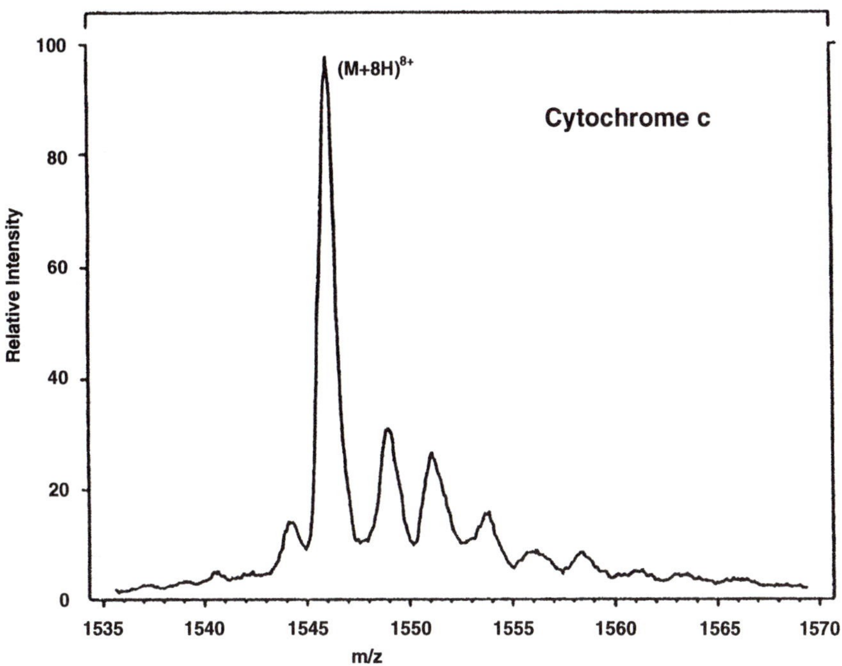

Figure 8.5. The ESI-mass spectrum of the $(M + 8H)^{8+}$ region from horse heart cytochrome c in the presence of cationic species. (© 1993, Finnigan MAT, San Jose, California. Reproduced from reference 14 with permission.)

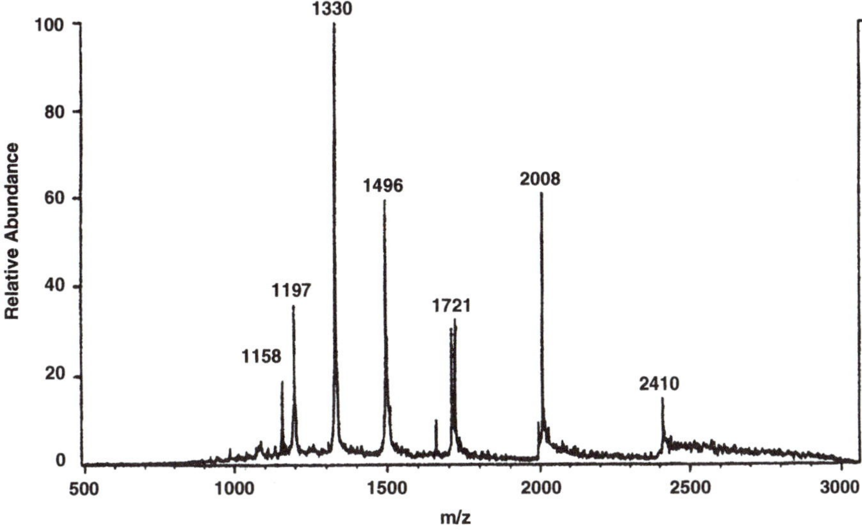

Figure 8.6 Negative-ion ESI-mass spectrum of a mixture of holo- and apo- rabbit parvalbumin protein in water. (© 1994, American Chemical Society. Reprinted from reference 15 with permission.)

PROBLEM 8.3 Figure 8.5 presents the ESI-mass spectrum of the 8+ envelope of peaks, including metal ion adducts for horse heart cytochrome c. The figure shows the $(M + 8H)^{8+}$ peak and five smaller peaks at higher masses.[14] Determine the adduct formulae for the proton–alkali and proton–mixed metal alkali species. Note: Balancing the formula for total charge in this protein yields approximately a 1-Da error in the bookkeeping of hydrogen.

PROBLEM 8.4 Figure 8.6 provides the ESI-mass spectrum of the holo- and apo-rabbit parvalbumin protein without added calcium salt in the buffer solution of the protein.[15] Provide the mass analysis of both parvalbumins and show that the difference of the masses yields two calcium ions.

REFERENCES

1. Wu, Q.; Liu, C.; Smith, R.D. *Rapid Commun. Mass Spectrom.* **1996,** *10,* pp 835-838.
2. Wu, J.; Pawliszyn, J. *Anal. Chem.* **1995,** *67,* pp 2010–2014.
3. Roboz, J.; Yu, Q.; Meng, A.; Soest, R. *Rapid Commun. Mass Spectrom.* 1994, *8,* pp 621– 626.
4. Kay, I.; Mallet, A.I. *Rapid Commun. Mass Spectrom.* **1993,** *7,* pp 744–746.
5. Stoney, K.; Nugent, K. In *Techniques in Protein Chemistry VI;* Crabb, J.W., Ed.; Academic Press: San Diego, CA, 1995, pp 277–284.
6. Liu, C.; Wu, Q.; Harms, A.C.; Smith, R.D. *Anal. Chem.* 1996, 68, pp 3295–3299.
7. Liu, C.; Muddiman, D.C.; Tang, K.; Smith, R.D. *J. Mass Spectrom.* **1997,** 32, pp 425–431.
8. Labowsky, M.; Whitehouse, C.; Fenn, J.B. *Rapid Commun. Mass Spectrom.* 1993, *7,* pp 71–84.
9. Covey, T.R.; Bonner, R.F.; Shushan, B.I.; Henion, J. *Rapid Commun. Mass Spectrom.* **1988,** *2,* pp 249–256.
10. Wang, J.; Ke, F.; Guevremont, R.; Siu, K.W.M. *Proc. 43rd ASMS Conference on Mass Spectrometry and Allied Topics* **1995,** p 379.
11. Covey, T.R.; Bonner, R.F.; Shushan, B.I. SCIEX HyperMass Application Note, No. 15489; SCIEX: Canada.
12. Hu, P.; Buckel, S.D.; Whitton, M.M.; Loo, J.A. *Proc. 43rd ASMS Conference on Mass Spectrometry and Allied Topics* **1995,** p 322.
13. Hu, P.; Buckel, S.D.; Whitton, M.M.; Loo, J.A. *Eur. Mass Spectrom.* **1996,** *2,* pp 69–76.
14. Chapman, J.R. *Practical Organic Mass Spectrometry; A Guide for Chemical and Biochemical Analysis*, 2nd Edition; John Wiley & Sons: Chichester, U.K., 1993; Chapter 6, pp 182–220.
15. Hu, P.; Ye, Q.; Loo, J.A. *Anal. Chem.* 1994, *66,* pp 4190–4194.

ANSWERS

PROBLEM 8.1

m/z 607:

$$\begin{array}{r} 607 \\ \times\ 2 \\ \hline 1214 \\ -1060 \\ \hline 154 \end{array}$$

$$\begin{array}{l} 156 = 4 \times 39 \\ -2H \\ \hline 154 \end{array}$$

$n = -2, m = 4$

$(M - 2H + 4K)^{2+}$

m/z 626:

$$\begin{array}{r} 626 \\ \times\ 2 \\ \hline 1252 \\ -1060 \\ \hline 192 \end{array}$$

$$\begin{array}{l} 195 = 5 \times 39 \\ -3H \\ \hline 192 \end{array}$$

$(M - 3H + 5K)^{2+}$

m/z 645:

$$\begin{array}{r} 645 \\ \times\ 2 \\ \hline 1290 \\ -1060 \\ \hline 230 \end{array}$$

$$\begin{array}{l} 234 = 6 \times 39 \\ -4H \\ \hline 230 \end{array}$$

$(M - 4H + 6K)^{2+}$

m/z 664:

$$\begin{array}{r} 664 \\ \times\ 2 \\ \hline 1328 \\ -1060 \\ \hline 268 \end{array}$$

$$\begin{array}{l} 273 = 7 \times 39 \\ -5H \\ \hline 268 \end{array}$$

$(M - 5H + 7K)^{2+}$

m/z 683:

$$\begin{array}{r} 683 \\ \times\ 2 \\ \hline 1366 \\ -1060 \\ \hline 306 \end{array}$$

$$\begin{array}{l} 312 = 8 \times 39 \\ -6H \\ \hline 306 \end{array}$$

$(M - 6H + 8K)^{2+}$

m/z 702:

$$\begin{array}{r} 702 \\ \times\ 2 \\ \hline 1404 \\ -1060 \\ \hline 344 \end{array}$$

$$\begin{array}{l} 351 = 9 \times 39 \\ -7H \\ \hline 344 \end{array}$$

$(M - 7H + 9K)^{2+}$

PROBLEM 8.2 For the first peak at m/z 1211.1,

$$\begin{array}{r} 1211.1 \\ \times\ 3 \\ \hline 3633.3 \\ -3630.3 \\ \hline 3.0 \end{array}$$

$(M + 3H)^{3+}$

and for the next peak,

$$\begin{array}{r} m/z\ 1216.8 \\ \times\ 3 \\ \hline 3650.4 \\ -3630.3 \\ \hline 20.1 \end{array}$$

Three positive charges must yield a mass of 20.1 Da. One sodium at 22.9 Da and the removal of 3H ions from the polypeptide yields $22.9 - 3 = 19.9 \simeq 20.1$. However, this results in $-3H + Na = -2$ net charge. Thus, sodium cannot be included in this adduct. The ammonium ion at 18 Da + 2H yields a mass of 20, and this is accomplished by a total of three positively charged species. Therefore, the ion at m/z 1216.8 is $(M + 2H + NH_4)^{3+}$.

For the third peak,

$$\begin{array}{rl} m/z\ 1218.4 & \\ \times\ 3 & \\ \hline 3655.2 & \\ -\ 3630.3 & \\ \hline 24.9 & = 22.9 + 2 \end{array}$$

This example is relatively straightforward and the formula is $(M+2H+Na)^{3+}$.

$$\begin{array}{rl} m/z\ 1224.0 & \\ \times\ 3 & \\ \hline 3672 & \\ -\ 3630.3 & \\ \hline 41.7 & \\ -\ 22.9 & Na^+ \\ \hline 18.8 & \\ -\ 18 & NH_4^+ \\ \hline 0.8 & H^+ \end{array}$$

Successive subtraction of ion masses from the 41.7-Da total adduct mass yields an adduct formula of $(M+H+Na+NH_4)^{3+}$.

$$\begin{array}{rl} m/z\ 1225.7 & \\ \times\ 3 & \\ \hline 3677.1 & (M+H+2Na)^{3+} \\ -\ 3630.3 & \\ \hline 46.8 & \\ -\ 45.8 & 2Na^+ \\ \hline 1.0 & H^+ \end{array}$$

PROBLEM 8.3 The mass of the $(M + 8H)^{8+}$ peak of cytochrome c in Figure 8.5 is $1546.1 \times 8 = 12368.8 - 8$ protons $= 12,360.8$ Da. An analysis of the next five peaks follows:

$$\begin{array}{rl} 1549 & \\ \times\ 8 & \\ 12,392.0 & \text{This value includes the eight protons} \\ -12,360.8 & \\ \hline 31.2 & \\ -\ 8 & \\ \hline 23.2 & \\ -\ 22.9 & Na^+ \\ \hline 0.3 & \text{remainder} \end{array}$$

However, this analysis results in one extra charge on the adduct; namely, 8H + Na = 9+. Thus, one less hydrogen is found on the adduct $(M+7H+Na)^{8+}$, and a 1.3-Da differential remains.

$$
\begin{array}{rrl}
m/z\ 1551 \times 8 = & 12{,}408.0 & \\
 & -\ \underline{12{,}360.8} & \\
 & 47.2 & \\
 & \underline{-\ 39.1} & K^{+}\ \text{ion} \\
 & 8.1 &
\end{array}
$$

To yield eight charges, only seven hydrogen ions are required: $(M+7H+K)^{8+}$, and 1.1 Da remains unaccounted.

$$
\begin{array}{rrl}
m/z\ 1553.75 \times 8 = & 12{,}430.0 & \\
 & \underline{-\ 12{,}360.8} & \\
 & 69.2 & \\
 & \underline{-\ 22.9} & Na^{+} \\
 & 46.3 & \\
 & \underline{-\ 39.1} & \\
 & 7.2 &
\end{array}
$$

Only six of the seven charges are required in the form of hydrogen ions. Thus, the adduct is $(M + 6H + Na + K)^{8+}$ with 1.2 Da remaining unaccounted.

$$
\begin{array}{rrl}
m/z\ 1556.2 \times 8 = & 12{,}449.6 & \\
 & \underline{-\ 12{,}360.8} & \\
 & 88.8 & \\
 & \underline{-\ 45.8} & 2\ Na^{+} \\
 & 43.0 & \\
 & \underline{-\ 39.1} & K^{+} \\
 & 3.9 & \simeq 4H^{+}
\end{array}
$$

The alkali and hydrogen ion yield a total of 7+ charges; therefore, one more hydrogen ion (1 Da) must be added to yield 8+ charges: $(M + 5H + 2Na + K)^{8+}$

$$
\begin{array}{rrl}
m/z\ 1558.3 \times 8 = & 12{,}466.4 & \\
 & -12{,}360.8 & \\
\hline
 & 105.6 & \\
 & -\ 78.2 & 2K^+ \\
\hline
 & 27.4 & \\
 & -\ 22.9 & Na^+ \\
\hline
 & 4.5 & \simeq 5H^+
\end{array}
$$

$$(M + 5H + Na + 2K)^{8+}$$

Most of the peaks are either ± 1 Da in variance with the charge state in the proton balance part of the protein adducts. Since the base peak yielded an accurate average mass of cytochrome c, the above analyses may be due to a ± 0.1-Da error in measuring the exact mass location of each peak in the spectrum. A ± 0.1-Da error is magnified 8× in the calculation of the zero charge mass, or ±0.8 Da, which is very close to the mass of one hydrogen ion.

PROBLEM 8.4

m_2	m_1	Δm	Approx. n_2	n_2	M
2,410	2,008	402	4.99	5.0	12,045
2,008	1,721	287	5.99	6.0	12,042
1,721				7.0	12,040
M_r = 12,042.3 ± 2.0					
1,496	1,330	166	8.01	8.0	11,960
1,330	1,197	133	8.99	9.0	11,961
1,197				10.0	11,960
M_r = 11,960.3 ± 0.5					

The difference of the two values yields 82 ± 2.5, and within experimental error yields the mass of two calcium ions. Note that a subtraction of 4 Da (four hydrogen ions) is not necessary because that is already reflected in the 5-, 6-, and 7-m/z values.

9

Interpretation of Peptide and Protein Mass Spectra

Proteinaceous substances are probably represented to a greater degree in the biological science literature than other biological molecules. In addition to proteins and enzymes, proteinaceous matter can be found in glycoproteins, phosphoproteins, and lipoproteins, and they have important roles in cellular biochemical reactions, communication and interaction recognition events, and structural elements. A perusal of the literature provides a comprehensive treatment of proteins, enzymes, and protein-containing biological substances with analytical instrumental methods. Proteins and their derivatives have certainly dominated biological utility and applications with mass spectrometry interrogative techniques.

Most of the protein structural elucidation methods and procedures with mass spectrometry were discovered mainly by fast atom bombardment (FAB) and liquid secondary-ion mass spectrometry (LSIMS) in the 1980s and early 1990s[1–6]. Electrospray ionization owes a great deal to these pioneering mass spectrometry ionization methods in providing clearly outlined protocols in the many aspects of protein structural elucidation.

An extensive series of ESI-product-ion mass spectra will be presented in a progressive fashion, detailing many of the structural nuances and variations in the characterization of protein biological molecules. In this chapter, proteins/enzymes that do not contain modifications with or additions of other classes of biological molecules will be considered. The latter "hybrid protein" species will be addressed in subsequent chapters. Mass spectral phenomena associated with select amino acids at certain positions in a peptide will be investigated.

Protein Structure and Fragmentation Pathways

An important aspect in the understanding of how proteins and peptides produce lower mass fragments resides in the knowledge of the different fragmentation pathways and mechanisms available to a peptide. The larger the peptide, the more potential sites of protein cleavage. Thus, from a purely statistical point of view, the signal of a simple peptide precursor can be spread into many m/z fragment ion positions and hence decrease the overall intensity of the individual product-ion species. In practice, the actual number of peptide fragments is smaller than the theoretical maximum, and certain amino acid–amino acid bonds are either more prone or more resistant to cleavage than bonds between amino acids in general. This is manifested as increased signal intensities or the absence of a signal at predetermined mass values.

Basically, protein fragmentation can take place in the CID cell of a mass spectrometer in the electronvolt (low-energy) or kiloelectronvolt (high-energy) regimes. Both energy deposition extremes provide ions generated from the same fragmentation pathways, except that in the high-energy regime, additional types of ions can be expected to appear. Amide bonds characterize the polymeric backbone of a peptide (*vide infra*), and it is these bonds that are susceptible to cleavage under low- as well as high-energy conditions. The high-energy regime provides further fragmentation on the pendant side chain of amino acid residues in the peptide analyte. Instruments that can be characterized as providing a relatively low-energy deposition into peptides include linear quadrupole, quadrupole ion-trap, Fourier transform (FT), and time-of-flight mass spectrometers. High-energy conditions are found in magnetic and electric mass spectrometers, or sector mass spectrometers. Low-energy peptide fragmentation information will be considered first in this chapter.

Peptide Nomenclature

Figure 9.1 presents the structures of the 20 commonly occurring amino acids, including their three- and one-letter abbreviations. These abbreviations are very important, because they allow for a compact way of expressing a relatively long peptide/protein molecule on the written page. Note that the amino acids are presented in the form of a residue structure as found in a peptide, where the N-terminal side of the molecule has a hydrogen missing and the carboxyl side has an hydroxyl group deliberately left out. Table 9.1 provides average mass information for each amino acid.[1,7–11] The residue mass and the free or intact amino acid mass are presented, and the free amino acid mass equals the residue mass plus a water molecule (18.1053 Da in average mass units). The nominal mass of an amino acid residue and that of the intact, free form is merely the whole-number portion of each numerical entry in Table 9.1. Four significant figures for the amino acid mass values are presented in Table 9.1. For the low-energy quadrupole mass spectrometer analyses, one significant figure in tenths of a dalton is the best that can be reported, while for the sector and FT mass spectrometers, up to three and four significant figures, respec-

Table 9.1 Average mass of amino acids in free and residue state[a]

	Codes		M_r	
Amino acid	3-letter	1-letter	Residue	Free
Alanine	Ala	A	71.0786	89.0938
Arginine	Arg	R	156.1870	174.2022
Asparagine	Asn	N	114.1036	132.1188
Aspartic acid	Asp	D	115.0884	133.1036
Cysteine	Cys	C	103.1386	121.1538
Glutamine	Gln	Q	128.1304	146.1456
Glutamic acid	Glu	E	129.1152	147.1304
Glycine	Gly	G	57.0518	75.0670
Histidine	His	H	137.1408	155.1560
Isoleucine	Ile	I	113.1590	131.1742
Leucine	Leu	L	113.1590	131.1742
Lysine	Lys	K	128.1736	146.1888
Methionine	Met	M	131.1922	149.2074
Phenylalanine	Phe	F	147.1762	165.1914
Proline	Pro	P	97.1164	115.1316
Serine	Ser	S	87.0780	105.0932
Threonine	Thr	T	101.1048	119.1200
Tryptophan	Trp	W	186.2128	204.2280
Tyrosine	Tyr	Y	163.1756	181.1908
Valine	Val	V	99.1322	117.1474

[a]The nominal mass of the residue and free amino acid is the whole-number portion of the rational-number value.

tively, can be achieved for the mass values of mass spectral signals. This has important ramifications in peptide structural elucidations for certain amino acid residues with very similar masses. For example, note the masses of Gln and Lys in Table 9.1. A quadrupole mass spectral determination could not differentiate between the two but this is possible with sector and FT instruments. Quadrupole instruments can easily distinguish between two compounds that possess a 1-mass-unit difference. For example, in Table 9.1, Asn and Asp have a 1-Da difference that originates from the amine ($-NH_2$) and hydroxy (–OH) moiety, respectively (Figure 9.1). Thus, a mass spectral analysis can differentiate between the m/z 133 and 134 protonated mass values of Asn and Asp, respectively. However, care must be taken when in two otherwise identical peptides of relatively high mass, the only difference is that one peptide has an Asn and the other has an Asp (Chapters 5 and 6). The operator must be confident that the instrument is tuned and properly calibrated in order to distinguish between 1-mass-unit difference at a relatively high mass; for example, where two peptides have masses of 2114 and 2115 Da. These values can represent two peptides that have identical residues with a mass of 2000 Da plus an extra internal amino acid residue, where the lighter peptide has Asn and the slightly heavier peptide has the Asp residue. Table 9.2 presents a complete list of mass differences of the common amino acids for up to 4 Da.[12]

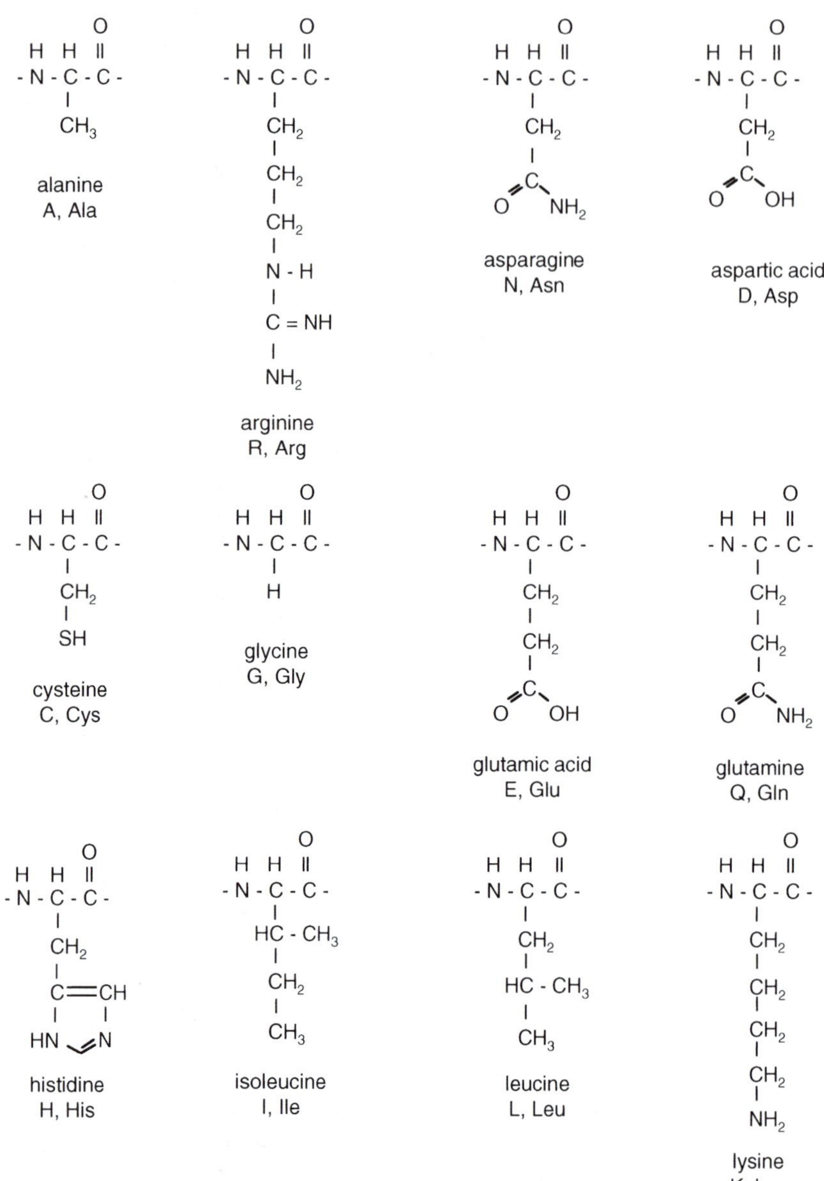

Figure 9.1 Commonly occurring amino acid residues.

Figure 9.2a is the heart of all protein structural determinations. Developed by Roepstorff and Fohlman,[13] and modified and expanded by Biemann et al.,[2] this systematic structural accounting of the backbone of a peptide sequence represents the focal point for protein fragment bookkeeping and a guide for the mass spectral "synthesis" of a peptide sample. It is customary to write a

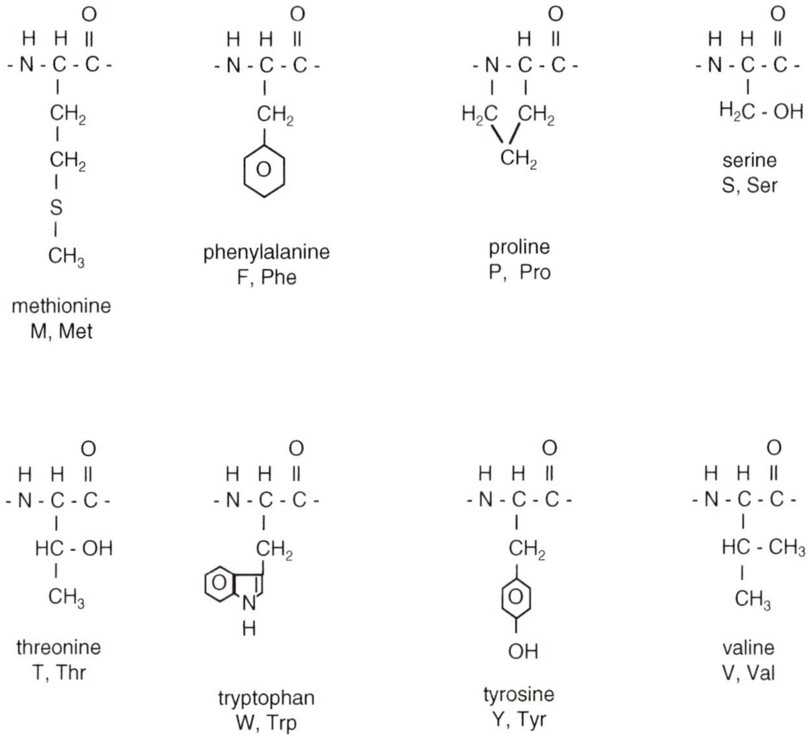

methionine
M, Met

phenylalanine
F, Phe

proline
P, Pro

serine
S, Ser

threonine
T, Thr

tryptophan
W, Trp

tyrosine
Y, Tyr

valine
V, Val

Figure 9.1 *(continued)*

Table 9.2 Nominal mass differences between standard amino acid residues

Δ Da	Masses	Respective residues
0	113–113	Leu–Ile
	128–128	Lys–Gln
1	114–113	Asn–Ile
	114–113	Asn–Leu
	115–114	Asp–Asn
	129–128	Glu–Gln
	129–128	Glu–Lys
2	99–97	Val–Pro
	101–99	Thr–Val
	103–101	Cys–Thr
	115–113	Asp–Leu
	115–113	Asp–Ile
	131–129	Met–Glu
3	131–128	Met–Gln
	131–128	Met–Lys
4	101–97	Thr–Pro
	103–99	Cys–Val

(Reprinted by kind permission of Elsevier Science-NL, Sara Burgerhartsraat 25, 1055 KV, Amsterdam, The Netherlands.)

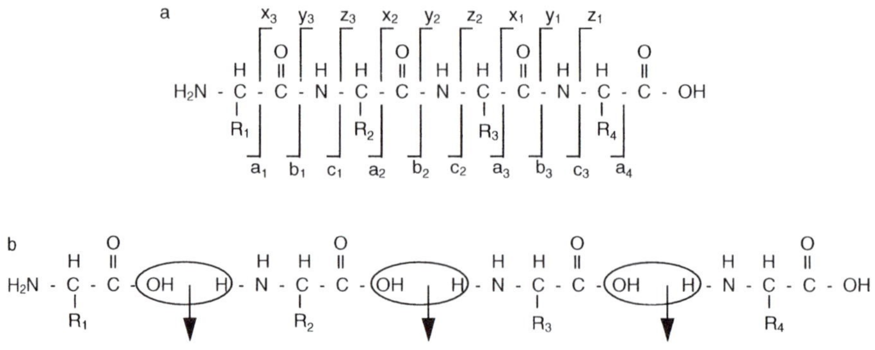

Figure 9.2 (a) Structural notation of a peptide backbone. (b) Dehydration reaction showing the synthesis of amino acids into a peptide.

peptide sequence with the N-terminal residue at the left-hand side. Figure 9.2b shows how the peptide in Figure 9.2a is derived from the free, individual amino acids. A pair of amino acids undergoes a dehydration reaction where a loss of water allows the generation of a peptide bond between the two amino acids. This peptide bond is an amide bond in organic chemistry nomenclature. Peptides can fragment at three different places in the amide backbone and these are descriptively noted at the a, x; b, y; and c, z sites in Figure 9.2a. The differences between each pair of labels is whether the bond cleavage produces a fragment that retains either the amino-terminal or the carboxyl-terminal portion of the peptide. For example, for the same bond at b_1 and y_3 in Figure 9.2a, the mass of the b_1 N-terminal fragment is considerably less than that of the y_3 C-terminal fragment. The numerical subscripts on the letters indicate the number of amino acid residues contained in that particular ion from either the N- or the C-terminal position. Thus, it seems reasonable to assume that if two adjacent ions of the same letter designation, e.g.— for example b_3 and b_4—are present in a mass spectrum, the difference of the two ions can yield a number that is equivalent to an amino acid residue mass. This is a valuable clue to the identity of the amino acid residue attached to b_3 in order to yield the b_4 ion.

Throughout this book, the a, b, c, x, y, z nomenclature will be expressed in lowercase letters when referring to the separate, independent backbone cleavages of a peptide, as suggested by Biemann et al.[2] Thus, the a, b, c ions describe bond-cleavage fragments that retain the amino or N-terminal side of the peptide, and the x, y, z, ions describe bond-cleavage fragments that retain the carboxyl or C-terminal side of the peptide.[7,8,14–16] These ions are displayed in more detail in Figure 9.3. Note that the mass of the a, x, b, and z ions can be deduced directly from the Figure 9.2a cleavage points. However, the c and y ions both have two additional hydrogen atoms; hence, the double prime symbol on each letter.[17,18] An important point to keep in mind is that the mass of a c or y ion is 2 Da greater than that of the backbone core, yet the ions have only a 1+ net charge. It is understood that $y_2'' = y_2$, $y_3'' = y_3$, and that y_2 refers to the 1+ charge state. Thus, a y_2^{2+} ion possesses two positive charges, and it has an additional hydrogen

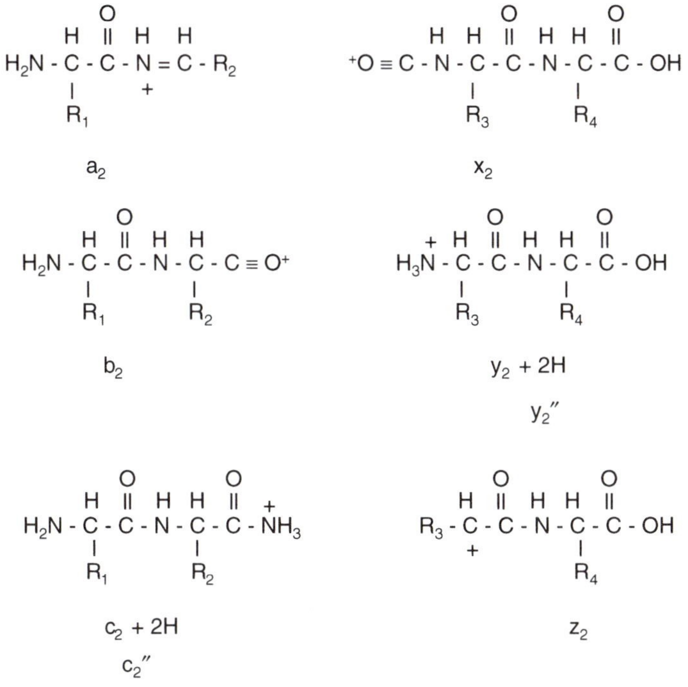

```
         O                                    O           O
    H    ||   H    H                  H   H   ||   H   H  ||
H2N - C - C - N = C - R2       +O ≡ C - N - C - C - N - C - C - OH
      |       +                           |           |
      R1                                  R3          R4

        a2                                  x2

          O                                   O           O
     H    ||   H    H                 +  H    ||   H   H  ||
H2N - C - C - N - C - C ≡ O+        H3N - C - C - N - C - C - OH
      |           |                       |           |
      R1          R2                      R3          R4

          b2                                  y2 + 2H
                                                y2''

          O           O                       O           O
     H    ||   H   H  ||  +              H    ||   H   H  ||
H2N - C - C - N - C - C - NH3        R3 - C - C - N - C - C - OH
      |           |                       +           |
      R1          R2                                  R4

          c2 + 2H                                 z2
           c2''
```

Figure 9.3 Complementary fragment ions of a peptide.

atom above the core mass of a y species, as shown in Figure 9.2a; namely, $(y + 3H)^{2+}$ or $(y''')^{2+}$.

Scheme 9.1 provides mechanistic details on the formation of b and a ions, as well as their neutral counterparts.[16,19–21] Note that there are two different pathways to form an "a" ion, and they are supported by detailed FAB-tandem mass spectrometry investigations carried out by Biemann et al.[16] The b and a ions are commonly observed in protein fragmentation mass spectra. Scheme 9.2 provides mechanistic information on the generation of c ions.[22] The origin of the two extra hydrogen atoms is depicted, where one hydrogen is the protonating hydrogen, and the other hydrogen comes from the β-carbon on one of the side-chain (R) groups. This latter hydrogen provides for a neutral amide functional group, and note that the c ion retains the original amino terminus of the peptide. In general, c ions are rarely observed in protein fragmentation mass spectra. Scheme 9.3 provides a detailed mechanism of the generation of the y ion. The y ion is commonly observed, and the origin of the two extra hydrogen atoms is shown. One hydrogen is a proton that adds onto a backbone nitrogen atom, while the other hydrogen originates from the adjacent (N-terminal side) backbone α-carbon at the peptide cleavage point. The b and y ions are the mass spectral information that usually provide most of the information necessary to reconstruct or "synthesize" a peptide analyte.

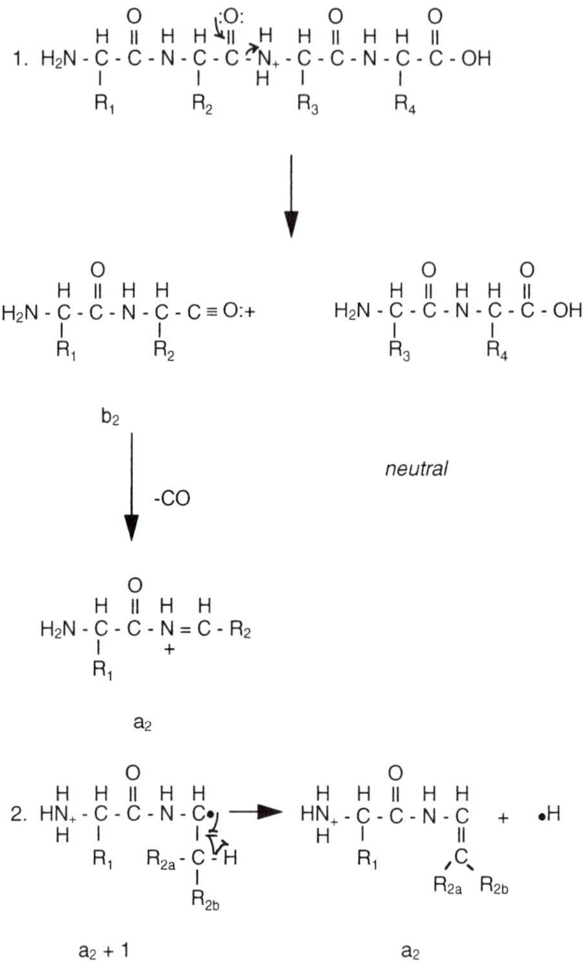

1. $H_2N-CH(R_1)-C(=O)-NH-CH(R_2)-C(=O)-N^{+}H_2-CH(R_3)-C(=O)-NH-CH(R_4)-C(=O)-OH$

$H_2N-CH(R_1)-C(=O)-NH-CH(R_2)-C\equiv O{:}^{+}$ $\quad$ $H_2N-CH(R_3)-C(=O)-NH-CH(R_4)-C(=O)-OH$

b_2 $\quad$ *neutral*

-CO

$H_2N-CH(R_1)-C(=O)-NH=C^{+}H-R_2$

a_2

2. $H_2N^{+}H-CH(R_1)-C(=O)-NH-C^{\bullet}H-C(H)(R_{2a})(R_{2b})$ → $H_2N^{+}H-CH(R_1)-C(=O)-NH-CH=C(R_{2a})(R_{2b})$ + $^{\bullet}H$

$a_2 + 1$ $\quad$ a_2

Scheme 9.1 Mechanisms of formation of b and a ions from peptide fragmentation.

The a, b, x, and z ions can be directly obtained by the Figure 9.2a peptide backbone depiction, while the y and c ions require an additional 2 Da with respect to the peptide backbone structure. However, the mass of an a, x, and z ion can also be obtained by using simple formulae, as shown in Table 9.3.[14,19,23,24] Using Figure 9.2a as a guide, an x_n ion is equal to the respective y_n ion with an addition of 28 Da for the carbonyl group and a subtraction of the two additional hydrogen atoms. The a_n ion is just the opposite, where 28 Da is subtracted from its respective b_n ion. The z_n ion is equal to its respective y_n ion by subtracting both the NH backbone moiety (−15 Da) and the two hydrogen atoms (−2 Da). The internal fragment ions are another class of fragmentation product and are delineated below.

Table 9.3 Equations for calculation of low-energy fragment ions

$$x_n = (y''_n + 28) - 2$$
$$a_n = b_n - 28$$
$$z = (y''_n - 15) - 2$$

Internal fragment ions ($n > 1$): $-(\text{residue}_n)-H^+$ or $-(\text{residue}_n)-2H^+$

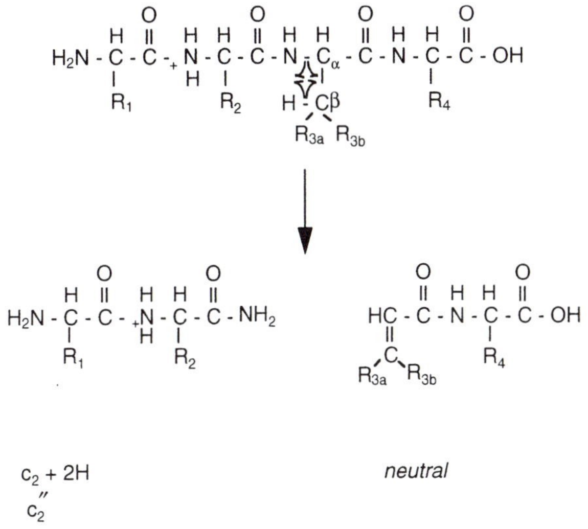

$c_2 + 2H$

c''_2

neutral

Scheme 9.2 Proposed mechanism of formation of c ions from peptide fragmentation.

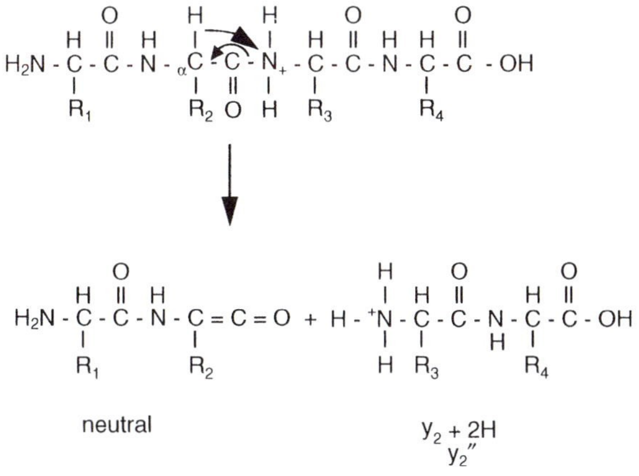

neutral

$y_2 + 2H$

y_2''

Scheme 9.3 Mechanism of formation of y ions.

Mass Spectral Determination of Peptide Fragmentation: Low-Energy CID

There are two fundamental ways of approaching the interpretation process of a biological-compound fragmentation pattern in a mass spectrum. The relatively easier situation is where the structure of the analyte is already known. Thus, the mass spectral interpretation is template-driven by the known sequence. When the sample analyte's sequence is not known, this provides a relatively more difficult challenge. The mass spectrum is the template and must draw on the information provided above in the figures and tables, and in the further information below. Further enzyme biochemical and chemical reactions of the sample are usually required in order to fragment a large protein into smaller, manageable peptides. Under tandem mass spectrometry conditions, these smaller peptides can provide important amino acid residue sequence information for the original, unmodified peptide analyte.

When the sequence is not known, and depending on the amount of mass spectral information and modifications to the amino acid residues, mass spectral interpretation may be relatively straightforward or it may be very time-consuming.

Tryptic Fragment of Myoglobin

Depending upon the size of the protein analyte, it may either be introduced directly into the mass spectrometer or undergo a series of processing and handling procedures prior to submittal for a comprehensive mass spectral sequence analysis. Sperm whale holo-myoglobin is a very large protein (M_r of 17,199.91 Da[25,26]) and must be cleaved into pieces or digested by an enzyme. The first example is a small sequence of amino acid residues from the sperm whale myoglobin protein.[27] In this investigation, the intact protein was isolated by sodium dodecyl sulfate (SDS)-polyacrylamide gel electrophoresis (PAGE), and the isolated myoglobin was electrotransferred to a nitrocellulose membrane. Trypsin was added to the membrane, and this enzyme cleaves a protein or peptide on the C-terminal side of every Lys or Arg residue. Further details on the action of trypsin can be found in Table 9.4.

On the nitrocellulose membrane, the myoglobin protein is cleaved into smaller fragments, where each fragment has either a Lys or an Arg amino acid residue as the carboxyl-terminal amino acid residue. The peptide mixture was extracted and subjected to reversed phase, high-pressure liquid chromatography (RP-HPLC). Figure 9.4 shows such a separation for myoglobin. Table 9.5 provides a listing of the various peptides (tryptic peptides) that were predicted based on the knowledge of the complete amino acid sequence with a particular focus on the location of the Arg and Lys residues. Table 9.5 also presents the observed peptides from the nitrocellulose method. Note that a few trypsin tryptic fragments are also observed; that is, the trypsin enzyme cleaved itself to a limited extent. This must be taken into account when digesting a sample analyte with trypsin,[31] and trypsin enzyme-to-substrate ratios are sensitive to the degree of trypsin autolysis products in a sample digest. It has been stated that it would be embarrassing to analyze the tryptic digest of a protein of unknown sequence when trypsin peptide fragments are also considered as part of the sample protein's sequence.[31]

Figure 9.5a shows a mass spectral analysis of the T2 tryptic fragment of myoglobin, and the product ions are labeled with the appropriate b and y nomenclature. We know that the C-terminal residue must be either an Arg or a Lys, and the T2 peptide sequence in Figure 9.5b shows that to be true. The $(M + H)^+$ for the T2 fragment is 1593.7 Da, and the mass spectral analysis originated from the CID of the doubly charged $(M + 2H)^{2+}$ ion at 797.4 Da. Recall from Chapter 6 that a doubly charged ion can be deduced by noting two peaks within a 1-Da interval, where the second peak is the ^{13}C isotope of the primary ^{12}C peak. The series of numbers below the sequence of amino acid residue abbreviations in Figure 9.5b are the b ions, and that listed above the sequence is the y ion series. All calculated values of both ion series are listed, and the underlined values are those observed experimentally in the spectrum. When the sequence of the analyte is already known, calculation of the masses of all the b and y ions is the first major step. Then, a comparison between the calculated values and the observed mass spectral peak values takes place in order to determine how many of the calculated masses actually appear in the mass spectrum. This initial survey serves as a good check in case some unexpected modifications or changes occurred in the sample, because these phenomena can happen, depending on the source of the sample. In the calculation of the various b_n and y_n masses, where the subscript n refers to any of the amino acid positions in the peptide, certain procedures should be followed. The average mass of the b_1 ion is equal to the residue mass (Table 9.1) plus the mass of a hydrogen atom (1.00797 Da). Note that for the valine residue, this yields $99.1 + 1.0 = 100.1$ Da. Subsequent b_n ion values are calculated by adding the residue mass of each respective amino acid (Figure 9.5b). The mass of the b ion containing the last amino acid residue (b_f), where this last residue is the C-terminal b residue, is not observed in the spectrum (Figure 9.5a). The b_{15} or b_f ion can be considered as a dehydrated ion from the $(M + H)^+$ ion. Thus a loss of water (HOH) is involved where the OH originates from the C-terminal hydroxyl moiety of the peptide, and the hydrogen comes from the proton on the protonated parent peptide $(M + H)^+$. This yields the equation $b_f = (M + H)^+ - H_2O$. The b series does not produce an ion equal to the average mass or M_r of the peptide.

For the y_n series, y_1 is equal to the mass of the C-terminal, intact, free amino acid plus one hydrogen atom. Reference to Figure 9.3 provides insight into this reasoning, where the amino acid residue portion of y_1 has the structure –NH–CHR–COOH. Addition of one hydrogen yields the neutral, intact amino acid, and the addition of a second hydrogen provides the protonated amino acid species. All subsequent residues have their residue mass added to the y_1 mass to yield the y_n series of ions. The last or highest mass y ion is obtained by addition of the N-terminal residue mass, as shown in Figure 9.5b.[27] The resulting value of the last y ion (which is y_f or y_{15}) is equal to the mass of the $(M + H)^+$. Therefore, we have $b_f + 17 + 1 = y_f^+$, where the f subscript refers

Table 9.4 Enzymes and their cleavage sites in proteins and peptides

Enzyme	Amino acid residue	Reference(s)
Carboxypeptidase A	Cleaves C-terminal residue. Tyr, Phe, Trp, Leu, Ile, Met, Ala, Val, His, Gln, Thr, and homoserine (hSer) are released easily. Lys, Asn, Ser, and Met sulfone are released slowly. Gly and acidic residues are released very slowly. Pro, Arg and hydroxyproline are not affected.	28–30
Carboxypeptidase B	Cleaves C-terminal Lys and Arg.	28
Carboxypeptidase Y	Cleaves C-terminal residue. Hydrophobic residues are released easily and charged residues are removed slowly. When Gly is next to the C-terminal residue, the latter is removed slowly.	28
Endoproteinase Glu-C; *Staphylococcus aureus* V8 protease	Produces peptides from proteins with a Glu or Asp (acidic residue) at the C-terminus unless Glu or Asp was originally at the protein's C-terminus. Will not cleave if Glu is close to N-terminus. pH 4.0 Cleaves C-terminus side of Glu. pH 7.8 Cleaves C-terminus side of Glu and Asp.	1, 31–35
Endoproteinase Asp-N	Cleaves at the N-terminal side of Asp and sometimes at Glu.	35, 36
Trypsin	(a) Produces peptides from proteins with a Lys or Arg (basic residue) at the C-terminus, unless it was originally at the protein's C-terminus. Acetylated Lys will not cleave. (b) Lys–Pro bond not cleaved; Arg–Pro bond may or may not be cleaved. (c) Double basic residue results in a mixture of peptides which have one or both basic residues in the C-terminal position. (d) A tryptic-like peptide is a peptide with a Lys or Arg at the C-terminus.	1, 31, 35–37
Endoproteinase Lys-C	Only cleaves Lys at the C-terminus in a protein except for Lys–Pro bonds.	31, 33, 36
α-Chymotrypsin	(a) Cleaves the C-terminal side of aromatic residues (Phe, Trp, Tyr), as well as Leu, Met, and His. Does not cleave X–Pro bonds when X = any of the six amino acid residues. (b) Does not cleave X–Ile bonds.	8, 31, 32, 35, 36, 38, 39
Asparaginyl endopeptidase	Cleaves the C-terminal side of Asn, with a few exceptions.	35, 40
Aspartyl-specific protease	Cleaves the N-terminal side of Asp and Lys residues.	31

(*continued*)

Table 9.4 (*continued*)

Enzyme	Amino acid residue	Reference(s)
PCAse or pyrrolidone carboxyl protease or pyroglutamyl peptidase	Removes the N-terminal pyroglutamic acid.	31, 41–43
Endoproteinase Arg-C	Cleaves the C-terminal side of Arg residues, except for Arg–Pro.	31, 33, 35, 36
Clostripain	Cleaves Arg-X residues, including Arg–Pro.	31, 33, 44
Pepsin	Cleaves the C-terminal side of Phe, Trp, Leu, and Met when they are adjacent to hydrophobic residues.	31, 35
Thermolysin	Cleaves the N-terminal side of Ile, Leu, Val, Phe. Cleaves other bonds to a lesser extent: Met, His, Tyr, Ala, Asp, Ser, Thr, Gly, Lys, Glu.	31, 33, 35, 36, 39, 45
Papain	Cleaves Arg–X and Lys–X bonds rapidly and more slowly where Arg, Lys are replaced by Gln, His, Gly, Tyr.	31, 33, 46
Pronase and subtilisin	Nonspecific cleavage of backbone amide bonds.	31

to the number of amino acid residues in the peptide analyte. The 1 represents the mass of a hydrogen ion, and the 17 value represents the mass of an hydroxyl moiety.

For the T2 myoglobin peptide (Figure 9.5a), almost all of the calculated b and y ions are observed. In addition to the doubly charged $(M + 2H)^{2+}$ species, two more doubly charged ions are observed. The y_{13}^{2+} ion is observed at $(1365.5 + 1)/2 = 683.3$ Da, and the y_{14}^{2+} ion is observed at $(1494.6 + 1)/2 = 747.8$ Da. A few of the calculated b and y ion values are not observed in

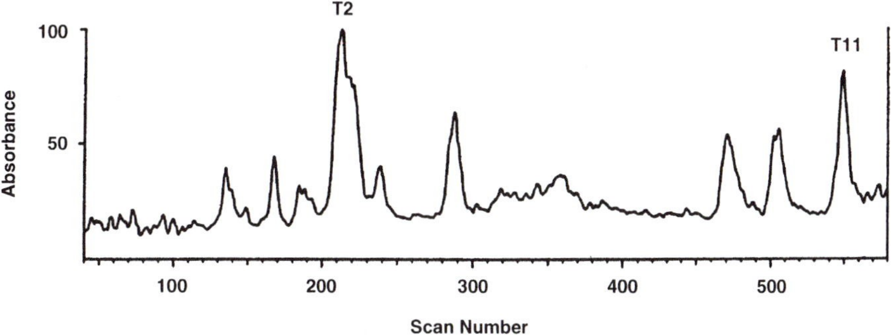

Figure 9.4 An RP-HPLC chromatogram of a tryptic digest of sperm whale myoglobin. The T2 and T11 peptides are noted. (© 1990. Reprinted from reference 31 by kind permission of Elsevier Science-NL, Sara Burgerhartstraat 25, 1055 KV Amsterdam, The Netherlands.)

Table 9.5 Tryptic fragments from sperm whale myoglobin

Predicted fragments	Their masses	Observed $(M + H)^+$
T1	1937.2	
T2	1593.7	1593.7
T3	407.5	
T4–T5	1359.5	1359.5
T6–T7	672.8	
T8–T9	1352.5	
T11	1393.7	1393.7
T14	1855.1	1855.1
T15–T16	736.0	
T17	1928.3	
T18	1516.6	1516.6
T19–T20	877.0	877.0
T21–T22	808.9	
T23	666.7	
Trypsin tryptic fragments observed		
T5		2164.3
T9		1154.2
T14		2274.6

the mass spectrum, and these instances will be elaborated upon. The b_f or b_{15} peptide is not observed in the spectrum. A b_1 ion is not observed, but a b_2 ion is present. Because the mass of b_2 is greater than that of any single amino acid, b_2 must be composed of at least two residues. This latter idea can be stated as such, even if the sequence was not known a priori. The *m/z* 229.3 is actually the sum of a number of residues + 1 (N-terminal hydrogen). Subtracting the hydrogen yields 228 Da. This value can be compared to that in Table 9.6, where the various combinations of residue masses are tabulated.[11] For a mass of 228 Da, a number of two, three, and even four residue combinations can be considered for the N-terminally located amino acids in the b ion series. This same problem occurs with the absence of the y_{14} ion. The difference between the y_{15} or $(M + H)^+$ and y_{13} ions is 228.2 Da. Either the b or the y ion series poses the same dilemma as to the identity of the N-terminal b ions or the high-mass y ions located at the N-terminal position. This is where the y_{14}^{2+} ion plays an important role. Conversion of *m/z* 747.8 (observed) to that of the singly charged species yields a mass of 1494.6 Da, which is not observed in the spectrum. The difference between 1494.6 Da and 1365.5 Da is 129.1 Da, and is equal to a Glu residue (Table 9.1). There is only one combination of residues in Table 9.6 that has 228 Da and that can include a Glu, and that is Val, Glu. Thus, the remaining residue is Val (99.1 + 1 Da), and this is at the b_1 position.

A similar occurrence is caused by the absence of the b_3 ion. Thus, the difference of b_4 and b_2 is 186 Da. Since this is a difference between internal residues, the 186 Da can be used directly in the interrogation of Table 9.6, without subtracting the mass of an N-terminal hydrogen. The 186 Da can

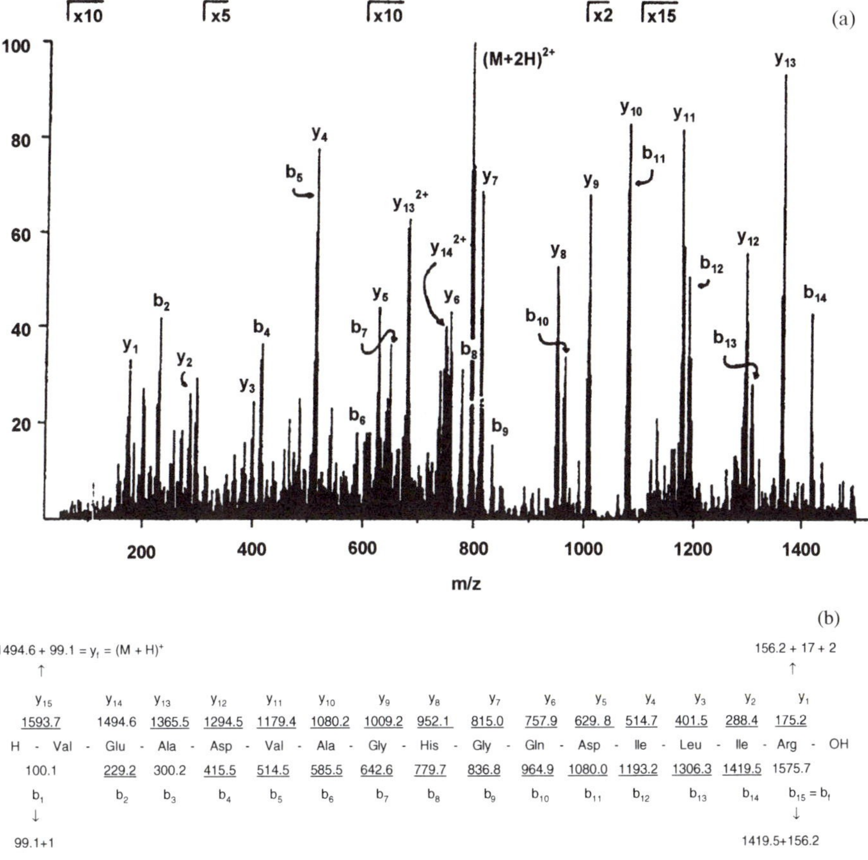

Figure 9.5 (a) The ESI-tandem mass spectrum of the $(M + 2H)^{2+}$ ion from the T2 fragment at *mz* 797.4^{2+}. (b) Amino acid sequence and ion identification of the tandem mass spectrum in (a). Experimentally observed ions are underlined. (© 1990. Reprinted from reference 31 by kind permission of Elsevier Science-NL, Sara Burgerhartstraat 25, 1055 KV Amsterdam, The Netherlands.)

equal the Trp residue or select combinations of residues. However, the y_{11} and y_{12} ions establish a difference of 115.1 Da, which is that of an Asp residue, and the difference between the y_{13}^{2+} and y_{12} ions (1365.5 – 1294.5 = 71 Da) establishes the presence of Ala at the N- terminal end of y_{13}. The Ala, Asp combination is one of the candidates for the 186-Da value in Table 9.6, and the above analysis identifies the third and fourth amino acid residues as Ala and Asp, respectively.

Continuing as though the peptide structure was not known a priori under the low-energy quadrupole conditions used, the residue at the C-terminus in the b_{10} ion, which is also the N-terminus for the y_6 ion, could be either Gln or Lys. Both residues have the same nominal mass at 128 Da; however, this can be solved by subjecting a portion of the sample to acetylation. As Table 9.7 shows, the acetylation reaction derivatizes the Lys residue, but not Gln. Since

Table 9.6 Nominal residue masses of single amino acids and small polypeptides

Mass	Amino acids[a]	Mass	Amino acids[a]	Mass	Amino acids[a]
57	Gly	71	Ala	87	Ser
97	Pro	99	Val	101	Thr
103	Cys	113	Xle	114	Asn
					Gly, Gly
115	Asp	128	Gln	129	Glu
			Lys		
			Gly, Ala		
131	Met	137	His	144	Gly, Ser
147	Phe	156	Arg	161	CMCys
			Gly, Val		
163	Tyr	168	Ala, Pro	170	Gly, Xle
					Ala, Val
171	Gly, Asn	172	Gly, Asp	174	Ala, Cys
	Gly, Gly, Gly		Ala, Thr		Ser, Ser
184	Ala, Xle	185	Gly, Gln	186	Trp
	Ser, Pro		Gly, Lys		Gly, Glu
			Ala, Asn		Ala, Asp
			Gly, Gly, Ala		Ser, Val
188	Gly, Met	190	Ser, Cys	194	Gly, His
	Ser, Thr				Pro, Pro
196	Pro, Val	198	Pro, Thr	199	Ala, Gln
			Val, Val		Ala, Lys
					Gly, Ala, Ala
200	Ala, Glu	210	Ser, Asn	202	Ala, Met
	Ser, Xle		Gly, Gly, Ser		Ser, Asp
	Val, Thr				Val, Cys
	Cys, Pro				Thr, Thr
204	Gly, Phe	206	Cys, Cys	208	pCys
	Thr, Cys				Ala, His
210	Pro, Xle	211	Pro, Asn	212	Pro, Asp
			Gly, Gly, Pro		Val, Xle
213	Gly, Arg	214	Val, Asp	215	Ser, Gln
	Val, Asn		Thr, Xle		Ser, Lys
	Gly, Gly, Val				Thr, Asn
	Ala, Ala, Ala				Gly, Gly, Thr
					Gly, Ala, Ser
216	Ser, Glu	217	Cys, Asn	218	Ala, Phe
	Thr, Asp		Gly, Gly, Cys		Ser, Met
	Cys, Xle				Cys, Asp
220	Gly, Tyr	224	Ser, His		Gly, CMCys
				225	Pro, Gln
					Pro, Lys
226	Pro, Glu	227	Ala, Arg		Gly, Ala, Pro
	Xle, Xle		Val, Gln	228	Pro, Met
			Val, Lys		Val, Glu
			Xle, Asn		Xle, Asp
			Gly, Gly, Xle		Asn, Asn
			Gly, Ala, Val		Gly, Gly, Asn
					Gly, Gly, Gly, Gly

(continued)

Table 9.6 (*continued*)

Mass	Amino acids[a]	Mass	Amino acids[a]	Mass	Amino acids[a]
229	Thr, Gln	230	Val, Met	231	Cys, Gln
	Thr, Lys		Thr, Glu		Cys, Lys
	Asn, Asp		Asp, Asp		Gly, Ala, Cys
	Gly, Gly, Asp				Gly, Ser, Ser
	Gly, Ala, Thr				
	Ala, Ala, Ser	234	Ala, Tyr	236	Val, His
232	Thr, Met		Ser, Phe		
	Cys, Glu		Pro, His		
	Ala, CMCys		Cys, Met		
238	Thr, His	239	Ala, Ala, Pro	240	Cys, His
241	Xle, Gln	242	Xle, Glu	243	Gly, Trp
	Xle, Lys		Asn, Gln		Ser, Arg
	Gly, Ala, Xle		Asn, Lys		Asn, Glu
	Gly, Ser, Pro		Gly, Gly, Gln		Asp, Gln
	Ala, Ala, Val		Gly, Gly, Lys		Asp, Lys
			Gly, Ala, Asn		Gly, Gly, Glu
			Gly, Gly, Gly, Ala		Gly, Ala, Asp
					Gly, Ser, Val
					Ala, Ala, Thr
244	Pro, Phe	245	Asn, Met	246	Val, Phe
	Xle, Met		Gly, Gly, Met		Asp, Met
	Asp, Glu		Gly, Ser, Thr		
			Ala, Ala, Cys		
			Ala, Ser, Ser		
247	Gly, Ser, Cys	248	Ser, CMCys	250	Ser, Tyr
			Thr, Phe		Cys, Phe
253	Arg, Pro	255	Arg, Val	256	Lys, Lys
	Gly, Pro, Val		Gly, Val, Val		Gln, Lys
			Gly, Pro, Thr		Gln, Gln
			Ala, Ala, Xle		Ala, Gly, Gln
			Ala, Ser, Pro		Ala, Gly, Lys
					Ala, Ala, Asn
					Ala, Ala, Gly, Gly
258	Glu, Glu				
	Pro, CMCys				
	Gly, Gly, Gly, Ser				
276	Xle, Tyr				
	Glu, Phe				
	Asp, CMCYs				

[a]Xle is used for Leu or Ile, CMCys for *S*-carboxymethyl cysteine, and pCys for *S*-pyridylethyl cysteine. (© 1995, *Mass Spectrometry Reviews*. Reprinted by permission of John Wiley & Sons, Inc.)

Table 9.7 Chemical modifications to protein amino acid residues for sequence information

Amino acid residue	Reagent or peptide modification	Product structure	Product residue	Peptide mass change	Reference(s)	Footnote
Cysteine $-NH-CH(CH_2SH)-C(=O)-$	Performic acid $H_2O_2/HCOOH$	$-NH-CH(CH_2SO_3H)-C(=O)-$	Cysteic acid	+48	28	
	Iodoacetic acid	$-NH-CH(CH_2-S-CH_2-CO_2H)-C(=O)-$	*S*-carboxymethyl cysteine	+58	28, 60–63	a
	Iodoacetamide	$-NH-CH(CH_2-S-CH_2-C(=O)-NH_2)-C(=O)-$	*S*-carboxamidomethyl cysteine	+57	28	
$-NH-CH(R)-C(=O)-NH-CH(CH_2SH)-C(=O)-$	NTCB 2-nitro-5-thiocyanobenzoic acid (NO_2, CO_2H, $SC^{*} \equiv N^{*}$ substituted benzene)	$-NH-CH(R)-CO_2^-$ + $H_2N^{*+}=C^{*}$ ring (S, HN, C(H)–C(=O)–)	2-iminothiazolidine-5-carboxylic acid (cys)	Backbone cleavage	5, 35, 64–66	

	Alkylation with 4-vinylpyridine $H_2C=CH-C_5H_4N$	$-NH-CH(CH_2-S-H_2C-CH_2-C_5H_4N)-C(=O)-$	*S*-pyridylethyl cysteine	+105	28, 31	
Methionine $-NH-{}^{\alpha}CH(\beta CH_2-\gamma CH_2-S-CH_3)-C(=O)-$	CNBr, cyanogen bromide	$-NH-CH(H_2C-H_2COH)-C(=O)-NH-CH(R)-C(=O)- \underset{OH^-}{\overset{H^+}{\rightleftharpoons}} -NH-CH-C(=O)$ (lactone) + $H_2N-CH(R)-C(=O)-$	Homoserine ⇌ homoserine lactone	−30 ⇌ Backbone cleavage	5, 31, 32, 46, 61, 65, 67–79	b
	Performic acid	$-NH-CH(CH_2-CH_2-S(=O)-CH_3)-C(=O)-$	Methionine sulfoxide	+16	46, 70, 71	
	Performic acid	$-NH-CH(CH_2-CH_2-S(=O)_2-CH_3)-C(=O)-$	Methionine sulfone	+32	70, 71	c

(*continued*)

Table 9.7 *(continued)*

Amino acid residue	Reagent or peptide modification	Product structure	Product residue	Peptide mass change	Reference(s)	Footnote
Cystine 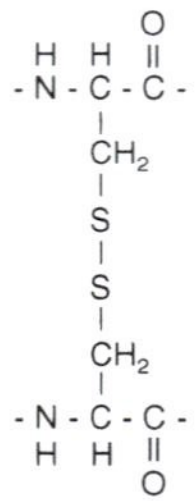	1,4-dithiothreitol DTT Cleland's reagent	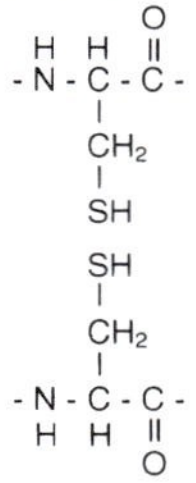	(Reduced) cysteine amino acid residues	Intrachain +2 Interchain 0	70, 72, 73	
	DTT, 4-vinylpyridine	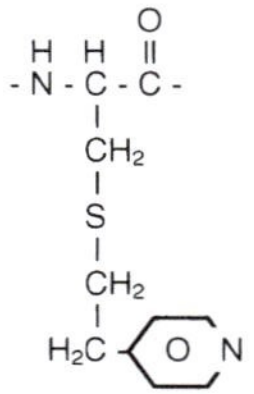	Two *S*-pyridylethyl cysteine residues	Intrachain +210 Interchain 0	28, 31, 61,74	
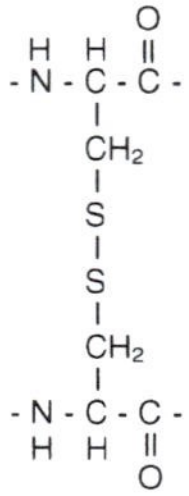	$Na_2S_2O_3$ ⇌ σ-iodosobenzoic acid	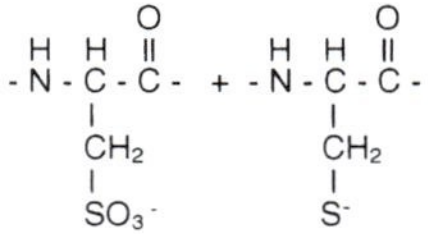	Sulfite and sulfide	Intrachain +48 Interchain 0	75, 76	

N-terminal residue $H_2N-CH(R)-C(=O)-$	PTM	$H-C(=O)-NH-CH(R)-C(=O)-$	*N*-formyl	+28	41	d
	Acetylation on free amine group, MeOH/acetic anhydride, PTM	$H_3C-C(=O)-NH-CH(R)-C(=O)-$	Acetyl	+42	1, 41	
N-terminal glutamine $H-NH-CH(CH_2-CH_2-C(=O)-NH_2)-C(=O)-$	Deamidation, PTM	H-N-C(H)-C(=O)- (cyclic, ring C=O)	Pyrrolidone carboxylic acid, PCA	−17	28, 42, 71, 77	
N-terminal glutamic acid $H-NH-CH(CH_2-CH_2-COOH)-C(=O)-$	Cyclization, PTM	H-N-C(H)-C(=O)- (cyclic, ring C=O)	Pyroglutamic acid pGlu, PCA	−18	28, 42, 71, 77	

(continued)

Table 9.7 (*continued*)

Amino acid residue	Reagent or peptide modification	Product structure	Product residue	Peptide mass change	Reference(s)	Footnote
C-terminal Glue and Gln -N(H)-C(H)(-CH₂-CH₂-C(=O)-NH₂ or -CH₂-CH₂-COOH)-C(=O)-OH	Cyclization	-N-C(H)-C(=O)-OH, ring closed with O= (pyroglutamyl ring)	Pyroglutamic acid, pGlu, PCA	−17 or−18	77	
-N(H)-C(H)(R)-C(=O)-OH	C-terminal amidation, PTM	-N(H)-C(H)(R)-C(=O)-NH₂	Amide of peptide	−1	37	
	Esterification, HCl/MeOH	-N(H)-C(H)(R)-C(=O)-OCH₃	Methyl ester	+14	37, 78	
Lysine -N(H)-C(H)(-CH₂-CH₂-CH₂-CH₂-NH₂)-C(=O)-	Acetylation, MeOH/acetic anhydride	-N(H)-C(H)(-(CH₂)₄-N(H)-C(=O)-CH₃)-C(=O)-	ε-Acetylaminolysyl, acetyllysyl	+42	1	e

Serine, threonine O H H ‖ -N-C-C- | CH_2 | HC-OH | CH_3	Methylation, HCl/MeOH	O H H ‖ -N-C-C- | CH_2 | HC-OCH_3 | CH_3	*O*-methylseryl	+14	70, 79
Serine	O-acetylation, MeOH/acetic anhydride	O H H ‖ -N-C-C- | CH_2 O | ‖ HC-O-C-CH_3 | CH_3	*O*-acetylseryl	+42	37
Tyrosine O H H ‖ -N-C-C- | CH_2 (ring) OH	O-acetylation, MeOH/acetic anhydride and pyridine	O H H ‖ -N-C-C- | CH_2 (ring) O-C-CH_3 ‖ O	*O*-acetyltyrosine	+42	70, 80
Glutamic acid O H H ‖ -N-$_\alpha$C-C- | βCH_2 | γCH_2 | COOH	γ-carboxylation, H^+/H_2O	O H H ‖ -N-C-C- | CH_2 | HC-COOH | COOH	γ-Carboxyglutamic acid residue	+44	28

(continued)

Table 9.7 (*continued*)

Amino acid residue	Reagent or peptide modification	Product structure	Product residue	Peptide mass change	Reference(s)	Footnote
	Methyl esterification, HCl/MeOH	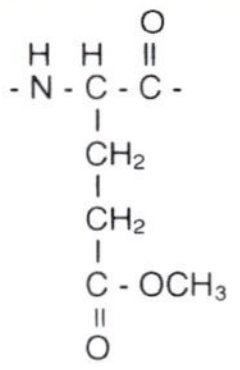	*O*-methyl-glutamyl	+14	37, 48	
Asp or Asn 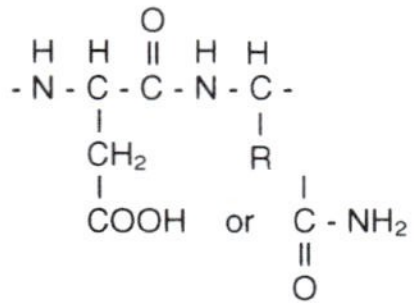	Isomerization, mild H^+	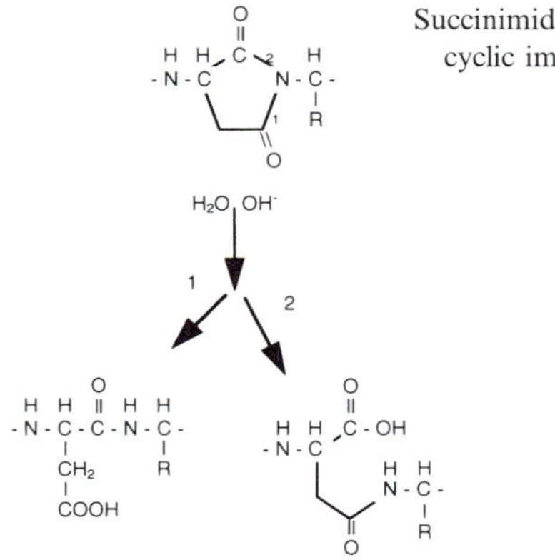 H_2O, OH^- ; 1: Original Asp α-aspartyl residue; 2: β-aspartyl residue, isoaspartyl residue	Succinimide, cyclic imide	−18 or −17 $-H_2O$, $-NH_3$	7, 24, 75, 81–83	

Asparagine or Gln O H H ‖ - N - C - C - I CH_2 I C - NH_2 ‖ O	Esterification, HCl/MeOH	O H H ‖ - N - C - C - I CH_2 I C - OCH_3 ‖ O	Methyl ester	+15	70
Peptide backbone	Mild HCl or oxalic acid at 100°C, 8 h	Random backbone amide bond cleavage; Asp–Pro bond cleaved before most other bonds	Fragments and amino acids		32 46

[a]His, Lys, and Met may also be modified.

[b]Met–Ser and Met–Thr cleave slowly; CNBr can oxidize free Cys–SH; therefore, they must be protected by alkylation (H-vinylpyridine) before reaction with CNBr.

[c]Performic acid also oxidizes tryptophan.

[d]PTM = post-translational modification. This is a cellular processing event on proteins and peptides.

[e]Asn and Gln do not acetylate.

Gln does not acetylate and Lys is not present in the peptide, the resulting peptide will be identical to the original peptide; therefore, no shift in mass of any of the b or y ions would occur.

The Leu and Ile residues at positions 12–14 of the peptide are another matter. By treating the peptide as an unknown, the difference in mass of the b_{11}–b_{12}, b_{12}–b_{13}, b_{13}–b_{14}, y_1–y_2, y_2–y_3, and y_3–y_4 ion pairs display absolute value differences of 113 Da, and, under low-energy conditions, Lxx would necessarily be the residue at positions 12–14, where Lxx stands for the uncertainty of the Leu–Ile pair. However, an enzymatic procedure can be used to distinguish between the two isomers. α-Chymotrypsin digestion (Table 9.4) cleaves peptides at the C-terminal side of the three aromatic amino acids, as well as Met, His, and Leu. The enzyme does not cleave X–Ile bonds, where X is any amino acid residue. Thus, the bond between residues 13 and 14 should be cleaved, as well as the bond in His^8–Gly^9. The former would experimentally provide for the presence of Leu at position 13. Pepsin could be considered (Table 9.4), because it cleaves the C-terminal side of selected amino acids, including Leu, but not Ile.

Mention should be made of the doubly charged ions. The peptide in Figure 9.5b displays an $(M + 2H)^{2+}$ ion in Figure 9.5a, and the protonation sites are most likely the Arg and N-terminal amine, as related in Chapter 4. The y_{13}^{2+} and y_{14}^{2+} ions most likely are charged on the Arg residue, and the second charge site may be either at the respective N-terminus of the y ions or at the His-8 residue (Chapter 4). Since there are no doubly charged ions in the y_8–y_{12} region (which would have the nitrogen-containing His-8 residue), it is likely that the protonation site is indeed the N-terminus of the y_{13} and y_{14} fragment ions. Table 9.8 delineates the amino acid residues based on their polarity and the ability of the side-chain R-group to ionize.[47]

Bovine α-s1-Phosphocasein Tryptic Peptide

The next example of peptide sequencing is a larger tryptic peptide from a trypsin digest of a phosphocasein protein.[27] The various peptides were separated by

Table 9.8 Amino acid polarity

1. Nonpolar or hydrophobic R group
Ala Val Leu Ile Pro Phe Trp Met
2. Polar, uncharged R group
Gly[a] Ser Thr Cys Tyr Asn Gln
3. Positively charged R group
Lys Arg His
4. Negatively charged R groups
Asp Glu

[a]Gly is considered polar because the polar nature of the amine and carboxyl groups make up a major portion of the molecule.

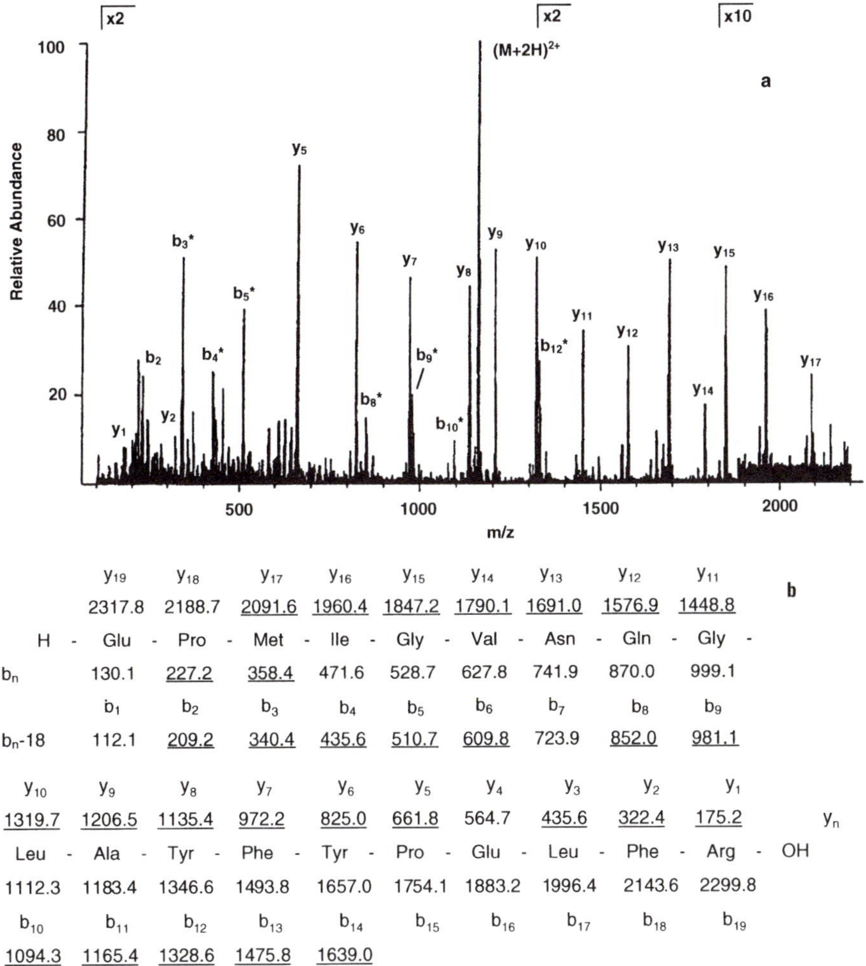

Figure 9.6 (a) The ESI-tandem mass spectrum of the $(M + 2H)^{2+}$ ion from a 19 residue tryptic peptide of bovine-α-sl-phosphocasein at m/z 1159.4. (b) Amino acid sequence and ion identification of the tandem mass spectrum in (a). Experimentally observed ions are underlined. (© 1991. Reprinted from reference 27 by kind permission of Elsevier Science-NL, Sara Burgerhartstraat 25, 1055 KV Amsterdam, The Netherlands.)

HPLC and electrosprayed into a tandem quadrupole mass spectrometer. The $(M + 2H)^{2+}$ ion, a 19 residue peptide, was selected at m/z 1159.4 $[(M + H)^+ = 2317.8$ and $M = 2316.8]$ and subjected to tandem mass spectrometry (Figure 9.6a). The sequence with the b_n and y_n series is shown in Figure 9.6b. Very few b ions are observed; however, many b_n-18 ions are displayed, which arise from the loss of a water molecule from the respective b_n ion (Figure 9.6b). Following the same rules to arrive at the b and y masses as used in Figure 9.5b, Figure 9.6b presents the calculated masses. Because only a few b ion values appear in the spectrum, it becomes obvious that about half of the b-ion information is actually in the dehydrated form, and these values are also listed in Figure 9.6b.

Relying only on the b_n series yields a few residues, while the b_n-18 series yields almost the first, or N-terminal, two-thirds of the peptide sequence. There are some uncertainties here because the b_1-18 and b_7-18 ions are not observed. Thus, the nature of the first two residues can be gleaned from the $b_2 - 1$ (subtraction of the N-terminal hydrogen) = 226 Da calculation. Table 9.6 shows this as either a Pro–Glu or Xle–Xle residue pair. The b_n-18 series provides no further insight into the identity of the first two residues, and this can also be said of the equivalent y_{18} and y_{19} N-terminal residues. Other than the $(M + 2H)^{2+}$, no other doubly charged ion is present. The difference between the b_6-18 and b_8-18 ion establishes a difference of 242.2 Da, and this can range from 2–4 amino acid residues (Table 9.6). The y_n series, however, establishes this region as the Asn–Gln residue pair. Another site of ambiguity, if the sequence was unknown, is y_4–y_5. The mass difference is 226.2 Da, which is identical to the ambiguity seen in the b_1–b_2 sequence determination. In addition, the b_{15-19} series is not observed, and thus cannot lend support to an assignment. If a small portion of the peptide sample is subjected to a methyl esterification chemical reaction, as shown in Table 9.7, the carboxylic acid R-group of Glu would undergo methylation.[37,48] This, in turn, would shift the mass of the y_5 ion to 661.8 + 14 = 675.8 Da, the y_6 ion to 825.0 + 14 = 839.0 Da, etc. This reaction would also increase the b_2 and b_3 ions by 14 Da, and the b_n-18 series by 14 Da. From this methylation reaction, both the b_1–b_2 and y_4–y_5 regions can be assigned as the Pro–Glu and not the Xle–Xle (or Lxx–Lxx) residue pair. However, this still does not establish the sequence or order of these two residues at both sites in an unknown situation. In this case, synthetic standards could be generated and their tandem mass spectra compared with that of the sample. An alternative possibility is to digest the peptide with *S. aureus* V8 protease (Table 9.4) at a pH of 4.0. It was established that Asp is not present, and thus digestion should cleave the peptide between the Glu^{16}–Leu^{17} bond and not at the Glu^1–Pro^2 bond. The resulting smaller peptides can be checked for their mass correspondence to the Figure 9.6b template for sequence purposes. It can be shown that Gln-8 fails to acetylate. Therefore, Lys can be eliminated as a candidate, because both Glu and Lys have the same nominal mass (Table 9.1). Finally, the Leu–Ile identities at the 4, 10, and 17 residue positions can be investigated with the α-chymotrypsin enzyme (Table 9.4).

PROBLEM 9.1 Figure 9.7 shows the ESI-tandem mass spectrum of the T11 tryptic peptide from sperm whale myoglobin (Table 9.5). The tandem mass spectrum was generated from the $(M + 2H)^{2+}$ species.[49] Useful ion masses are tabulated in Table 9.9. Deduce the sequence taking into account further useful chemical enzymatic tests to refine any residue identity/order ambiguities. Keep in mind that the list in Table 9.9 consists only of the observed ion masses.

PROBLEM 9.2 The ESI-tandem mass spectrum in Figure 9.8 is that of a tryptic peptide of β-lactoglobulin from an HPLC separation of the trypsin digest peptide mixture. The $(M + 2H)^{2+}$ at m/z 596.6 was used as the parent ion. Table

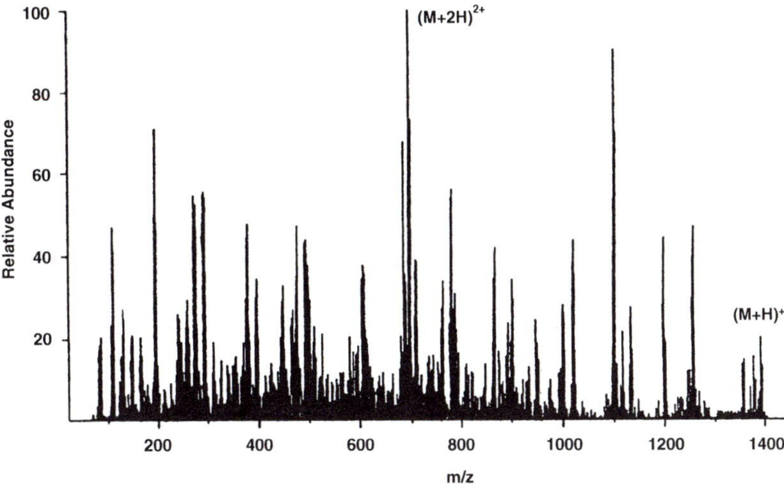

Figure 9.7 The ESI-tandem mass spectrum of the $(M + 2H)^{2+}$ ion from the T11 peptide from a tryptic digest of sperm whale myoglobin at *mz* 697.4. (© 1992. Reprinted from reference 49 by permission of Academic Press, Inc., Orlando, Florida.)

9.10 presents a tabulation of the essential observed masses. Two of the listed ions are a-type ions, and there is a signal that results from the dehydration of a certain ion type. Deduce the sequence of the peptide.[50–52]

PROBLEM 9.3 The ESI-tandem mass spectrum of glufibrinopeptide B[53] is shown in Figure 9.9, where the precursor ion is the $(M + 2H)^{2+}$ at *m/z* 786.0. This complete peptide can be considered as a tryptic-like peptide. Deduce the sequence of the peptide.

Another strategy for the interpretation of the product-ion mass spectrum of a peptide can be found in the literature,[54] especially when the C-terminus of the peptide is known to contain a specific amino acid. This method can be valuable if the peptide was produced from the digestion of a protein by a specific enzyme. Thus, by labeling that C-terminus residue as y_1, that mass value can be checked for its appearance in the product-ion mass spectrum. Whether or not the y_1 mass is present, the y_2 mass value can be arrived at by systematically calculating all 20 amino acid additions onto y_1 and noting the presence/absence of the calculated value in the product-ion mass spectrum.

Table 9.9 Observed ions from the ESI-tandem mass spectrum (Figure 9.7) of the T11 tryptic peptide of sperm whale myoglobin

147.2	195.2	260.4	294.3	373.5	395.4	444.6
494.6	501.6	607.7	614.8	697.4^{2+}	708.8	779.9
787.0	893.1	900.2	950.1	999.3	1021.2	1100.4
1134.4	1199.5	1256.6	1393.7			

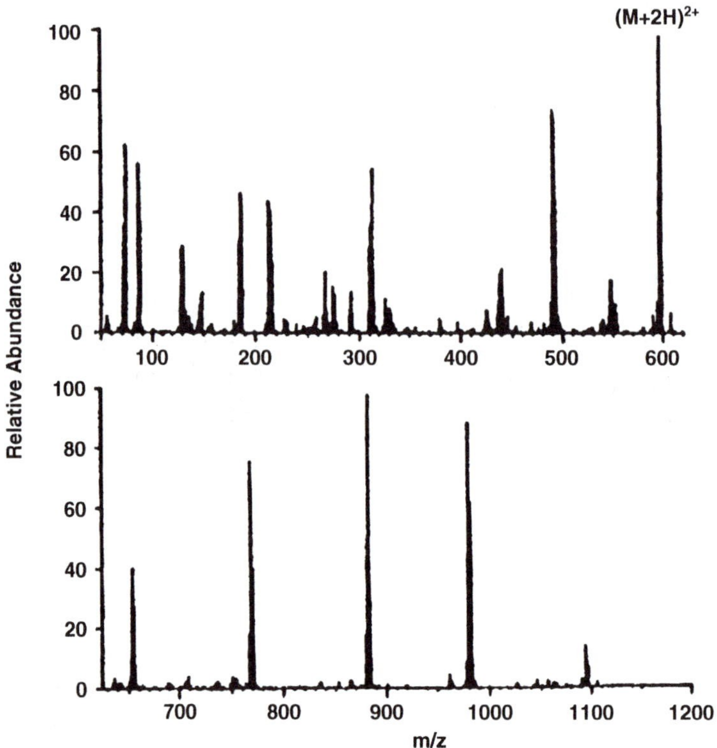

Figure 9.8 The ESI-tandem mass spectrum of the $(M + 2H)^{2+}$ ion from a tryptic peptide of β-lactoglobulin at m/z 596.6. (© 1991. Reprinted from reference 52 by permission of Plenum Publishing Company, New York.)

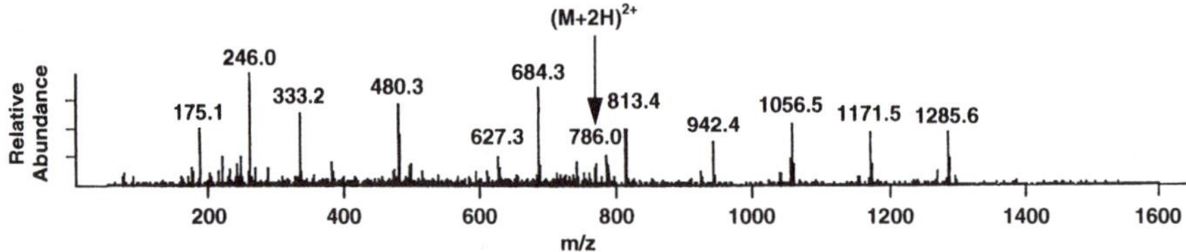

Figure 9.9 The ESI-tandem mass sepctrum of the $(M + 2H)^{2+}$ ion from glufibrinopeptide B at m/z 786.0. (© 1995. American Chemical Society. Reprinted from reference 53 with permission.)

Table 9.10 Observed ions from the ESI-tandem mass spectrum of the T(92–101) tryptic peptide of β-lactoglobulin

72.0	129.2	147.2	185.0	213.2	275.4	312.3	425.5
438.5	442.0^{2+}	491.5^{2+}	548.1^{2+}	553.6	596.6^{2+}	654.7	769.8
883.0	982.1	1095.3					

2+ indicates a doubly charged ion.

If only one value is actually found in the mass spectrum, then the value is labeled y_2. If two of the calculated values are observed, then both resulting amino acids must be considered until further mass spectral evidence dismisses one possibility, such as the b-ion series calculations. This method is continued until all of the observed masses are accounted. Missing b or y ions can result in complications with this sequence process.

Tryptic Peptide of Equine Cytochrome c

Figure 9.10 shows the complete sequence of equine cytochrome c. Note that some of the residues are not included in a tryptic peptide, and virtually all of these residues are Lys. Figure 9.11 shows an infusion sample introduction of the entire tryptic peptide mixture directly into the mass spectrometer, and some of the peptides are doubly charged.[55] Figure 9.12a presents the ESI-tandem mass

5 10 15
Ac- Gly- Asp- Val- Glu- Lys-| Gly- Lys-| Lys-| Ile- Phe- Val- Gln- Lys-| Cys- Ala-
T_1 | T_2 | | T_3 |

20 25 30
Gln- Cys- His- Thr- Val- Glu- Lys-| Gly- Gly- Lys-| His- Lys-| Thr- Gly- Pro-
T_4 +heme | T_5 | T_6 |

35 40 45
Asn- Leu- His- Gly- Leu- Phe- Gly- Arg-| Lys-| Thr- Gly- Gln- Ala- Pro- Gly-
T_7 | |

50 55 60
Phe- Thr- Tyr- Thr- Asp- Ala- Asn- Lys-| Asn- Lys-| Gly- Ile- Thr- Trp- Lys-|
T_8 | T_9 | T_{10} |

65 70 75
Glu- Glu- Thr- Leu- Met- Glu- Tyr- Leu- Glu- Asn- Pro- Lys-| Lys-| Tyr- Ile-
T_{11} | |

80 85 90
Pro- Gly- Thr- Lys-| Met- Ile- Phe- Ala- Gly- Ile- Lys-| Lys- Lys-| Thr- Glu-
T_{12} | T_{13} | | T_{14}

95 100
Arg-| Glu- Asp- Leu- Ile- Ala- Tyr- Leu- Lys-| Lys-| Ala- Trp- Asn- Glu
| T_{15} | | T_{16}

Figure 9.10 Complete amino acid sequence of equine cytochrome c and a tryptic digest map of the residue sequences. (© 1991. Reprinted from reference 55 by permission of Academic Press, Inc., Orlando, Florida.)

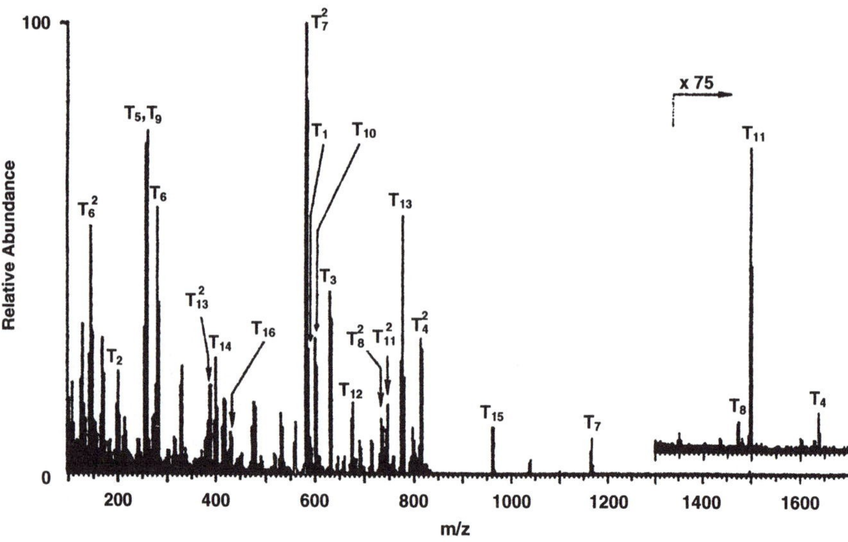

Figure 9.11 Infusion ESI-mass spepctrum of the tryptic digest of cytochrome c in Figure 9.10. (© 1991. Reprinted from reference 55 by permission of Academic Press, Inc., Orlando, Florida.)

spectrum of the T7 fragment, and Figure 9.12b shows the analysis. The precursor ion was the $(M + 2H)^{2+}$ at m/z 585. The y_n series is observed except for the last y ion, $y_f = 1169.3$ Da, which equals the $(M + H)^+$. Instead, this value can be calculated from the $(M + 2H)^{2+}$ species. No b ions are observed; however, two other ions can be noted—namely, ions designated PN and PNL. These are internal fragment ions as noted in Table 9.3. The mass of an internal fragment ion is equal to the residue masses plus either one or two hydrogen atoms, and their origin and structure are delineated in Scheme 9.4.[11,56,57] Since no b ions are observed in this example, the nomenclature in Scheme 9.4 cannot be used; however, the one-letter residue abbreviations of the fragment ions will suffice. The internal fragment ion PN, corresponding to the third and fourth residues, has a mass of 211 or 212 depending on whether there are one or two additional hydrogens, respectively. A similar observation is made from the ion of minor intensity at m/z 324 or 325, which is generated from the PNL internal fragment.

PROBLEM 9.4 Figure 9.13 shows the ESI-tandem mass spectrum of the T15 tryptic fragment of equine cytochrome c; the precursor ion was the $(M + H)^+$ at m/z 965.1. No doubly charged species were observed in the tandem mass spectrum. Derive the structure of the tryptic peptide fragment using the masses listed in Table 9.11. Suggest possible experiments to deal with ambiguities that result from the mass spectral analysis.

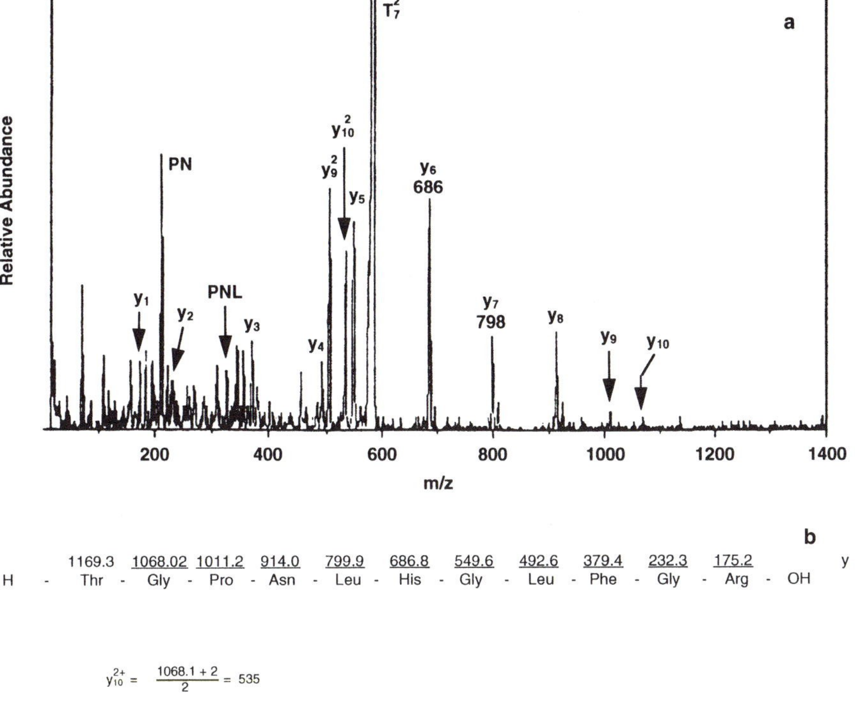

$$y_{10}^{2+} = \frac{1068.1 + 2}{2} = 535$$

$$y_9^{2+} = \frac{1011.1 + 2}{2} = 506.5$$

$$(M{+}2H)^{2+} = \frac{1168.1 + 2}{2} = 585$$

Figure 9.12 (a) The ESI-tandem mass spectrum of the $(M + 2H)^{2+}$ ion from the tryptic peptide T7 in Figures 9.10 and 9.11 at m/z 585 (T_7^2). (b) Amino acid sequence and ion identification of the mass spectrum in (a). Experimentally obseved ions are underlined. (© 1991. Reprinted from reference 55 by permission of Academic Press, Inc., Orlando, Florida.)

Protein Modifications

The examples of protein sequencing by ESI-mass spectral methods have been relatively straightforward where the analytes were composed of unmodified amino acid residues. Ambiguities arose in mass spectral interpretation, however, because of the absence of mass spectral features and the isomeric nature of certain amino acid residues. Where ambiguity existed about the identity of isomeric residues, chemical and enzymatic methods were suggested that selectively modify and cleave, respectively, specific targeted residues in the peptides.

Many proteins of interest contain modifications to the 20 amino acids, and these posttranslational modifications will be introduced and elaborated. The term posttranslational modification is derived from the fact that after the messenger-RNA is translated into a linear sequence of amino acids by the transfer-

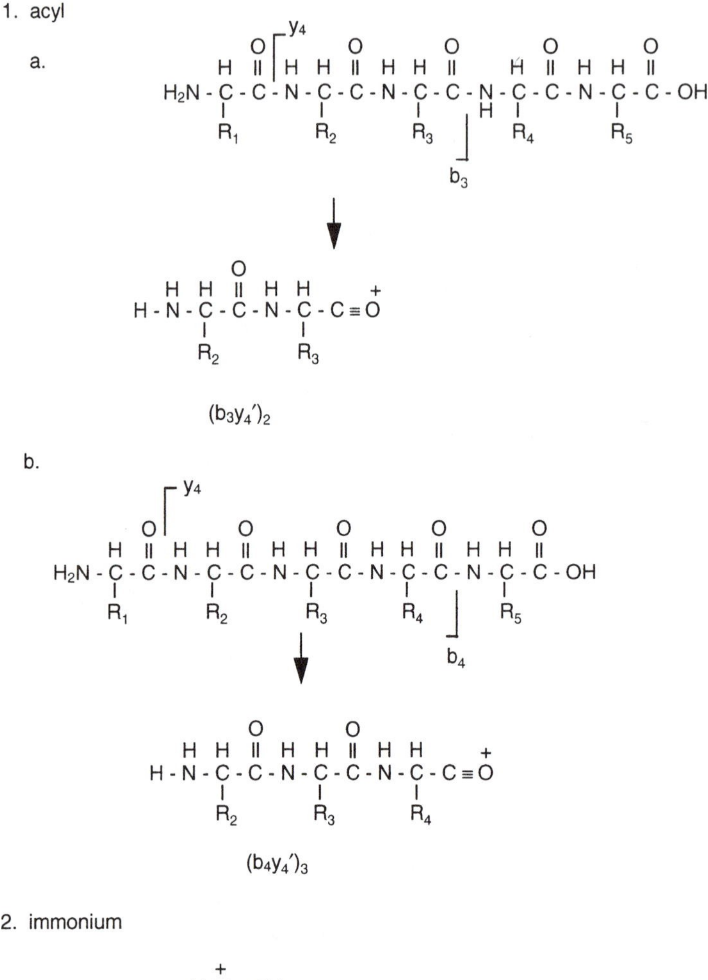

Scheme 9.4 Internal fragment ions from peptide CID.

RNA and ribosome apparatus, further cellular processing or modification(s) of the nascent peptide/protein chain can take place.[37,58,59]

Rat Cerebellar Protein

This example of peptide sequencing deals with a peptide that has a particular modification. The modification can be found as an acetamide derivative of the N-terminal amine moiety. The equivalent chemical reaction modification is acetylation of the N-terminus moiety (Table 9.7).

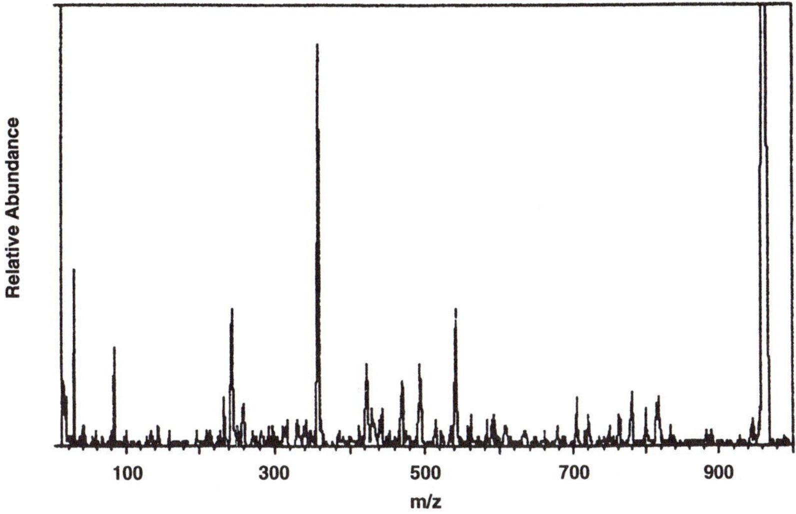

Figure 9.13 Infusion ESI-tandem mass spectrum of the $(M + H)^+$ ion from the T15 tryptic peptide in Figures 9.10 and 9.11 at *m/z* 965.1. (© 1991. Reprinted from reference 55 by permission of Academic Press, Inc., Orlando, Florida.)

Processing of the peptide consisted of subjecting a crude rat extract to two-dimensional PAGE, which separated the mixture into more than 400 different protein spots.[84] One of the protein spots, the calcium-binding protein calmodulin, was removed, purified, and cleaved with trypsin enzyme. The enzyme digest was then submitted to HPLC-ESI-tandem mass spectrometry. Figure 9.14a shows the mass spectrum of the T1 tryptic peptide from the calmodulin protein, and Figure 9.14b shows the sequence and mass analysis. The precursor ion was the $(M + 2H)^{2+}$ ion at *m/z* 782.5, yielding an $(M + H)^+$ and M of 1564 and 1563 Da (average masses rounded to the nearest whole number), respectively. In the determination of the sequence from the spectrum, starting anywhere in the mass spectrum yields differences between various peaks that can be attributed to amino acid residues when interrogating both the b and y series. However, when the lowest masses are investigated—in particular, b_1 at *m/z* 114—the inclination is to identify it as *m/z* 113 + 1 (either a Leu or Ile residue). In reality, the b_1 ion is an acetylated alanine residue (acetylalanine or Ac-Ala), which causes it to have the same nominal mass (71 Da + 43 Da) as that of Xle (113 Da + 1H).

Table 9.11 Observed ions from the ESI-tandem mass spectrum of the T15 tryptic peptide of cytochrome c

147.2	245.2	260.3	358.3	423.5	471.5	494.6
542.6	607.1	705.8	720.9	818.9	965.1	

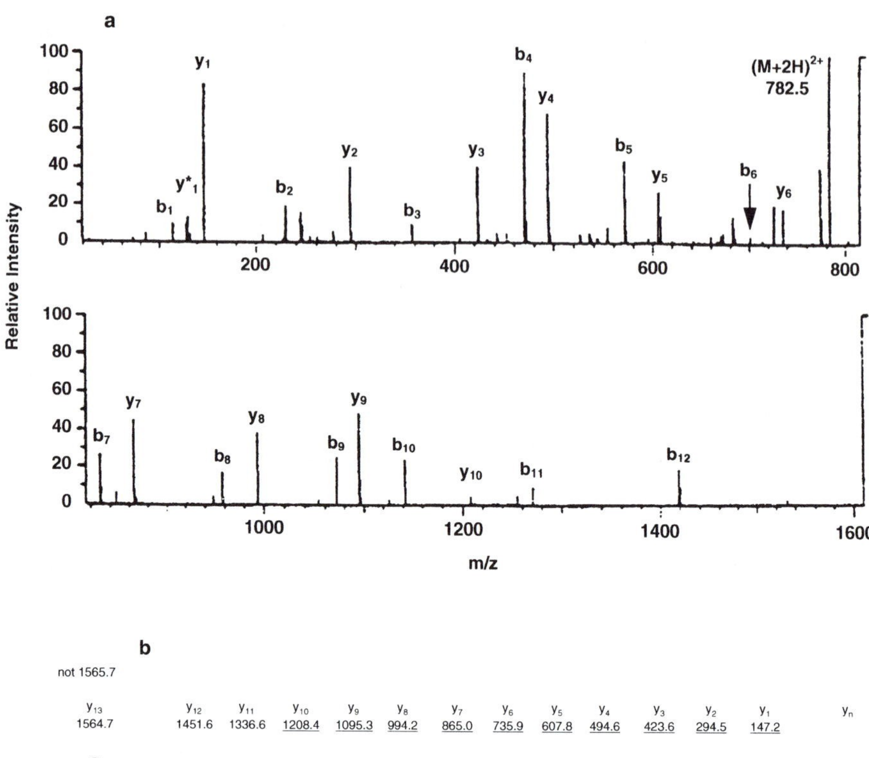

O
||
H3C - C - Ala - Asp - Gln - Leu - Thr - Glu - Glu - Glu - Ile - Ala - Glu - Phe - Lys - OH

b_n	114.1	229.0	357.1	420.3	571.4	700.5	829.6	957.8	1070.9	1142.0	1271.1	1418.3	1546.5
	b_1	b_2	b_3	b_4	b_5	b_6	b_7	b_8	b_9	b_{10}	b_{11}	b_{12}	b_{13}

$(y''_1)^* = y_1^* = y_1 - H_2O = 147.2 - 18 = \underline{129.2}$

Figure 9.14 (a) The ESI-tandem mass spectrum of the $(M + 2H)^{2+}$ ion from the T1 tryptic peptide of rat cerebellar calmodulin protein at m/z 782.5. (b) Amino acid sequence and ion identification of the mass spectrum in (a). Experimentally observed ions are underlined. (© 1996. Reprinted from reference 84 by kind permission of Elsevier Science-NL, Sara Burgerhartstraat 25, 1055 KV Amsterdam, The Netherlands.)

Fortunately, there is a complete b_n series where b_1–b_{12} are observed and b_{13} (b_f here) can be deduced from the $(M + 2H)^{2+}$ ion at m/z 782.5. This yields a rounded average mass M_r of 1563 Da (or 1563.5 Da average molecular mass in Figure 9.14b without rounding to a whole number). The y-type ions differ from each other by an amino acid residue and are calculated on that basis except for the y_1 ion. The y_1 ion is calculated from the mass of the free amino acid plus one hydrogen ion, where the latter is the origin of the charge on the ionized y_1 species. The y_1 structure can be dissected further into the residue mass, the OH carboxyl-terminus moiety, the ionizing proton, and the N-terminal hydrogen (Figure 9.1). These four y_1 subunits are present throughout the entire y_n series until the y_f residue is reached. Even though y_{11} and y_{12} are not present in the spectrum, they can be deduced by inference to the b_n series, and these values are found in Figure 9.14b. Remember that without prior knowledge of the N-terminal amino acid modification, this position was labeled as Xle.

There is no N-terminal hydrogen, yet the residue mass is all that is required to yield the $(M + H)^+ = y_f$ ion, and the acetylalanyl and leucyl–isoleucyl residues have the same mass. One of the methyl hydrogens of the acetyl group can conceptually act as the "N-terminal" hydrogen.

If one is resourceful, the very first cycle of the classic Edman degradation process can expose an altered first residue in the peptide. Edman degradation[31] is a time-consuming chemical method that derivatizes the N-terminal amino acid, followed by cleavage of the derivatized N-terminal amino acid with subsequent identification by reversed-phase HPLC. This process is repeated for each residue. However, if the N-terminus is blocked (i.e., absence of a free amine moiety), the Edman reaction will not proceed, and the N-terminal amino acid will not be released. This would be the case with the present peptide and would signify a blocked N-terminus. The absence of a released amino acid would then indicate that m/z 114.1 consists of a residue and blocking group. Proline has a single free N-terminal hydrogen, and this reacts with the Edman derivatization reagent. Table 9.12 lists a number of uncommon amino acid residues. Hydroxyproline and norleucine have the same nominal residue mass as Ac-Ala, but both of these uncommon amino acids have a free amine group. Thus, one cycle of the Edman degradation process can shed insight on the nature of the first residue. However, the mass spectrum would not necessarily prompt this line of investigation.

It has been reported that an acylpeptide hydrolase can effectively remove an N-terminal acyl or formyl blocking group.[85] The use of this enzymatic procedure can yield a peptide with an $(M + H)^+$ that is 43 Da lower than the original, blocked peptide shown in Figure 9.14b.

PROBLEM 9.5 Synthetic solid-phase synthesized peptide. This problem deals with a peptide that is blocked at the N-terminal residue, and a modification exists at the C-terminal residue. In solving this problem, provide careful bookkeeping of the mass values to within 1 Da. The spectrum with pertinent mass values[58] is given in Figure 9.15. The observed precursor ion for the product-ion mass spectrum is the $(M + 2H)^{2+}$ at m/z 581.2. However, with a knowledge of the sequence, a more accurate $(M + 2H)^{2+}$ mass value is 581.6 Da; thus, it is prudent in this learning experience to use that $(M + 2H)^{2+}$ value in order to arrive at unambiguous residue assignments. Also, not all masses that are labeled in the spectrum (Figure 9.15) are necessarily useful a-, b-, c-, x-, y-, z-type ions. Some ions may be irrelevant (noise). Other ions may have experienced a loss of H_2O (−18Da) or NH_3 (−17 Da), thus resulting in a_n-18, a_n-17, b_n-18, b_n-17, y_n-18, y_n-17 type ions, and so on. Initially, one should try to identify ions that are spaced 17 or 18 Da apart. This can signify b_n-17, b_n; b_n-18, b_n; y_n-17, y_n ion pairs, and so on. Also note that a_n, b_n pairs can exist

Table 9.12 Uncommon amino acid residues

Name	Abbreviation	Residue structure	Residue M_r
Sarcosine, *N*-methyl glycine	Sar	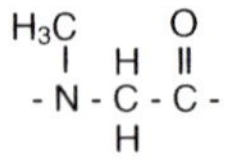 H_3C O I H II - N - C - C - H	71.08
Homoserine lactone	Hsl	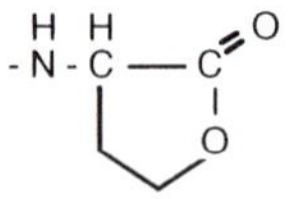 H H O - N - C — C I O	100.10
Homoserine	Hse	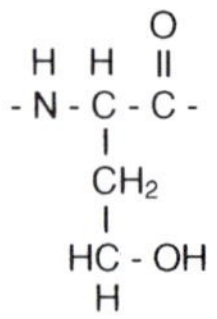 O H H II - N - C - C - I CH_2 I HC - OH H	101.10
Pyroglutamic acid	pGlu Glp < Glu	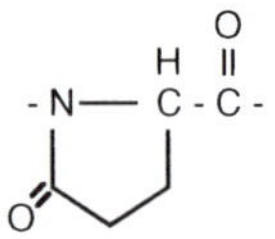 O H II - N — C - C - O	111.11
4-Hydroxyproline	HOPro	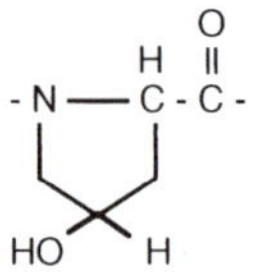 O H II - N — C - C - HO H	113.12
Norleucine	Nle	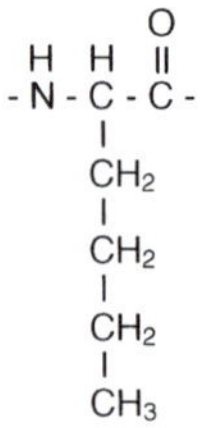 O H H II - N - C - C - I CH_2 I CH_2 I CH_2 I CH_3	113.16
Ornithine	Orn	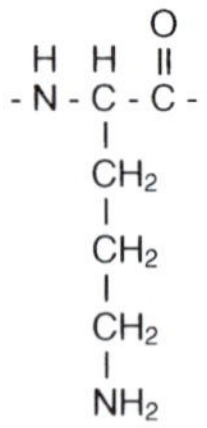 O H H II - N - C - C - I CH_2 I CH_2 I CH_2 I NH_2	114.15

(*continued*)

Table 9.12 (*continued*)

Name	Abbreviation	Residue structure	Residue M_r
Methionine sulfoxide	SxMet	O H H ‖ - N - C - C - | CH_2 | CH_2 | S = O | CH_3	147.20
Cysteic acid	Cya	O H H ‖ - N - C - C - | CH_2 | SO_3H	151.14
S-carboxymethyl cysteine	CMCys	O H H ‖ - N - C - C - | CH_2 | S | CH_2 | C - OH ‖ O	161.18
Methionine sulfone	SMet	O H H ‖ - N - C - C - | CH_2 | CH_2 | O = S = O | CH_3	163.20
S-(4-pyridylethyl)-cysteine	pCys	O H H ‖ - N - C - C - | CH_2 | S | CH_2 | CH_2 | N	208.28

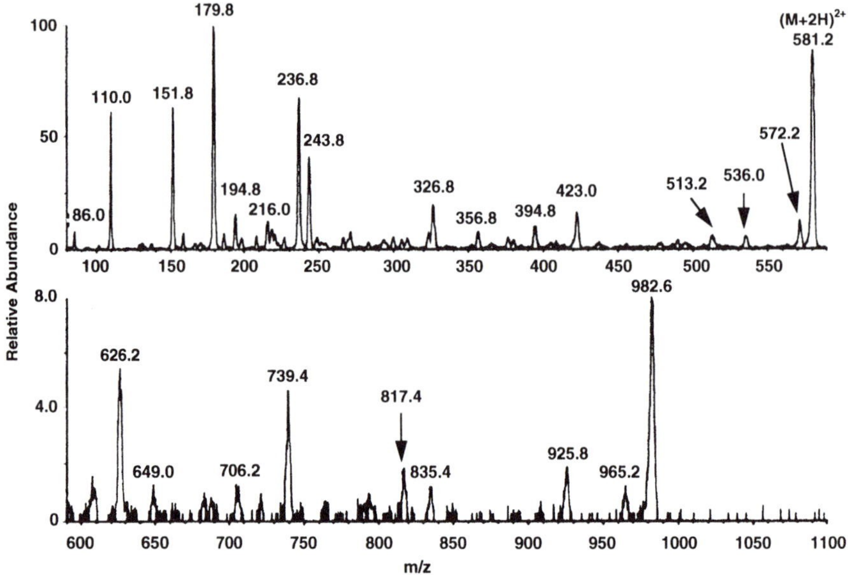

Figure 9.15 The ESI-tandem mass spectrum of the $(M + 2H)^{2+}$ ion from a synthetic solid-phase synthesized peptide at *m/z* 581.2. (© 1991, American Chemical Society. Reprinted from reference 58 with permission.)

where $b_n - 28 = a_n$; therefore one would look for b_n-28, b_n pairs (Figure 9.2) in the mass spectrum (Figure 9.15). The same analogy can be applied to y_n-28, y_n ion pairs where $y_n - 28 = x_n$. In addition, another class of ion that is called an immonium ion[11,19,23,56,57] is present in Figure 9.15. Scheme 9.4 presents the structure of this ion, and, basically, an immonium ion is the equivalent of a b_1 amino acid residue minus the CO moiety. Table 9.13 lists the masses of the immonium ion for each amino acid residue; certain residues have additional lower and higher mass fragments from their respective immonium ion.[11,23,56,57] Table 9.14 lists immonium ions of some uncommon amino acids.[11,23,56,57] The point must be stressed that a residue does not have to be at the b_1 position to yield an immonium ion. A residue can be internal to the peptide and provide an immonium ion in the mass spectrum. It is these ions that could cause confusion in assignment of the b, y series. Thus, the low-mass region is not the best place to start the sequencing procedure.

PROBLEM 9.6 The next protein analysis study with ESI-tandem mass spectrometry provides an interesting observation.[87] A particular voltage (3 kV) was placed on the ESI needle and, using a common peptide analyte, Figure 9.16a was obtained under tandem mass spectrometry conditions. When the voltage was raised to 5–6 kV, a change was noticed in the mass spectrum of the peptide (Figure 9.16b). What could be the nature of this change and exactly where in the peptide did this on-line, real-time modification occur?

Table 9.13 Mass of immonium and other fragment ions of amino acid residues

Amino acid	Immonium ion mass	Other fragment masses	Comments
Gly	30		
Ala	44		
Ser	60		High intensity
Pro	70		This high-intensity mass also originates from an Arg fragment mass. If other Arg fragment ions are missing, then there is a high probability that *m/z* 70 originates from Pro
Val	72		
Ac-Gly[a]	72		
Thr	74		
Cys	76		Low intensity
Leu	86	72	Very low intensity
Ile	86	72	Very low intensity
Asn	87	70	Both ions very low intensity
Asp	88		Ion is absent when C-terminal
Gln	101	84, 129	*m/z* 129 usually very low intensity
Lys	101	70, 84, 112, 129	*m/z* 70 and 101 usually very low intensity
Glu	102		Very low intensity when C-terminal
Met	104	61	*m/z* 104 of very low intensity
His	110	82, 121, 123, 138, 166	Fragment ions of very low intensity; high-intensity immonium ion
Phe	120	91	*m/z* 120 of high intensity and *m/z* 91 is of very low intensity
Arg	129	59, 70, 73, 87, 100, 112	*m/z* 73 and 129 of very low intensity
Tyr	136	91, 107	High-intensity immonium and very low intensity fragment ions
εAc-Lys[b]	143	126	
Ac-Glu[c]	144	102	N-terminal origin
Trp	159	117, 130, 170, 171	High-intensity immonium and fragment ions except *m/z* 117 is low intensity
AcTyr[d]	178	136	N-terminal origin

Immonium ion of the following residues:
[a]Acetylglycyl.
[b]Lysyl residue acetylates at the ε-amine group.
[c]Acetylglutamyl.
[d]Acetyltyrosyl.
(Information taken from references 56, 57, and 86.)

PROBLEM 9.7 Sometimes immonium and internal fragment ions can be found in peptide mass spectra, and in other cases they are not observed. The analyte in Figure 9.17 provides a significant amount of immonium and internal fragment ions.[88] Keep in mind that an a_1 ion is an immonium ion (Scheme 9.4 and Figure 9.2). Look for ions separated by 17, 18, and 28 Da. Deduce the identity of the peptide.

Table 9.14 Mass of immonium and other fragment ions of uncommon amino acid residues

Amino acid	Immonium ion mass	Other fragment masses	Comments
pGlu	84		N-terminal pGlu present
HO Pro	86		
Nle	86		
CMCys	134		

Aspartic Acid

Aspartic acid is prone to spontaneous cyclization under mild acid conditions (Table 9.7). This happens especially when Asp is followed by a Gly residue. Once in the cyclized state, aqueous conditions can cause either amide bond to open up. As Table 9.7 shows, an isomeric mixture of both the original Asp and the isoaspartyl residue is formed.

A recombinant protein, hirudin variant 1, or r-hirudin, is expressed from *Saccharomyces cerevisiae*. This protein is an antagonist of human α-thrombin, and the structure of r-hirudin is shown in Figure 9.18.[82] The Asp residue occurs at a number of places in the sequence, and the isomerization reaction was investigated with this protein. The cyclization to a succinimide ring structure of the Asp residue is spontaneous. The protein underwent a radiolabeled reaction which opened up the ring; this will be discussed below.

The structure of the protein shows that a number of cystine bridges are present. These disulfide bridges were reduced with DTT (Table 9.7) and subsequently subjected to S-alkylation with 4-vinylpyridine (β-pyridylethylation) as shown in Table 9.7. This procedure is one of a number of methods used to identify cystine disulfide residues in a protein. The cystine-reduced and alkylated protein was then digested with *Staphylococcus aureus* V8 protease, or Glu-C (Table 9.4), at pH 7.8. This procedure cleaves the C-terminal end of Glu residues (E), and a number of peptides were generated in the enzymatic reaction.

Two peptides were investigated for the Asp cyclization phenomenon. The first peptide, V_{44-61} or Q4, consists of the residues 44–61 shown in Figure 9.18. Residue 53 is the Asp under investigation. Note that cysteines are absent; thus, the pyridylethylation reaction did not affect this peptide product of r-hirudin. Also, the Asp-55 residue did not show evidence of cyclization. Figure 9.19 provides structural details of the Asp region under investigation. The initial radiolabel experiment consisted of splitting the protein solution into two vials and allowing the intact r-hirudin protein to equilibrate in H_2O and $H_2{}^{18}O$. The ring-opening hydrolysis reaction (Table 9.7) takes place in both reagents, and both α- and β-aspartyl residues are generated. One ^{18}O is incorporated as a hydroxyl moiety to form the γ-carboxyl R-group. This results in a scission of the cyclic succinimide. After subsequent derivatization and digestion of the intact protein, the Q4 peptide was analyzed, and Figures 9.20a and b present

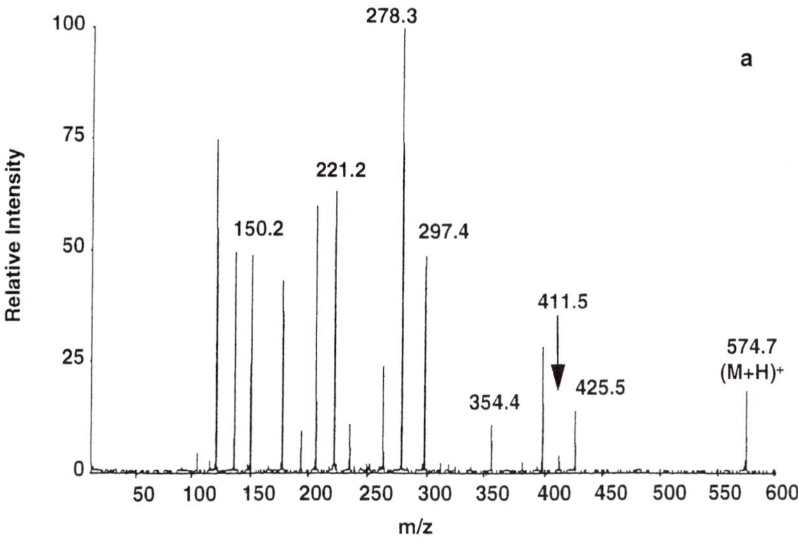

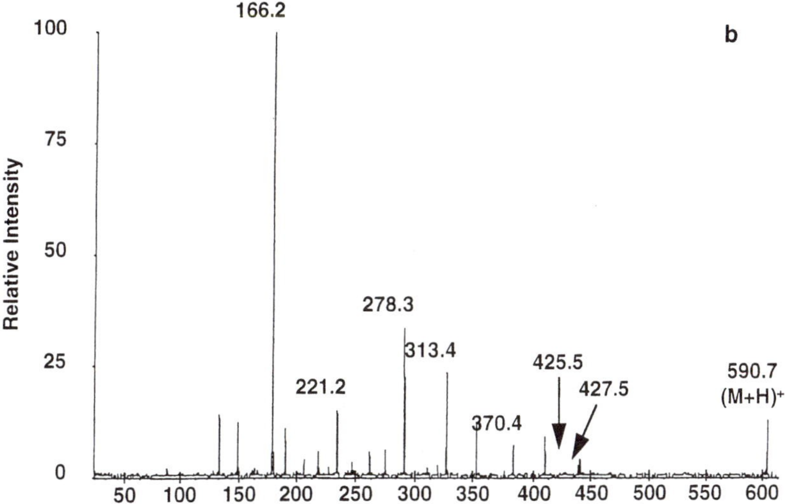

Figure 9.16 (a) The ESI-tandem mass spectrum of the $(M + H)^+$ ion from a peptide at *m/z* 574.7. The electrospray needle voltage was set at 3 kV. (b) The ESI-tandem mass spectrum of the $(M + H)^+$ ion from a modified form of the peptide at *m/z* 590.7, where the needle voltage was raised to 5–6 kV. (© 1993. Reproduced from reference 87 by permission of John Wiley & Sons, Limited.)

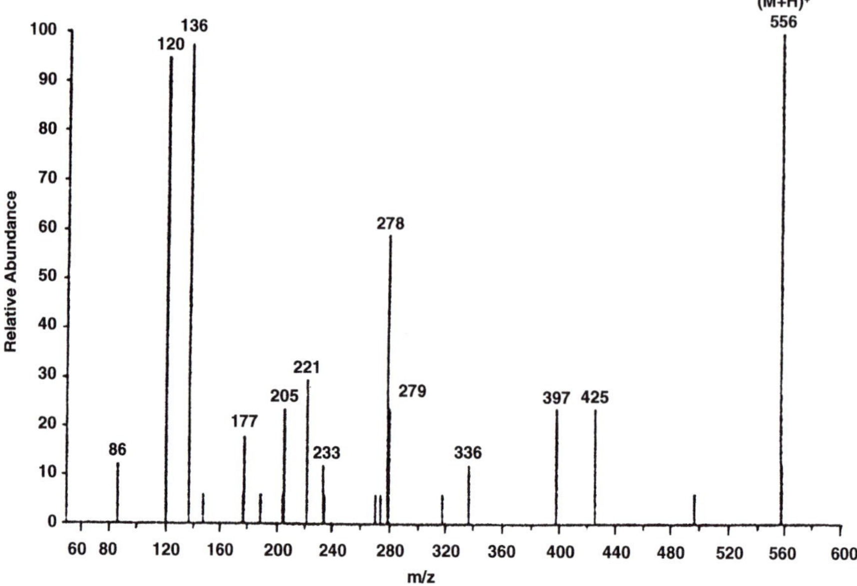

Figure 9.17 The ESI-tandem mass spectrum of the $(M + H)^+$ ion from a peptide at m/z 556.0. (© 1989. Reprinted from reference 88 by kind permission of Elsevier Science-NL, Sara Burgerhartstraat 25, 1055 KV Amsterdam, The Netherlands.)

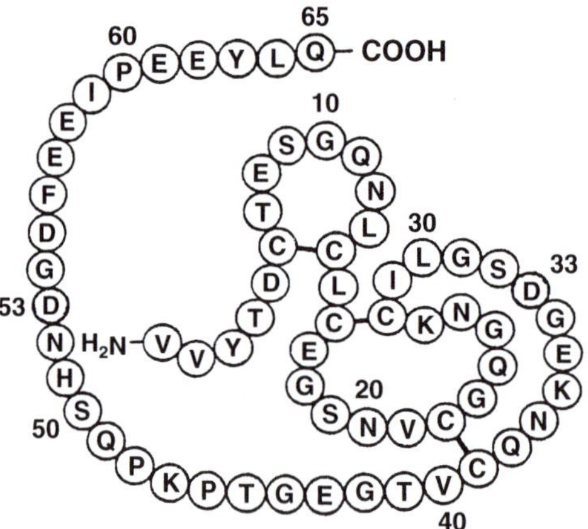

Figure 9.18 Amino acid sequence of recombinant hirudin variant 1. (© 1993. Reproduced from reference 82 by permission of John Wiley & Sons, Limited.)

H2N - G T P K P Q S - N - **H** - C - N - **N** - C - N - C - C - N - **G** - C -

	b_8	b_9	b_{10}
H_2O	833.4	947.2	1062.3
$H_2{}^{18}O$	833.5	947.3	1064.2

b_{15}^{2+}	b_{15}	b_{16}^{2+}	b_{16}
820	1639	877	1753
821	1641	878	1755

Figure 9.19 Structural details and ion identification information for the amino acid region of the Asp-53 residue in the Q4 peptide of r-hirudin variant 1. Radiolabeled water hydrolysis of cyclized Asp-53 is also shown.

the ESI-tandem mass spectra of the H_2O and $H_2{}^{18}O$ reaction, respectively.[82] The essential part of this study is portrayed in Figure 9.19. Note that the b_8 and b_9 fragments (Figures 9.19 and 9.20) have not changed in mass between the H_2O–$H_2{}^{18}O$ solvent equilibration. However, the b_{10} ion increases by 2 Da in the $H_2{}^{18}O$ reagent.

The b_{15} and b_{16} singly and doubly charged ions are of significant intensity in Figure 9.20b, and they are also affected by the $H_2{}^{18}O$ hydrolysis of Asp-53. The same mass shift is observed for these sets of ions as that of b_{10}. Overall, Figure 9.20 does not provide much structural information for the Q4 peptide; however, it provides the crucial information necessary to verify the $H_2{}^{18}O$ reaction on Asp-53.

PROBLEM 9.8 With respect to the *S. aureus* V8 protease digestion of r-hirudin, the possibility of succinimide formation and hydrolysis in the V_{18-35} or Q5 enzyme digestion product were investigated. In this peptide, there is an Asp-33 residue, and there are two cysteine residues (Figure 9.18) that were S-alkylated with 4-vinylpyridine. Refer to the section on "Aspartic Acid" for details on the handling and preparation of the protease peptide products. Figures 9.21a and b present the tandem mass spectra of the Q5 peptide when equilibrated with H_2O and $H_2{}^{18}O$, respectively.[82] Given the sequence of the peptide in Figure 9.18, provide b and y labels for the observed ions in each of the spectra. Highlight the b and y ion pairs; that is, a "b" ion and a "y" ion in Figures 9.21a and b, which shift in mass due to the radiolabel incorporation. Are there b, y ions in the spectra that specifically identify the residue responsible for the radiolabeled oxygen incorporation?

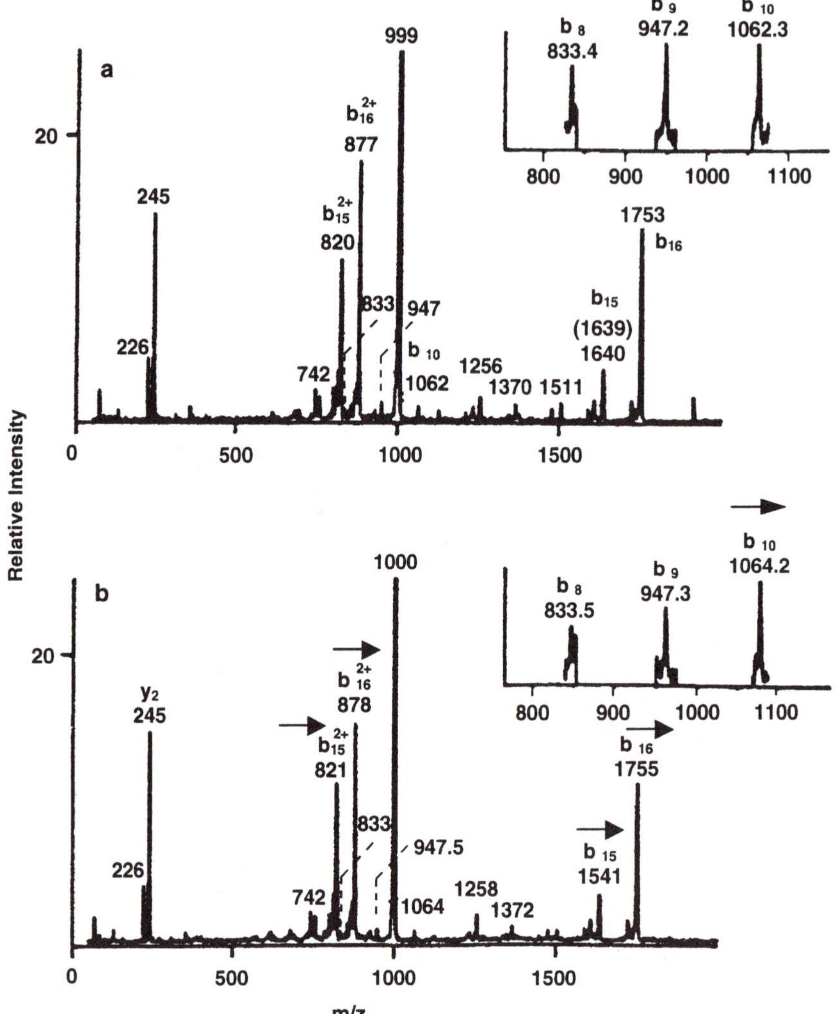

Figure 9.20 The ESI-tandem mass spectra of the Q4 peptide taken after water hydrolysis, derivatization, and digestion of r-hirudin. Spectra are from the hydrolysis in (a) H_2O and (b) $H_2{}^{18}O$. The $(M + 2H)^{2+}$ was used for tandem mas spectrometry at (a) *m/z* 999 and (b) *m/z* 1000. (© 1993. Reproduced from reference 82 by permission of John Wiley & Sons, Limited.)

Syntheses of peptides are important for many uses and applications. They are used as standards and control substances to ascertain an isolated unknown from biological systems. Biomedical, clinical, analytical and biochemical research routinely use synthetic peptides. A peptide was synthesized[61,81] as a test substance. It is common to find various amounts of by-products after the synthesis of a peptide. Preparative HPLC methods are used to separate the

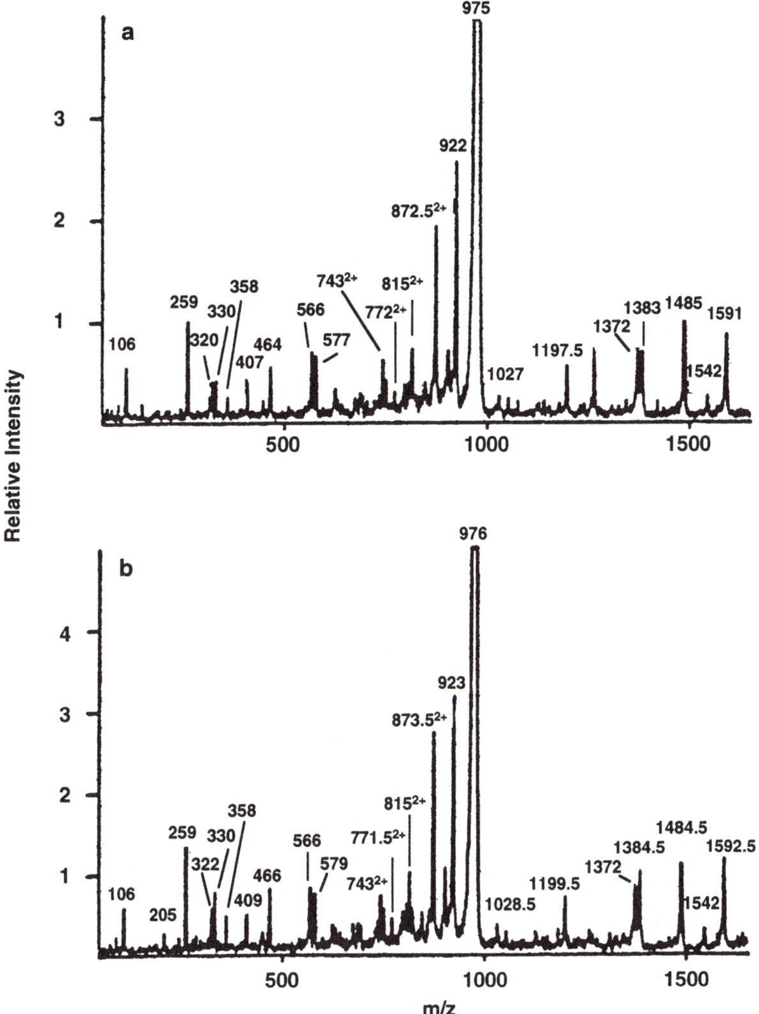

Figure 9.21 The ESI-tandem mass spectra of the Q5 peptide taken after water hydrolysis, derivatization, and digestion of r-hirudin. Spectra are from the hydrolysis in (a) H_2O and (b) $H_2{}^{18}O$. The $(M + 2H)^{2+}$ ions used for tandem mass spectrometry were (a) m/z 975 and (b) m/z 976. (© 1993. Reproduced from reference 82 by permission of John Wiley & Sons, Limited.)

various products. During the synthesis, oftentimes certain side-chain R-groups are reacted with chemical reagents to block or protect reactive moieties—such as hydroxyl or carboxyl groups, because they may react in an undesirable fashion during the peptide synthesis procedure. For example, Asp residues are usually protected where the carboxyl hydrogen is replaced with a benzyl group, and, for the Tyr hydroxyl group, the reactive hydrogen is protected with a bromo-ben-

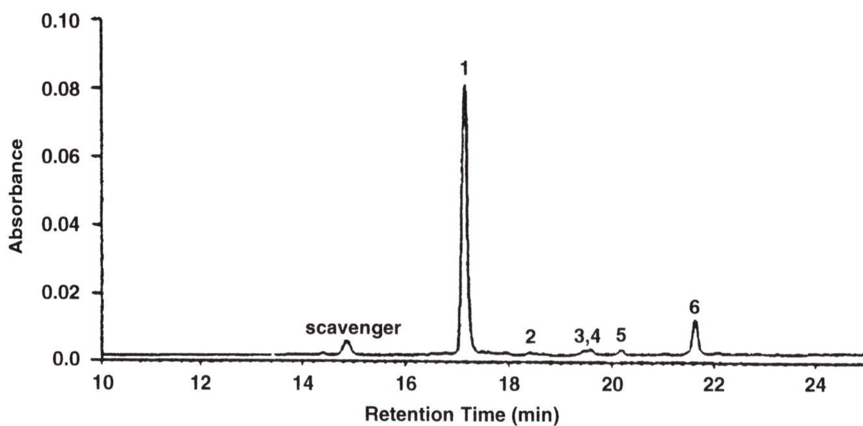

Figure 9.22 An RP-HPLC analysis of the unpurified solution of synthesized acyl carrier protein (ACP) (65–74) (peak 1). UV absorbance was at 220 nm. Peaks 2–6 represent by-products of peak 1. (© 1992. Reprinted from reference 81 by permission of Academic Press, Inc., Orlando, Florida.)

zyloxycarbonyl group. The Asn and Gln side groups are usually more refractory and less reactive to synthetic procedures, and, thus, are usually unprotected. Figure 9.22 shows an analytical reversed-phase HPLC analysis of the resultant solution of a synthesized peptide. Peak number 1 is the peptide of interest and peaks 2–6 are low-level amounts of unwanted by-products. The next series of problems (Problems 9.9 through 9.12) investigates these peaks.

PROBLEM 9.9 Figure 9.23 presents the ESI-tandem mass spectrum of the peptide product at peak 1 in Figure 9.22, and an $(M + H)^+$ is observed at m/z 1064. Deduce the sequence of the peptide.

PROBLEM 9.10 Peak 2 in Figure 9.22 has an $(M + H)^+$ of 1047 Da. Its origin in the synthetic procedure is uncertain; however, it displays a different $(M + H)^+$ species than that of the ACP(65–74) peak number 1 compound. Figure 9.24 presents the tandem mass spectrum of peak 2. Based on the change in $(M + H)^+$ mass, show which amino acid residue experiences the change and what the nature of the change is.

PROBLEM 9.11 Peak 3 is another by-product in Figure 9.22 in the synthesis of ACP(65–74), and the ESI-tandem mass spectrum of peak 3 is presented in Figure 9.25. The $(M + H)^+$ is observed at 1046 Da. Describe the position and nature of the modification/change. There is one c ion in Figure 9.25.

PROBLEM 9.12 Peak 6 in Figure 9.22 provides a tandem mass spectrum in Figure 9.26 with an $(M + H)^+$ of 1155 Da. Taking into consideration the various steps used in the synthesis, describe the position and nature of the modification to the peptide.

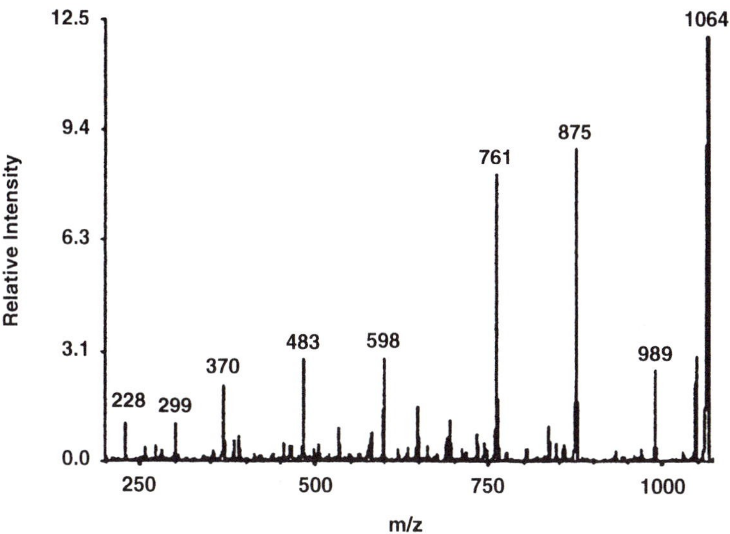

Figure 9.23 The ESI-tandem mass spectrum of peak 1 in Figure 9.22 at an $(M + H)^+$ of m/z 1064. (© 1992. Reprinted from reference 81 by permission of Academic Press, Inc., Orlando, Florida.)

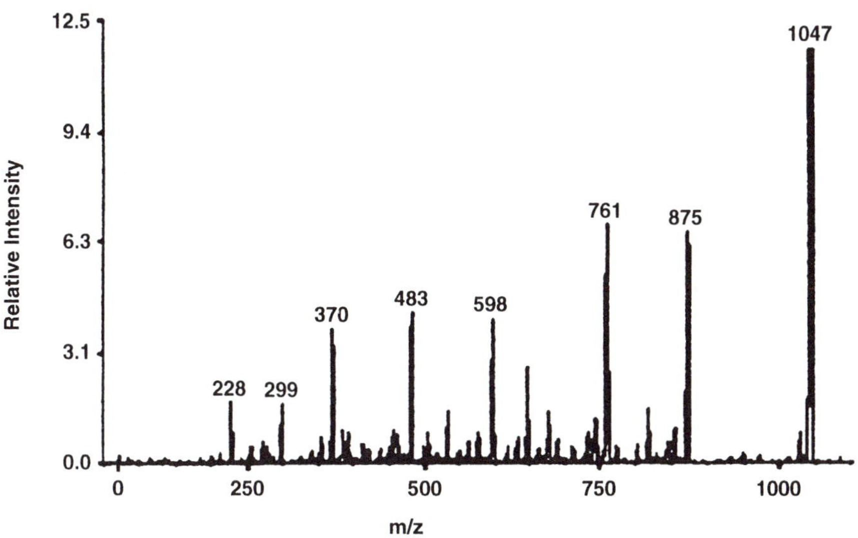

Figure 9.24 The ESI-tandem mass spectrum of peak 2 in Figure 9.22 at an $(M + H)^+$ of m/z 1047. (© 1992. Reprinted from reference 81 by permission of Academic Press, Inc., Orlando, Florida.)

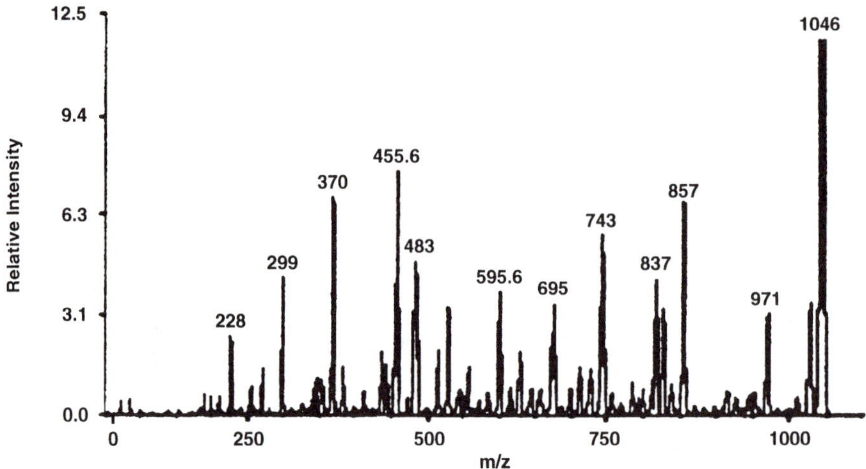

Figure 9.25 The ESI-tandem mass spectrum of peak 3 in Figure 9.22 at an $(M + H)^+$ of m/z 1046. (© 1992. Reprinted from reference 81 by permission of Academic Press, Inc., Orlando, Florida.)

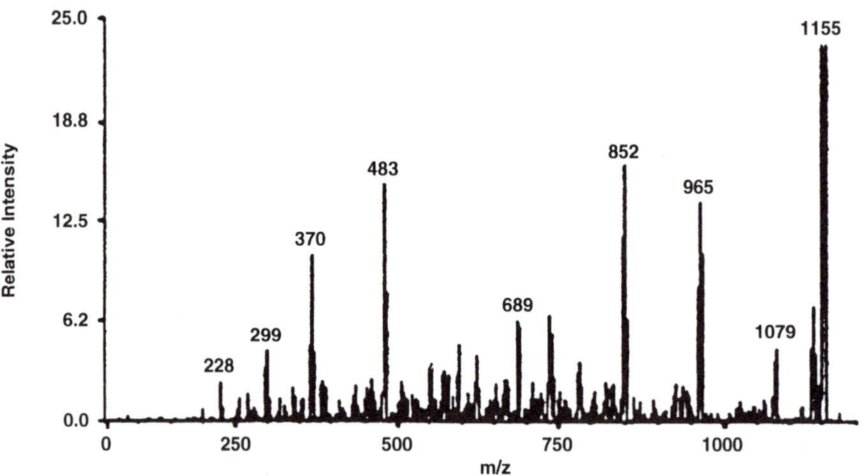

Figure 9.26 The ESI-tandem mass spectrum of peak 6 in Figure 9.22 at an $(M + H)^+$ of m/z 1155. (© 1992. Reprinted from reference 81 by permission of Academic Press, Inc., Orlando, Florida.)

Proline provides an interesting mass spectral display of relative intensities depending on its N-terminal neighbor. If the residue in the sequence X-Pro is of a charged nature—for example, Arg, Asp, or Glu—the ion that results from cleavage of the bond is of a low intensity. However, if the residue is neutral in nature, then an intense signal is observed.[15,17,56,89–97] Figure 9.27 shows an example of luteinizing hormone-releasing hormone (LH-RH) and an identical peptide except that the Arg-8 residue is replaced with a neutral Gln-8 residue. The ions that impact this phenomenon are the b_8 and/or y_3 ions, and the mass spectra are shown in Figure 9.28. As a note of interest, this peptide displays a blocked N-terminus and modified C-terminus. The b_8 ion shows an intense signal for the modified LH-RH [(Gln^8)-LH-RH], while the b_8 signal for LH-RH is of negligible abundance.

This phenomenon (proline effect) is observed for substance P(2–11) peptide whose structure is shown in Figure 9.29, and the tandem mass spectrum of the $(M + H)^+$ is shown in Figure 9.30a.[90] The b_2 ion is formed by the cleavage of the amide bond between the K–P residue, and the b_2 ion displays a high abundance due to the neutral lysine (K) residue. The nominal mass of 226 Da can actually be that of the b_2 ion as well as the KP′ internal fragment (Figures 9.29 and 9.30). A fairly intense y_8 ion is also observed, and this is also due to the neutral nature of the Lys residue. Observation of Figure 9.30b provides a different analysis. Figure 9.30b is the ESI-tandem mass spectrum of the $(M + 2H)^{2+}$ species of substance P(2–11). Note the decreased abundance of b_2 and the negligible y_8 intensity. The significant decrease in abundance of both b_2 and y_8 fragments is most likely due to the lysine side group accepting a proton, and the N-terminal amine attaching a proton. These two sites are the most likely to accept a proton (Chapter 4). Thus, in the $(M + H)^+$ species, a

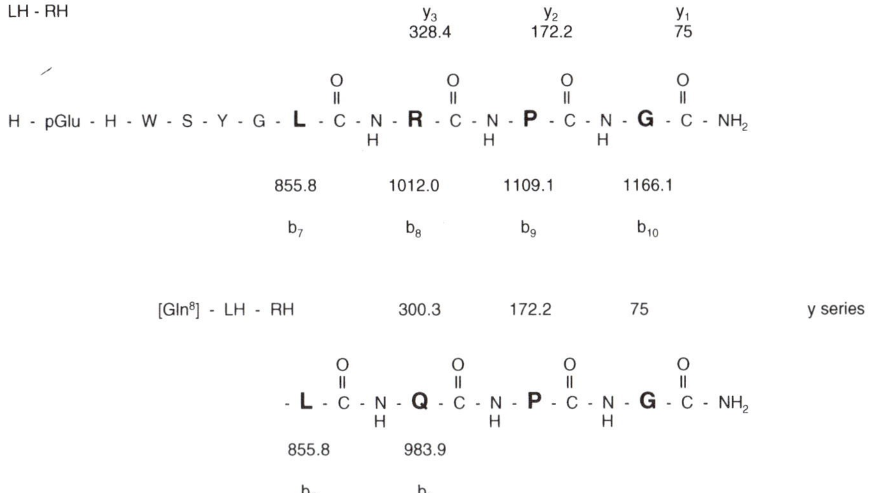

Figure 9.27 Structure and partial mass assignment of the LH-RH and [Gln^8]-LH-RH peptides.

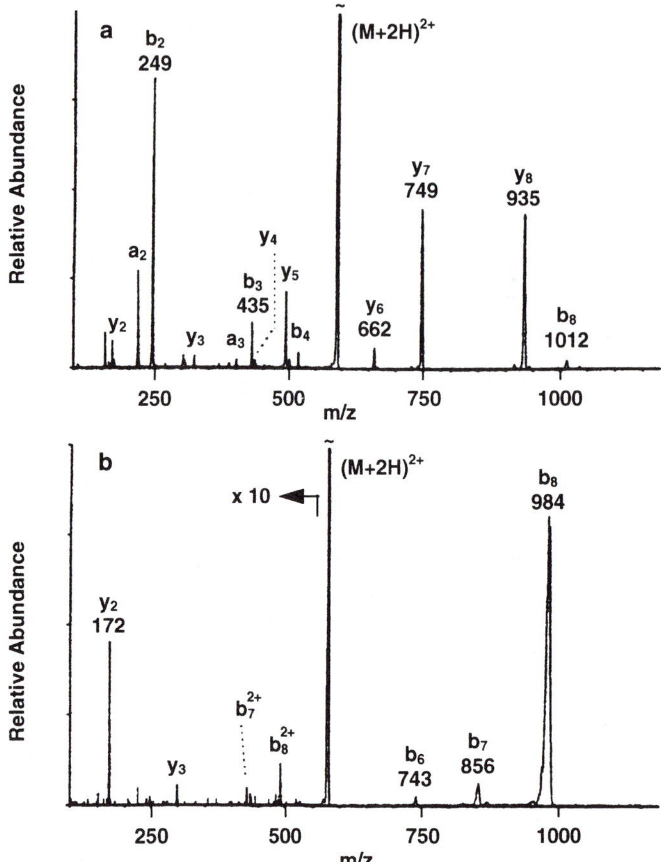

Figure 9.28 The ESI-tandem mass spectra of the $(M + 2H)^{2+}$ ions from (a) LH-RH at m/z 592 and (b) [Gln^8]-LH-RH at m/z 578. (© 1993, American Chemical Society. Reprinted from reference 90 with permission.)

y_{10} y_9 y_8

$(M+H)^+$ = 1192.4 1095.3 967.2

H_2N - **P** - C(=O) - NH - **K** - C(=O) - NH - **P** - C(=O) - NH - **Q** - Q - F - F - G - L - M - C(=O) - NH_2

98.1 226.3 323.4

b_1 b_2 b_3

$b_2 = H_2N$ - P - C(=O) - NH - K - C ≡ O^+ or K P′ = $(b_3\, y_9')_2$ Internal Fragment

Figure 9.29 Structure and partial mass assignment of substance P(2–11).

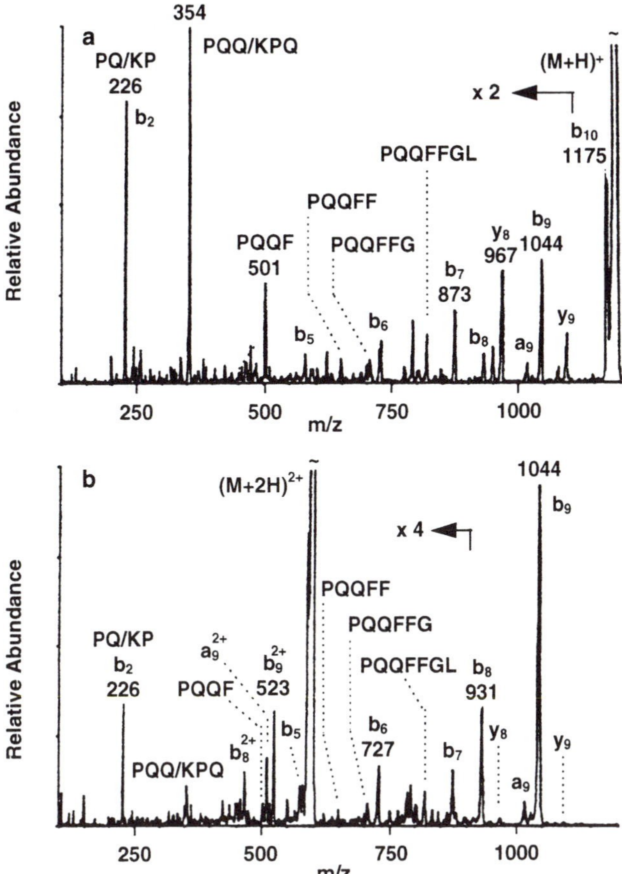

Figure 9.30 The ESI-tandem mass spectra of substance P(2–11) ions at (a) $(M + H)^+$ of m/z 1192.4 and (b) $(M + 2H)^{2+}$ of m/z 596.7. (© 1993, American Chemical Society. Reprinted from reference 90 with permission.)

neutral lysine is found, while in the $(M + 2H)^{2+}$ species, a charged lysine is produced. It is the charged nature of lysine in the K–P residue sequence that can cause the significant decrease in intensity of the b_2 and y_8 ions.

In the $(M + H)^+$ species, the amine terminus is the most likely site for protonation, and this leaves the Lys residue in the neutral state. However, in the $(M + 2H)^{2+}$, the Lys-2 residue must be the second site for protonation and provides a charge to the residue. This would cause the significant decrease in observed intensity for the b_2 and y_8 ions (Figure 9.30b). The above line of reasoning for cleavage at the N-terminal side of a Pro residue has another interesting caveat. When the Pro residue is near (second residue in peptide) the N-terminal residue, Pro is next to a charged species. For an N-terminal residue, the charge most likely is on the amine moiety, and this confers a charged, polar status to the residue. Thus, the X–P bond in ^+H_3N-X–Pro, where X is the N-terminal residue, may be influenced in a similar fashion as though X were a residue with a charged side-chain R-group. This appears to be the case, as Figures 9.31 and 9.32a and b show for the tryptic peptides[90] of Met-human growth hormone (Met-hGH)(1–9) and hGH(1–8). The amino acid sequence of Met-hGH(1–9) is MFPTIPLSR and that of hGH(1–8) is FPTIPLSR, and Figure 9.31 shows the structures where Pro is the third and second residue from the N-terminus, respectively. In Figure 9.32a, intense y_7 and y_7^{2+} ions are observed because of the cleavage of the F–P bond. The b_2 ion is present, albeit at a relatively low intensity. The intense y_7 and y_7^{2+} ions can be characterized by the neutral Phe next to the Pro residue. Note that these y ions are considerably more intense than the y_6 ion. However, in Figure 9.32b, the Pro is next to the charged N-terminus. The b_1 ion is absent, and the y_7 and y_7^{2+} ions are much lower in intensity than that of the Met-hGH species. As a note of interest, the y_6 ion at m/z 687 is much more intense in Figure 9.32b than that in Figure 9.32a.

PROBLEM 9.13 In the sequence of γ-endorphin, YGGFMTSEKSQTPLVTL, which complementary set of ions would be expected to be of either low or high relative abundance in a tandem mass spectrum of the $(M + 2H)^{2+}$ precursor?

PROBLEM 9.14 In the sequence EPMIGVNQELAYFYPELFR, a tryptic peptide, there are two Pro residues. What would be the relative intensities of at least one ion in each b, y ion pair for the amide bonds on the N-terminal side of each Pro residue?

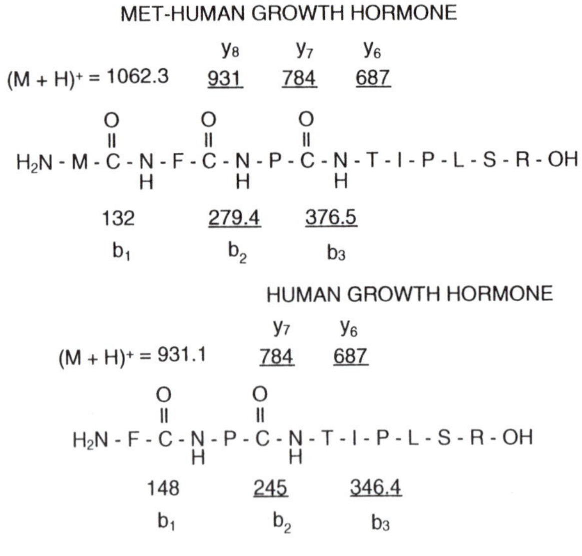

MET-HUMAN GROWTH HORMONE

y_8 y_7 y_6

$(M + H)^+ = 1062.3$ 931 784 687

$H_2N-M-C(=O)-NH-F-C(=O)-NH-P-C(=O)-NH-T-I-P-L-S-R-OH$

132 279.4 376.5

b_1 b_2 b_3

HUMAN GROWTH HORMONE

y_7 y_6

$(M + H)^+ = 931.1$ 784 687

$H_2N-F-C(=O)-NH-P-C(=O)-NH-T-I-P-L-S-R-OH$

148 245 346.4

b_1 b_2 b_3

Figure 9.31 Structures and partial mass analyses of Met-human growth hormone (Met-hGH) and human growth hormone (hGH).

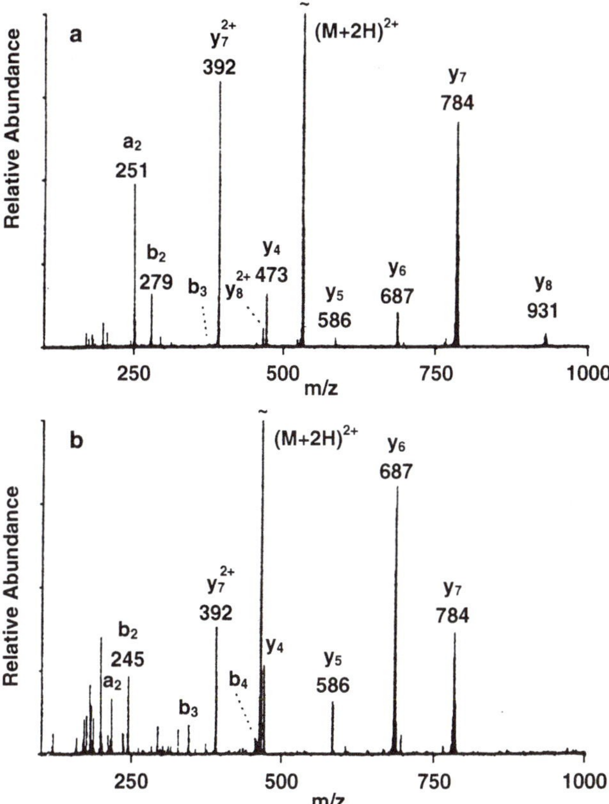

Figure 9.32 The ESI-tandem mass spectra of the $(M + 2H)^{2+}$ ions from (a) Met-hGH(1–9) at *m/z* 531.5 and (b) hGH(1–8) at *m/z* 466. (© 1993, American Chemical Society. Reprinted from reference 90 with permission.)

There is a phenomenon known as charge-site-remote fragmentation.[17,18,58,99] This is based on the observation that—given a stable charge in a molecule—the farther away a bond cleavage occurs from the charge site, the relatively higher in intensity is that product ion with respect to product ions which fragment closer to and retain the charged site.[15–18,58,93,100,101] This was originally observed in fatty acids,[102] but it has important ramifications in the ESI-tandem mass spectra of proteins. Doubly charged ions in a conventional ESI-mass spectrum are usually the result of protonation on the N-terminus and on a Lys/Arg/His residue. If these basic residues are at or near the C-terminus (e.g., tryptic peptide), the higher mass y_n ions exhibit pronounced intensities, because their proton-charged N-terminus is farther away from the protonated C-terminus with respect to the lower mass y_n ions. In addition, doubly charged y ions (y_n^{2+}) of the high-mass series are also observed. If the basic amino acids are near the N-terminus, then the higher mass b_n-ion series can dominate in the tandem mass spectrum.[17,58,103]

When a strongly basic residue is either in the middle of a peptide, generating an $(M + 2H)^{2+}$ ion, or not present in a peptide, yielding only an $(M + H)^{+}$ ion, the tandem mass spectrum does not usually produce a bias as to a particular ion series with respect to intensity. This arises by charge-site-initiated fragmentation and is characterized by the formation of complementary b and y ion pairs.[17,56,94] Alternatively, when the proton charge is on the N-terminus, to yield an $(M + H)^{+}$, the tandem mass spectrum yields predominantly b_n series ions in the high-mass region of the spectrum (*vide supra*).

Examples of this phenomenon can be observed for the pronounced high-mass y_n series (Figures 9.5a, 9.6a, 9.8, 9.12a, 9.14a, 9.15, 9.29, 9.53, and 9.55a), b_n series (Figures 9.23, 9.24, and 9.26), and for relatively equal y and b ion presence and intensity distributions (Figures 9.14a, 9.25, and 9.33a). The presence of proline in the peptide can alter the above-mentioned trends, and an example can be found in Figure 9.28b.

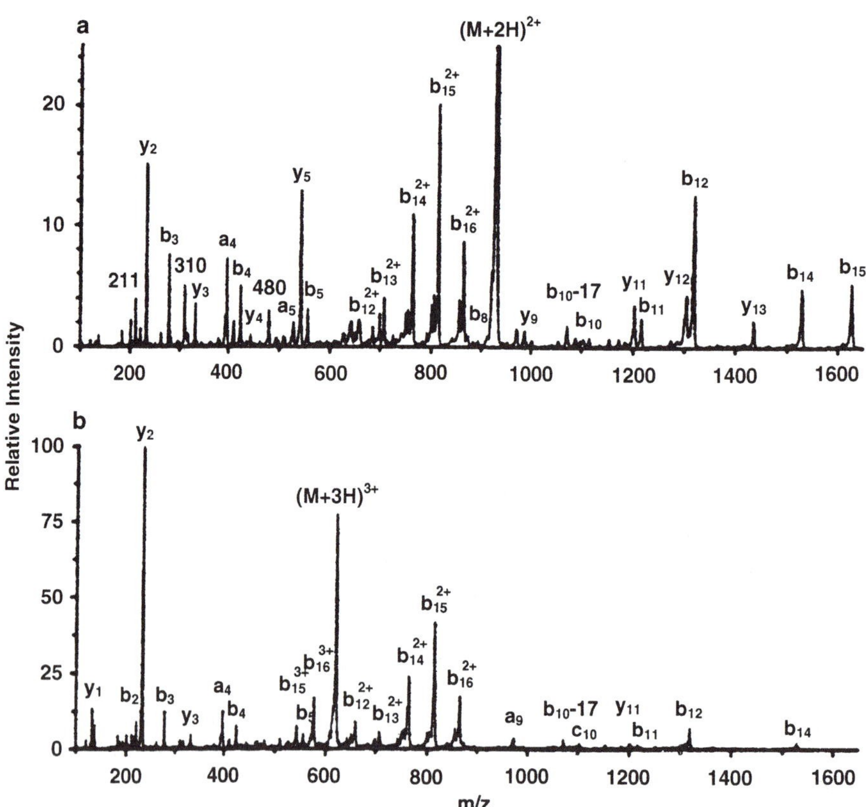

Figure 9.33 The ESI-tandem mass spectra of γ-endorphin for the following ions: (a) $(M + 2H)^{2+}$ at m/z 930.6 and (b) $(M + 3H)^{3+}$ at m/z 620.7. (© 1993, American Chemical Society. Reprinted from reference 17 with permission.)

PROBLEM 9.15 Figure 9.34 provides an ESI-mass spectrum of a peptide.[11] Provide a sequence for the substance and rationalize any proline effects. There may be useful, yet unlabeled, masses in Figure 9.34, and there are unlabeled doubly charged ions.

Mass Spectral Determination of High-Mass Protein Fragmentation: Low-Energy CID

The information presented up to this point deals with relatively small, sequence-manageable peptides. This section deals with the larger, much higher amino acid residue-number proteins. Usually, a protein is enzymatically or chemically degraded to small manageable peptides, while the intact protein is not subjected to CID. It has been stated[58,103–108] that it is difficult to reliably sequence peptide ions that are greater than 2500–3000 Da. Reasons for this include the rapid distribution of vibrational energy within the relatively large molecule, which leads to the dispersal of ion current over many fragment ions that have relatively low intensities. In addition, the charge is usually spread out over a number of multiply charged precursor ions (Chapter 4), as opposed to the peptide having only one or two charges. In this section, studies and experiments will be shown where, indeed, intact high-mass proteins are subjected to CID.

High-mass proteins offer higher charge state $(M + nH)^{n+}$ precursor ions in comparison with the $(M + H)^+$ and $(M + 2H)^{2+}$ ions of relatively low-mass peptides. Figure 9.35 shows ESI-tandem mass spectra of the $(M + 3H)^{3+}$ to $(M + 6H)^{6+}$ precursor ions (marked with an asterisk) of melittin, a protein with 26 amino acid residues and an average mass of 2846.5 Da.[91,109] In all four spectra, y- and b-type ions dominate. Abundant b ions appear to be fragments close to the N-terminus, and the y-type ions yield fragments that extend to

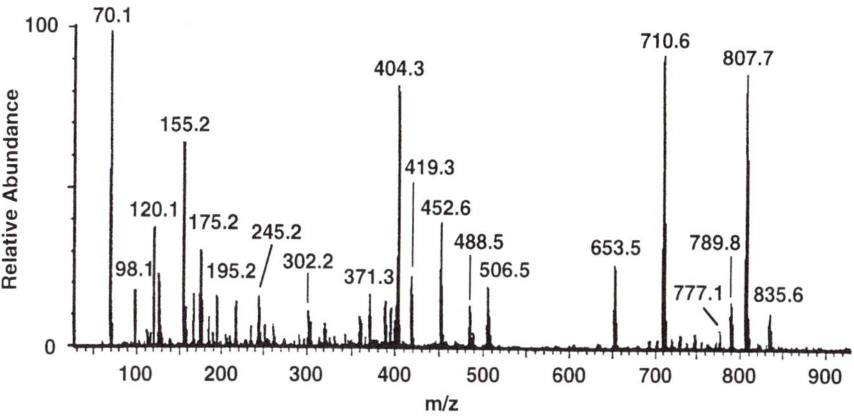

Figure 9.34 The ESI-tandem mass spectrum of a peptide from its $(M + 2H)^{2+}$ ion at m/z 452.6. (© 1995, *Mass Spectrometry Reviews*. Reprinted from reference 11 by permission of John Wiley & Sons, Inc.)

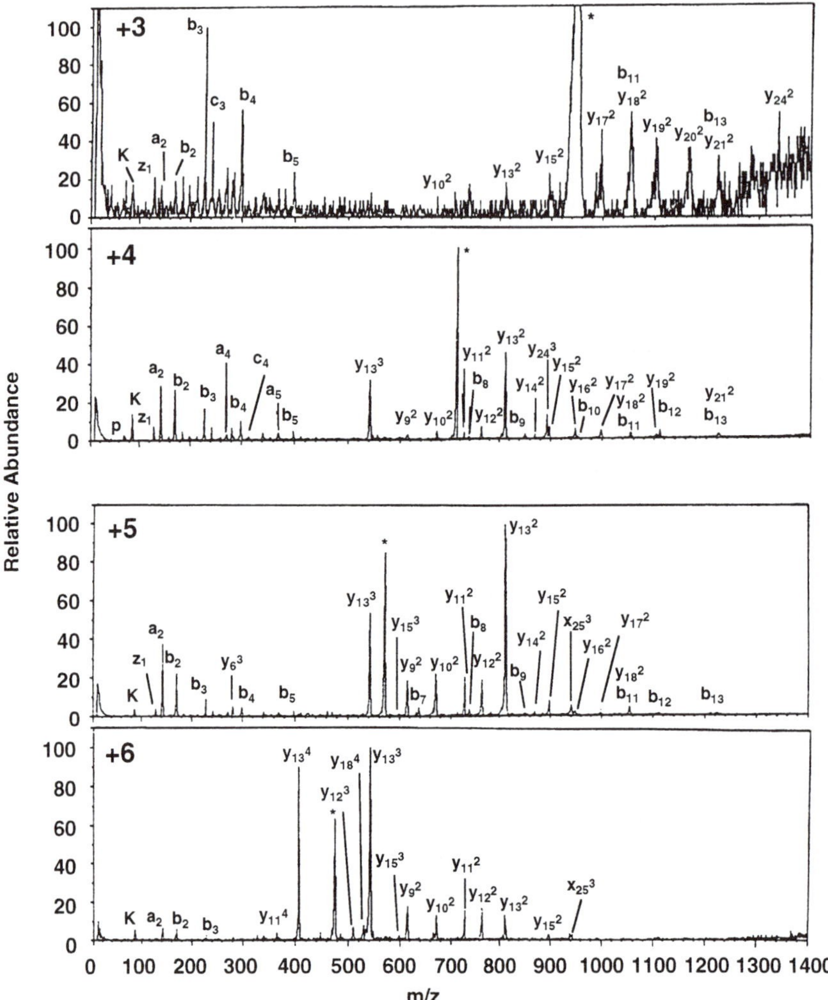

Figure 9.35 The ESI-tandem mass spectra of the 3+, 4+, 5+, and 6+ ESI-mass spectral peaks of melittin (cf. figure 9.36a). *, Precursor ion. Superscripts refer to the multiply charged positive-ion state. (© 1989. Reproduced from reference 91 by permission of John Wiley & Sons, Limited.)

the middle and close to the N-terminus. The b- and y-type ions that are indicative of the region near the C-terminus are absent; thus, only partial sequencing capabilities can be obtained.

Figures 9.36a–c once again feature melittin, but, in this case, fragmentation of the protein is promoted by heating the capillary metal tube (Figure 1.6) that transports and dries the ions from the ESI source to the first quadrupole.[109–112] As the tube is heated to higher temperatures, fragmentation of all the multiply charged precursor ions (Figure 9.36a) yields a series of y_n ions, as shown in Figure 9.36c.

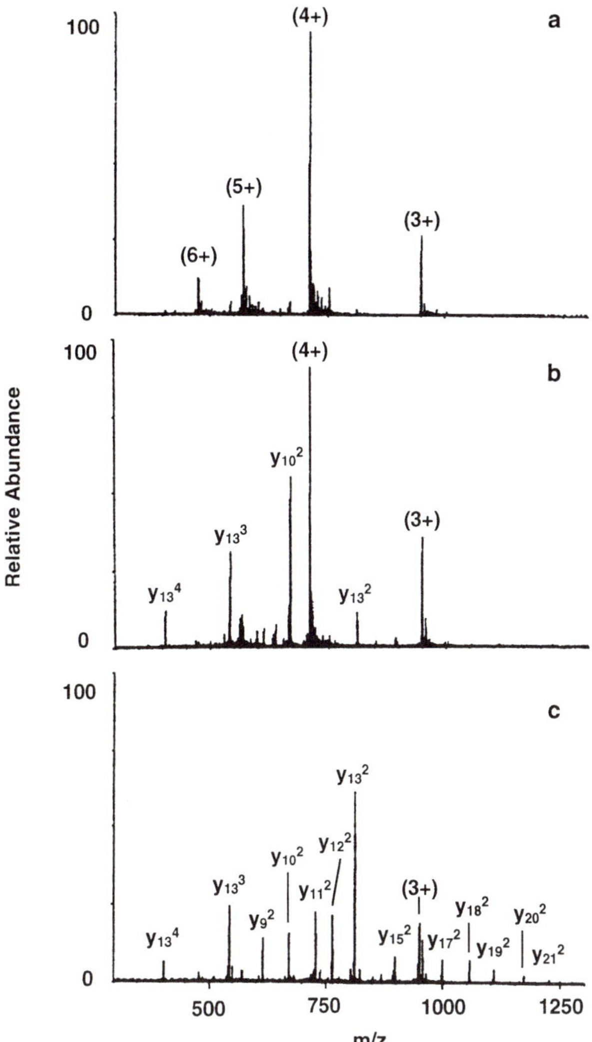

Figure 9.36 Thermally induced dissociation of the melittin multiply charged mass spectral peaks by heating the capillary metal tube between the ESI source and mass analyzer. Tube temperatures (heater power in watts) were (a) 200°C (25 W) with essentially no fragmentation, (b) 300°C (37 W), and (c) 400°C (49 W). (© 1991. Reproduced from reference 110 by permission of John Wiley & Sons, Limited.)

Bovine pancreatic ribonuclease A (RNase A) is a relatively large protein with an average mass M_r of 13,682 Da, 124 amino acid residues, and four disulfide bonds. Figures 9.37a and b show tandem mass spectra of the protein from the $(M + 12H)^{12+}$ and $(M + 13H)^{13+}$ precursor ions, respectively. Figures 9.37c and d show the same spectra as Figures 9.37a and b except that the four disulfide bonds were reduced. This procedure allows a protein to "open up" or

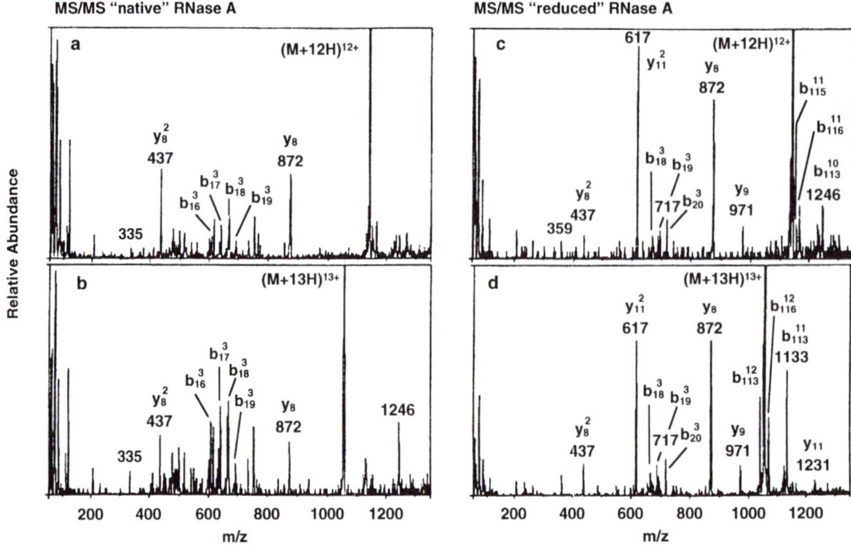

Figure 9.37 The ESI-tandem mass spectra of the (a) $(M + 12H)^{12+}$ and (b) $(M+13H)^{13+}$ mass spectral peaks of the intact RNase A protein and the (c) $(M + 2H)^{12+}$ and (d) $(M + 13H)^{13+}$ mass spectral peaks of the disulfide bond-reduced RNase A protein. Superscripts refer to the multiply charged positive-ion state. (© 1990, American Association for the Advancement of Science. Reprinted from reference 113 with permission.)

provide for a greater portion of the protein to be exposed to the solution. Essentially, very little sequence information is observed.[109,113] The b and y ions provide only a glimpse for relatively few residues near the N- and C-terminal ends of the protein.

Sheep serum albumin is a protein of M_r 66,385 Da and has approximately 580 amino acid residues. Figure 9.38 shows the extremely daunting task of sequencing the protein when using it as an analyte in the intact state.[114,115] Here, the protein was reduced with DTT (Table 9.7). Figure 9.38 shows the conventional multiply charged spectrum with a nozzle-skimmer potential difference (ΔNS) of 110 V. All charge states subsequently fragment by raising the nozzle-skimmer voltage to effect up-front CID (Figures 1.4, 1.9, and 1.10). Once again, only N-terminal fragments are found, and they are of the b-type ion series.

A question at this point can be raised about how the labeling of the mass spectral signals can be attained. Essentially, a computer program calculates all possible mass values, given the detailed sequence information. The program calculates the mass values for a complete set of 1+, 2+, 3+ product ions, and so on. These series of values are compared with the experimental record to find the best matches. Attempting to manually calculate and interpret spectra such as these would be nothing short of Herculean.

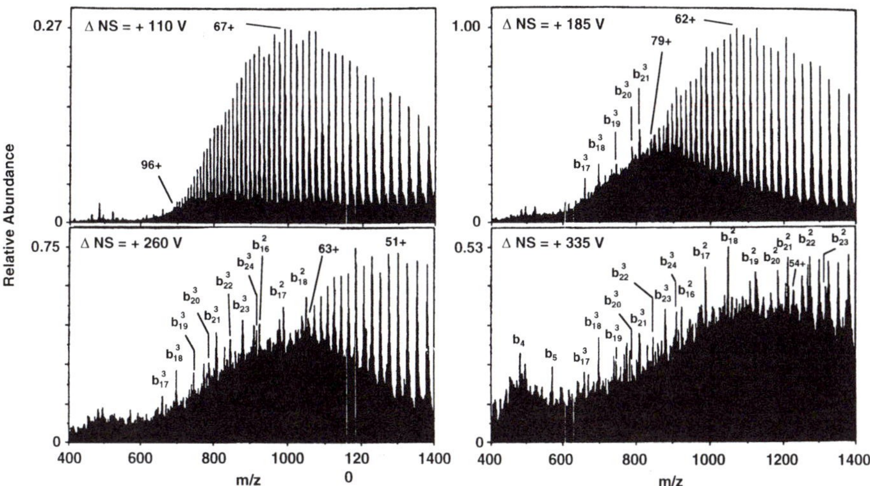

Figure 9.38 The ESI-up-front dissociation (CID) mass spectra of the multiply charged states of disulfide bond-reduced sheep serum albumin at the given nozzle-skimmer potential differences (ΔNS). Superscripts refer to the multiply charged positive-ion state. (© 1991, American Chemical Society. Reprinted from reference 114 with permission.)

Figure 9.38 provides tandem mass spectra for DTT-reduced sheep serum albumin that were derived from all the multiply charged precursor ions. Figures 9.39a and b present tandem mass spectra of, respectively, the $(M + 49H)^{49+}$ and $(M + 68H)^{68+}$ ions of the disulfide bond-reduced intact protein species. It is known that dissociation and fragmentation become more efficient as the charge state increases for a given molecule. The problem with this efficient fragmentation is that many more ions can be created as the charge state increases, and this tends to spread the ion current into many low-intensity charge states. Thus, the law of diminishing returns can be observed here. Figure 9.39b presents an example of this phenomenon, where the higher charge state provides essentially no useful tandem mass spectrum.[114] The $(M + 49H)^{49+}$ data provide a limited N-terminal sequencing capability. Figure 9.40 takes this one step further. The up-front CID causes the generation of a number of fragment ions (Figure 9.39a). The b_{21}^{3+} and b_{18}^{3+} ions were selected in the first quadrupole and subjected to CID. The spectra (Figure 9.40) indicate some structure and possible N-terminal sequencing capabilities.[114] A comparison of the ESI-up-front CID mass spectra of dog, sheep, and horse serum albumin is shown in Figures 9.41a–c, respectively.[114] All charge states of the molecular species participate in the fragmentation process, and the proteins had their four disulfide bonds intact. Doubly and triply charged ions of the same b_n value are present for the sheep serum albumin. The 5+, 4+, and 3+ b_n species are observed for horse serum albumin. However, only b ions are observed, and they are near the N-terminus.

Human hemoglobin (Hb) consists of four protein subunits, where two are the α-chain of 141 residues (M_r of 15,126.4 Da) and two are the β-chain (M_r of

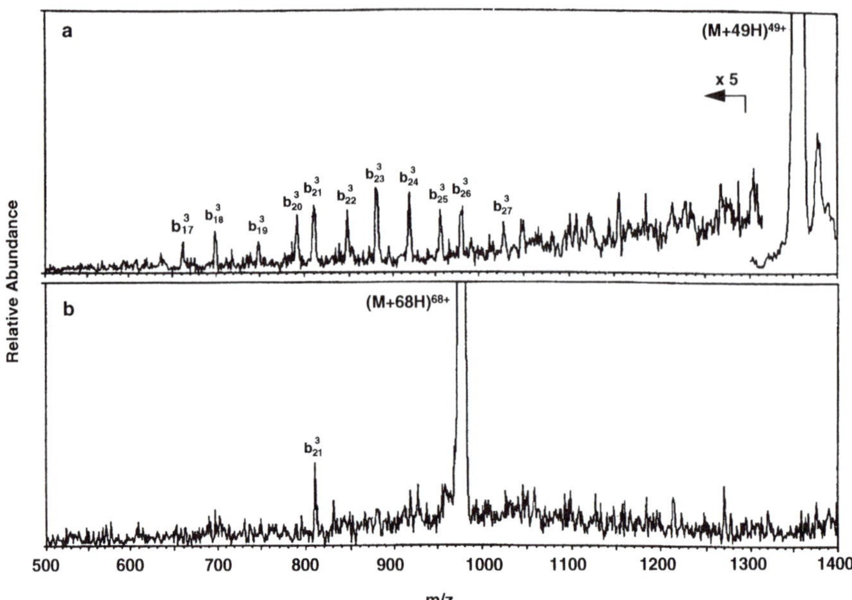

Figure 9.39 The ESI-tandem mass spectra of the (a) $(M + 49)^{49+}$ and (b) $(M + 68)^{68+}$ precursor ions from the disulfide bond-reduced sheep serum albumin. Superscripts refer to the multiply charged positive-ion state. (© 1991, American Chemical Society. Reprinted from reference 114 with permission.)

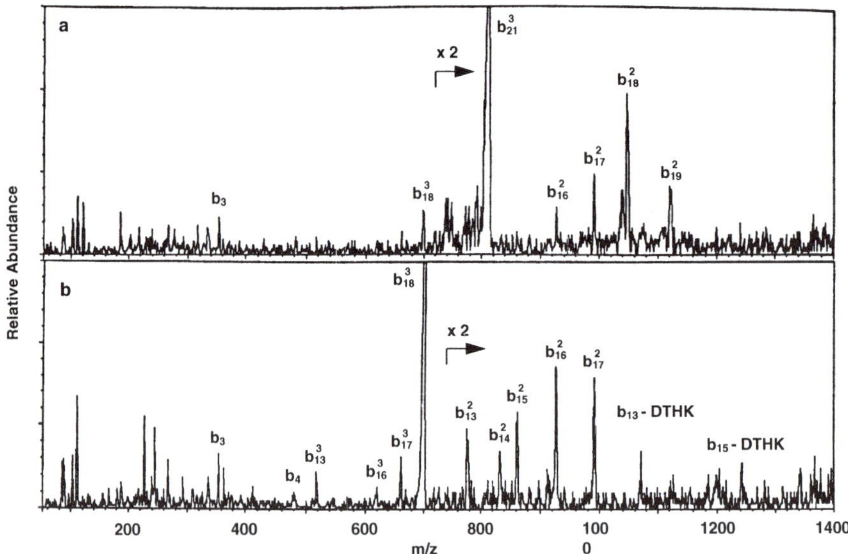

Figure 9.40 The ESI-tandem mass spectra of the up-front CID generated (a) b_{21}^{3+} and (b) b_{18}^{3+} ions from Figure 9.39a. ΔNS = +335 V, and two internal fragments can be observed at b_{13}-DTHK and b_{15}-DTHK. Superscripts refer to the multiply charged positive-ion state. (© 1991, American Chemical Society. Reprinted from reference 114 with permission.)

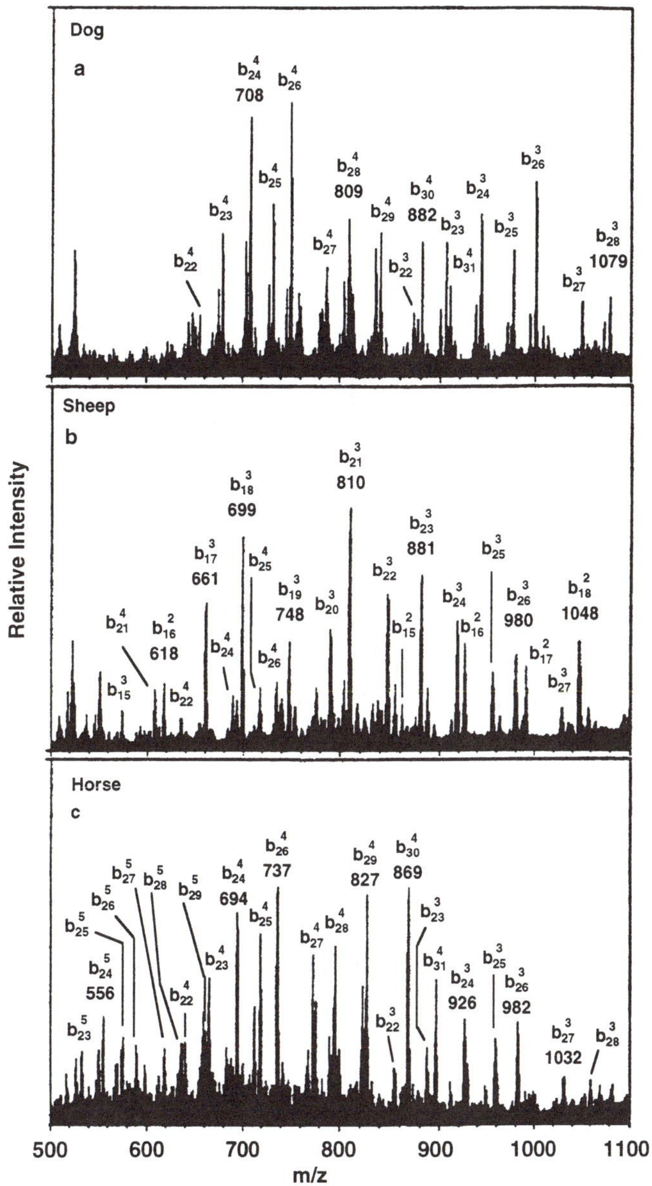

Figure 9.41 The ESI-up-front CID mass spectra of the entire multiply charged mass spectral ion envelope of the native, nonreduced serum albumin of (a) dog, (b) sheep, and (c) horse. Superscripts refer to the multiply charged positive-ion state. (© 1991, American Chemical Society. Reprinted from reference 114 with permission.)

15,867.2 Da). An FT-MS system was used to obtain tandem mass spectra (Figure 9.42) of four multiply charged precursor ions of the α-chain of apo-Hb. The apo designation refers to the protein without the heme group. A comparison of Figures 9.42a and b with Figures 9.42c and d shows quite different fragmentation processes.[116,117] The 11+ and 12+ spectra show sporadically

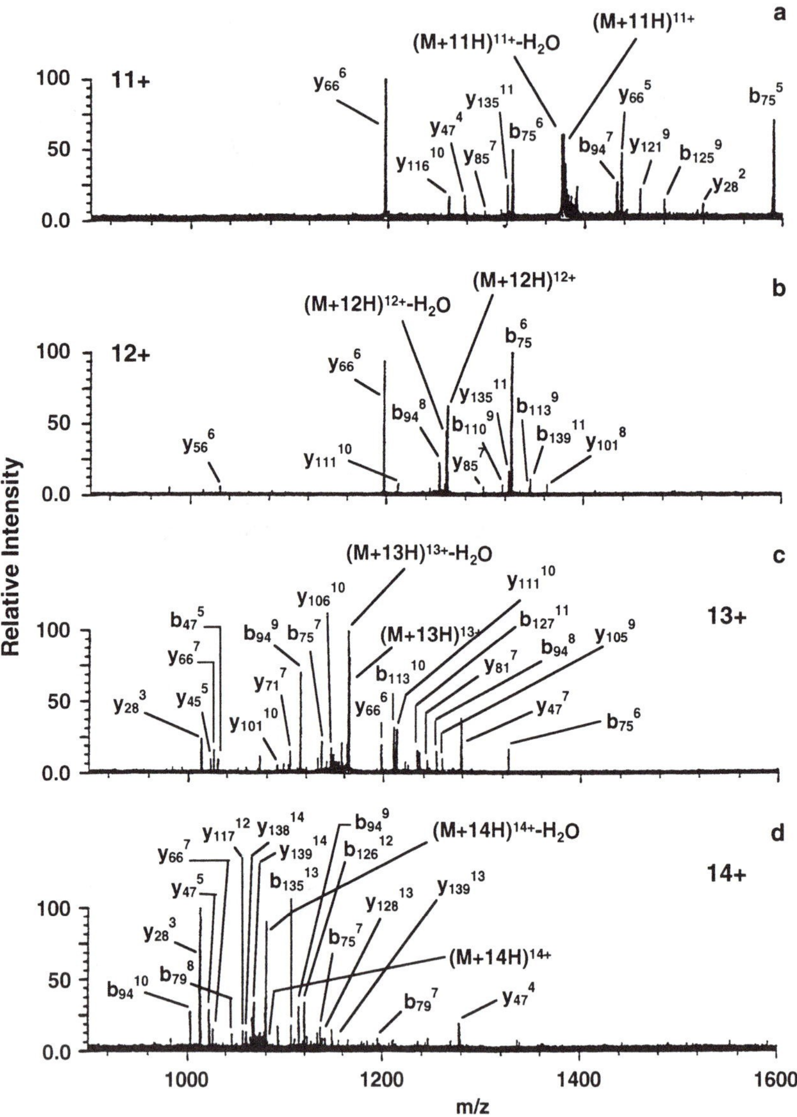

Figure 9.42 The FT-MS product-ion mass spectra of the α-chain of apo-human hemoglobin for the following ions: (a) $(M + 11H)^{11+}$, (b) $(M + 12H)^{12+}$, (c) $(M+ 13H)^{13+}$, and (d) $(M + 14H)^{14+}$. Superscripts refer to the multiply charged positive-ion state. (© 1994. Reproduced from reference 117 by permission of John Wiley & Sons, Limited.)

placed fragmentation throughout the protein. The same can be said of Figures 9.42c and d, except that a different fragmentation distribution is observed with respect to Figures 9.42a and b. Clearly, ion-type information, let alone extensive amino acid residue sequence information, cannot be reliably attained at the present time with intact, high-mass proteins. Only isolated portions of very large proteins can be reliably analyzed with tandem mass spectrometry.[90,91,114]

Mass Spectral Determination of Peptide Fragmentation: High-Energy CID

This chapter has presented tandem mass spectra of peptides and proteins obtained under relatively low-energy conditions. This is characterized by energy deposition in the precursor protonated molecules in the 100 eV energy region. Fragmentation generally can be ascribed to occur at various places along the peptide/protein polyamide backbone. The low-energy regions of energy deposition can be found to occur in quadrupole instruments, including linear and ion-trap devices, and in time-of-flight and FT-MS instruments. Ions that can be formed are of the a, b, c, x, y, z, immonium, and internal fragment type, as well as dehydration, NH_3, and CO losses.[16,56]

Sector instruments, which include electric and magnetic designs, can produce all ions that are observed in low-energy systems, as well as three more types of ions. Sector instruments typically supply energies to ions in the kilo-electron-volt range, and these energy domains are necessary in order to fragment the carbon–carbon bonds found in the side chains. Figure 9.43 presents the structure of these three types of ions, labeled d, w, and v. A common feature of each ion

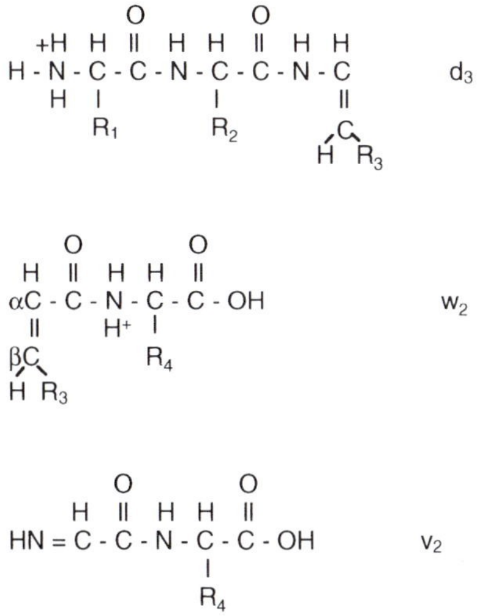

```
              O           O
   +H  H  ||  H  H  ||  H  H
 H - N - C - C - N - C - C - N - C          d3
     H   |           |           ||
         R1          R2         C
                               / \
                              H   R3

       O           O
 H    ||   H  H   ||
 αC - C - N - C - C - OH                    w2
 ||       H+  |
 βC           R4
 / \
H   R3

         O           O
     H   ||   H  H   ||
 HN = C - C - N - C - C - OH                v2
                  |
                  R4
```

Figure 9.43 Peptide fragment ions formed from sector high-energy collision: d, w, and v ions.

type is that cleavage and fragmentation occur on the side-chain (R) group of the amino acid residue.[3,4,16,19,23,58] Note that the d and w ions are complementary in that the d ion is produced from the β–γ carbon cleavage of a side-chain R-group and the d ion retains the N-terminal residue. This is shown in Scheme 9.5 for Leu and Ile.[3,4,15,118–120] The Leu residue can form only one d ion, but Ile can form two types of d ions. Note that the structures of the d_3 ion for Leu and the d_{3a} and d_{3b} ions for Ile in Scheme 9.5 are all different, and hence have different masses. Thus, high-energy CID can distinguish between the Leu and Ile isomers! These fragmentation schemes can be constructed for the other amino acids. Scheme 9.6 shows the same information as Scheme 9.5 except that a peptide ion is obtained

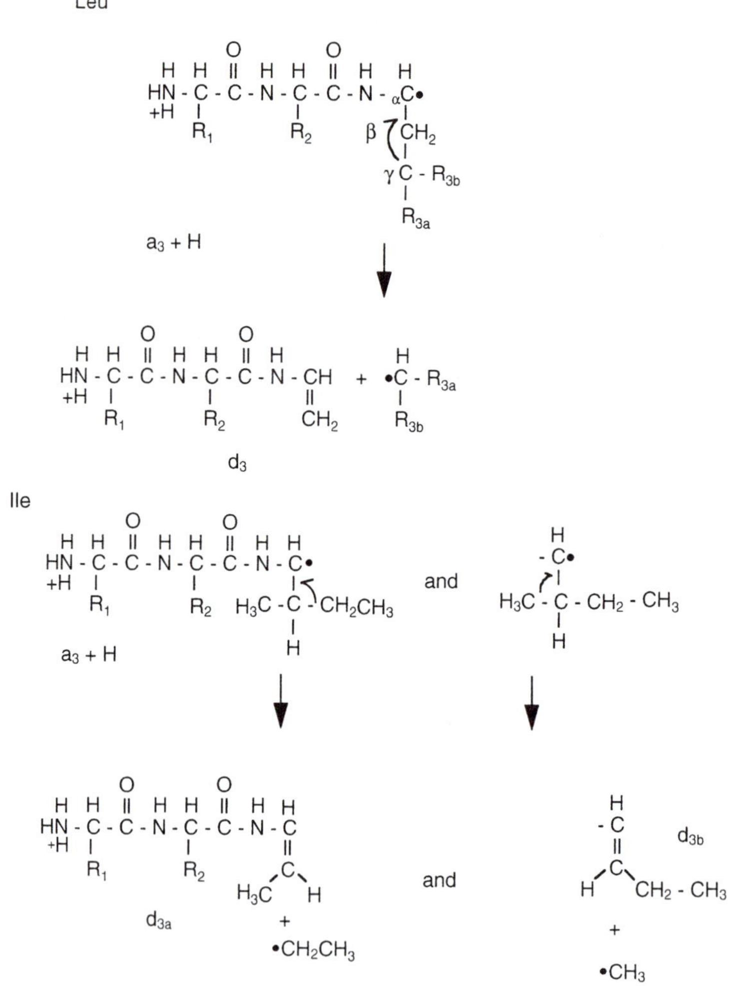

Scheme 9.5 Two examples of the mechanisms of formation of d ions from peptide fragmentation.

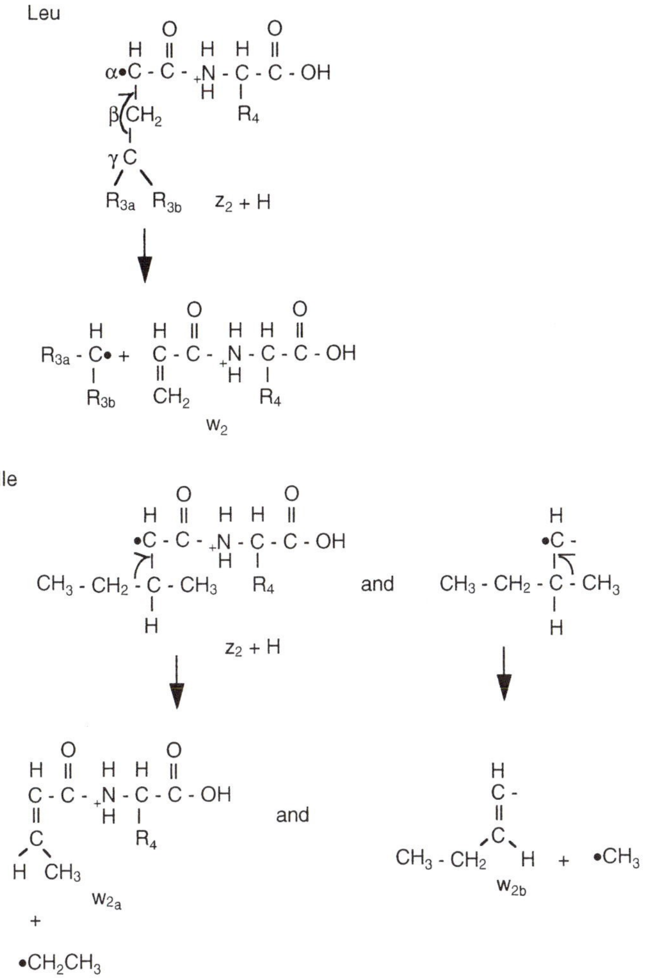

Leu

α•C(H) - C(=O) - $_{+}$N(H) - C(H)(R_4) - C(=O) - OH

β CH_2

γ C(R_{3a})(R_{3b})

z_2 + H

R_{3a} - C•(H)(R_{3b}) + C(H)(=CH_2) - C(=O) - $_{+}$N(H) - C(H)(R_4) - C(=O) - OH

w_2

Ile

•C(H) - C(=O) - $_{+}$N(H) - C(H)(R_4) - C(=O) - OH

CH_3 - CH_2 - C(H) - CH_3

z_2 + H

and

H •C -

CH_3 - CH_2 - C(H) - CH_3

C(H)(=C(H)CH_3) - C(=O) - $_{+}$N(H) - C(H)(R_4) - C(=O) - OH

w_{2a}

+ •CH_2CH_3

and

H C - = C(CH_3 - CH_2)(H) + •CH_3

w_{2b}

Scheme 9.6 Mechanisms of formation of w ions from peptide fragmentation.

that has the C-terminal residue, and this d ion complementary fragment is labeled a w ion. The same differentiation of Leu and Ile applies as in Scheme 9.5. The d ions originate from $(a_n + H)^+$ ions, and the w ion is formed from a $(z_n + H)^+$ ion. The last fragment, the v ion, arises from a y″ ion, and is formed by a cleavage of the entire R-group from a residue (Scheme 9.7). Thus, this ion retains the C-terminal residue portion of the precursor peptide. Apparently, there is no complement to the v-type ion, which retains the N-terminal sequence. Table 9.15 provides compact mathematical expressions for these ions to aid in the interpretation of high-energy sector ESI-tandem mass spectra of peptides.[16]

There are documented exceptions where d ions are observed in the low-energy quadrupole regime.[121] Intrachain/intraionic acid–base proton transfers

```
          O           O
  H  H    ||  H  H    ||
H - N - C - C - N - C - C - OH
  +  |    |           |
  H    R3             R4

            y2''

             ↓

                  O           O
         H   H    ||  H  H    ||
HR3 + HN = C - C - N - C - C - OH
         +                |
                          R4

              v2
```

Scheme 9.7 Mechanism of formation of v ions from peptide fragmentation.

from an N-terminal Arg and an interior acidic residue can provide a low-energy pathway for the formation of the d ion. The resulting d ion consists of a protonated N-terminal Arg, where the proton resides on the side-chain amine group. The acidic residue (glutamic, aspartic, or cysteic acid) loses carbon dioxide to form an unsaturated side chain.

Mass Spectral Analysis of the Signal Recognition Particle

The example presented here originates from the signal-recognition particle (SRP), which is composed of seven proteins and RNA.[14,122] This particle is necessary to cause certain proteins to arrive at the endoplasmic reticulum membrane of cells. The proteins were isolated by SDS-PAGE, and one of the proteins was digested with trypsin in the gel. The resulting peptide mixture was extracted from the gel and subjected to HPLC-ESI-mass spectrometry. Figure 9.44 shows the HPLC analysis, and the peptide chosen for tandem mass spectrometry is marked with an arrow. Figure 9.45 shows that this peak is composed of two peptides (SRP-66 and SRP-68) of 1003.09 and 1501.88 Da average mass, respectively. The 1004.1-Da $(M + H)^+$ peak was subjected to tandem mass spectrometry, and the spectrum is displayed in Figure 9.46. The sequence of the

Table 9.15 Equations for the calculation of fragment ions[a]

$$z_n + 1 = (y''_n - 15) - 1$$
$$w_n = (y_n - 15) - 1 - (\beta, \gamma \text{ moiety})$$
$$= z_n + 1 - (\beta, \gamma \text{ moiety})$$
$$a_n + 1 = (b_n - 28) + 1$$
$$d_n = (b_n - 28) + 1 - (\beta, \gamma \text{ moiety})$$
$$= a_n + 1 - (\beta, \gamma \text{ moiety})$$
$$v_n = (y''_n - 1) - \text{R group}$$

[a] β, γ moiety refers to the entire side-chain R group removed by scission of the β–γ bond.

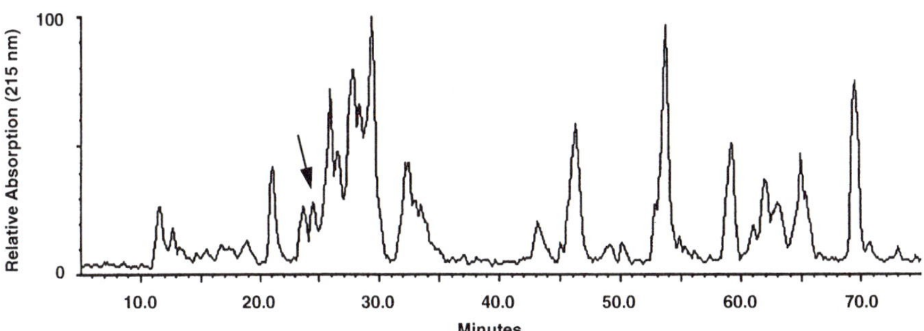

Figure 9.44 Absorption spectrum at 215 nm of an RP-HPLC elution of a trypsin digestion of two unresolved proteins (SRP-66 and SRP-68) from the yeast signal-recognition particle (SRP). The numbers 66 and 68 indicate the mass of the SRP proteins in kilodaltons. (© 1995. Reprinted from reference 122 by permission of Academic Press, Inc., Orlando, Florida.)

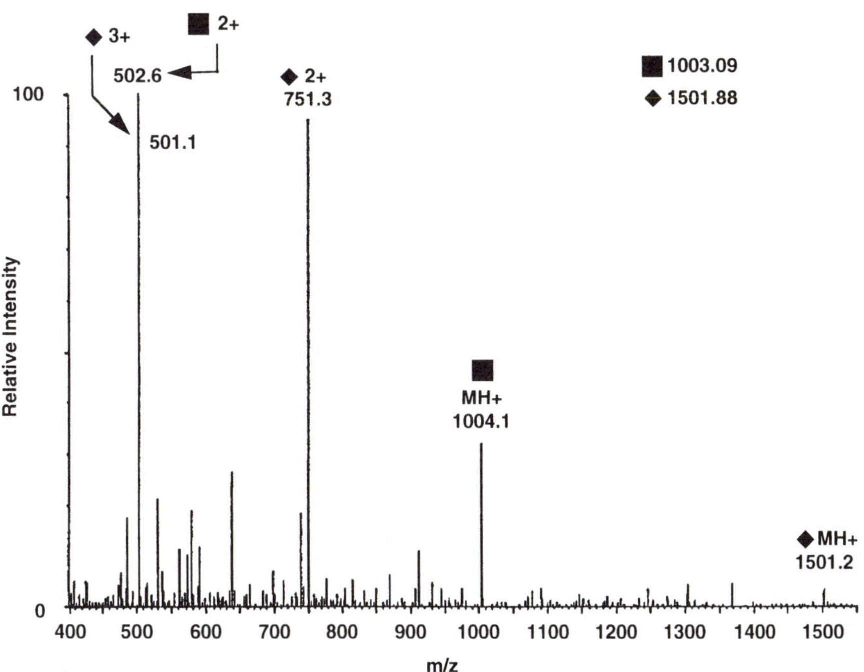

Figure 9.45 The ESI-mass spectrum of the indicated RP-HPLC tryptic peptide peak (arrow) in Figure 9.44. (© 1995. Reprinted from reference 122 by permission of Academic Press, Inc., Orlando, Florida.)

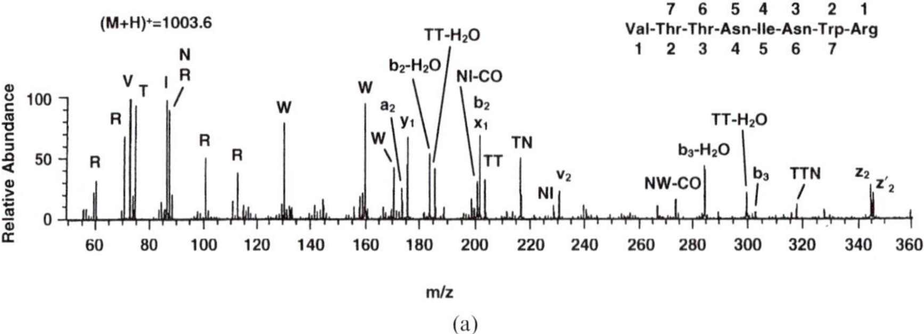

(a)

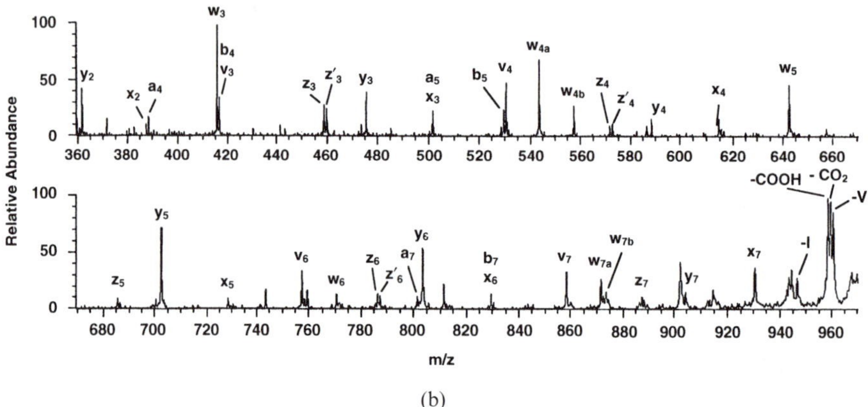

(b)

Figure 9.46 High-energy product-ion mass spectrum of the SRP-66 tryptic peptide in Figure 9.45 at an $(M + H)^+$ of m/z 1004.1. (© 1995. Reprinted from reference 122 by permission of Academic Press, Inc., Orlando, Florida.)

octapeptide is Val-Thr-Thr-Asn-Ile-Asn-Trp-Arg, and the backbone cleavages are detailed in Figure 9.47. The complete sequence can be obtained from the spectrum and is mainly due to the y series of ions. The a, b and x, y, z mass values that appear in the spectrum are listed below and above the peptide sequence, respectively. Low-mass immonium ions are observed for Arg, Val, Thr, Ile, and Trp (Figure 9.46). A couple of internal fragment ions can be observed, as well as dehydration fragments from backbone and internal fragment ions. Internal fragment ions with loss of CO can be noted. Also, some of the z_n ions appear to be in the z and the z′ form, where the z retains one of the two hydrogen atoms ($z' = z + H$) from the y''_n precursor (compare Tables 9.3 and 9.15). Importantly, in this sector tandem mass spectrum, w and v ions can be observed. The following discussion details the calculations of the ions following the equations shown in Table 9.15.

931.0	829.8	728.7	614.7	501.5	387.4	201.2	x
905.0	803.8	702.7	588.7	475.5	361.4	175.2	y
888.0	787.8	685.7	573	459.5	345.4		z′
	786.8		572	458.5	344.4		z

y_7 y_6 y_5 y_4 y_3 y_2 y_1

H₂N - **V** - C(=O) - N(H) - **T** - C(=O) - N(H) - **T** - C(=O) - N(H) - **N** - C(=O) - N(H) - **I** - C(=O) - N(H) - **N** - C(=O) - N(H) - **W** - C(=O) - N(H) - **R** - C(=O) - OH

b_2 b_3 b_4 b_5 b_7

	b_2	b_3	b_4	b_5	b_7
a	173		388.4	501.6	801.9
b	201.2	302.3	416.4	529.6	829.9

Figure 9.47 Amino acid sequence of the tryptic peptide of SRP-66 (marked with an arrow in Figure 9.44) and an $(M + H)^+$ of 1004.1 Da (Figure 9.45). Ion identification and mass information are presented.

In each calculation for a w_n and v_n ion, the high-mass value is that of the equivalent y_n ion.

v_2
361.4
−1
360.4
−130 Trp
230.4

v_3
475.5
−1 Asn
474.5
−58 $H_2C{-}C(=O){-}NH_2$
416.5

v_4
588.7
−1
587.7 Ile
−57.0 $HC(CH_3)(CH_2CH_3)$
530.7

v_6
803.9
−1
802.9 Thr
−45 $HC(CH_3)OH$
757.9

v_7
905.0
−1
904.0
−45 Thr
859.0

w_3
475.5
−15
460.5
−1.0
459.5 $= z_3 + 1 = z'_3$
−44 Asn
415.5

w_4
588.7
−15
573.7
−1 Ile
572.7
−29 $H_2C{-}CH_3$
543.7 w_{4a}

572.7
−15 CH_3
557.7 w_{4b}

w_5
702.8
−15
687.8
−1
686.8 Asn
−44 $O{=}CNH_2$
642.8

w_6		w_7	
803.9		905	
−15		−15	
788.9 Thr		890 Thr	
−1		−1	
787.9	787.9	889	889
−17 OH	−15 CH_3	−17	−15
770.9 w_{6a}	772.9 w_{6b} negligible	872 w_{7a}	874 w_{7b}

The spectrum in Figure 9.46 also shows that a number of ions have the same nominal mass: NI–CO, b_2, $x_1 = 201$; b_4, $v_3 = 416$; a_5, $x_3 = 501$; b_7, $x_6 = 829$. Also, some ions are separated by 1 Da: z_2, z_2'; x_2, a_4; w_3, (b_4, v_3); z_3, z_3'; b_5, v_4; z_4, z_4'; z_6, z_6'. The peptide is a tryptic peptide, and, as such, it displays intense y_n ions (y_6 and y_5) in the high-mass region.

Mass Spectral Analysis of Human Pancreatic Polypeptide

In a previous section, quadrupole mass spectrometry was shown to provide incomplete spectra with respect to protein sequencing when the molecule became relatively high in mass. The present example shows that, with sector instruments, it is possible to generate an ESI-tandem mass spectrum of a 36-amino acid residue protein that yields enough information to provide a complete sequence. Figures 9.48a–d provide the mass spectrum of the $(M + 5H)^{5+}$ precursor ion at m/z 837.3 of human pancreatic polypeptide (M_r of 4181.7 Da). Figure 9.49 identifies the ions that are observed, and Figure 9.50 provides the numerical values for the y- and b-series ions.[123] Note that the protein has a modified C-terminus. In addition, selected d, w, and v ion mass values are calculated. The interpretation of this spectrum yields ions from 1+ to 4+ charge states. The deconvolution and charge-state assignment of the peaks in the spectrum (Figure 9.48) were accomplished by setting the resolution on the third quadrupole to 3600 full width at half-maximum peak height. Baseline resolution could not be achieved because of relatively low ion counts, which produce less-than-desired ion statistics. However, this resolution was sufficient for counting the number of isotope peaks in a 1-Da increment. With a knowledge of the sequence already in hand, and the charge states of the product ions, the sequencing procedure was analogous to that of a mass spectrum from an $(M + H)^+$ species. The generous number of peaks available in the spectrum for sequencing purposes takes its toll on the peak intensities. The signal is spread out to the extent that many peaks have very low relative intensities. Figure 9.49 provides an insightful view of the multiple-charging phenomenon. No lysine residues are in the protein, but a number of arginine residues are present. Let us focus on these Arg species and keep track of the charge. Starting from the C-terminus, the y ions (y_1, y_2) are singly charged, including the Arg-35 residue. When Arg-33 is encountered, the ions become doubly charged. As the y-ion fragments increase in mass, only the

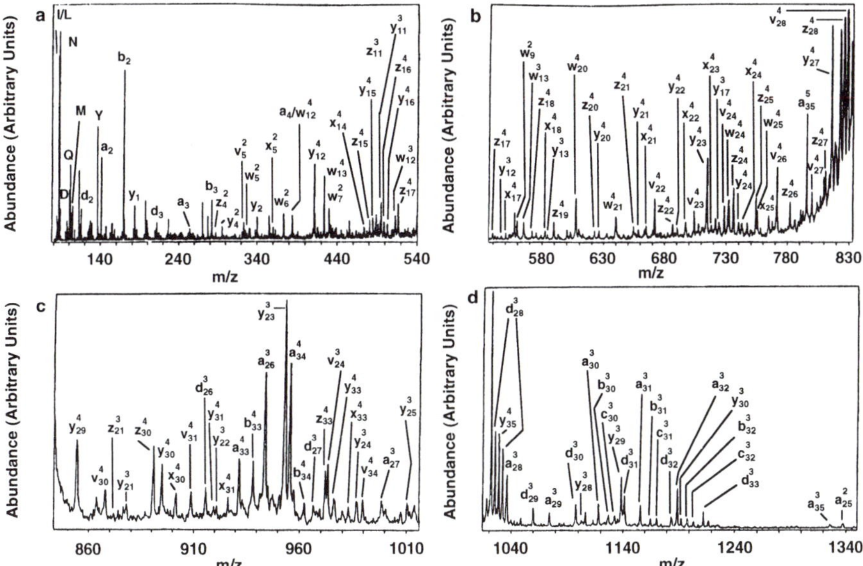

Figure 9.48 The ESI-high-energy tandem mass spectrum of the $(M + 5H)^{5+}$ ion at m/z 837.3 of human pancreatic polypeptide. Superscripts refer to the multiply charged positive-ion state. (© 1995, American Society for Mass Spectrometry. Reprinted from reference 123 by permission of Elsevier Science, Inc.)

doubly charged forms are observed up to Arg-26. Then, we see one set of ions that are triply charged (y_{11}^3 and z_{11}^3). The next residue is another arginine at Arg-25, and this residue produces 4+ charge states in the ensuing fragments. All the rest of the y and related series of ions, from amino acid residue 25 to 2, are quadruply charged. The entire peptide then becomes quintuply charged because of the N-terminus residue. This trend is also observed in the b and related series of ions. Even the Arg^{25}–Arg^{26} residues each have a proton; this strongly suggests that the proton affinity of an Arg residue is energetically favorable with respect to the electrostatic repulsion of the adjacent proton-charged residues.[123]

In interpreting the information given in Figure 9.50, a conservative mathematical approach was used, where one starts with a singly charged species at a specific ion label and residue value (n), and the final step is conversion to the multiply charged species. For example, in deducing the mass of the w_{24}^4 ions, the initial mass value is the equivalent y_{24} value. By following the equation in Table 9.15, the singly charged values are obtained. These values then can be converted into the 4+ charge-state value. Note that Thr can yield two w ions; however, the 4+ charge state yields two fragments that are separated by only 0.5 Da. The d_3 ion follows another equation (Table 9.15) starting from the b_3 value. Note that for the calculation of a "d" or "w" ion, the initial b and y ion masses do not have to be present in the tandem mass spectrum, and a similar reasoning can be applied to a v ion. The equations in Table 9.15 are merely aids in arriving at an observed ion. The Ile-28 residue provides an interesting analysis of the comple-

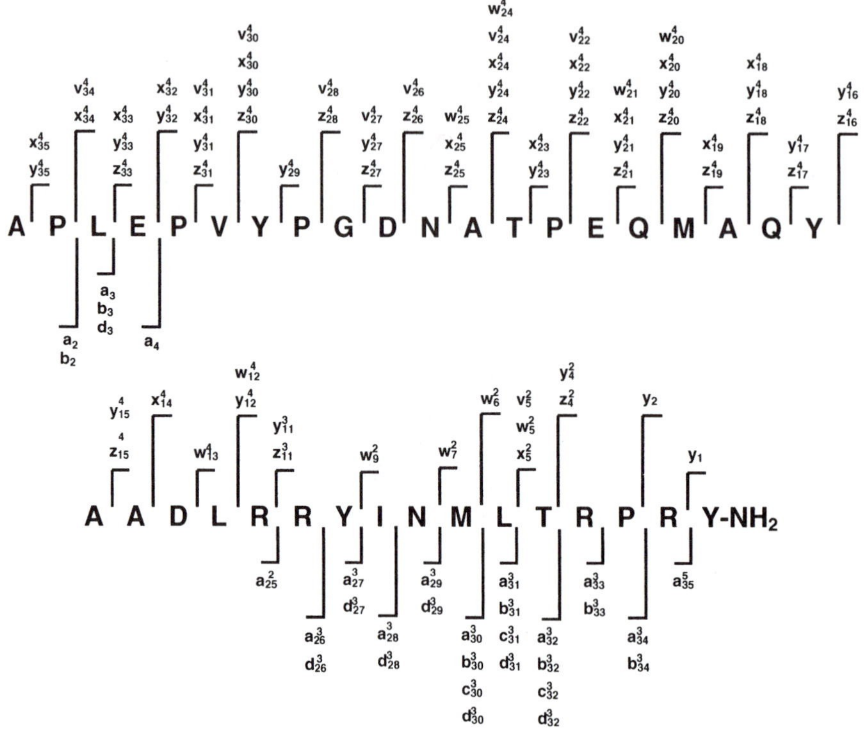

Figure 9.49 Complete sequence of human pancreatic polypeptide and structural nomenclature from Figure 9.48. Superscripts refer to the multiply charged positive-ion state. (© 1995, American Society for Mass Spectrometry. Reprinted from reference 123 by permission of Elsevier Science, Inc.)

mentary w_9, d_{28} ion pair. The d species is observed for both ethyl and methyl fragmentations (d^3_{28a} and d^3_{28b}), but only the lower mass (loss of ethyl) is observed for the w ion; namely, w^2_{9a}.

PROBLEM 9.16 Using Figures 9.48–9.50 as guides, calculate the mass values of the following v ions: v^2_5, v^4_{22}, v^2_{28}, v^4_{30}, and v^4_{31}. Calculate the masses of x^4_{23}, y^3_{17}, v^4_{24} and w^4_{24}, v^4_{31} and compare their values.

Mass Spectral Analysis of Tyr-corticotropin Releasing Factor

Figures 9.51a–d present high-energy tandem mass spectra of the $(M + 6H)^{6+}$ precursor ion at *m/z* 821.1 of Tyr-corticotropin releasing factor (human recombinant); this has an M_r of 4920.7 Da and comprises 42 amino acid residues.[123] Figure 9.52 provides the sequence and the observed ions in the mass spectrum. Surprisingly, enough mass spectral information is present in order to provide a

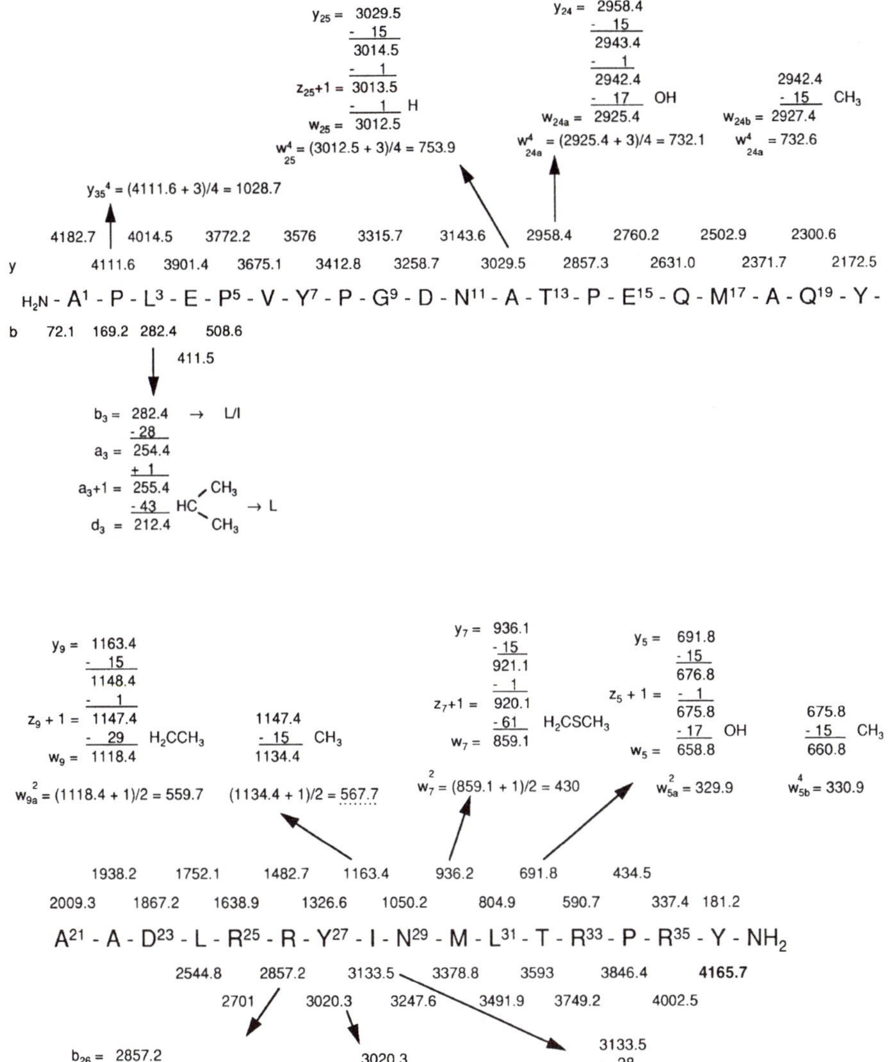

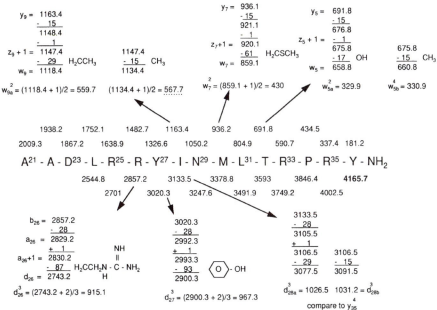

Figure 9.50 Sequence, residue mass values, and selected fragment ion calculations for the peptide in Figure 9.49. (© 1995, American Society for Mass Spectrometry. Reprinted from reference 123 by permission of Elsevier Science, Inc.)

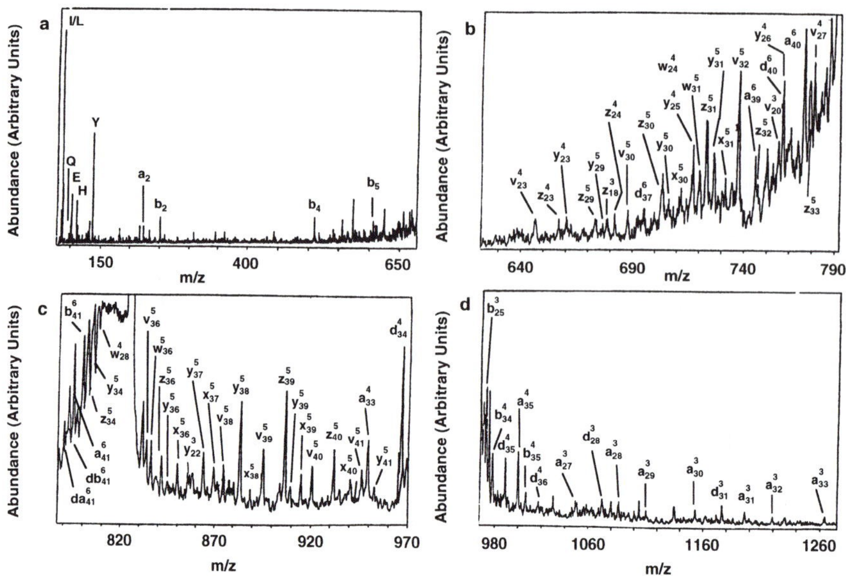

Figure 9.51 The ESI-high-energy tandem mass spectrum of the $(M + 6H)^{6+}$ ion at m/z 821.1 of human recombinant Tyr-corticotropin releasing factor. Superscripts refer to the multiply charged positive-ion state. (© 1995, American Society for Mass Spectrometry. Reprinted from reference 123 by permission of Elsevier Science, Inc.)

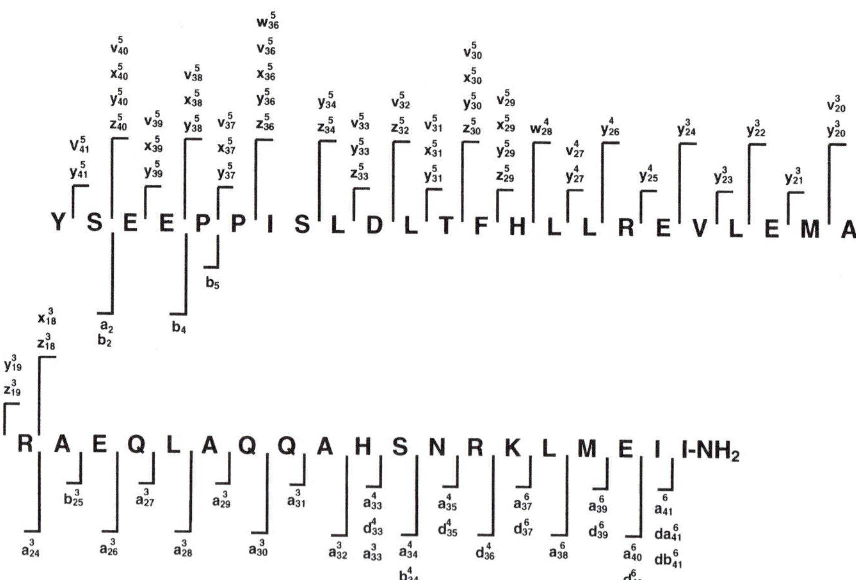

Figure 9.52 Sequence and structural nomenclature from Figure 9.51 of human Tyr-corticotropin releasing factor. (© 1995, American Society for Mass Spectrometry. Reprinted from reference 123 by permission of Elsevier Science, Inc.)

nearly complete sequence. The arginine, lysine, and histidine residues roughly provide a match between their distribution and the transition of different charge states on the product ions.

PROBLEM 9.17 (a) Given the observed ions in Figure 9.52, can the Ile^7–Ser^8 residues be unambiguously identified and sequenced? If not, rationalize as to the extent this region can be sequenced.
(b) Compare the mass differences of the following ions: d^6_{41a}, d^6_{41b}, a^6_{41}, z^5_{34}, y^5_{34}, and w^4_{28}.

Supplementary Problems

It is the intention of this addendum to Chapter 9 to provide supplementary problems and further examples of the sequencing of proteins based on ESI-tandem mass spectral characteristics. Very little guidance is provided, and some of the answers are not provided. Instead, the reader is referred to the literature for the solution to the following problems. This approach provides a good exercise, along with a greater familiarity with the literature and reasons and applications for peptide investigations.

PROBLEM 9.S1 Figures 9.S1(a–e) have a common residue in the peptide sequences, as well as a common observation with respect to fragmentation patterns. In addition, this residue has a propensity to highlight a particular ion series. The $(M + 2H)^{2+}$ and $(M + H)^+$ ions were the precursor ions in Figure 9.S1(a–d) and Figure 9.S1e, respectively, and immonium and internal fragment ions can be found. Deduce the sequence of all five peptides, and then provide the information as to the residue and fragmentation pattern characteristic. The explanation for the observed fragmentation pattern with respect to the amino acid residue in question can be found in Chapter 9. References 18 and 124 provide solutions for Figures 9.S1(a–c) and 9.S1d, respectively. Figure 9.S1e can be found in reference 125.

PROBLEM 9.S2 Figure 9.S2 provides a mass spectrum of a peptide opiate agonist.[88] Provide the sequence of the peptide.

PROBLEM 9.S3 Figure 9.S3 shows the mass spectrum of a small peptide[125] which is a hormone that has a very significant influence in the human body. Despite the relatively simple spectrum and low mass, the sequencing process is difficult. As a matter of fact, the methyl derivative of this peptide was one of two Symposium Test Peptides-4 (STP-4) of the Protein Society.[126]

PROBLEM 9.S4 Figure 9.S4 presents an ESI-tandem mass spectrum of a peptide with three Pro residues.[52] Deduce the sequence, making use of Table 9.6, and apply the "proline effect" to the spectrum. Rationalize the results. There is more than one proline effect in the mass spectrum.

PROBLEM 9.S5 Figure 9.S5 presents an ESI-tandem mass spectrum of a peptide.[18] Deduce the sequence and rationalize any proline effect.

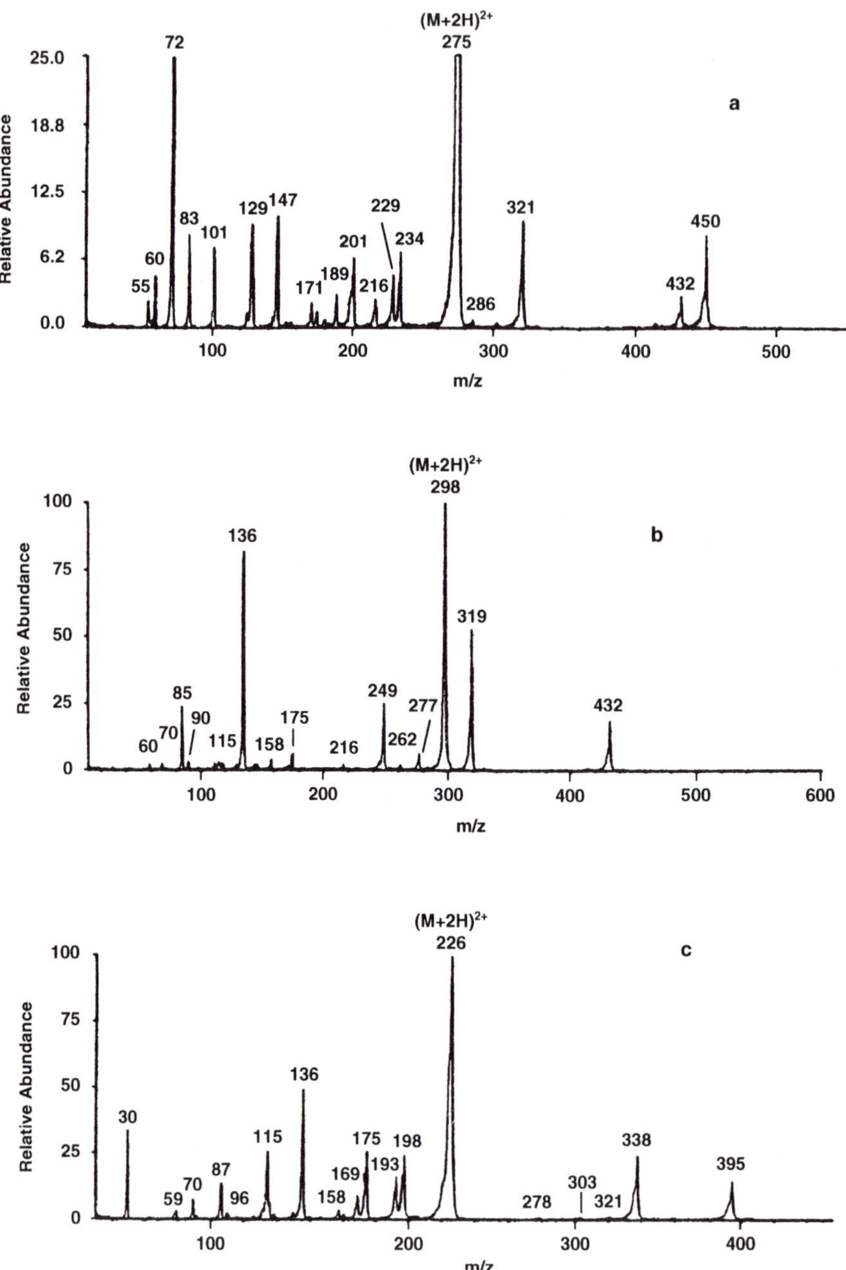

Figure 9.S1 The ESI-product-ion mass spectra of the (a–d) $(M + 2H)^{2+}$ ions and (e) $(M + H)^{+}$ ion of various tryptic-like peptides. [(a–c) © 1992. Reproduced from reference 18 by permission of John Wiley & Sons, Limited. (d) © 1990. Reprinted from reference 124 by permission of Academic Press, Inc., Orlando, Florida. (e) © 1991. Reprinted from reference 125 by kind permission of Elsevier Science-NL, Sara Burgerhartstraat 25, 1055 KV Amsterdam, The Netherlands.]

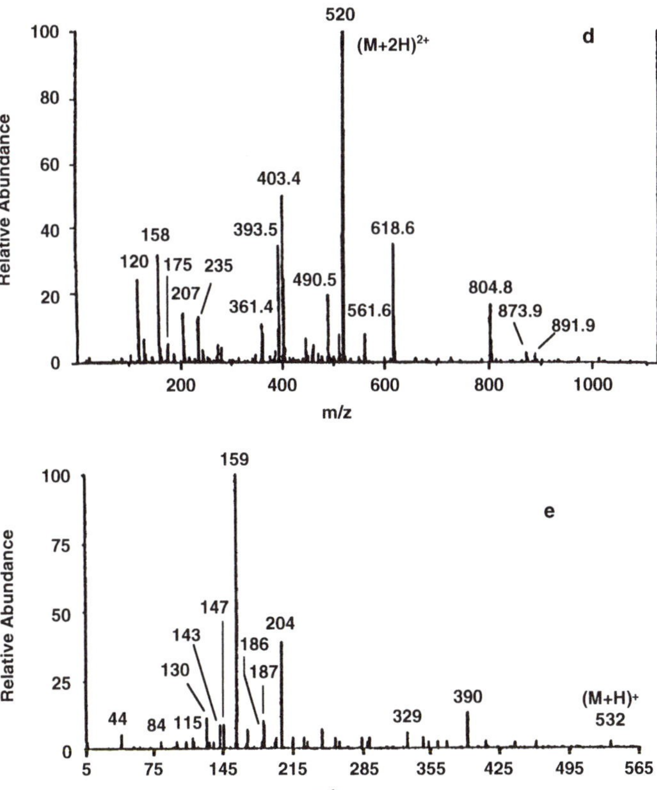

Figure 9.S1 d and e (*continued*)

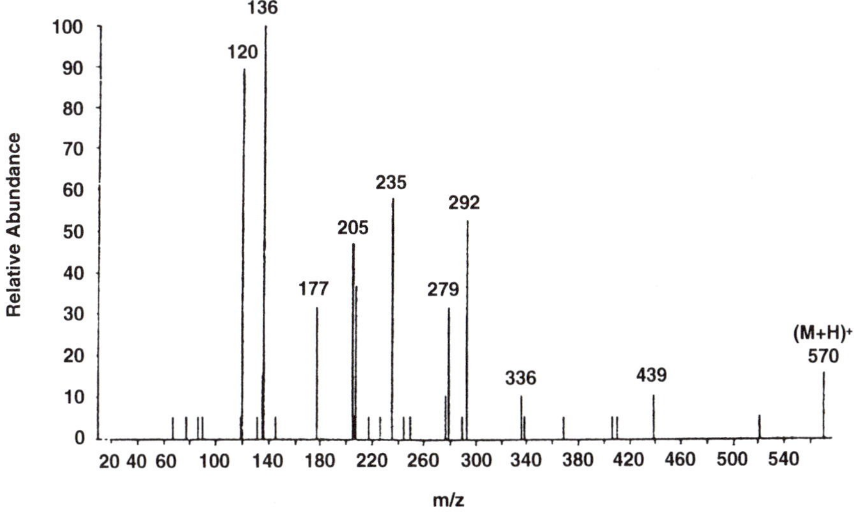

Figure 9.S2 The CZE-ESI-tandem mass spectrum of the $(M + H)^+$ ion of a peptide. (© 1989. Reprinted from reference 86 by kind permission of Elsevier Science-NL, Sara Burgerhartstraat 25, 1055 KV Amsterdam, The Netherlands.)

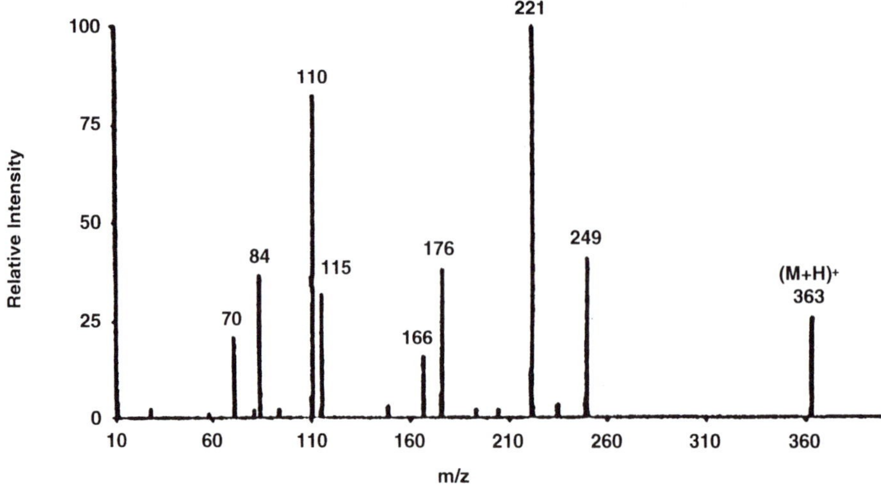

Figure 9.S3 The CZE-ESI-tandem mass spectrum of the $(M + H)^+$ ion of a small peptide. (© 1991. Reprinted from reference 125 by kind permission of Elsevier Science-NL, Sara Burgerhartstraat 25, 1055 KV Amsterdam, The Netherlands.)

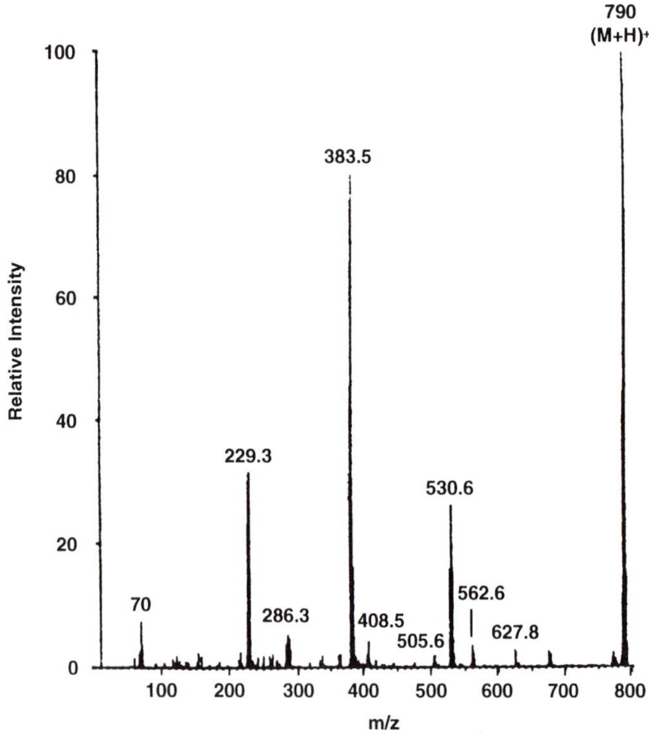

Figure 9.S4 The CZE-ESI-tandem mass spectrum of the $(M + H)^+$ ion of a peptide. (© 1991. Reprinted from reference 52 by permission of Plenum Publishing Company, New York.)

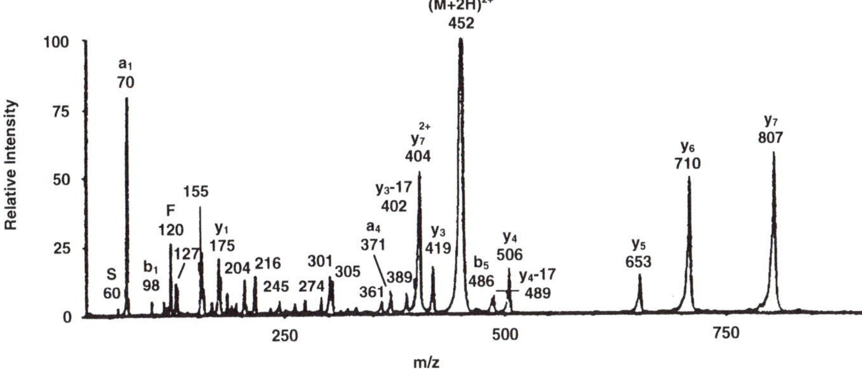

Figure 9.S5 The ESI-product-ion mass spectrum of the $(M + 2H)^{2+}$ ion of a tryptic-like peptide. (© 1992. Reproduced from reference 18 by permission of John Wiley & Sons, Limited.)

PROBLEM 9.S6 Figure 9.S6 shows the ESI-tandem mass spectrum of m/z 442.3^{2+} from 550 attomoles of an angiotensin peptide, RVYVHPI.[127,128] Rationalize the proline effect with respect to the signal intensities.

Answer The b_5 and/or the y_2 ion can experience a proline effect. Its neighbor, His, is a neutral residue. Thus, these two ions should be enhanced in intensity. They are the two most abundant peaks in the spectrum.

PROBLEM 9.S7 Figure 9.S7 presents the tandem mass spectrum by Hunt et al.[129] of the second of two peptides in the STP-4 mixture[126] (see Problem 9.S3).

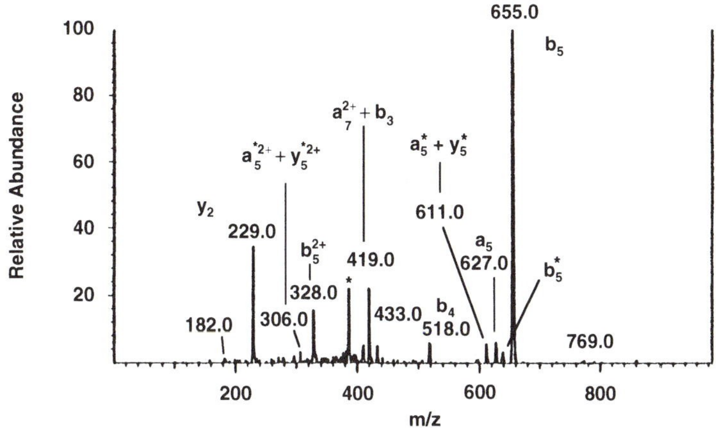

Figure 9.S6 The ESI-tandem mass spectrum of the $(M + 2H)^{2+}$ ion at m/z 442.3^{2+} of a small peptide. Neglect the peak marked *.

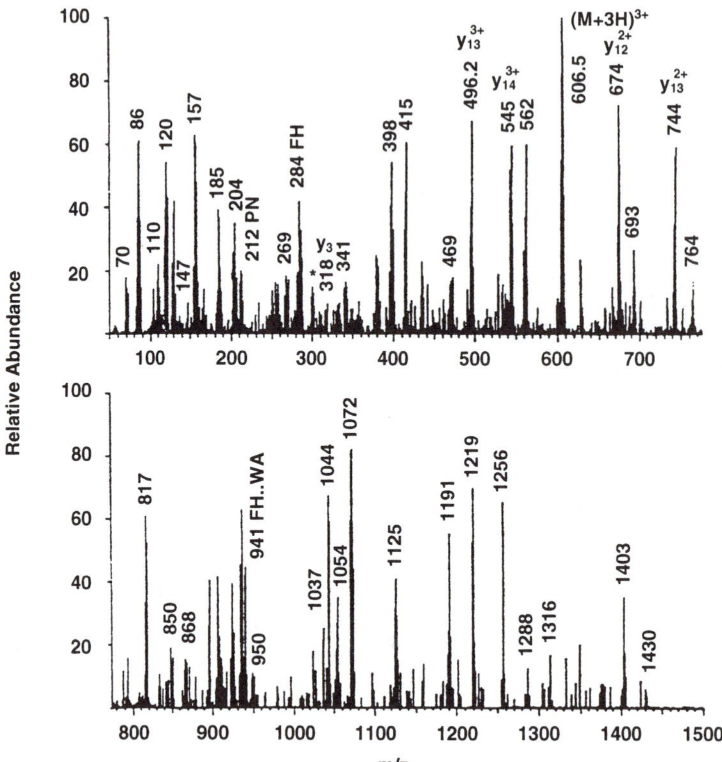

Figure 9.S7 The ESI-tandem mass spectrum of the $(M + 3H)^{3+}$ ion at m/z 606.5 of a tryptic-like peptide. A few internal fragments are labeled on the respective masses. (© 1991. Reprinted from reference 129 by permission of Academic Press, Inc., Orlando, Florida.)

There are many internal fragment ions ranging from low to very high mass. Also present are immonium ions, triply and doubly charged fragment ions, and a couple of signals of respectable intensity that are not part of the peptide (and/or are unidentifiable). Deduce as much of the sequence as possible of the peptide based on the mass spectrum. A few internal fragment assignments are given.

PROBLEM 9.S8 Figure 9.S8a presents an ESI-tandem mass spectrum of a peptide, as well as an expanded region (Figure 9.S8b) of m/z 336–358.[11] Most of the peptide can be sequenced and there is more than one doubly charged mass in the spectrum.

PROBLEM 9.S9 Figures 9.S9a and b provide ESI-tandem mass spectra[11] of a peptide that contains a Pro residue. Figure 9.S9b provides greater detail of the low-mass region. Provide a sequencing rationale for the peptide and rationalize any proline effects.

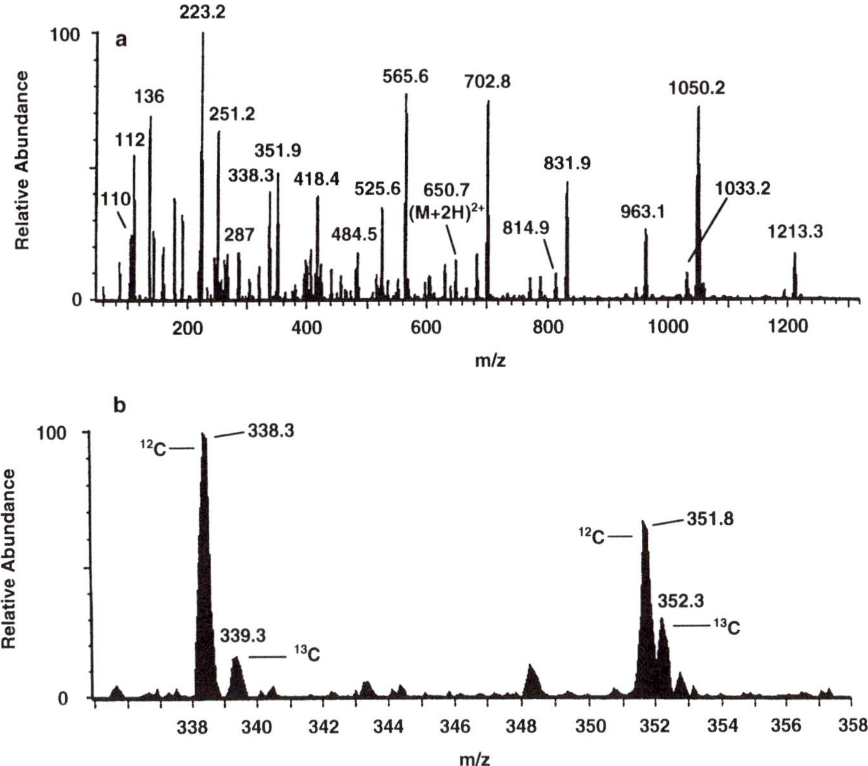

Figure 9.S8 (a) The ESI-tandem mass spectrum of the $(M+2H)^{2+}$ ion at m/z 650.7 of a peptide, (b) expansion of the m/z 336–358 region in (a). (© 1995, Mass Spectrometry Reviews. Reprinted from reference 11 by permission of John Wiley & Sons, Inc.)

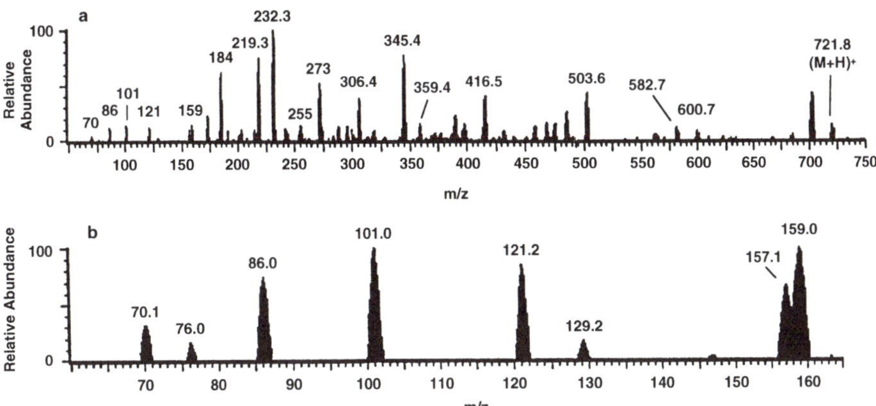

Figure 9.S9 (a) The ESI-tandem mass specptrum of the $(M+H)^{+}$ ion at m/z 721.8 of a peptide, (b) expansion of the mass spectral region below m/z 160 in (a). (© 1995, Mass Spectrometry Reviews. Reprinted from reference 11 by permission of John Wiley & Sons, Inc.)

Answer Reference 11 provides the spectral interpretation, and a proline effect (Pro-6) is observed for the intense signals of the b_5 and y_2 ions, because the Pro has a polar, uncharged Ser residue (Table 9.8) on the N-terminal side.

REFERENCES

1. Biemann, K. *Annu. Rev. Biochem.* **1992**, *61*, pp 977–1010.
2. Biemann, K. *Biomed. Environ. Mass Spectrom.* **1988**, *16*, pp 99–111.
3. Vath, J.E.; Biemann, K. *Int. J. Mass Spectrom. Ion Processes* **1990**, *100*, pp 287–299.
4. Johnson, R.S.; Martin, S.A.; Biemann, K.; Stults, J.T.; Watson, J.T. *Anal. Chem.* **1987**, *59*, pp 2621–2625.
5. Biemann, K. In *Methods in Protein Sequence Analysis*; Imahori, K.; Sakiyama, F., Eds.; Plenum Press: New York, 1993; pp 119–126.
6. Hunt, D.F.; Zhu, N.Z.; Shabanowitz, J. *Rapid Commun. Mass Spectrom.* **1989**, *3*, pp 122–124.
7. Biemann, K. In *Mass Spectrometry of Biological Materials*; McEwen, C.N.; Larsen, B.S., Eds.; Marcel Dekker: New York, 1990; Chapter 1, pp 3–24.
8. Biemann, K. In *Biological Mass Spectrometry: Present and Future*; Matsuo, T.; Caprioli, R.M.; Gross, M.L.; Seyama, Y., Eds.; John Wiley & Sons: Chichester, U.K., 1994; Chapter 3.1, pp 275–297.
9. Biemann, K. In *Methods in Enzymology: Mass Spectrometry*; McCloskey, J.A., Ed.; Academic Press: San Diego, CA, 1990; Volume 193, Appendix A6, p 888.
10. Feng, R.; Konishi, Y.; Bell, A.W., *J. Am. Soc. Mass Spectrom.* **1991**, *2*, pp 387–401.
11. Papayannopoulos, I.A. *Mass Spectrom. Rev.* **1995**, *14*, pp 49–73.
12. Wada, Y.; Matsuo, T. In *Biological Mass Spectrometry: Present and Future*; Matsuo, T.; Caprioli, R.M.; Gross, M.L.; Seyama, Y., Eds.; John Wiley & Sons: Chichester, U.K., 1994; Chapter 3.7, pp 369–399.
13. Roepstorff, P.; Fohlman, *J. Biomed. Mass Spectrom.* **1984**, *11*, p 601.
14. Medzihradszky, K.F.; Burlingame, A.L. *METHODS: A Companion to Methods in Enzymology* **1994**, *6*, pp 284–303.
15. Biemann, K. In *Methods in Enzymology: Mass Spectrometry*; McCloskey, J.A., Ed.; Academic Press: San Diego, CA, 1990; Volume 193, Chapter 25, pp 455–479.
16. Johnson, R.S.; Martin, S.A.; Biemann, K. *Int. J. Mass Spectrom. Ion Processes* **1988**, *86*, pp 137–154.
17. Tang, X.J.; Thibault, P.; Boyd, R.K. *Anal. Chem.* **1993**, *65*, pp 2824–2834.
18. Tang, X.J.; Boyd, R.K. *Rapid Commun. Mass Spectrom.* **1992**, *6*, pp 651–657.
19. Ashcroft, A.E.; Derrick, P.J. In *Mass Spectrometry of Peptides*; Desiderio, D.M., Ed.; CRC Press: Boca Raton, FL, 1991; Chapter 7, pp 121–138.
20. Yalcin, T.; Csizmadia, I.G.; Peterson, M.R.; Harrison, A.G. *J. Am. Soc. Mass Spectrom.* **1996**, *7*, pp 233–242.
21. Downard, K.M.; Biemann, K. *Int. J. Mass Spectrom. Ion Processes* **1995**, *148*, pp 191–202.
22. Downard, K.M.; Biemann, K. *J. Am. Soc. Mass Spectrom.* **1993**, *4*, pp 874–881.
23. Biemann, K. In *Methods in Enzymology: Mass Spectrometry*; McCloskey, J.A., Ed.; Academic Press: San Diego, CA, 1990; Volume 193, Appendix A5, pp 886–887.
24. Biemann, K.; Scoble, H.A. *Science* **1987**, *237*, pp 992–998.
25. Jardine, I. In *Methods in Enzymology: Mass Spectrometry*; McCloskey, J.A., Ed.; Academic Press: San Diego, CA, 1990; Volume 193, Chapter 24, pp 441–455.

26. Zaia, J.; Annan, R.S.; Biemann, K.; *Rapid Commun. Mass Spectrom.* **1992**, *6*, pp 32–36.
27. Griffin, P.R.; Coffman, J.A.; Hood, L.E.; Yates, J.R. III, *Int. J. Mass Spectrom. Ion Processes* **1991**, *111*, pp 131–149.
28. Allen, G., Ed.; *Sequencing of Proteins and Peptides*; Elsevier Science Publishers B.V.: Amsterdam, The Netherlands, 1989; Volume 9, Chapter 2, pp 19–71.
29. Ambler, R.P. In *Methods in Enzymology: Enzyme Structure*; Hirs, C.H.W.; Timasheff, S.N., Eds.; Academic Press: New York, 1972; Volume 25, Part B, Chapter 10, pp 143–154.
30. Ambler, R.P. In *Methods in Enzymology: Enzyme Structure*; Hirs, C.H.W.; Timasheff, S.N., Eds.; Academic Press: New York, 1972; Volume 25, Part B, Chapter 21, pp 262–272.
31. Shively, J.E.; Paxton, R.J. In *Mass Spectrometry of Biological Materials*; McEwen, C.N.; Larsen, B.S., Eds.; Marcel Dekker: New York, 1990; Chapter 2, pp 25–86.
32. Smith, D.L.; Zhou, Z. In *Methods in Enzymology: Mass Spectrometry*; McCloskey, J.A., Ed.; Academic Press: San Diego, CA, 1990; Volume 193, Chapter 20, pp 374–389.
33. Smith, B.J. In *Methods in Molecular Biology: Basic Protein and Peptide Protocols*; Walker, J.M., Ed.; Humana Press: Totowa, NJ, 1994; Chapter 32, pp 289–296.
34. Drapeau, G.R. In *Methods in Enzymology: Enzyme Structure*; Hirs, C.H.W.; Timasheff, S.N., Eds.; Academic Press: New York, 1977; Volume 47, Part E, Chapter 21, pp 189–191.
35. Lee, T.D.; Shively, J.E. In *Methods in Enzymology: Mass Spectrometry*; McCloskey, J.A., Ed.; Academic Press: San Diego, CA, 1990; Volume 193, Chapter 19, pp 361–375.
36. Kraft, R. In *Protein Structure Analysis: Preparation, Characterization & Microsequencing*; Kamp, R.M.; Choli-Papadopoulou, T.; Wittman-Leybold, B., Eds.; Springer-Verlag: New York; Chapter 5, pp 61–71.
37. Krishna, R.G.; Wold, F. In *Methods in Protein Sequence Analysis*; Imahori, K.; Sakiyama, F., Eds.; Plenum Press: New York, 1993; pp 167–172.
38. Nguyen, D.N.; Becker, G.W.; Riggin, R.M. *J. Chromatogr., A* **1995**, *705*, pp 21–45.
39. Carr, S.A.; Roberts, G.D.; Hemling, M.E. In *Mass Spectrometry of Biological Materials*; McEwen, C.N.; Larsen, B.S., Eds.; Marcel Dekker: New York, 1990; Chapter 3, pp 87–136.
40. Ishii, S.; Abe, Y.; Mitta, M.; Matsushita, H.; Kato, I. In *Methods in Protein Sequence Analysis*; Imahori, K.; Sakiyama, F., Eds.; Plenum Press: New York, 1993; pp 95–100.
41. Tsunasawa, S.; Hirano, H. In *Methods in Protein Sequence Analysis*; Imahori, K.; Sakiyama, F., Eds.; Plenum Press: New York, 1993; pp 45–53.
42. Podell, D.N.; Abraham, G.N. *Biochem. Biophys. Res. Commun.* **1978**, *81*, pp 176–185.
43. Doolittle, R.F. In *Methods in Enzymology: Enzyme Structure*; Hirs, C.H.W.; Timasheff, S.N., Eds.; Academic Press: New York, 1972; Volume 25, Part B, Chapter 18, pp 231–244.
44. Mitchell, W.M. In *Methods in Enzymology: Enzyme Structure*; Hirs, C.H.W.; Timasheff, S.N., Eds.; Academic Press: New York, 1977; Volume 47, Part E, Chapter 18, pp 165–170.
45. Heinrikson, R.L. In *Methods in Enzymology: Enzyme Structure*; Hirs, C.H.W.; Timasheff, S.N., Eds.; Academic Press: New York, 1977; Volume 47, Part E, Chapter 20, pp 175–189.

46. Allen, G., Ed.; *Sequencing of Proteins and Peptides*; Elsevier Science Publishers B.V.: Amsterdam, The Netherlands, 1989; Volume 9, Chapter 3, pp 73–104.
47. Conn, E.E.; Stumpf, P.K., Eds.; *Outlines of Biochemistry*, 3rd Edition; John Wiley & Sons: New York, 1972; Chapter 4, pp 69–101.
48. Hunt, D.F.; Shabanowitz, J.; Yates, J.R., III; Zhu, N.Z.; Russell, D.H.; Castro, M.E. *Proc. Natl. Acad. Sci. U.S.A* **1987**, *84*, pp 620–623.
49. Griffin, P.R.; Furer-Jonscher, K.; Hood, L.E.; Yates, J.R., III; Schwartz, J.; Jardine, I. In *Techniques in Protein Chemistry III*; Angeletti, R.H., Ed.; Academic Press: San Diego, CA, 1992; pp 467–476.
50. Weigt, C.; Meyer, H.E.; Kellner, R. In *Microcharacterization of Proteins*; Kellner, R.; Lottspeich, F.; Meyer, H.E., Eds.; VCH Verlagsgesellschaft mbH: Weinheim, Germany, 1994; Chapter V.3, pp 189–205.
51. Hunt, D.F.; Alexander, J.E.; McCormack, A.L.; Martino, P.A.; Michel, H.; Shabanowitz, J.; Sherman, N.; Moseley, M.A.; Jorgenson, J.W.; Tomer, K.B. In *Techniques in Protein Chemistry II*; Villafranca, J.J., Ed.; Academic Press: San Diego, CA, 1991; Chapter 43, pp 441–454.
52. Hunt, D.F.; Shabanowitz, J.; Moseley, M.A.; McCormack, A.L.; Michel, H.; Martino, P.A.; Tomer, K.B.; Jorgenson, J.W. In *Methods in Protein Sequence Analysis*; Jörnvall, H., Höög, J.-O.; Gustavsson, A.-M., Eds.; Birkhäuser Verlag: Basel, Switzerland, 1991; pp 257–266.
53. Covey, T. In *Biochemical and Biotechnological Applications of Electrospray Ionization Mass Spectrometry*; Snyder, A.P., Ed.; ACS Symposium Series 619; American Chemical Society: Washington, DC, 1995; Chapter 2, pp 21–59.
54. Johnstone, R.A.W.; Rose, M.E., Eds.; *Mass Spectrometry for Chemists and Biochemists*; 2nd Edition; Cambridge University Press: Cambridge, U.K., 1996; Chapter 11, pp 397–426.
55. Edmonds, C.G.; Loo, J.A.; Ogorzalek Loo, R.R.; Smith, R.D. In *Techniques in Protein Chemistry II*; Villafranca, J.J., Ed.; Academic Press: San Diego, CA, 1991; Chapter 47, pp 487–495.
56. Hunt, D.F.; Yates, J.R., III; Shabanowitz, J.; Winston, S.; Hauer, C.R. *Proc. Natl. Acad. Sci. U.S.A.* **1986**, *83*, pp 6233–6237.
57. Falick, A.M.; Hines, W.M.; Medzihradszky, K.F.; Baldwin, M.A.; Gibson, B.W. *J. Am. Soc. Mass Spectrom.* **1993**, *4*, pp 882–893.
58. Carr, S.A.; Hemling, M.E.; Bean, M.F.; Roberts, G.D. *Anal. Chem.* **1991**, *63*, pp 2802–2824.
59. Metzger, J.W.; Eckerskorn, C. In *Microcharacterization of Proteins*, Kellner, R.; Lottspeich, F.; Meyer, H.E., Eds., VCH Publishers: New York, 1994; Volume 2, pp 167–187.
60. Gurd, F.R.N. In *Methods in Enzymology: Enzyme Structure*; Hirs, C.H.W.; Timasheff, S.N., Eds.; Academic Press.: New York, 1972; Volume 25, Part B, Chapter 34a, pp 424–438.
61. Hudson, D. *J. Org. Chem.* **1988**, *53*, pp 617–624.
62. Lundblad, R.L., Ed.; *Techniques in Protein Modification*; CRC Press: Boca Raton, FL, 1995; Chapter 6, pp 63–89.
63. Lundblad, R.L., Ed.; *Techniques in Protein Modification*; CRC Press: Boca Raton, FL, 1995; Chapter 8, pp 97–103.
64. Papayannopoulos, I.A.; Biemann, K. *Protein Sci.* **1992**, *1*, pp 278–288.
65. Smith, B.J. In *Methods in Molecular Biology: Basic Protein and Peptide Protocols*; Walker, J.M., Ed.; Humana Press: Totowa, NJ, 1994; Chapter 33, pp 297–309.

66. Smith, B.J. In *The Protein Protocols Handbook*; Walker, J.M., Ed.; Humana Press: Totowa, NJ, 1996; Chapter 66, pp 385–387.
67. Carr, S.A. *Adv. Drug Delivery Rev.* **1990**, *4*, pp 113–147.
68. Smith, B.J. In *The Protein Protocols Handbook*; Walker, J.M., Ed.; Humana Press: Totowa, NJ, 1996; Chapter 63, pp 369–373.
69. Gross, E. In *Methods in Enzymology: Enzyme Structure*; Hirs, C.H.W., Ed.; Academic Press: New York, 1967; Volume 11, Chapter 27, pp 238–255.
70. Hunt, D.F.; Krishnamurthy, T.; Shabanowitz, J.; Griffin, P.R.; Yates, J.R., III; Martino, P.A.; McCormack, A.L.; Hauer, C.R. In *Mass Spectrometry of Peptides*; Desiderio, D.M., Ed.; CRC Press: Boca Raton, FL, 1991; Chapter 8, pp 139–158.
71. Mant, C.T.; Hodges, R.S. In *High Performance Liquid Chromatography of Peptides and Proteins: Separation, Analysis, and Conformation*; Mant, C.T.; Hodges, R.S., Eds.; CRC Press: Boca Raton, FL, 1991; pp 906–915.
72. Smith, R.D.; Loo, J.A.; Ogorzalek Loo, R.R.; Busman, M.; Udseth, H.R. *Mass Spectrom. Rev.* **1991**, *10*, pp 359–451.
73. Konigsberg, W. In *Methods in Enzymology: Enzyme Structure*; Hirs, C.H.W.; Timasheff, S.N., Eds.; Academic Press: New York, 1972; Volume 25, Part B, Chapter 13, pp 185–188.
74. Carne, A.F. In *Methods in Molecular Biology: Basic Protein and Peptide Protocols*; Walker, J.M., Ed.; Humana Press: Totowa, NJ, 1994; Chapter 34, pp 311–320.
75. Ling, V.T.; Eng, M.L.; Lee, P.J.; Keck, R.G.; Keyt, B.A.; Canova-Davis, E. In *New Methods in Peptide Mapping for the Characterization of Proteins*; Hancock, W.S., Ed.; CRC Press: Boca Raton, FL, 1996; Chapter 1, pp 1–30.
76. Lundblad, R.L., Ed.; *Techniques in Protein Modification*; CRC Press: Boca Raton, FL, 1995; Chapter 7, pp 91–96.
77. Waugh, R.J.; Bowie, J.H.; Hayes, R.N. *Org. Mass Spectrom.* **1991**, *26*, pp 250–256.
78. Hunt, D.F.; Shabanowitz, J.; Yates, J.R.; Griffin, P.R. In *Mass Spectrometry of Biological Materials*; McEwen, C.N.; Larsen, B.S., Eds.; Marcel Dekker: New York, 1990; Chapter 5, pp 169–195.
79. Roepstorff, P. In *Mass Spectrometry in the Biological Sciences: A Tutorial*; Gross, M.L., Ed.; Kluwer Academic Publishers: Dordrecht, The Netherlands, 1992; pp 213–227.
80. Riordan, J.F.; Vallee, B.L. In *Methods in Enzymology: Enzyme Structure*; Hirs, C.H.W.; Timasheff, S.N., Eds.; Academic Press: New York, 1972; Volume 25, Part B, Chapter 42, pp 500–506.
81. Schnölzer, M.; Jones, A.; Alewood, P.F.; Kent, S.B.H. *Anal. Biochem.* **1992**, *204*, pp 335–343.
82. Grossenbacher, H.; Märki, W.; Coulot, M.; Müller, D.; Richter, W.J. *Rapid Commun. Mass Spectrom.* **1993**, *7*, pp 1082–1085.
83. Smith, B.J. In *The Protein Protocols Handbook*; Walker, J.M., Ed.; Humana Press: Totowa, NJ, 1996; Chapter 67, pp 389–392.
84. Nakayama, H.; Uchida, K.; Shinkai, F.; Shinoda, T.; Okuyama, T.; Seta, K.; Isobe, T. *J. Chromatogr., A* **1996**, *730*, pp 279–287.
85. Jones, W.M.; Manning, J.M. *Biochem. Biophys. Res. Commun.* **1985**, *126*, pp 933–940.
86. Thibault, P.; Pleasance, S.; Laycock, M.V.; MacKay, R.M.; Boyd, R.K. *Int. J. Mass Spectrom. Ion Processes* **1991**, *111*, pp 317–353.
87. Morand, K.; Talbo, G.; Mann, M. *Rapid Commun. Mass Spectrom.* **1993**, **7**, pp 738–743.

88. Mück, W.M.; Henion, J.D. *J. Chromatogr.* **1989**, *495*, pp 41–59.
89. Schwartz, B.L.; Bursey, M.M. *Biol. Mass Spectrom.* **1992**, *21*, pp 92–96.
90. Loo, J.A.; Edmonds, C.G.; Smith, R.D. *Anal. Chem.* **1993**, *65*, pp 425–438.
91. Barinaga, C.J.; Edmonds, C.G.; Udseth, H.R.; Smith, R.D. *Rapid Commun. Mass Spectrom.* **1989**, *3*, pp 160–164.
92. Huang, E.C.; Henion, J.D. *J. Am. Soc. Mass Spectrom.* **1990**, *1*, pp 158–165.
93. Martin, S.A.; Biemann, K. *Int. J. Mass Spectrom. Ion Processes* **1987**, *78*, pp 213–228.
94. Biemann, K.; Martin, S.A. *Mass Spectrom. Rev.* **1987**, *6*, pp 1–76.
95. Schwartz, B.L.; Erickson, B.W.; Bursey, M.M.; Marbury, G.D. *Org. Mass Spectrom.* **1993**, *28*, pp 113–122.
96. Tang, X.J.; Thibault, P.; Boyd, R.K. *Int. J. Mass Spectrom. Ion Processes* **1992**, *122*, pp 153–179.
97. Harvan, D.J.; Hass, J.R.; Wilson, W.E.; Hamm, C.; Boyd, R.K.; Yajima, H.; Klapper, D.G. *Biomed. Environ. Mass Spectrom.* **1987**, *14*, pp 281–287.
98. Wu, Z.; Fenselau, C. *Rapid Commun. Mass Spectrom.* **1992**, *6*, pp 403–405.
99. Vachet, R.W.; Asam, M.R.; Glish, G.L. *J. Am. Chem. Soc.* **1996**, *118*, pp 6252–6256.
100. Mueller, D.R.; Eckersley, M.; Richter, W.J. *Org. Mass Spectrom.* **1988**, *23*, pp 217–222.
101. Kenny, P.T.M.; Nomoto, K.; Orlando, R.; *Rapid Commun. Mass Spectrom.* **1992**, *6*, pp 95–97.
102. Jensen, N.J.; Tomer, K.B.; Gross, M.L. *J. Am. Chem. Soc.* **1985**, *107*, pp 1863–1868.
103. Smith, R.D.; Loo, J.A.; Edmonds, C.G.; Barinaga, C.J.; Udseth, H.R. *Anal. Chem.* **1990**, *62*, pp 882–899.
104. Neumann, G.M.; Derrick, P.J. *Org. Mass Spectrom.* **1984**, *19*, pp 165–170.
105. Neumann, G.M.; Sheil, M.M.; Derrick, P.J. *Z. Naturforsch* **1984**, *39a*, pp 584–592.
106. Ganem, B.; Li, Y.T.; Hsieh, Y.L.; Henion, J.D.; Kaboord, B.F.; Frey, M.W.; Benkovic, S.J. *J. Am. Chem. Soc.* **1994**, *116*, pp 1352–1358.
107. Carr, S.A.; Bean, M.F.; Hemling, M.E.; Roberts, G.D. In *Biological Mass Spectrometry: Proceedings of the Second International Symposium on Mass Spectrometry in the Health and Life Sciences*; Burlingame, A.L.; McCloskey, J.A., Eds.; Elsevier Science Publishers B.V.: Amsterdam, The Netherlands, 1990; pp 621–652.
108. Gross, M.L.; Tomer, K.B.; Cerny, R.L.; Giblin, D.E. In *Mass Spectrometry in the Analysis of Large Molecules*; McNeal, C.J., Ed.; John Wiley & Sons: Chichester, U.K., 1986; pp 171–190.
109. Smith, R.D.; Loo, J.A.; Edmonds, C.G. In *Mass Spectrometry: Clinical and Biomedical Applications, Volume 1*; Desiderio, D.M., Ed.; Plenum Press: New York, 1992; Chapter 2, pp 37–98.
110. Rockwood, A.L.; Busman, M.; Udseth, H.R.; Smith, R.D. *Rapid Commun. Mass Spectrom.* **1991**, *5*, pp 582–585.
111. Busman, M.; Rockwood, A.L.; Smith, R.D. *J. Phys. Chem.* **1992**, *96*, pp 2397–2400.
112. Rockwood, A.L.; Busman, M.; Smith, R.D. *Int. J. Mass Spectrom. Ion Processes* **1991**, *111*, pp 103–129.
113. Loo, J.A.; Edmonds, C.G.; Smith, R.D. *Science* **1990**, *248*, pp 201–204.
114. Loo, J.A.; Edmonds, C.G.; Smith, R.D. *Anal. Chem.* **1991**, *63*, pp 2488–2499.
115. Loo, J.A.; Edmonds, C.G.; Ogorzalek Loo, R.R.; Udseth, H.R.; Smith, R.D. In *Experimental Mass Spectrometry*; Russell, D.H., Ed.; Plenum Press: New York, 1994; Chapter 7, pp 243–286.

116. Smith, R.D.; Bruce, J.E.; Wu, Q.; Cheng, X.; Hofstadler, S.A.; Anderson, G.A.; Chen, R.; Bakhtiar, R.; Van Orden, S.O.; Gale, D.C.; Sherman, M.G.; Rockwood, A.L.; Udseth, H.R. In *Mass Spectrometry in the Biological Sciences*; Burlingame, A.L.; Carr, S.A., Eds.; Humana Press: Totowa, NJ, 1996; pp 25–68.
117. Bakhtiar, R.; Wu, Q.; Hofstadler, S.A.; Smith, R.D. *Biol. Mass Spectrom.* **1994**, *23*, pp 707–710.
118. Summerfield, S.G.; Cox, K.A.; Gaskell, S.J. *J. Am. Soc. Mass Spectrom.* **1997**, *8*, pp 25–31.
119. Hall, S.C.; Smith, D.M.; Clauser, K.R.; Andrews, L.E.; Walls, F.C.; Webb, J.W.; Tran, H.M.; Epstein, L.B.; Burlingame, A.L. In *Mass Spectrometry in the Biological Sciences*; Burlingame, A.L.; Carr, S.A., Eds.; Humana Press: Totowa, NJ, 1996; pp 171–202.
120. Stults, J.T.; Watson, J.T. *Biomed. Environ. Mass Spectrom.* **1987**, *14*, pp 583–586.
121. Summerfield, S.G.; Whiting, A.; Gaskell, S.J. *Int. J. Mass Spectrom. Ion Processes* **1997**, *162*, pp 149–161.
122. Burlingame, A.L.; Medzihradszky, K.F.; Clauser, K.R.; Hall, S.C.; Maltby, D.A.; Walls, F.C. In *Biochemical and Biotechnological Applications of Electrospray Ionization Mass Spectrometry*; Snyder, A. P., Ed.; ACS Symposium Series 619; American Chemical Society: Washington, DC, 1995; Chapter 24, pp 472–511.
123. Kolli, V.S.K.; Orlando, R. *J. Am. Soc. Mass Spectrom.* **1995**, *6*, pp 234–241.
124. Hail, M.; Lewis, S.; Zhou, J.; Jardine, I.; Whitehouse, C. In *Current Research in Protein Chemistry: Techniques, Structure, and Function*; Villafranca, J.J., Ed.; Academic Press: San Diego, CA, 1990; Chapter 10, pp 105–116.
125. Johansson, I.M.; Huang, E.C.; Henion, J.D.; Zweigenbaum, J. *J. Chromatogr.* **1991**, *554*, pp 311–327.
126. Lee, T.D.; Ronk, M. In *Techniques in Protein Chemistry II*; Villafranca, J.J., Ed.; Academic Press: San Diego, CA, 1991; Chapter 45, pp 467–475.
127. Bier, M.E.; Schwartz, J.C.; Zhou, J.; Syka, J.E.P.; Taylor, D.; Land, A.P.; James, M.; Fies, B. *Proc. 43rd ASMS Conference on Mass Spectrometry and Allied Topics* **1994**, p 988.
128. Bier, M.E.; Schwartz, J.C.; Zhou, J.; Taylor, D.; Syka, J.; James, M.; Fies, B.; Stafford, G. *Proc. 43rd ASMS Conference on Mass Spectrometry and Allied Topics* **1994**, p 1117.
129. Hunt, D.F.; Alexander, J.E.; McCormack, A.L.; Martino, P.A.; Michel, H.; Shabanowitz, J. In *Techniques in Protein Chemistry II*; Villafranca, J.J., Ed.; Academic Press: San Diego, CA, 1991; Chapter 44, pp 455–465.

ANSWERS

PROBLEM 9.1 The $(M + 2H)^{2+}$ ion is shown in Figure 9.7 and listed in Table 9.9 as m/z 697.4. This yields an $(M + H)^+$ of $(697.4 \times 2) - 1 = 1393.8 \simeq 1393.7$ Da, and this value is also observed in Table 9.9. Therefore, these two ions can be labeled as such in Table 9.16a. Remember that the $(M + H)^+ = y_f$. Furthermore, $M = 1393.7 - 1 = 1392.7$ Da. This mass cannot be observed, because it is the mass of the neutral peptide. This value has a dashed underline in Table 9.16a, signifying its absence in the mass spectrum. In the absence of an $(M + H)^+$ ion, the mass of the analyte can still be found by the $(M + 2H)^{2+}$ species in the conventional mass spectrum. If the difference between the $(M + H)^+$ value (whether it is present or not in the tandem mass spectrum)

Table 9.16 Reconstruction of the T11 tryptic peptide of sperm whale myoglobin from the b_n and y_n ions in Figure 9.7

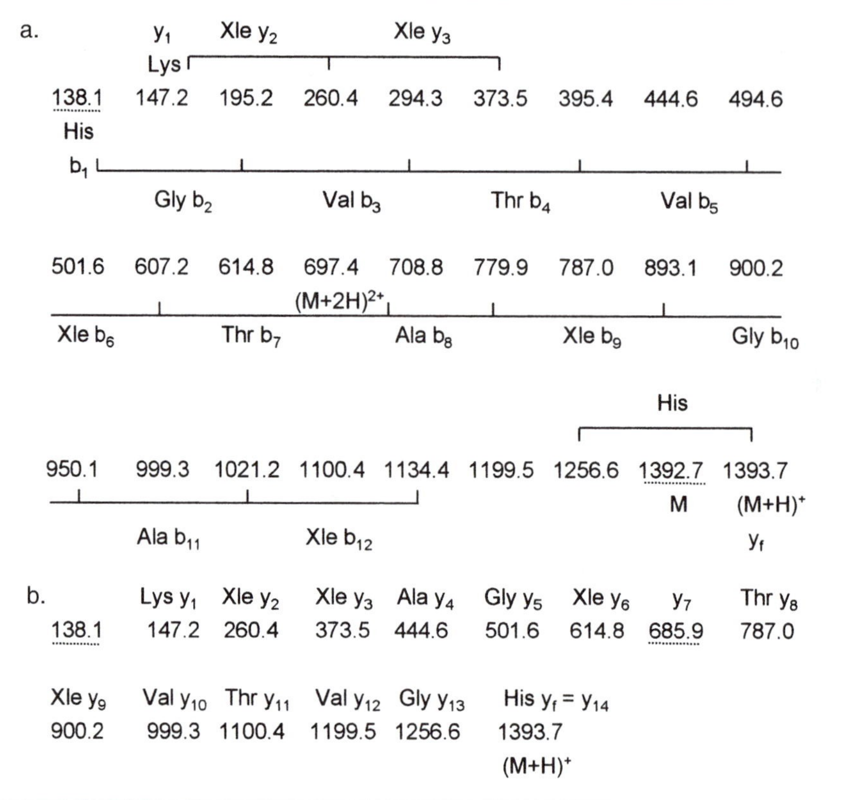

a.

y_1 Lys, Xle y_2, Xle y_3

138.1	147.2	195.2	260.4	294.3	373.5	395.4	444.6	494.6

His b_1; Gly b_2; Val b_3; Thr b_4; Val b_5

501.6	607.2	614.8	697.4	708.8	779.9	787.0	893.1	900.2
			$(M+2H)^{2+}$					

Xle b_6; Thr b_7; Ala b_8; Xle b_9; Gly b_{10}

His

950.1	999.3	1021.2	1100.4	1134.4	1199.5	1256.6	1392.7	1393.7
							M	$(M+H)^+$
								y_f

Ala b_{11}; Xle b_{12}

b.

	Lys y_1	Xle y_2	Xle y_3	Ala y_4	Gly y_5	Xle y_6	y_7	Thr y_8
138.1	147.2	260.4	373.5	444.6	501.6	614.8	685.9	787.0

Xle y_9	Val y_{10}	Thr y_{11}	Val y_{12}	Gly y_{13}	His $y_f = y_{14}$
900.2	999.3	1100.4	1199.5	1256.6	1393.7 $(M+H)^+$

and a lower mass fragment is equal to an amino acid residue, then there is a good probability that this ion is the first residue of the peptide sequence. In the present case, we have 1393.7 − 1256.6 = 137.1 Da = His residue. Thus, His must be the first residue in the peptide. One can then look for that same mass value in the low-mass range since the b_1 ion is the same residue as that which terminates the y_n series. The b_1 value of His would be 137.1 + 1 = 138.1 Da, and this value is not observed in the spectrum. However, it appears that this analysis has produced mass anchor points that consist of the first residue (or unobserved b_1) and y_f, where the first mass (or unobserved b_1) is 138.1 Da (dashed underline in Table 9.16a), and the last mass is 1393.7 Da.

Differences between masses are sought that yield amino acid residue mass values. It is usually prudent to start in the middle of the mass spectral features and work toward higher and lower masses. This will become clearer as we proceed through this chapter. When two masses have a difference that is much smaller than the value of the lowest mass residue (Gly at 57 Da), then this is a prompt to signify that one ion may be a b-type and the other may be a y-type ion.

For example, *m/z* 607.7 and 614.8 are clearly ions that originate from two different series, as well as the *m/z* pairs 260.4, 294.3; 373.5, 395.4; 494.6, 501.6; 779.9, 787.0; 893.1, 900.2; 999.3, 1021.2; 1100.4, 1134.4 (Table 9.9). Therefore, in each pair, one may be a b-type and the other may be a y-type ion to a first approximation. Elements in each nearest pair can then be subtracted to yield certain residue masses (Table 9.1). For the two lowest pairs of masses:

$$\begin{array}{r} 373.5 \\ -260.4 \\ \hline 113.1 \end{array}\ \text{Xle} \qquad \begin{array}{r} 373.5 \\ -294.3 \\ \hline 79.2 \end{array} \qquad \begin{array}{r} 395.4 \\ -260.4 \\ \hline 135.0 \end{array} \qquad \begin{array}{r} 395.4 \\ -294.3 \\ \hline 101.1 \end{array}\ \text{Thr}$$

Note that all four masses distinguish themselves such that two of the resulting values do not yield residue masses. Xle refers to either Ile or Leu because they both have identical masses (isomers). Table 9.16a and b show that structural aids can be developed such as:

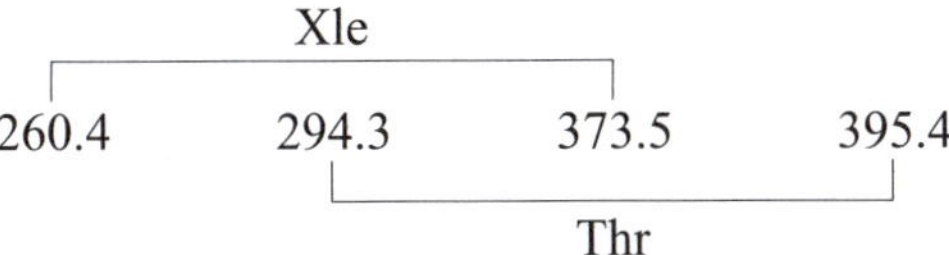

The Xle and Thr can next be examined for their relationship to the two observed lower mass values and the absent b_1 value:

$$\begin{array}{r} 260.4 \\ -195.2 \\ \hline 65.2 \end{array} \qquad \begin{array}{r} 260.4 \\ -147.2 \\ \hline 113.2 \end{array} \qquad \begin{array}{r} 260.4 \\ -138.1 \\ \hline 122.3 \end{array}$$

In Table 9.16a, a connection is drawn between *m/z* 147.2 and 260.4, which represents another Xle residue. Here *m/z* 147.2 appears to be the end point for the Xle–Xle series. If it ended at *m/z* 138.1, which is the b_1 = His residue, Xle–Xle would be b_2 and b_3. Since Xle–Xle cannot "link up" with b_1, they must be in the y series, and *m/z* 147.2 is low enough in mass to be a residue. Note carefully that *m/z* 147.2 cannot be Phe (Table 9.1)! Also, 147.2 Da could not be the b_1 ion, because the mass would have to consist of a residue of 146.2 Da. No amino acid has a residue mass of 146.2 Da. Thus, *m/z* 147.2 must be the y_1 ion; therefore, $(147.2 - 17) - 2 = 128.2$ Da, and this is equal to the residue mass of either Lys or Gln. Since the analyte is a tryptic peptide, the residue must be Lys. The *m/z* 147.2 is the y_1 ion and is the C-terminal or last amino acid residue in the peptide. Above each series of numbers in Table 9.16b, the y ions are interrogated by working in the C-terminal to N-terminal direction ("backwards"). Below each series of numbers in Table 9.16a, the b ions are interrogated by working in the N-terminal to C-terminal direction ("forwards").

Proceeding forward from the His b_1 residue (Table 9.16a): $195.2 - 138.1 = 57.1$ = Gly, and b_2 = 195.2 Da. At this point it appears that *m/z* 195.2 and 294.3 should be adjacent residues since *m/z* 260.4 clearly belongs to the y series; therefore, 294.3 – 195.2 = 99.1 = Val. Thus, b_3 = 294.3 Da. Continuing in this pattern for the b series:

$$\begin{array}{r} 395.4 \\ -294.3 \\ \hline 101.1 \end{array} = b_4 \text{ Thr} \qquad \begin{array}{r} 444.6 \\ -395.4 \\ \hline 49.2 \end{array} \qquad \begin{array}{r} 494.6 \\ -395.4 \\ \hline 99.2 \end{array} = b_5 \text{ Val}$$

This relationship is shown in Table 9.16a, and

$$\begin{array}{r} 607.7 \\ -494.6 \\ \hline 113.1 \end{array} = b_6 \text{ Xle} \qquad \begin{array}{r} 614.8 \\ -494.6 \\ \hline 120.2 \end{array}$$

For the b_7 ion, m/z 614.8 and 697.4 are neglected, and

$$\begin{array}{r} 708.8 \\ -607.7 \\ \hline 101.1 \end{array} = b_7 \text{ Thr} \qquad \begin{array}{r} 779.9 \\ -607.7 \\ \hline 172.2 \end{array}$$

Continuing in this fashion,

$$\begin{array}{r} 779.9 \\ -708.8 \\ \hline 71.1 \end{array} = b_8 \text{ Ala} \qquad \begin{array}{r} 787.0 \\ -708.8 \\ \hline 78.2 \end{array}$$

$$\begin{array}{r} 893.1 \\ -779.9 \\ \hline 113.2 \end{array} = b_9 \text{ Xle} \qquad \begin{array}{r} 900.2 \\ -779.9 \\ \hline 120.3 \end{array}$$

$$\begin{array}{r} 950.1 \\ -893.1 \\ \hline 57.0 \end{array} = b_{10} \text{ Gly} \qquad \begin{array}{r} 999.3 \\ -893.1 \\ \hline 106.2 \end{array}$$

$$\begin{array}{r} 999.3 \\ -950.1 \\ \hline 49.2 \end{array} \qquad \begin{array}{r} 1021.2 \\ -950.1 \\ \hline 71.1 \end{array} = b_{11} \text{ Ala} \qquad \begin{array}{r} 1100.4 \\ -950.1 \\ \hline 150.3 \end{array}$$

$$\begin{array}{r} 1100.4 \\ -1021.2 \\ \hline 79.2 \end{array} \qquad \begin{array}{r} 1134.4 \\ -1021.2 \\ \hline 113.2 \end{array} = b_{12} \text{ Xle}$$

$$\begin{array}{r} 1199.5 \\ -1134.4 \\ \hline 65.1 \end{array} \qquad \begin{array}{r} 1256.5 \\ -1134.4 \\ \hline 122.2 \end{array} \qquad \begin{array}{r} 1392.7 \\ -1134.4 \\ \hline 258.3 \end{array} = M$$

Between the b_{12} ion and the b_f or M value, there must be a residue that is absent in the tandem mass spectrum because none of three numbers (i.e., 65.1, 122.2, and 258.3 Da) match an amino acid residue (Table 9.1). However, the difference between the m/z values of 1392.7 and 1134.4 is 258.3 Da and 17 Da (C-terminal hydroxyl) must be subtracted to yield the residue value, which is equal to 241.3. Table 9.6 shows that there are five possibilities. This problem can be addressed by the y series of ions, and is immediately solved in that the y_1 and y_2 ions provide the answer to this question. The last two residues, b_{13} and $b_f = b_{14}$ comprise (b_{12} + Xle) and (b_{13} + Lys), in that order. Therefore, the mass for b_{13} would be 1134.4 + 113.1 (Xle) = 1247.5 (absent in the mass spectrum), and to cross-check: 1247.5 + 128.2 (Lys) = 1375.7 and 1375.7 + 17 (OH) = 1392.7 = M.

Table 9.16b is an abbreviated version of the full set of ions, because only the y_n series is shown. Thus, it appears that every mass merely accounts for a particular residue, and product-ion mass spectral analyses should be relatively straightforward, in general, unless a y ion happens to be missing in the series (*vide infra*). Successive subtractions should yield the reverse of the b_n series of ions, and each residue can be checked with the established b_n series of ions. Between m/z 614.8 = y_6 and m/z 787, the difference is 172.2 Da, which does not equal a residue. Table 9.6 indicates that this value can originate from either a Gly, Asp or an Ala, Thr. The b_n series identifies this region as the Ala, Thr residues. Therefore, the y_7 ion is missing (<u>685.9</u>), where the underline defines a predicted (yet experimentally missing) mass, and m/z 787.0 represents the y_8 ion. The rest of the sequence mirrors the established b_n series, and both series are labeled in their proper places in the ESI-tandem mass spectrum shown in Figure 9.53. Table 9.17 provides the solution[49] in compact form for the b_n and y_n ions, where the nonunderlined mass values are absent in Figures 9.7 and 9.53.

The identity of the Xle residues cannot be determined by the mass spectrum. However, some of these residues may be able to be assigned by cleaving the peptide with α-chymotrypsin and analyzing which bonds cleave. Table 9.4 shows that α-chymotrypsin cleaves the C-terminal side of Phe, Trp, Tyr, Leu, Met, and His. The peptide does not contain a Phe, Trp, Tyr, or Met residue. Thus, cleavages could result in the severing of the following amide bonds: His^1–Gly^2; Leu^6–Thr^7; Leu^9–Gly^{10}; Leu^{13}–Lys^{14}; and the Ile^{12}–Leu^{13} bond should not be cleaved. It is possible that this enzymatic reaction could provide information on some or all three residues at positions 6, 9, and 13 as Leu while leaving position 12 as Ile.

PROBLEM 9.2 Figure 9.54 presents the sequence and ion type values, along with the identities of the doubly charged y peaks. This peptide is the T(92–101) tryptic fragment of β-lactoglobulin. Knowledge of the doubly charged species allows the calculation of the singly charged species, and these latter values can be searched in the tandem mass spectrum for their presence or absence. The Xle residues (X in Figure 9.54) at positions 2 and 4 cannot be discerned from the mass spectrum, and thus are either Leu or Ile. Both residues are known to be Leu. The peptide is a tryptic peptide, and the C-terminal mass is either an Arg or a Lys. The y_1 value of 147 Da allows the Lys determination to be made. The C-terminal residue could not be the Gln residue, which nevertheless has the same nominal mass as Lys. However, the ninth residue could be either Lys or Gln because of the nominal 128-Da difference between y_1 and y_2. The acetylation reaction (Table 9.7) can be used in order to identify this residue.

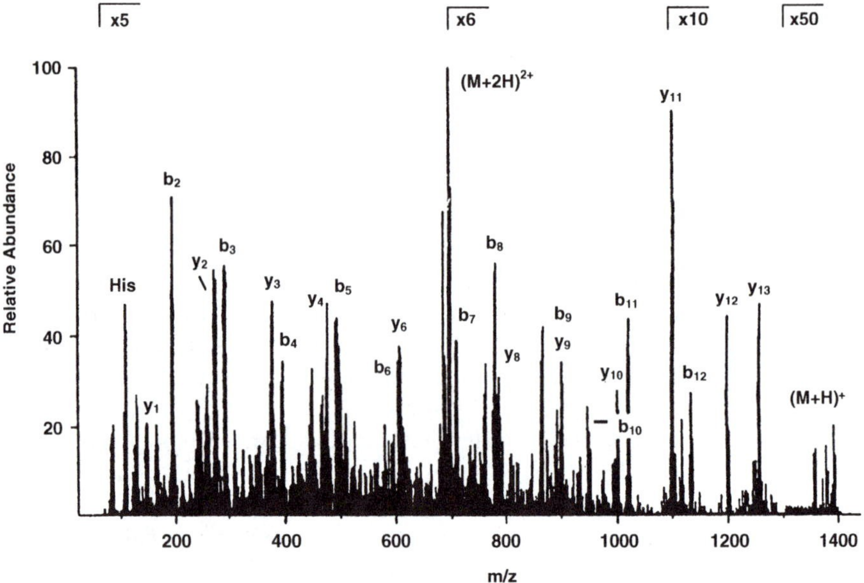

Figure 9.53 The ESI-tandem mass spectrum of the $(M + 2H)^{2+}$ ion from the T11 peptide of sperm whale myoglobin at m/z 697.4. The peaks are labeled with the appropriate nomenclature. (© 1992. Reprinted from reference 49 by permission of Academic Press, Inc., Orlando, Florida.)

Table 9.17 Predicted fragment ions for the T11 tryptic myoglobin fragment (Figure 9.53)

b_n	Residue	b value	y value	y_n
1	His	138.1	1393.7	14
2	Gly	195.2	1256.6	13
3	Val	294.3	1199.5	12
4	Thr	395.4	1100.4	11
5	Val	494.6	999.3	10
6	Leu	607.7	900.2	9
7	Thr	708.8	787.0	8
8	Ala	779.9	685.9	7
9	Leu	893.1	614.8	6
10	Gly	950.1	501.6	5
11	Ala	1021.2	444.6	4
12	Ile	1134.4	373.5	3
13	Leu	1247.5	260.4	2
14	Lys	1375.7	147.2	1

$b_{14} + 17 = 1392.7 = M$.
$y_{14} + 1393.7 = (M + H)^+$.

y_9	y_8	y_7	y_6	y_5	y_4	y_3	y_2	y_1
<u>1095.3</u>	<u>982.1</u>	883.0	<u>769.8</u>	<u>654.7</u>	<u>553.6</u>	<u>438.5</u>	<u>275.4</u>	145.2 + 2 = <u>147.2</u>

H - N - **V** - C(=O) - N(H) - **X** - C(=O) - N(H) - **V** - C(=O) - N(H) - **X** - C(=O) - N(H) - **D** - C(=O) - N(H) - **T** - C(=O) - N(H) - **D** - C(=O) - N(H) - **Y** - C(=O) - N(H) - **K** - C(=O) - N(H) - **K** - C(=O) - OH

a_1 <u>72</u>; b_1 100; a_2 <u>185</u>; b_2 <u>213.2</u>; b_3 312.3; b_4 <u>425.5</u>

<u>147.2</u> - 18 = <u>129.2</u> loss of H_2O
y_1 $\quad$ y_1^*

$$548.1 = \frac{1095.3 + 1}{2} = \frac{y_9 + 1}{2} = y_9^{2+} = (y_9'')^{2+}$$

$$491.5 = \frac{982.1 + 1}{2} = \frac{y_8 + 1}{2} = y_8^{2+} = (y_8'')^{2+}$$

$$442.0 = \frac{883.0 + 1}{2} = \frac{y_7 + 1}{2} = y_7^{2+} = (y_7'')^{2+}$$

Figure 9.54 Amino acid sequence, structural notation, and ion identification of the tandem mass spectrum in Figure 9.8. Experimentally observed ions are underlined. (©. 1991. Reprinted from reference 52 by permission of Plenum Publishing Company, New York.)

PROBLEM 9.3 A tryptic-like peptide signifies that either an Arg or a Lys residue is present at the C-terminal end. The m/z 175.2 can be labeled the y_1 ion. As it turns out, no b_n species are observed. Figure 9.55a shows the mass spectrum complete with the y_n labels, and Figure 9.55b shows the analysis. The M_r of the peptide is $(786 \times 2) - 2 = 1570$ Da; the $(M + H)^+$ is 1571 Da. Once the

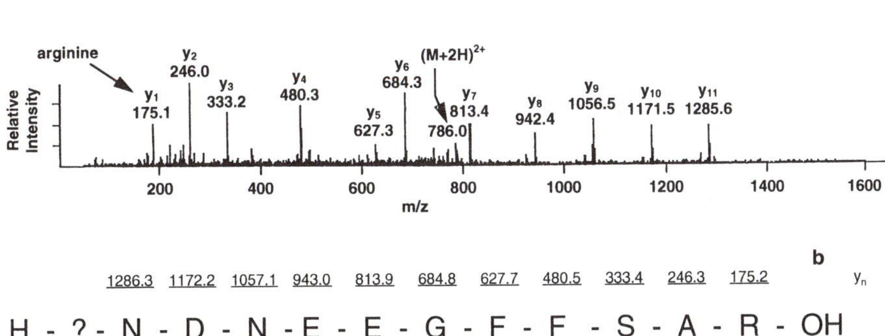

Figure 9.55 (a) The ESI-tandem mass spectrum of the $(M + 2H)^{2+}$ ion from glufibrinopeptide (Figure 9.9) with ion identification labels. (b) Amino acid sequence and ion identification of the tandem mass spectrum in (a). Experimentally observed ions are underlined. (© 1995, American Chemical Society. Reprinted from reference 53 with permission.)

sequence is obtained from the spectrum, there is a marked difference between the last observed ion, y_{11} at m/z 1286.3, and the $(M + H)^+$ value of 1571 Da. This difference is equal to 284.7 Da and can be due to any number of residue combinations. The correct sequence of the first three residues is Glu-Gly-Val. Chemical modification methodologies on these three residues may or may not be detected in the observed y_n series, such as esterification of the Glu residue. Other than the $(M + 2H)^{2+}$ ion, no other doubly charged species is observed. Thus, the charges must reside on the N-terminal amine and the C-terminal Arg residue. The small differences between respective masses in Figures 9.9, 9.55a, and Figure 9.55b are due to experimental error.

PROBLEM 9.4 Figure 9.56a presents the labeled tandem mass spectrum, and Figure 9.56b shows the structural analysis. Indeed, the Leu and Ile residues must be labeled Xle in the context of this problem. The sequence assignment is straightforward except for the first two residues. Both b_2 and y_7 ions are missing; therefore, a mass of 245.2 (subtracting the end H yields a mass of 244.2) must be consulted in Table 9.6. A residue mass of 244 Da can belong

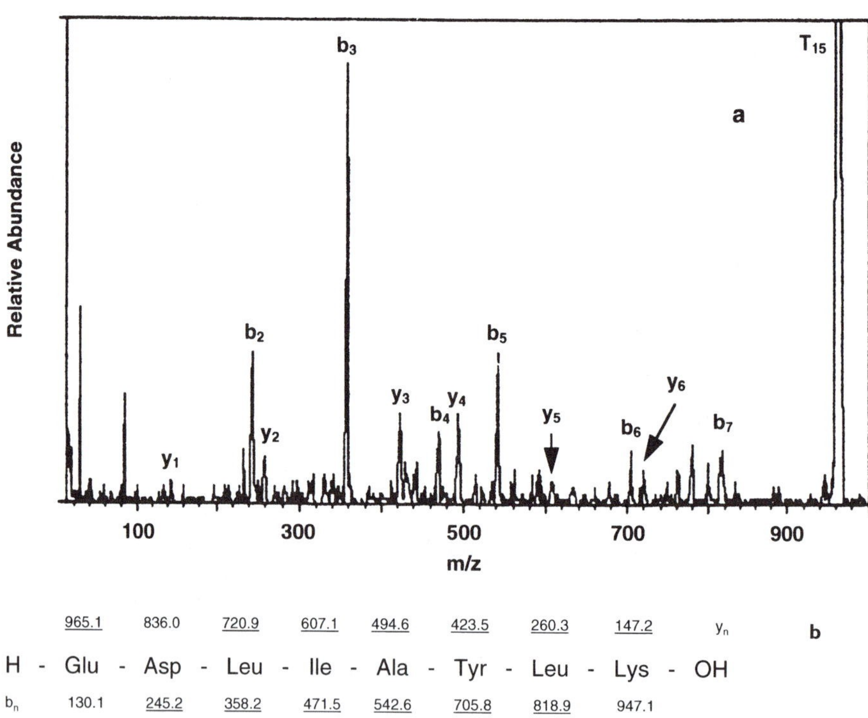

Figure 9.56 (a) The ESI-tandem mass spectrum of the $(M + H)^+$ ion from the equine cytochrome c T15 tryptic peptide at m/z 965.1 with ion identification labels. (b) Amino acid sequence and ion identification of the mass spectrum in (a). Experimentally observed ions are underlined. (© 1991. Reprinted from reference 55 by permission of Academic Press, Inc., Orlando, Florida.)

to three pairs of residues. A combination of chemical and/or enzymatic procedures may shed light on the identity of the residue pair. Methyl esterification will methylate both the Glu[1] and Asp[2] residues. An increase in 28 Da for the b_2 and $(M + H)^+$ ions should be observed, and this can confirm the presence of the Glu–Asp pair among the three possibilities for 244 Da in Table 9.6. However, the order of the two residues cannot be obtained because of the absence of the b_1 and y_7 ions. With this information, treatment of the original peptide with *S. aureus* V8 protease can be performed, first at pH 4.0 and then at pH 7.8. At pH 4.0, the N-terminal Glu will *not* cleave to yield a peptide with an $(M + H)^+$ of m/z 836.0 (Table 9.4). This can signify that Glu is indeed the N-terminal amino acid residue. The same reaction at pH 7.8 should yield peptides that have an $(M + H)^+$ at m/z 836 and 720.9, where cleavages occur at the C-terminal side of both the Asp and Glu residues. This also can establish Asp as the second amino acid in the peptide.

PROBLEM 9.5 First, construct a table of m/z values (Table 9.18) similar to that in Tables 9.9, 9.10, and 9.11. The $(M + 2H)^{2+}$ precursor yields $(M + H)^+$ and M values of 1161.4 and 1160.4 Da, respectively. However, more accurate values of the $(M + H)^+$ and M mass values are 1162.2 and 1161.2 Da. The y_f or $(M + H)^+$ ion is not observed in the spectrum (Figure 9.15). Unfortunately, the following calculations:

$$\begin{array}{r} 1162.2 \\ -965.2 \\ \hline 197.0 \end{array} \qquad \begin{array}{r} 1162.2 \\ -925.8 \\ \hline 236.4 \end{array} \quad \text{Val, His}$$

do not yield a single amino acid value; however, the first two amino acids, comprising b_1 and b_2, appear to be Val or His and Val, His, respectively (Table 9.6). Note that there is no combination of residues that yields a nominal mass of 197 Da. But, it was stated in the problem that both termini were changed and/or modified. Thus, the above analysis is invalid! However, the values 197.0 or 236.4 Da could reflect the first two amino acids with the modification, but which value is it? And what is the N-terminal modification? Obviously, Edman degradation will not release the N-terminal residue, but the modified N-terminal residue and the second residue can both be contained in either 197.0 or 236.4 Da. This is a riddle within a puzzle; therefore, we must look elsewhere.

We were told to look for pairs of masses separated by 17, 18, or 28 Da. This procedure can at least pinpoint a "b" or "y" ion, and they are 817.4, 835.4 (Δ18); 394.8, 423.0 (Δ28.2); 151.8, 179.8 (Δ28). Since both termini are altered from the usual residue structures, 86.0 and 110.0 Da (Figure 9.15) cannot be considered for sequence purposes at present. However, they do match that of the Xle and the His immonium ions, respectively. Thus, there is a strong possibility that these residues are present in the peptide.

From the above analysis, we have a high degree of confidence that m/z 835.4, 423.0, and 179.8 are residue values and either b or y ions. These pieces of information are labeled in Table 9.18a. Table 9.18b continues the analysis.

Table 9.18 Table of consecutive m/z values in Figure 9.19 for a synthetic peptide

a.	86.0 Xle	110.0 His	151.8 -28 ion	179.8 b or y	194.8	216.0	236.8	243.8		
	326.8	356.8	394.8 -28 ion	423.0 b or y	513.2	536.0	572.2	581.2 $(M+2H)^{2+}$		
	626.2	649.0	706.2	739.4	817.4 -18 ion	835.4 b or y	925.8	965.2	1161.2 M	1162.2 $(M+H)^+$ y_f
b.	86.0 Xle	110.0 His	151.8 -28 ion	179.8	194.8	216.0	236.8	243.8		
				residue		Gly				
			Trp*				Xle			
	326.8	356.8	394.8 -28 ion	423.0	513.2	536.0	572.2	581.2 $(M+2H)^{2+}$		
			Trp		Xle					
		Xle			Trp					
	626.2	649.0	706.2	739.4	817.4 -18 ion	835.4	925.8	965.2	1161.2 M	1162.2 $(M+H)^+$ y_f
	Xle	Gly			Glu					

Table 9.18b shows the determination of residues above the mass of 179.8 Da, and the N-terminal modification is mathematically part of m/z 179.8. This analysis is written below the mass series. Note that from m/z 236.8, one needs to proceed to m/z 423.0 to yield another difference that equals a residue. The m/z 394.8 is omitted from the analysis, because it is a decarbonylation ion. Residues are discovered until the following is obtained:

$$\begin{array}{r} 925.8 \\ -835.4 \\ \hline 90.4 \end{array} \qquad \begin{array}{r} 965.2 \\ -835.4 \\ \hline 129.8 \end{array} \neq \text{Glu}$$

Even though it is close, 129.8 is 0.7 Da from the mass of a Glu residue. The previous residue is Glu: 835.4 – 706.2 = 129.2, which is within 0.08 Da of the mass of a Glu residue. Thus, 0.7 Da is an unacceptable level of error, and m/z 925.8 and 965.2 cannot contribute to this ion series. Depending on whether this is a "b" or "y" series, there is a mass difference of 326.8 or 325.8 Da between m/z 835.4 and the $(M + H)^+$ ion or M mass, respectively. Thus, this series of ions cannot unequivocally supply either of the terminal and near-terminal residues. Including the modification, it is possible that m/z 179.8 represents a b_1 or y_1 ion. Being unsure of the status of m/z 194.8, differences are sought from masses in the middle of the sequence that reflect residue values, and are placed above the numerical series. Starting at m/z 513.2:

$$\begin{array}{rl} 513.2 & \\ -356.8 & \\ \hline 156.40 & \\ -156.19 & \text{Arg} \\ \hline \Delta 0.21 & \text{Da} \end{array} \qquad \begin{array}{rl} 513.2 & \\ -326.8 & \\ \hline 186.4 & \\ -186.21 & \text{Trp} \\ \hline \Delta 0.19 & \text{Da} \end{array}$$

Both masses yield values for residues, yet only one is correct. Having already established most of the other ion series, an Arg is not present, while a Trp is. Thus, this can be labeled Trp (Trp* in Table 9.18b). If the other series (b or y) was not addressed first, then this problem (i.e., either Arg or Trp as a residue) would indicate that it must be addressed first. It is possible that m/z 356.8 is not used at all and represents noise or an impurity. However, it is also possible that this represents an internal fragment ion, and can be revisited later.

Proceeding down in mass from 326.8 Da:

$$\begin{array}{r} 326.8 \\ -243.8 \\ \hline 83.0 \end{array} \qquad \begin{array}{r} 326.8 \\ -216.0 \\ \hline 110.8 \end{array} \qquad \begin{array}{r} 326.8 \\ -194.8 \\ \hline 130.0 \end{array} \neq \text{Met}$$

Thus, no b or y ion can be discovered from the mass series that lies below and includes m/z 326.8.

Proceeding above m/z 513.2:

$$\begin{array}{r} 572.2 \\ -513.2 \\ \hline 59.0 \end{array} \qquad \begin{array}{r} 626.2 \\ -513.2 \\ \hline 113.0 \end{array} \text{ Xle} \qquad \begin{array}{r} 739.4 \\ -626.2 \\ \hline 113.2 \end{array} \text{ another Xle}$$

At this point, an overlap of the b and y series is beginning to appear:

925.8	965.2
–739.4	–739.4
186.4 Trp	225.8 Pro, Glu or Xle, Xle

This choice appears to be Trp, because the other series (written below the mass values in Table 9.18b) does not have a Pro residue. However, the lower placed ion series of masses appears to have one Trp, while the upper series of mass assignments in Table 9.18b has two Trp residues. A Trp residue has the same nominal mass as a combination of other amino acids. Table 9.6 shows three combinations of two residues that have a nominal mass of 186 Da. Comparison of the lower series to the residue combinations from Table 9.6 makes the Gly, Glu pair very likely. Thus, before a final determination is made, let us summarize the sequence:

			Glu	Gly								
				Trp*		Xle			Xle		Trp	Gly
236.8	326.8	423.0	456.2	513.2	536.0	626.2	649.0	706.2	739.4	835.4	925.8	982.9
Gly		Trp			Xle		Xle	Gly		Glu		

The Trp* must actually be Glu–Gly in that order. Remember that one series is the reverse of the other. Thus, a "b" or "y" ion is missing in the spectrum, and this ion would have been 513.2 – 57.05 (Gly) = 456.2. A dashed underline mass value indicates its absence in the ESI-tandem mass spectrum.

Also, in the upper series, a Gly is on the C-terminal side of Trp because it is on the N-terminal side of Trp in the lower series; thus, 925.8 + 57.05 = 982.9. Subtraction of this value from y_f and M yields the following:

1161.2 M	1162.2 y_f
–982.9	–982.9
178.3	179.3

The value 179.3 Da, is close to 179.8 Da, as observed in Table 9.18a and b. The y_f originates from the N-terminal side, and it appears that *m/z* 179.3 is the b_1 ion in a modified form. Thus, the b_n series can be as follows:

1162.2							
y_f							
residue -	Gly -	Trp -	Xle -	Xle -	Gly -	Glu -	
b_1	b_2	b_3	b_4	b_5	b_6	b_7	M
						835.4	1161.2

Both sets of ions that belong to the last of the b series and the beginning of the y series are missing in the spectrum. Thus, the identities of the C-terminal residues appear to be in jeopardy. The difference between 1161.2 Da and 835.4 Da is 325.8 Da, and is exactly 1 Da less than *m/z* 326.8. The *m/z* 326.8 was established as a y ion. Thus, the last or C-terminal b residues have a combined mass of 325.8 Da, and the first or N-terminal y residues have a combined mass of 326.8 Da. This makes sense since the y series possesses the extra hydrogen with respect to the b series.

The residue at the N-terminus most likely can be the His residue in a modified form. His is made visible by the very important immonium ion at *m/z* 110. Thus, 179.8 – 137.14 = 42.7 = acetyl derivative. Thus, b_1 can be the acetylhistidyl residue.

Unfortunately, the literal absence of both b and y ions near and at the C-terminus does not provide for the sequence at this end of the peptide.

Note that *m/z* 356.8 is essentially equal to 57.05 (G) + 186.21 (W) + 113.16 (X) = GWX, and this ion is an internal fragment (Scheme 9.4) composed of GWX, where X is either Leu or Ile. Also, *m/z* 243.8 is essentially equal to 57.05 (G) + 186.21 (W) = GW, and appears to be another internal fragment ion. Apparently, no H transfer to the N-terminus has taken place for either ion.

Note that in Table 9.18b, ions such as *m/z* 194.8, 216.0, and 965.2 were not used in the analysis.

The identity of the complete peptide is Ac-H-G-W-I-L-G-E-H-G-D-NH_2, where the C-terminal residue is an amide modification of the carboxyl group, and this results in $y_f = y_{10}$ residue. In addition, the ions at *m/z* 151.8, 179.8 represent a_1, b_1; 394.8, 423.0 represent a_3, b_3; and 817.4, 835.4 are b_7-H_2O, b_7.

PROBLEM 9.6 Each spectrum can be sequenced based on the labeled peaks by using the strategies found in the previous problems. An important clue can be found immediately by comparing the $(M + H)^+$ values of the separate compounds. The difference yields 16 Da and, as such, can be either an oxygen or NH_2 species. Sequencing of the peptide in Figure 9.16a yields the YGGFM peptide, and this is Met-enkephalin. The y_1 or b_5 residue is the site of modification, and Table 9.7 should yield an important clue as to the nature of this modification. A Met residue can undergo oxidation to the methionine sulfoxide residue (+16 Da), and this is apparently what happened as the voltage in the ESI source was increased.

PROBLEM 9.7 Figure 9.57 provides the solution.[88] The compound is Leu-enkephalin and is closely related to Met-enkephalin, which was the subject of Problem 9.6. Note that the determination of the b and y ions can be confusing due to the many immonium and internal fragment ions. This example shows the importance of starting a sequence analysis toward the middle or upper mass region instead of the lower mass region.

Scheme 9.4 shows the nomenclature used to describe the internal fragment ions. However, in order to properly describe them symbolically, the b and y series assignments must be known. In this example, it is possible to make the assignments.

The $a_5y'_1$ is the immonium ion of Leu. The superscript on $a_5y'_1$ signifies the additional hydrogen. An alternative designation would simply be L, because this is a case where an immonium ion is an internal fragment.

The $(a_4y'_3)_2$ is equivalent to a GF designation. The $a_4y'_2$ is the immonium ion of Phe and $(a_4y_4)_3$ is GGF without an additional hydrogen atom. Note that internal fragments need not be only of the b_ny_n variety.

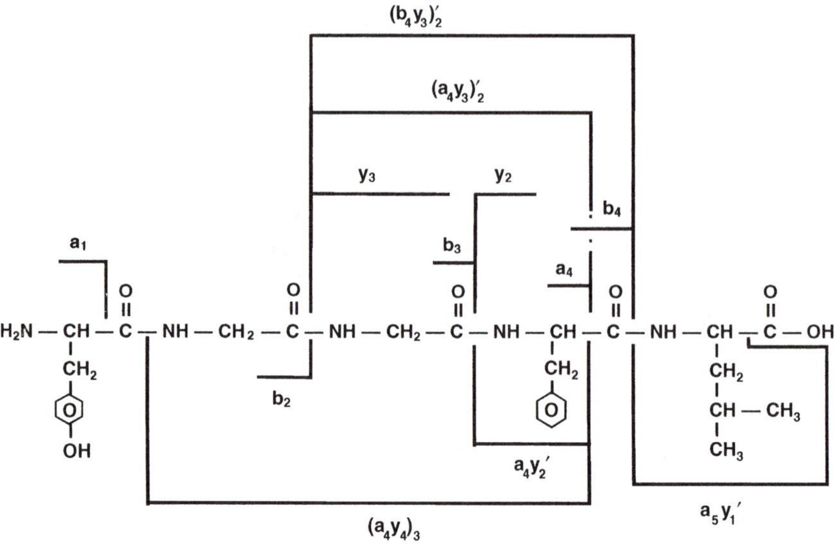

Figure 9.57 Interpretation of the tandem mass spectrum (Figure 9.17) of the Leu-enkephalin peptide YGGFL. (© 1989. Reprinted from reference 88 by kind permission of Elsevier Science-NL, Sara Burgerhartstraat 25, 1055 KV Amsterdam, The Netherlands.)

PROBLEM 9.8 Figures 9.58a and b provide the calculated b and y labels for the ion masses. This sequencing results from a systematic b and y analysis of the Q5 peptide. Remember to add 105 Da to both Cys residues (Table 9.7) due to the S-pyridylethylation reaction. The residues in the spectrum that display a 2-Da shift are all y_n ions of $n \geq 3$; namely, y_3, y_4, y_5, y_6, y_9, y_{11}, y_{13} and y_{14}. Only b_{16}^{2+} of the b series of ions shows a shift of 2 Da. The doubly charged ions are b_{13}^{2+}–b_{16}^{2+}, and the Asp-33 residue that incorporates the radiolabel is found on the b_{16}^{2+} and y_3 ions.

PROBLEM 9.9 The structure is that of a decapeptide with amine and carboxyl termini. Masses 228–989 Da are the b_2–b_9 residues, respectively. The peptide is known as ACP(65–74), or acyl carrier protein(65–74), and the sequence can be deduced except for the b_1 and b_2 residues. The residue mass of the b_1, b_2 pair is 227 Da, which reflects six different combinations of residues (Table 9.6). The correct sequence is VQAAXDYXNG, where X is an isoleucine residue in both cases.

PROBLEM 9.10 Figures 9.23 and 9.24 show the same b_2–b_8 ions. Figure 9.24 is missing the b_9 ion; therefore, the change must occur on either the b_9 or b_{10} residue; namely, Asn-73 or Gly-74. A difference of 17 Da is observed between the $(M + H)^+$ species in Figures 9.23 and 9.24, and this can be due to a loss of ammonia, which signifies the Asn residue. Table 9.7 depicts the cyclization of Asn into a succinimide with loss of ammonia (17 Da), and this is the cause of the change.

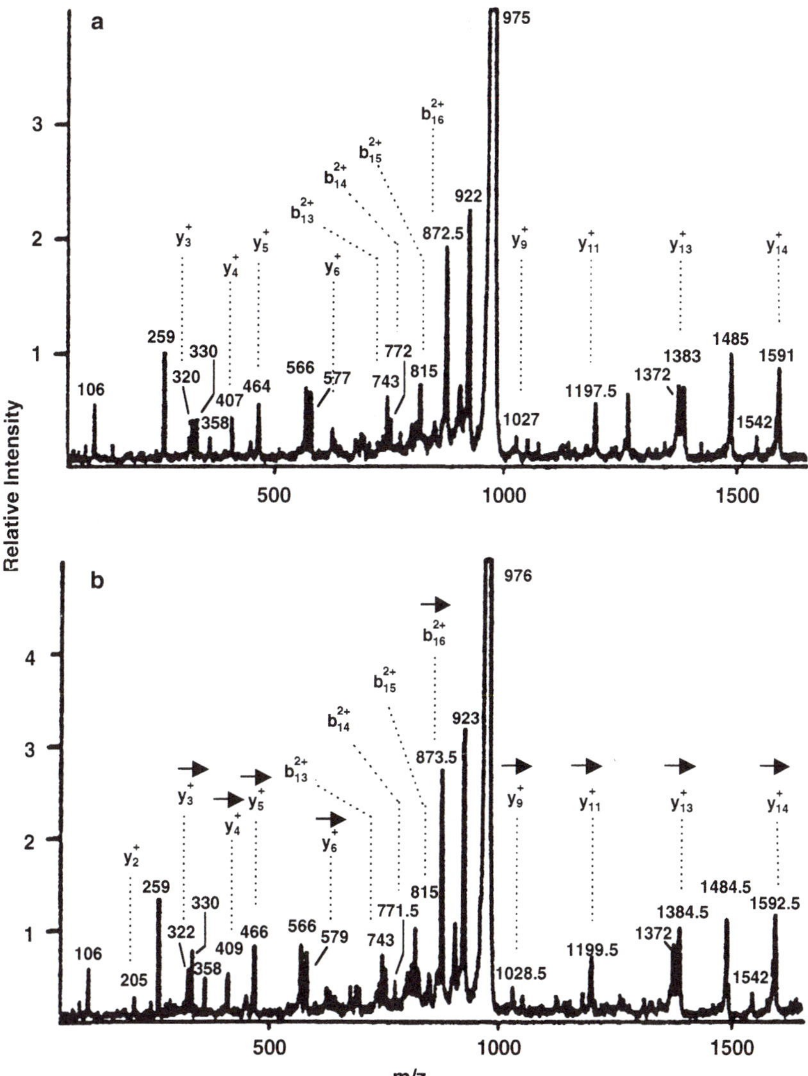

Figure 9.58 The ESI-tandem mass spectra of Figures 9.21a and b with the inclusion of the fragment ion nomenclature. (© 1993. Reproduced from reference 82 by permission of John Wiley & Sons, Limited.)

PROBLEM 9.11 An analysis of the masses (in daltons) listed in Figure 9.25 yields the following:

228, b_2; 299, b_3; 370, b_4; 455.6, a_5; 483, b_5; 595.6, c_6; 695, y_6; 743, b_7; 837, y_8; 857, b_8; 971, b_9.

The absence of a b ion for residue 70 (b_6 is not observed) and the appearance of a c_6 ion indicates a significant change in that particular residue. The $(M + H)^+$

Table 9.19 Sequence analysis of the peptide in Problem 9.15

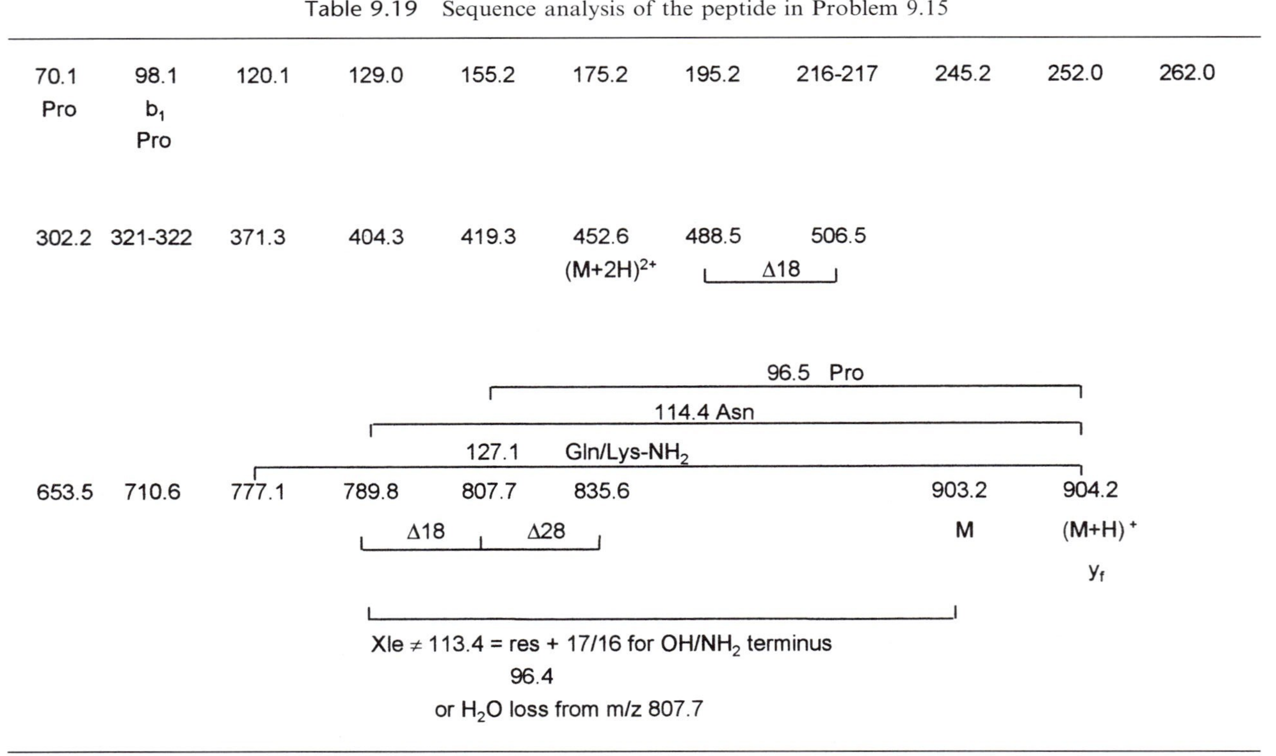

70.1	98.1	120.1	129.0	155.2	175.2	195.2	216-217	245.2	252.0	262.0
Pro	b_1									
	Pro									
302.2	321-322	371.3	404.3	419.3	452.6	488.5	506.5			
					$(M+2H)^{2+}$	Δ18				
				96.5 Pro						
			114.4 Asn							
		127.1 Gln/Lys-NH_2								
653.5	710.6	777.1	789.8	807.7	835.6			903.2	904.2	
			Δ18	Δ28				M	$(M+H)^+$	
									y_f	

Xle ≠ 113.4 = res + 17/16 for OH/NH_2 terminus

96.4

or H_2O loss from m/z 807.7

changes by a loss of 18 Da from the ACP(65–74) peptide. This can indicate the loss of a water molecule. Since a change occurs at Asp-70, formation of a succinimide ring can explain the change, in a similar fashion that loss of ammonia causes cyclization of the Asn-73 residue.

PROBLEM 9.12 The b_2–b_5 ions (m/z 228, 299, 370, and 483, respectively) are the same as that of the ACP(65–74) analysis; however, b_6 (m/z 689) is 91 Da higher than that of the unmodified peptide ($b_6 = m/z$ 598, Figure 9.23). This difference in mass from that of the unmodified peptide continues for the b_7–b_9 ions (m/z 852, 965, and 1079, respectively) and $(M + H)^+$ species. Thus, the b_6 or Asp-70 is again changed, but in a different way from the succinimide cyclization. It was stated earlier that Asp was protected with a benzyl group to guard against unwanted reactions. The benzyl moiety is 91 Da, and, as such, this is the nature of the modification. It occurs in the form of a benzyl ester at the γ-carboxyl group of Asp-70.

PROBLEM 9.13 Figure 9.33a shows the "proline effect" on the $(M + 2H)^{2+}$ of γ-endorphin, where the amide bond in Thr^{12}–Pro^{13} is preferentially cleaved to form the intense y_5, b_{12} complementary ions. Note that in Figure 9.33b, for the tandem mass spectrum of the $(M + 3H)^{3+}$ species, the b_{12} ion is greatly reduced in abundance, and y_5 is of negligible intensity. The sites of the two protons in the $(M + 2H)^{2+}$ species can be visualized on the N-terminus and Lys residue. Proton affinity considerations (Chapter 4 and references 17, 18, and 98) possibly allow for the third proton in the $(M + 3H)^{3+}$ species to be located on either the Gln-11 or the Pro-13 residue itself. The latter case would provide a reason as to the reduced relative intensities of the y_5 and b_{12} residues, because Pro itself would be charged.

PROBLEM 9.14 The answer is in Figure 9.6a. For the Pro-15 residue, the y_5 ion displays a strong abundance, and for Pro-2, the b_2 ion shows a much lower abundance because of its proximity to the proton-charged N-terminus. The b_{14} ion, of the b_{14}, y_5 complementary pair, is of low intensity because the acidic Tyr residue is located next to the Pro N-terminus. The y_{18} species, of the b_2, y_{18} complementary pair, is absent.

PROBLEM 9.15 An analysis of this problem can be found in reference 11 by Papayannopoulos. However, a detailed analysis follows, mainly because some additional pitfalls can be realized in the interpretation of the spectrum.

An initial analysis begins at the high-mass end since it is significantly less complex than in the low-mass range. Table 9.19 provides a mass framework; note that more masses are included than are labeled in Figure 9.34. The $(M + 2H)^{2+}$ ion yields the M and $(M + H)^+$ values in Table 9.18. A high-mass analysis yields the information in Table 9.19. Three possibilities exist for the b_1 ion, and a possibility exists for the y_1 ion. Because of the significant intensity of m/z 807.7, this value is most likely that of a y residue. This may be ascribed to a Pro residue. Since this appears to be the N-terminal residue in the y_f ion, this also is equal to the b_1 ion. A b_1 mass for Pro is 97.1 + 1 = 98.1 Da, and an m/z 70 Pro immonium ion is also noted. Note that m/z 789.8 could

Table 9.20 Sequence analysis of the peptide in Problem 9.15

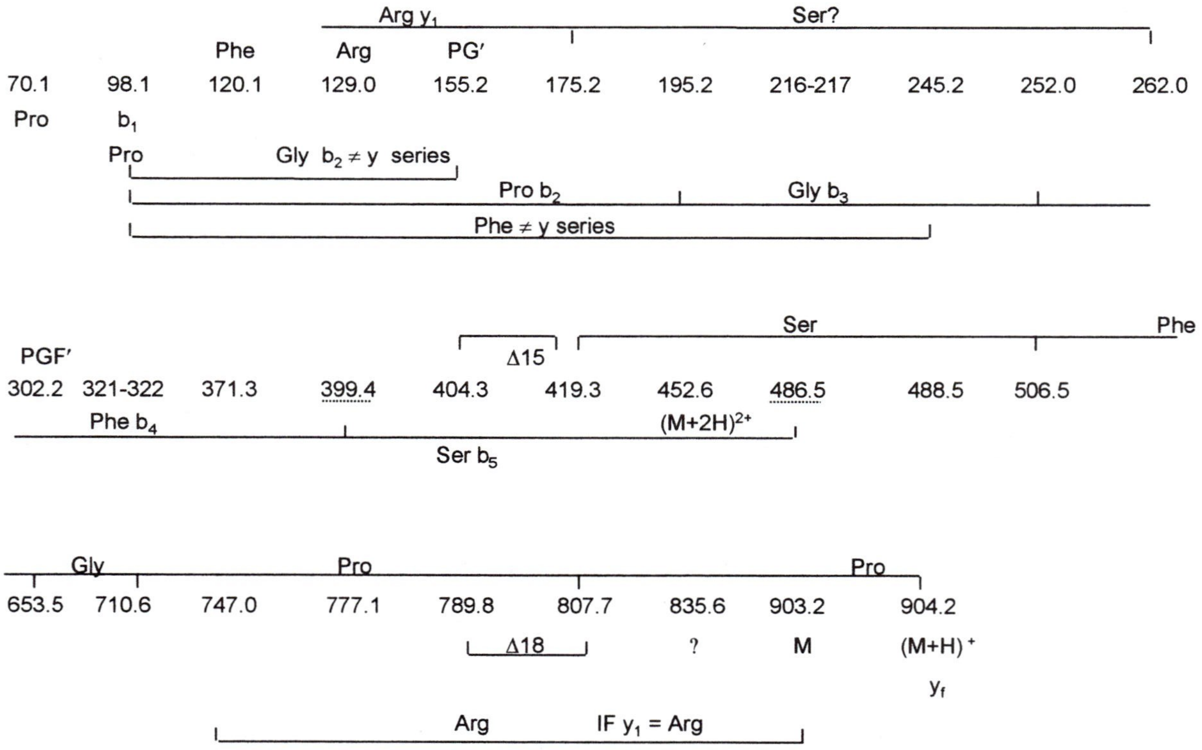

			Arg y_1				Ser?			
		Phe	Arg	PG′						
70.1	98.1	120.1	129.0	155.2	175.2	195.2	216-217	245.2	252.0	262.0
Pro	b_1									

Pro Gly $b_2 \neq y$ series

Pro b_2 Gly b_3

Phe $\neq$ y series

				Δ15		Ser				Phe
PGF′										
302.2	321-322	371.3	399.4	404.3	419.3	452.6	486.5	488.5	506.5	
	Phe b_4					$(M+2H)^{2+}$				
				Ser b_5						

Gly			Pro					Pro
653.5	710.6	747.0	777.1	789.8	807.7	835.6	903.2	904.2
				Δ18		?	M	$(M+H)^+$
								y_f

Arg IF y_1 = Arg

be the final b ion (b_f), because it appears to be the dehydration product of m/z 807.7. This must be temporarily put aside until more information is gathered. Given a Pro as the b_1 ion and the N-terminal residue in the y_f ion, a partial series of y ions becomes obvious (Table 9.20): H_2N-Pro-Pro-Gly-Phe-Ser. Note the intense nature of the high-mass y ions. Since these ions do not include the N-terminus, and the spectrum was generated from an $(M + 2H)^{2+}$ ion, it seems reasonable to believe that a strong proton acceptor is near the C-terminus, perhaps a Lys/Arg. Once the partial y series is obtained, the b_n ion mirror-image series can be sought. Note the possible confusion for the b_2 ion. Since Pro is b_1, a Gly, Pro, or Phe can be postulated as b_2 from the N-terminal low-mass analysis (Table 9.20). However, Pro is clearly shown as the terminal residues for the y_f and y_{f-1} ions. Thus, Pro must also be a b_2 ion. The b_3 ion is also present; however, the b_4 and b_5 ions are absent (dashed underline).

Note that m/z 129 can be the immonium ion of Arg or an immonium ion fragment of Lys. It most likely is not Gln, because m/z 129 is approximately 20% of the base peak (Figure 9.34), while that of Gln is of very low intensity (Table 9.13). The y_1 ion of Arg and Lys is 147.2 and 175.2 Da, respectively. The m/z 175.2 is present, and thus Arg can be considered the y_1 ion, as well as the C-terminal residue. A Gly, Val combination also appears feasible (Table 9.6) in place of Arg; however, it would be difficult to rationalize the appearance of the $(M + 2H)^{2+}$ and the high-intensity, high-mass y ions if a Lys/Arg were not present in this peptide. Thus, the protonation sites for the peptide reside on each terminus. The Arg also voids the possibility that the C-terminus residue is a Pro (Table 9.19). Note also the appearance of the two internal fragment ions PG′ (m/z 155.2) and PGF′ (m/z 302.2), and the Phe immonium ion at m/z 120. A Ser appears to be a possibility for y_2. Figure 9.59 condenses the results obtained in Tables 9.19 and 9.20.

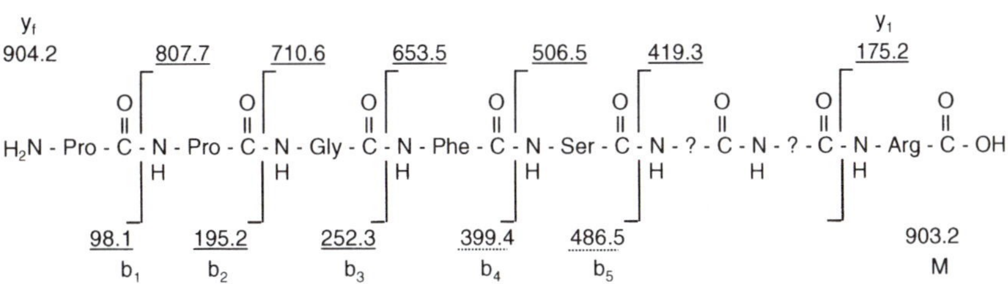

y_2 Possibilities	y_2 Mass	Presence in Spectrum
Pro-Arg-OH	272.3	No
Phe-Arg-OH	322.4	Yes
Xle-Arg-OH	288.4	No
Met-Arg-OH	306.4	No
Asp-Arg-OH	290.3	No
Glu-Arg-OH	304.3	No

Figure 9.59 Partial sequence and ion identification nomenclature for the peptide in Figure 9.34.

Note that m/z 404.3 has the same mass as that of the z ion of the m/z 419.3 y ion. However, its intense nature leads to speculation as to its relationship with the high-intensity, high-mass y ions. The m/z 404.3 ion can be the doubly charged ion of m/z 807.7.

If Ser is indeed y_2, it should be present in the analysis of the difference between y_1 and m/z 419.3. A residue difference of 244 Da (Table 9.6) does not yield Ser as a possibility for the y_2 ion. Figure 9.59 shows six different possibilities for the y_1, y_2 ion pair. Only m/z 322 can be observed in the spectrum. It appears that m/z 304 may be present, despite the absence of the written mass label. There appears to be no more information, except for the possibility that some of the unassigned mass values may yield internal fragments in this region; that is, for PF. The PF′ internal fragment has a mass of 245.3 Da and is present in the spectrum. Thus, the complete sequence of the peptide is Pro-Pro-Gly-Phe-Ser-Pro-Phe-Arg, and this is termed des-Arg[1]-bradykinin.

PROBLEM 9.16

$$
\begin{array}{rll}
 & v_5^2 & \\
y_5 = & 691.8 & \\
 & -1 & \\
\hline
 & 690.8 & \text{Thr-32} \\
 & -45 & HC(CH_3)OH \\
\hline
 & 645.8 & \\
\end{array}
$$

$(645.8 + 1)/2 = 323.4 = v_5^2$

$$
\begin{array}{rll}
 & v_{22}^4 & \\
y_{22} = & 2760.2 & \\
 & -1 & \\
\hline
 & 2759.2 & \text{Glu-15} \\
 & -73 & CH_2CH_2COOH \\
\hline
 & 2686.2 & \\
\end{array}
$$

$(2686.2 + 3)/4 = 672.3 = v_{22}^4$

$$
\begin{array}{rll}
v_{28}^4 & & \\
3315.7 & & \\
-1 & & \\
\hline
3314.7 & & \\
-1 & H & \text{Gly-9} \\
\hline
3313.7 & & \\
\end{array}
$$

$(3313.7 + 3)/4 = 829.2 = v_{28}^4$

$$
\begin{array}{rl}
v_{30}^4 & \\
3576 & \\
-1 & \\
\hline
3575 & \text{Tyr-7} \\
-107 & H_2C\text{–}C_6H_4\text{–}OH \\
\hline
3468 & \\
\end{array}
$$

$(3468 + 3)/4 = 867.8 = v_{30}^4$

$$
\begin{array}{rl}
v_{31}^4 & \\
3675.1 & \\
-1 & \\
\hline
3674.1 & \text{Val-6} \\
-43 & H_3C\text{–}HC\text{–}CH_3 \\
\hline
3631.1 & \\
\end{array}
$$

$(3631.1 + 3)/4 = 908.5$

$$
\begin{array}{rll}
 & x_{23}^4 & \\
y_{23} = & 2857.3 & \\
 & +28 & \\
\hline
y_{23} = & 2885.3 & \\
 & +3 & \text{Hydrogen ions} \\
\hline
 & 2888.3 & \\
\end{array}
$$

$2888.3/4 = 722.1$

$$
\begin{array}{rl}
 & y_{17}^3 \\
y_{17} = & 2172.5 \\
\end{array}
$$

$(2172.5 + 2)/3 = 724.8$

v^4_{24}	w^4_{24} from Figure 9.57:
2958.4	$w^4_{24a} = 732.1$; $w^4_{24b} = 732.6$
−1	
2957.4	
−45 Thr	
2912.4	
$(2912.4 + 3)/4 = 728.9$	

The four ions form a close series with respect to mass, and can be observed in Figure 5.55b.

PROBLEM 9.17 (a) The y^5_{36} and y^5_{34} delineate this region. Thus, $y_{36} = 4218.9$ Da and $y_{34} = 4018.7$ Da. The difference is 200.2 Da. Table 9.6 provides four different pairs of amino acids whose summed peptide residue masses equal 200 Da. They are Ala, Glu; Ser, Xle; Val, Thr; and Cys, Pro. Five ions related to the seventh residue are observed, and two ions related to the ninth residue are observed. Of these seven ions, only the w^5_{36} ion provides fragmentation directly related to the R-group. This is so because part of the R-group is retained and part of it is lost in the fragmentation reaction. Thus, the portion that is lost may be the key in deriving an *m/z* value to match that of the given value of w^5_{36}.

Since we know that Ile is in position 7, we can start here:

$y_{36} = 4218.9$
−15
4203.9
−1

$z_{36} + 1 = 4202.9$	4202.9
−29 H_2CCH_3	−15 CH_3
4173.9	4187.9
$(4173.9 + 4)/5$	$(4187.9 + 4)/5$
$w^5_{36a} = 835.6$	$838.4 = w^5_{36b}$
not observed	observed

Only one of the two potential mass values is observed in Figure 9.51. It is the mass of the leaving group that can differentiate and yield the correct value of 838.4 Da for the w^5_{36} ion.

Thus, the mass of the leaving group of each candidate residue can be interrogated:

Residue	Leaving group mass
Ile	15
Ser	17
Ala	1
Glu	59
Val	15
Thr	15, 17
Cys	46
Pro	—

There is a redundancy in the potential pairs of amino acids. The Ala–Glu and Cys–Pro combinations can be eliminated, but, with the mass spectral data given, the choice can be either Ile, Ser or Val, Thr. Both pairs have identical 15- and 17-Da leaving groups: therefore, they both yield the same w_{36}^5 ion mass. Less spectral congestion in the region of the w_{36}^5 ion could allow a difference to be obtained as to whether a 15- or 17-Da loss occurred from residue number seven. The w_{36b}^5 ion of an Ile residue is 838.4 Da, and the hydroxyl loss from Thr or Ser is

$$\begin{array}{rl} z_{36} + 1 = & 4202.9 \\ & \underline{-17} \\ & 4185.9 \\ (4185.9 + 4)/5 = & 838.0 \text{ Da} \end{array}$$

Only 0.4 Da separates the possibilities for residue number seven, and this degree of discrimination cannot be obtained by a visual analysis of Figure 9.51c.

(b)

d_{41}^6

$$\begin{array}{rl} b_{41} = & 4791.5 \\ & \underline{-28} \\ a_{41} = & 4763.5 \\ & \underline{+1} \\ & 4764.5 \\ & \underline{-29} \\ & 4735.5 \end{array} \qquad \begin{array}{l} 4764.5 \\ \underline{-15} \\ 4749.5 \end{array}$$

$(4735.5 + 5)/6 = 790.1 = d_{41a}^6$ $\quad (4749.5 + 5)/6 = 792.4 = d_{41b}^6$

a_{41}^6

$$\begin{array}{rl} y_{41} = & 4791.5 \\ & \underline{-28} \\ a_{41} = & 4763.5 \end{array}$$

$(4763.5 + 5)/6 = 794.8 = a_{41}^6$

$y_{34}^5 = (4018.7 + 4)/5 = 804.5$

z_{34}^5

$$\begin{array}{rl} y_{34} = & 4018.7 \\ & \underline{-15} \\ & 4003.7 \\ & \underline{-2} \\ z_{34} = & 4001.7 \end{array}$$

$(4001.7 + 4)/5 = 801.1$

w_{28}^4

$$\begin{array}{rl} y_{28} = & 3291.9 \\ & \underline{-15} \\ & 3276.9 \\ & \underline{-1} \\ & 3275.9 \\ & \underline{-43} \\ & 3232.9 \end{array}$$

$(3232.9 + 3)/4 = 809.0$

These ions are within a mass range of about 20 Da and occur on the rising slope of the $(M + 6H)^{6+}$ precursor ion signal.

10

Interpretation of Phosphoprotein Mass Spectra

Chapter 9 presents a comprehensive introduction to the analysis of proteins by electrospray ionization-mass spectrometry (ESI-MS). A recurring theme in Chapter 9 was the changes and modifications that can take place to the 20 basic amino acids. These modifications can be imposed by experimental design, or they can be found in vivo as part of the protein. "Artificial" changes to proteins are usually performed in order to assist in the characterization of the sequence and structure with respect to the interrogational analytical technique. In vivo modifications of proteins and peptides take place after their synthesis in association with ribosomes and a messenger-RNA (mRNA) template.[1] Chapter 9 portrayed many types of changes to the basic amino acid sequence, and these changes are termed posttranslational modifications (PTM). That is, after the mRNA is translated into a sequence of amino acids, subsequent cellular entities can modify the nascent peptide/protein at various cellular locations. This chapter deals with a specific type of PTM, that of phosphorylation. The majority of phosphoproteins are phosphorylated on three amino acids.[2] Phosphorylation is an addition of HPO_3 to the hydroxyl side group of Ser, Thr, and Tyr, and this results in an H_2PO_4 moiety attached to the side-group carbon atom. The phosphate group is added to a protein by a protein kinase, and a phosphate can be removed from a Ser, Thr, or Tyr side group by a protein phosphatase.[1,3]

Approximately one-third of mammalian proteins are phosphoproteins.[4,5] Some are relatively isolated with respect to interaction with other phosphoproteins; however, many phosphoproteins interact directly with other phosphoproteins. A majority of the phosphoproteins facilitate cellular events, such as signal transduction.[4,6–8] That is, phosphoproteins act as intermediaries between the triggering or activation of cell-surface receptors by extracellular stimuli (hormones; cytokines; neurotransmitters; light stimulation of photoreceptors;

antigen, protein, and cell-surface interactions) and intracellular processes.[7,9] Some of these pathways can be characterized as cascades, where various internal proteins and other entities interact with each other like "dominoes." Usually, when a protein or peptide is phosphorylated, it is considered as activated; and when it is dephosphorylated, it is inactivated. In addition, phosphoproteins regulate intracellular reactions, stimulated by cell-surface receptors, by reversible phosphorylation/dephosphorylation events.[4,10]

The most common type of protein kinase is the Ser/Thr kinase family of phosphotransferases; thus, phosphate addition is most commonly found on the Ser and Thr amino acid residue hydroxyl groups.[2,11] There can be more than one phosphate on a protein, and the phosphate moiety can occur on adjacent residue sites or on more widely spaced residues in the protein sequence. The next most common phosphate modification to a protein is effected by a protein tyrosine kinase. Other residues that can be modified with a phosphate include histidine, arginine, lysine, cysteine, aspartic acid, and glutamic acid.[2] In this chapter, only the Ser, Thr, and Tyr amino acids will be considered, and serine is found as a phospho-amino acid residue to a greater extent than the other amino acids.

Protein kinases usually derive the phosphate from the γ or terminal phosphate from the energy-rich adenosine triphosphate (ATP) molecule.[2,5,10] Kinases not only phosphorylate other proteins on the side-group hydroxyl sites, but they also catalyze phosphate addition on themselves (autophosphorylation).[10,12–15]

Table 10.1 presents the Ser, Thr, and Tyr phospho-amino acid residues and their fragmentation characteristics in the conventional, up-front CID, and the CID or tandem mass spectrometry modes of energy deposition. Depending on the residue and energy deposition process, different residue and phosphoryl fragments are produced, and each plays an important role in providing mass spectral information with respect to amino acid sequence characterization procedures. Table 10.1 acts as a convenient reference for the examples presented in this chapter.

Finding the Location of the Phosphoryl Moiety in a Peptide

This section presents a number of different techniques and methods to identify the protein section, protein fragment, or enzymatically generated peptide(s) (Table 9.4) that contain(s) the phosphate moiety. Direct infusion or LC separation methods of a sample into the ESI-mass spectrometer highlight these techniques.

Neutral Loss Experiments

The nicotinic acetylcholine nerve receptor (AChR) has a number of protein subunits. There are two sites of phosphorylation on the delta AChR subunit, and in this case the nerve receptor protein was obtained from the Torpedo electric ray. The subunit was digested with Lys-C enzyme, and an HPLC total ion current (TIC) spectrum is shown in Figure 10.1.[1] In this complex chromatogram, two peptides exist that are phosphorylated, and they are known to be found on serine residues. The positive-ion mode was interrogated, and since the neutral-loss mode uses tandem mass spectrometry with the central collision cell elevated in pressure with argon gas, the central section of Table 10.1 is referred to for interpretation purposes. Fragmentation of phospho-Ser/Thr residues results in losses of H_2O and HPO_3. Combined, these losses yield an H_3PO_4 molecule of 98 Da. However, the loss actually takes place in two steps, where the HPO_3 moiety is lost first and then the side chain undergoes a dehydration to yield an H_2O neutral loss. The residue becomes unsaturated by forming a double bond as shown in Table 10.1. Serine becomes a dehydroalanyl residue, and threonine becomes a dehydro-2-aminobutyryl residue. This can be shown using serine as an example: $H_2O_3P{-}O{-}CH_2{-}C\ (R,\ R'){-}H \rightarrow PO_3H + HO{-}CH_2{-}C\ (R,\ R'){-}H \rightarrow H_2O + H_2C{=}C\ (R,\ R')$ where R and R′ are the N- and C-terminal moieties on the Ser residue.

In a neutral-loss scan, the first and third quadrupoles are scanned at the same rate with an offset of 98 Da. Signals that result from sample introduction signify that a sample that enters the collision cell releases a fragment of 98 Da if it is singly charged. A doubly charged species must be scanned with the quadrupoles offset by 49 Da, and this latter value was used in order to discern the phosphopeptides in the Lys-C AChR digest. Figure 10.2 shows an experiment[20] where the complete Lys-C AChR digest was interrogated by HPLC neutral-loss mass spectrometry using 49 Da as the offset. Figure 10.2 shows two nearly coeluting peaks at approximately 50 min, and this is the same peak marked with an asterisk in Figure 10.1. These were the only peaks in the neutral-loss mass spectrum that indicated two peptides, each of which contained at least one phosphate group. The mass spectrum of each peak is shown in Figure 10.3, and the higher mass peptide elutes earlier than the lower mass peptide (Figure 10.2). Both display doubly and triply charged parent ions. The doubly charged mass spectral peaks provide higher intensities than the respective triply charged peaks. Other methods, such as Edman degradation, identified the details of phosphorylation on each peptide.

PROBLEM 10.1 In the above neutral-loss analysis, two phosphopeptides are observed. It was found that the amino acid sequence was identical for each peptide. The difference between the two resided within the presence of phosphate. Explain this observation in light of the different observed masses in Figure 10.3.

One other point is worth mentioning. The HPLC mode in this example[20] is that of a reversed-phase HPLC (RP-HPLC), and RP-HPLC preferentially retains nonpolar compounds to a greater extent with respect to polar compounds. Note that in Figure 10.2 the higher mass peptide with two phosphates elutes earlier than the lower mass peptide with one phosphate. Under RP-HPLC conditions, the higher polarity biphosphate species would be expected to elute earlier than the relatively lower polarity monophosphate peptide.

Table 10.1 Gas-phase decomposition of phosphate-containing amino acid residues for $(M + H)^+$ ions and $(M - H)^-$ ions

Spectrum	Residue	Mechanism	Ion type[a]	Reference(s)
Conventional mass spectrum	Tyr	HO-P(O)(OH)-O-⟨O⟩-CH₂-C(H)(N-H)(C=O) → HO-⟨O⟩-CH₂-C-H + HPO_3 80Da	±	1
		→ HO-⟨O⟩-CH₂-C-H + HPO_4 96Da	±	1
	Ser/Thr	HO-P(O)(OH)-O-CH₂-C(H)(N-H)(C=O) (Ser) → HO-CH₂-C-H + HPO_3 80 Da	±	1
		→ H—CH₂-C-H + HPO_4 96 Da	±	1
Tandem (CID) mass spectrum	Tyr	HO-P(O)(OH)-O-⟨O⟩-CH₂-C(H)(N-H)(C=O) → HO-⟨O⟩-CH₂-C + HPO_3	$(M + H)^+$	16
		→ no phosphoryl loss	±	1, 16, 17
		→ PO_3^- 79 Da	$(M - H)^-$	3

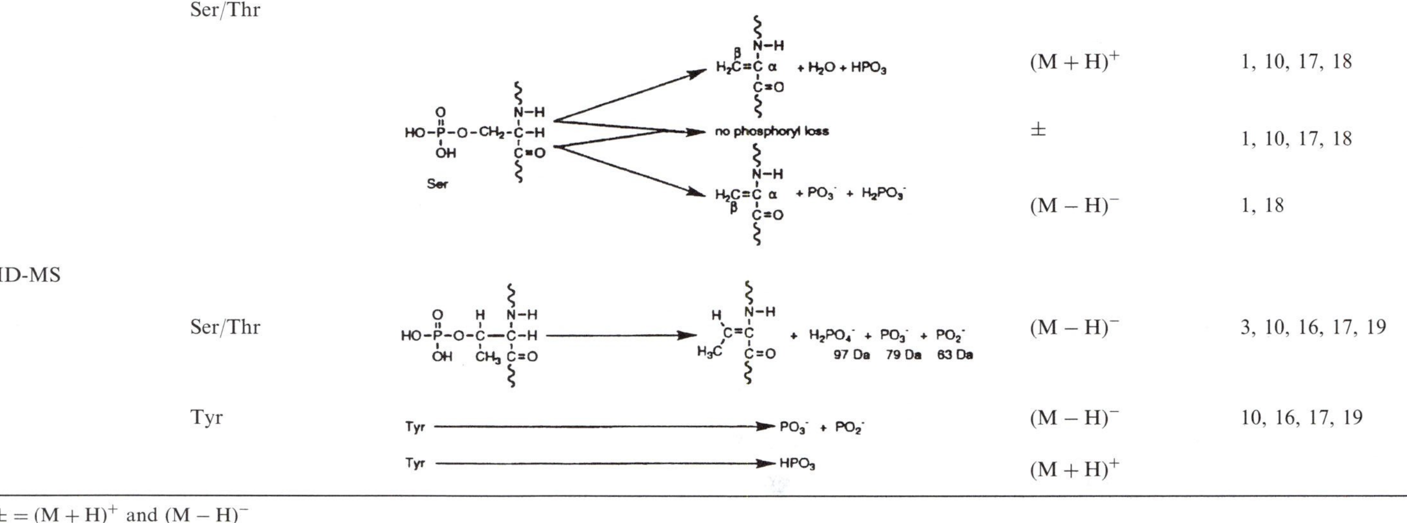

Method	Residue	Fragmentation	Mode	References
	Ser/Thr	(see diagram above)	$(M+H)^+$	1, 10, 17, 18
			±	1, 10, 17, 18
			$(M-H)^-$	1, 18
CID-MS	Ser/Thr		$(M-H)^-$	3, 10, 16, 17, 19
	Tyr		$(M-H)^-$	10, 16, 17, 19
			$(M+H)^+$	

[a] $\pm = (M+H)^+$ and $(M-H)^-$

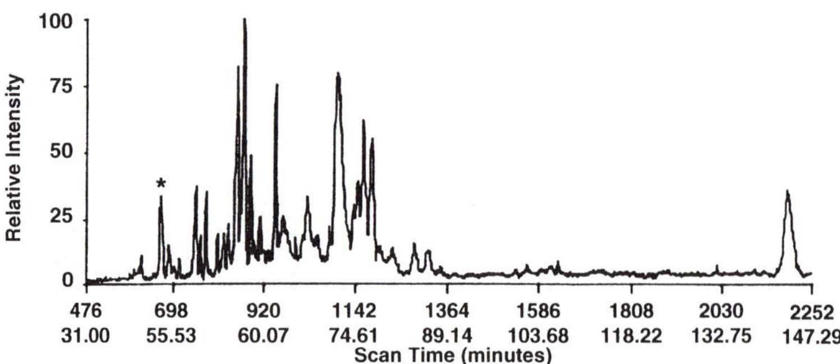

Figure 10.1 The ESI-LC mass spectral total ion current (TIC) of a Lys-C digest of the AChR delta subunit. (© 1990. Reprinted from reference 1 by permission of Plenum Publishing Company, New York.)

HPLC-Mass Spectral Identification

Another method of identifying a phosphopeptide in an enzymatic digest is shown in Figure 10.4 for a tryptic digest of profilaggrin.[21] Profilaggrin is a phosphoprotein that enters into the terminal or maturation stages of differentiation in mammal epidermal skin tissue.[22] In the differentiation process, profilaggrin undergoes a series of dephosphorylation steps, catalyzed by protein phosphatases, and is also enzymatically degraded to yield the filaggrin protein. Filaggrin is subsequently involved in the binding of cellular filaments. Figure 10.4 presents the RP-HPLC two-dimensional mass spectral representation of the elution profile. This is a contour projection of an HPLC plot such as that shown in Figure 10.1. The numerous dots and dashes in the plots all represent discrete mass spectral peaks. Note, in particular, the two circled areas. The upper circle contains three short horizontal streaks, and each streak is composed of a discrete mass which elutes over a finite time period. The difference in mass between each streak is approximately 26.7 Da (or 27 Da on the vertical scale), and each mass has a charge of 3+. Thus, each mass differs by 80 Da (HPO_3). A similar analysis

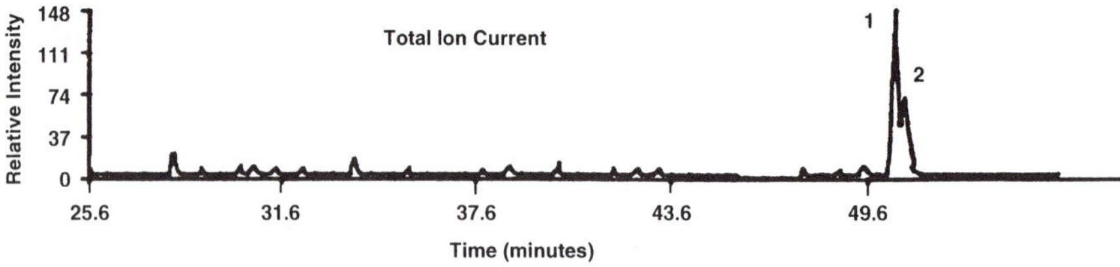

Figure 10.2 The ESI-LC neutral-loss mass spectrum of a Lys-C digest of the AChR-δ subunit. A mass difference of 49 Da was used, which represents the loss of H_3PO_4 from $(M + 2H)^{2+}$ peptides. (© 1991. Reprinted from reference 20 by permission of Plenum Publishing Company, New York.)

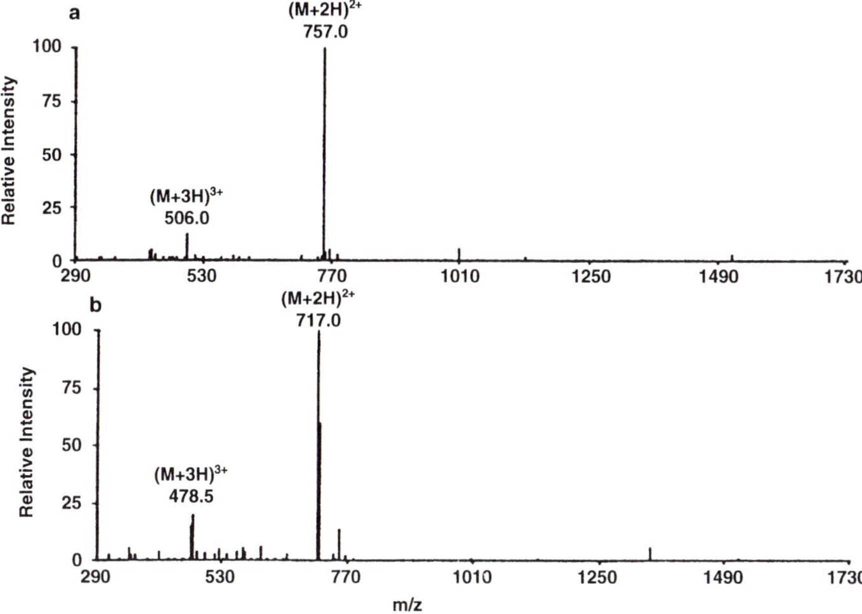

Figure 10.3 The ESI-LC mass spectra of the phosphorylated peptides observed in the Figure 10.2 neutral-loss mass spectra. The spectra represent the peptides labeled in Figure 10.2 as (a) 1 and (b) 2. (© 1991. Reprinted from reference 20 by permission of Plenum Publishing Company, New York.)

can be made of the lower circled streaks, where they each differ by 20 Da. Since there is a 4+ charge on these masses, they too differ by 80 Da. The earliest eluting peak in the upper and lower circle in Figure 10.4 is the 4+ and 3+ form of a diphosphopeptide, respectively, and the middle peak in the upper and lower circle represents the 4+ and 3+ form, respectively, of the same peptide in the monophosphate form. The later eluting peak in the upper and lower circle in Figure 10.4 represents the 4+ and 3+ form, respectively, of the same peptide in an unphosphorylated state. Thus, all six masses in the two circles represent the same peptide in different charge and phosphorylation states.

Up-Front CID

A technique was developed by Carr[3] (stepping voltage method) that uses CID-MS (up-front CID) and conventional scan mass spectra in the same single mass spectrum. This seemingly paradoxical statement separates the two mass spectral techniques in one scan using the element of time. The conditions were as follows: m/z 59–99 and m/z 400–2000 were scanned in one mass spectral acquisition in 0.25-m/z increments with a total scan time of 6 s. Here, 1642 Da/6000 ms = 0.27 Da/ms. The first 41-Da scan takes 41 Da/(0.27 Da/ms) = 152 ms, and a relatively high up-front CID voltage (−350 V) is applied. After this time period, and

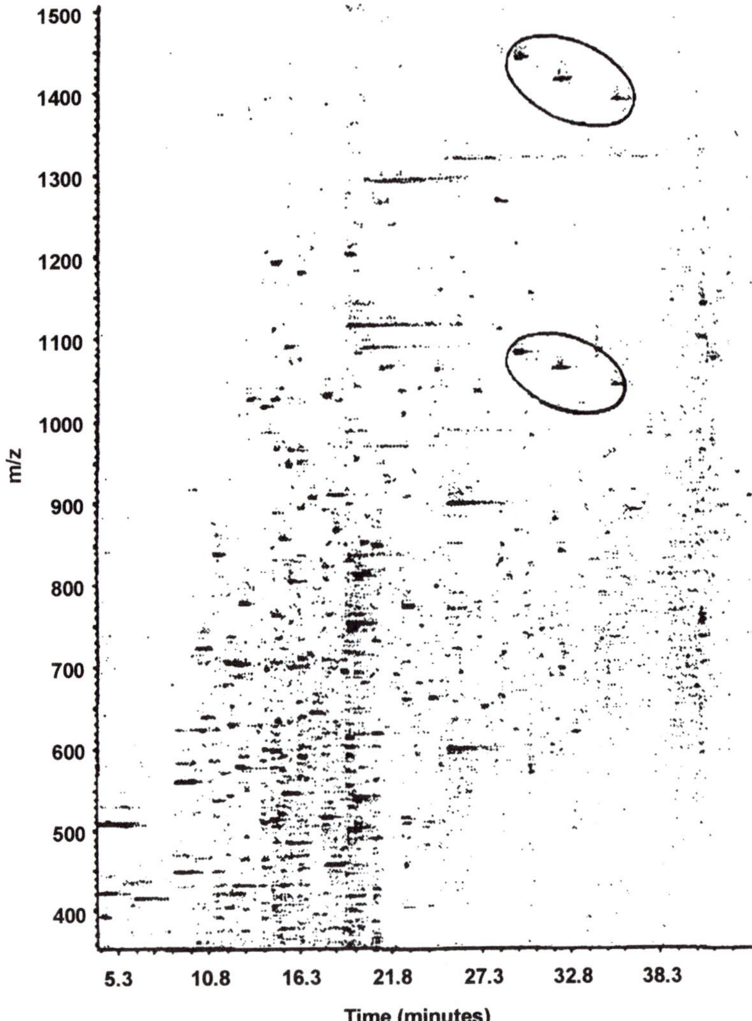

Figure 10.4 The HPLC-ESI mass spectral contour plot of the elution of a tryptic digest of epidermal profilaggrin protein. A 2-mm i.d. C_{18} Hypersil column and acetonitrile gradient was used. The top and bottom circles represent masses in an $(M + 3H)^{3+}$ and $(M + 4H)^{4+}$ protonated state, respectively. (© 1993. Reprinted from reference 21 by permission of Plenum Publishing Company, New York.)

for the remainder of the mass spectral scan (6000 − 152 = 5848 ms), the voltage is dropped (stepped-down) to −115 V.

A mixture of 16 peptides, where three of them (numbers 3, 4, and 14) are phosphopeptides, was eluted and interrogated with RP-HPLC-ESI-MS using the stepped-down, up-front CID collision voltage.[3] Figure 10.5 displays peaks for all the peptides (shaded peaks), because this is the total ion current (TIC). However, when only the 63-Da (PO_2^-) and 79 Da (PO_3^-) masses are involved

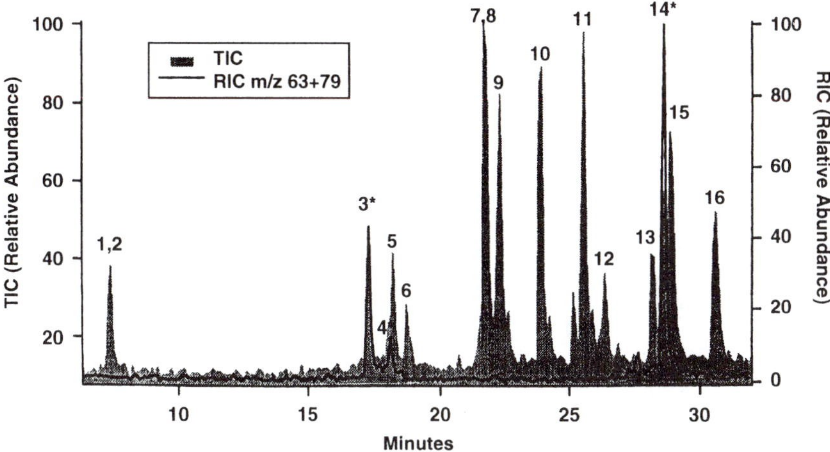

Figure 10.5 Overlay of two HPLC-ESI mass spectra of a mixture of 16 peptides. The mass spectral parameters exhibited a stepped-collisional mode for up-front CID in the *m/z* 59–99 region when using −350 V on the mass spectrometer interface plate while keeping the voltage on the quadrupole ion-focusing rods constant. An immediate lowering of the voltage to −115 V occurred in the remaining 400–2000-Da scan. This process was repeated for each mass spectrum and represents the TIC (shaded chromatogram). Three peptides were phosphopeptides, and the RIC of *m/z* 63 + 79 yielded the solid line that contains peptide numbers 3*, 4*, and 14*. (© 1993, American Society for Mass Spectrometry. Reprinted from reference 3 by permission of Elsevier Science, Inc.)

in the reconstructed ion chromatogram (RIC) in Figure 10.5 (dark line tracing), the three phosphopeptide signals are observed. Note, in particular, that phosphopeptide 4* is hidden in the peptide 5 peak; however, it is clearly observed in the 63- and 79-Da RIC trace. Peak 14* appears as a shoulder on the fast side of peptide 15 in the TIC, and it is clearly observed upon 63- and 79-Da RIC analysis in Figure 10.5. The mass spectrum[3] of each phosphopeptide with the stepped-down voltage approach is shown in Figure 10.6. The left-most part of each spectrum was scanned only in the range 59–99 Da in order to capture the 63-, 79-, and 97-Da PO_2^-, PO_3^-, and $H_2PO_3^-$ masses, respectively. Note that 79 Da is the most stable phosphoryl mass and occurs at a high up-front CID voltage. At the high voltage of −350 V, higher masses have very low intensity, because they fragment under high-voltage conditions (data not shown). At the lower voltage of −115 V, the precursor and adduct ions are observed in Figure 10.6. In Figures 10.6a–c, in addition to the $(M-H)^-$ ion for all three phosphopeptides, trifluoroacetic acid (TFA) adducts are observed. The TFA is usually added in very small amounts in the electrospray buffer as a modifier which aids in the sample ionization process.

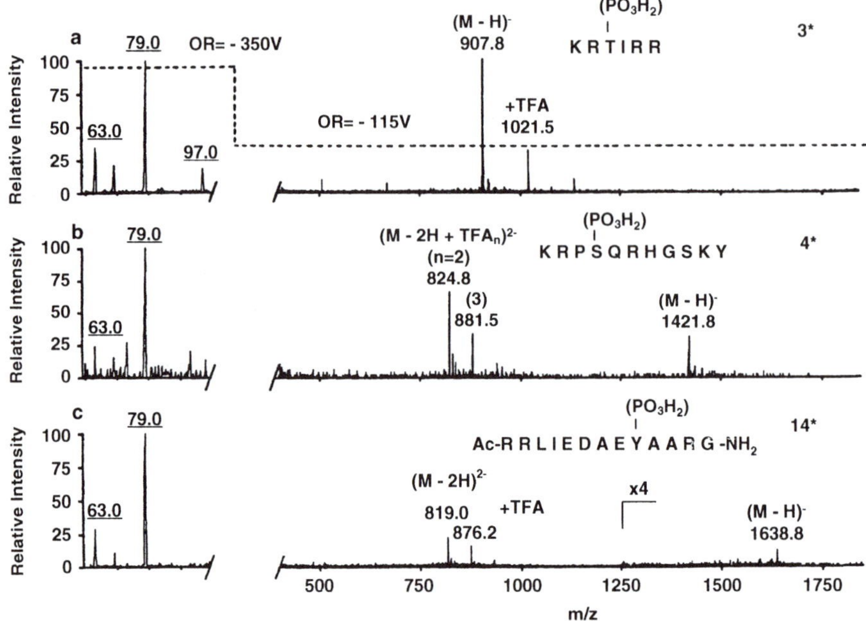

Figure 10.6 The ESI-mass spectra of peaks (a) 3*, (b) 4*, and (c) 14* in Figure 10.5. OR is the voltage on the mass spectrometer interface plate in the up-front CID region. Each scan is composed of a high (left side) and a low (right side) OR voltage mass spectral region in order to capture the phosphoryl and multiply charged, $(M - nH)^{n-}$ peaks of the phosphopeptides. (© 1993, American Society for Mass Spectrometry. Reprinted from reference 3 by permission of Elsevier Science, Inc.)

Immobilized Metal-Ion Affinity Chromatography

Many methods of characterization for an enzymatic digest of a protein invariably use separation methods, such as liquid chromatography or electrophoretic (Chapter 2) techniques, on the individual peptides. These techniques provide for the sequential analysis of many peptides in one experiment. Another type of chromatographic method is similar to the many types of biochemically useful affinity chromatography columns, and this is immobilized metal-ion affinity chromatography (IMAC). Immobilized ferric ions (Fe^{3+}) preferentially retain phosphorylated peptides, and the nonphosphorylated peptides interact weakly, yet are retained, on the IMAC column. An acetic acid wash easily removes the nonphosphorylated peptides, and the phosphopeptides are eluted with an ammonium acetate/ammonium hydroxide buffer.

A tryptic digest of bovine β-casein is used as an example. The digest contains 16 peptides, of which two are phosphopeptides. The T1,2 peptide contains four phosphoserine residues, and the T6 peptide has one phosphoserine residue. An infusion, without separation, by ESI-MS yielded the mass spectrum shown in Figure 10.7, where almost all of the tryptic peptides are observed, including

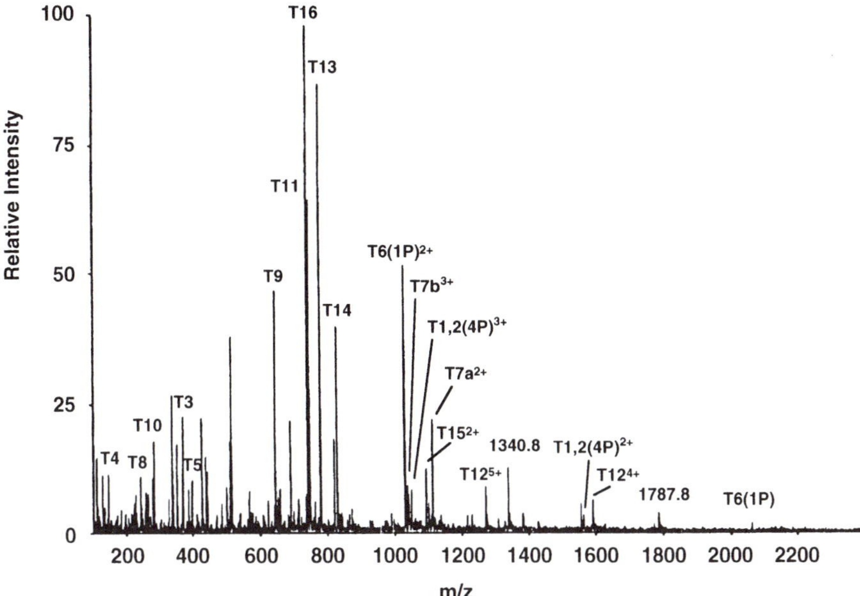

Figure 10.7 The ESI-mass spectrum of a direct infusion of a tryptic digest of bovine β-casein. Selected peptides are highlighted, including the two phosphopeptides. (© 1993, American Society for Mass Spectrometry. Reprinted from reference 8 by permission of Elsevier Science, Inc.)

the two phosphopeptides.[8] The T1,2 peptide is observed as doubly and triply charged ions, while T6 occurs as singly and doubly charged ions. The RP-HPLC analysis of the same digest (Figure 10.8) shows most of the peptides; however, T1,2 is absent[8] In RP-HPLC, highly polar analytes are not usually observed in that they elute quickly and are hidden or masked by the coeluting buffer compounds. An IMAC elution profile is shown in the inset to Figure 10.9, where a relatively sharp peak is observed.[8] The averaged mass spectrum of the 9-min peak in the inset in Figure 10.9 shows both phosphopeptides. They were initially retained on the IMAC column, and they subsequently eluted at the same time in the basic buffer elution conditions.

Mass Spectral Analysis of a Phosphopeptide

An analysis of a product-ion mass spectrum of a phosphopeptide will be presented here. In an ion-trap mass spectrometer, a phosphopeptide was isolated and allowed to undergo CID. An intense CID product ion was observed as the dephosphorylated, dehydration product: $(M - H_2O - HPO_3 + 2H)^{2+}$, and the phosphoryl loss occurred at a Thr residue.[23] This ion was isolated and subjected to further CID. Figure 10.10 provides the product-ion spectrum of the peptide Leu-Lys-Arg-Ala-Thr*-Leu-Gly-NH_2, and Scheme 10.1 shows the analysis fol-

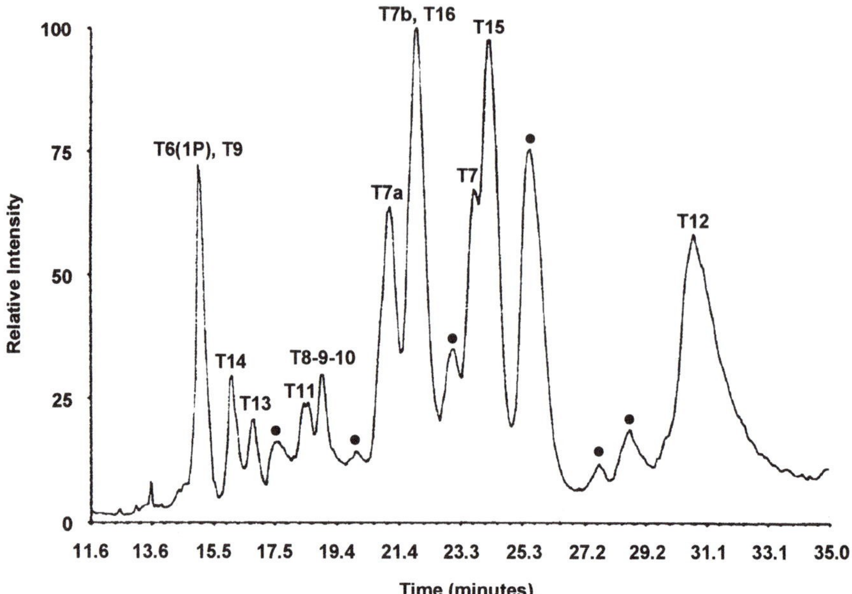

Figure 10.8 The HPLC-ESI mass spectral TIC of the same tryptic digest of bovine β-casein as in Figure 10.7. The T6 phosphopeptide is observed, but T1,2 is absent. •, nontryptic and modified peptides. (© 1993, American Society for Mass Spectrometry. Reprinted from reference 8 by permission of Elsevier Science, Inc.)

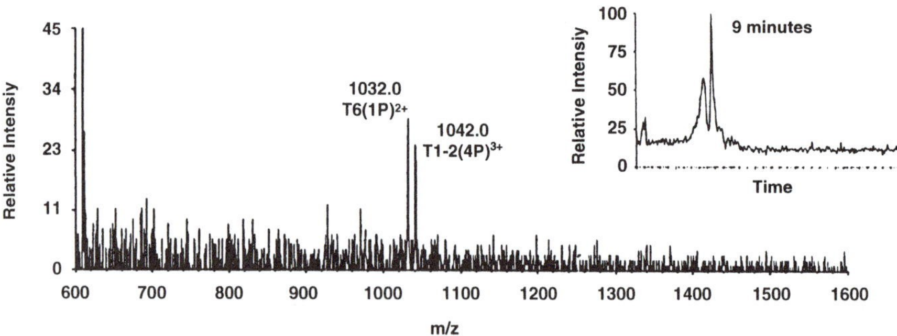

Figure 10.9 The IMAC-ESI mass spectrum of the retained T1,2 and T6 phosphopeptides from the tryptic digest of bovine β-casein. Inset shows the IMAC elution TIC of the nearly coeluting phosphopeptides in the mass spectrum. (© 1993, American Society for Mass Spectrometry. Reprinted from reference 8 by permission of Elsevier Science, Inc.)

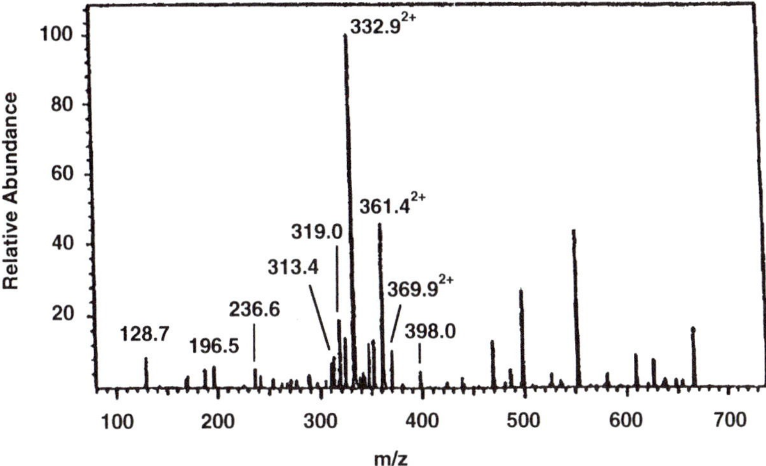

Figure 10.10 The ESI-product-ion spectrum of the $(M - H_2O - HPO_3 + 2H)^{2+}$ ion at *m/z* 370 Da of LKRApTLG − NH_2.

lowing a procedural format. Nominal mass units are used, and the $(M + 2H)^{2+}$ ion at 370 Da was used as the precursor mass. The underlined mass values are observed, and Thr* is the dehydroaminobutyric acid residue. The residue label in Scheme 10.1 refers to the unmodified form of the peptide; namely, the unphosphorylated standard residue of Thr in the peptide. The basics of protein sequence determination are presented in Chapter 9 and will be helpful in this chapter. Even though the phosphate group is not present in the dehydrated peptide, evidence of its prior existence in the peptide is seen in the b_5^* and b_6^* ions, where the asterisk indicates both dephosphorylation and dehydration. Note, in particular, the absence of the b_5 and y_6 ions. The b_5 and b_6 residue ions could not be present in the spectrum because the precursor ion is the

	Leu	Lys	Arg	Ala	Thr	Leu	Gly	NH_2	
		724	596	440	369				+80 (HPO_3)
	757	644	516	360	289	188	75		residue y
	739	626	498	342	271				-18 (H_2O)
	370								y^{*2+}
b^{2+}						333	361.5		
-18 (H_2O)					552	665		738	
b residue	114	242	398	469	570	683			
+80 (HPO_3)					650	763			

Scheme 10.1

dehydrated form of the Thr residue; namely, Thr*. Because a y_3^* ion is not present, subsequent y ions, if observed, will indicate the presence of a dehydration. Since y_5^* and y_6^* are observed, these ions indicate that a dephosphorylation and dehydration occurred somewhere between y_1 and y_4. Since this can occur only on the Thr residue, the Thr must have been phosphorylated with respect to the interpretation capability of the y_5^* and y_6^* ions. The b_7^* or M mass of the dehydrated peptide equals 738 Da, and the $(M + 2H)^{2+}$ ion, $(738 + 2)^{2+}$ at 370 Da, was the precursor ion. A neutral loss of the terminal amine group as ammonia (17 Da) yields the doubly charged fragment at $(738 + 2 - 17)/2 = 361.5^{2+}$, and this ion is observed in the spectrum in Figure 10.10 (361.4^{2+}). If the phosphate was present on the precursor ion, then the $y_3 \cdot P$ fragment would be observed at 369 Da (Scheme 10.1), where P = phosphate moiety. Note the mathematical aids in Scheme 10.1, which are used in the deduction of the mass values for the phosphopeptide. Further sequence characterization schemes will be formulated in the determination of phosphate position(s) in peptides.

Mass Spectral Analysis of a Synthetic Phosphopeptide

The synthetic phosphopeptide in Figure 10.6c (phosphopeptide 14* in Figure 10.5) was subjected to product-ion analysis in order to determine the site of phosphorylation.[16] Figure 10.11 presents the product-ion spectrum and sequence of the phosphopeptide. The precursor ion was the $(M + 2H)^{2+}$ species at 821 Da.

PROBLEM 10.2 Using Scheme 10.1 as a guide, and Table 10.1, determine where the single phosphorylation is located and calculate the mass values of the labeled peaks for the phosphopeptide in Figure 10.11.[16] Use Table 10.1 in order to

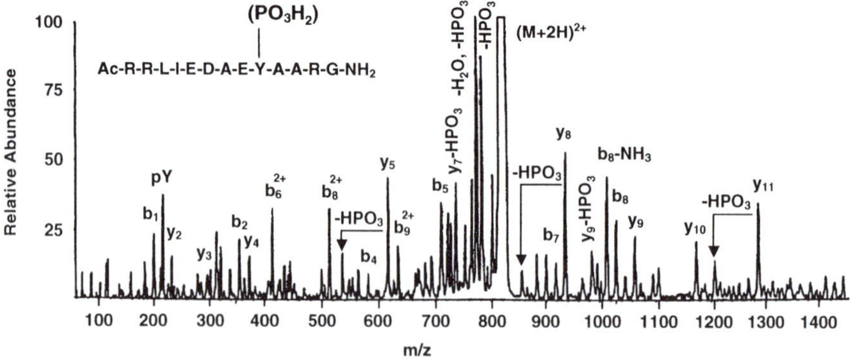

Figure 10.11 The ESI-product-ion mass spectrum of the $(M + 2H)^{2+}$ ion at 821 Da of the phosphopeptide number 14* in Figure 10.5 and 10.6c. (© 1994. Reprinted from reference 16 by permission of Academic Press, Inc., Orlando, Florida.)

correctly set up the calculations for the phosphopeptide. The HPO_3 loss at 80 Da has an exact mass of 79.98 Da, and that of H_2O for an 18-Da loss is 18.015 Da.

Mass Spectral Analysis of a Phosphopeptide from FGF-R1 Kinase

Cell growth and differentiation are controlled by independent sets of enzymes and proteins. There are at least seven different fibroblast growth factor proteins (FGF), which interact with the FGF cell receptor. One of them is a kinase labeled FGF-R1, and it consists of phosphotyrosine residues.[24,25] Purification of the kinase was accomplished by immunological methods and labeled with ^{32}P. The FGF-R1 phosphoprotein was subsequently immobilized onto beads and was hydrolyzed by trypsin while attached to the beads. Further column purification and RP-HPLC yielded six radioactive phosphopeptides, and all contained phosphotyrosine residues. Figure 10.12 provides the ESI-tandem mass spectrum of the $(M + 2H)^{2+}$ ion at 706.4 Da from one of the FGF-R1 tryptic phosphopeptides that has a sequence of DXHHXDYYKK. The X symbol refers to the presence of either Leu or Ile.

PROBLEM 10.3 In Figure 10.12, identify the site(s) of phosphorylation on the peptide using the product-ion mass spectrum as a guide for the residue masses from D^{647} to K^{656}.

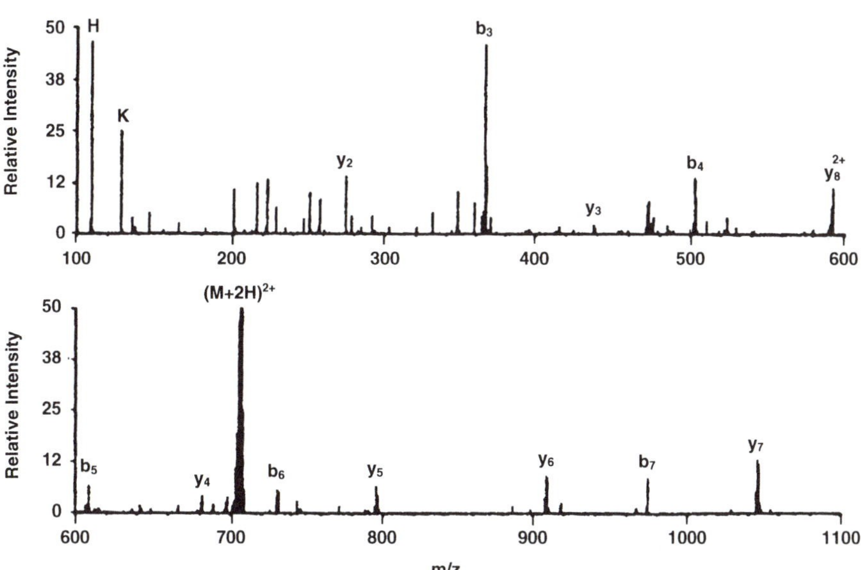

Figure 10.12 The ESI-product-ion mass spectrum of the $(M + 2H)^{2+}$ ion at 706.4 Da of a tryptic phosphopeptide from the FGF-R1 kinase. (© 1993. Reprinted from reference 24 by permission of Academic Press, Inc., Orlando, Florida.)

Mass Spectral Analysis of a Phosphopeptide from Rhodopsin Kinase

The eye retina tissue relies on a number of phosphoproteins that are intermediaries in the efficient functioning of light-induced enzyme processes. Rhodopsin becomes photoactivated by light in the photoreceptor cells in retina tissue, and to quench this excited, activated state, rhodopsin kinase (RK) phosphorylates the excited rhodopsin. Thus, phosphorylation deactivates the photoactivated rhodopsin. Phosphorylation occurs on multiple (4–9 sites) Ser and Thr residues and primarily in the C-terminus region of rhodopsin.[12,26] The RK itself is a phosphoprotein with multiple phospho-Ser and -Thr residues. The RK was isolated from the retina red-outer-segment (ROS) tissue with extensive clean-up procedures. The RK was digested with the Asp-N endoproteinase, and the peptides were subjected to radioactive ^{32}P labeling. The RP-HPLC analysis was performed with a scintillation counter, and Figure 10.13 shows the resulting radiochromatogram.[12] Three radioactive ^{32}P-labeled peaks are observed, and the intense phosphopeptide peak was studied further. This phosphopeptide could not be observed in an ESI-mass spectrum because of the contaminating Tween 80, which suppressed ionization of the sample (data not shown). A limited Lys-C digestion of the intense phosphopeptide peak in Figure 10.13 produced two peptides. One was a phosphopeptide and was subjected to CID of the $(M + 2H)^{2+}$ ion at 858.2 Da. The product-ion mass spectrum and sequence is shown in Figure 10.14.[12] The peptide portion of the RK enzyme is

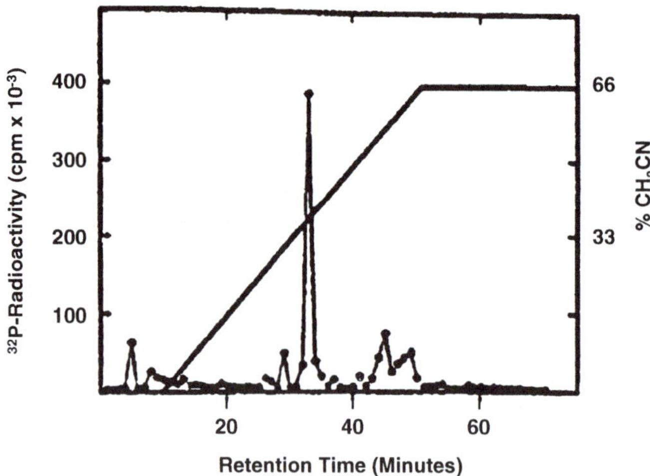

Figure 10.13 Radioactive detection (scintillation counter) of an Asp-N endoproteinase digest of ^{32}P-labeled eye retina rhodopsin kinase (RK). The HPLC column was a Partisil 5 ODS3 column (6.34 × 250 mm, 5-μm particle size). The mobile phase consisted of a 0–66% AcCN gradient with 0.1% TFA at 0.8 ml/min flow rate. (© 1992. Reproduced from reference 12 by permission of the American Society for Biochemistry and Molecular Biology, Bethesda, MD.)

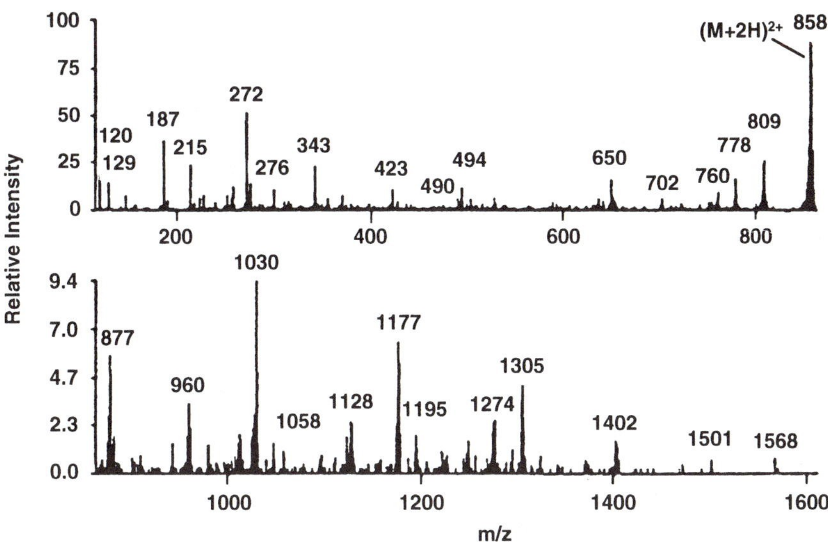

Figure 10.14 The intense radioactive phosphopeptide peak in Figure 10.13 was digested with Lys-C into two peptides. One was a phosphopeptide, and an ESI-tandem mass spectrum of its $(M + 2H)^{2+}$ ion at 858.2 Da is shown. (© 1992. Reproduced from reference 12 by permission of the American Society for Biochemistry and Molecular Biology, Bethesda, MD.)

from D-483 to K-497, and the X symbol refers to serine and/or threonine residues.

PROBLEM 10.4 Deduce the identities of the X-488 and X-489 residues in Figure 10.14, and determine whether one or both contain phosphate moieties. Identify all labeled mass peaks and use nominal mass units in the calculations. Remember that CID conditions can cause dehydration to occur after the loss of a phosphoryl group for Ser and Thr residues. Once the identities of the X residues are determined, consider all six different mass possibilities for each residue, assuming the presence of two phosphate groups.

Mass Spectral Analysis of a Phosphopeptide from Osteopontin

Osteopontin is a phosphoprotein that binds the osteoclast bone cells together. This rat phosphoprotein was expressed in recombinant fashion from Chinese hamster ovary cells and treated with trypsin. The RP-HPLC analysis was performed and a 63- and 79-Da RIC produced a number of phosphopeptides. The early eluting phosphopeptide that consists of the residues from I-287 to N-299 was subjected to ESI-tandem mass spectrometry, and the spectrum is presented in Figure 10.15 along with the peptide sequence.[19]

PROBLEM 10.5 Determine where the one phosphorylation site is located on the phosphopeptide in Figure 10.15, as well as the identity of the labeled ions. The X symbol is Leu/Ile. Note that the (*, b_9) and (*, b_{11}) pairs of peaks are close in mass. Table 10.2 provides a list of the labeled masses in Figure 10.15. An asterisk indicates a phosphate loss and water loss from the next highest-in-mass b_n ion, and this b_n ion is equal to the $b_n \cdot P$ ion. An ion indicated by an asterisk can also be thought of as a dehydration from the unmodified residue of the next highest-in-mass $b_n(= b_n \cdot P)$ species.

Mass Spectral Analysis of a Phosphopeptide from gD-2

Herpes virus relies on a glycoprotein (Chapter 12), termed gD-2, to cause cellular infection. The gD-2 is a structural glycoprotein with an attached phosphate, and upon viral infection the glycoprotein finds its way to the surface of the infected cell. This protein then triggers the host immune response.[27] The carbohydrate portion of gD-2 is elaborated upon in Chapter 12; however, upon RP-HPLC

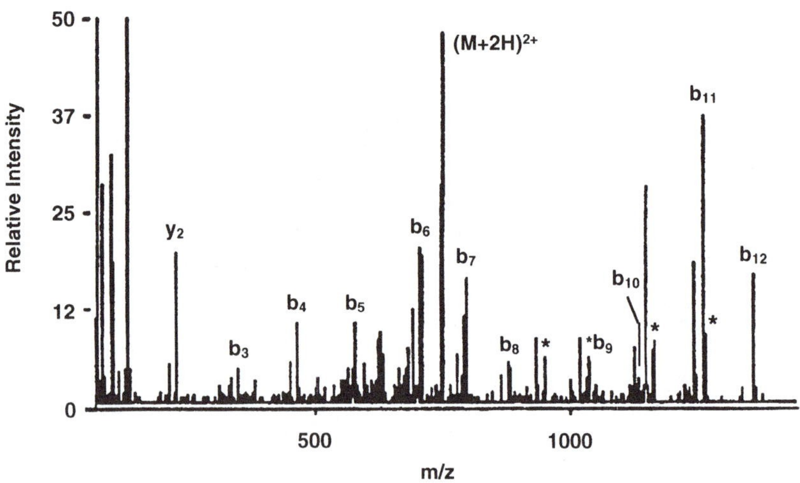

Figure 10.15 The ESI-product-ion mass spectrum of the $(M + 2H)^{2+}$ ion at 748 Da from the I-287 to N-299 tryptic phosphopeptide of osteopontin. The asterisk indicates a phosphoric acid loss (phosphoryl and water) from the next highest-in-mass $b_n(= b_n \cdot P)$ ion. (© 1995. Reprinted from reference 19 by permission of Academic Press, Inc, Orlando, Florida.)

Table 10.2 Listing of observed ions in Figure 10.15

232.25	952.98
338.38	1040.05
467.49	1050.97
580.65	1138.05
709.76	1169.17
748.0	1267.16
796.84	1268.31
883.91	1366.30

analysis (Figure 12.40c), a series of ions was observed for a particular tryptic peptide that had 80-Da spacings. This peptide appeared to be a glycopeptide with appendant phosphorylation site(s), and it eluted earlier as a shoulder on the equivalent nonphosphorylated glycopeptide. Upon removal of the carbohydrate portion from the tryptic peptide, the resulting (apparent) phosphopeptide still eluted earlier than the nonphosphorylated glycopeptide. This elution order can be explained by the fact that more polar compounds elute relatively faster than nonpolar compounds on RP-HPLC systems. Thus, the presence of phosphate on a glycopeptide causes an increase in the relative polarity in comparison with a nonphosphorylated peptide. The tryptic digest of the gD-2 glycoprotein with appendant phosphate moiety was treated with an enzyme that removes carbohydrate species, and an RP-HPLC chromatogram is shown in Figure 10.16. The phosphopeptide in this resulting digest was isolated by 63- and 79-Da RIC analysis.[27] Only one peak provided a signal for both 63 Da and 79 Da near a 40-min elution time (Figure 10.16a and b). Figure 10.17 presents the ESI-product-ion mass spectrum of that phosphopeptide, and the precursor ion was the $(M + 4H)^{4+}$ species at 1292 Da.[27] The peptide sequence is shown in Figure 10.17, and it spans the I-238 to D-284 C-terminal region of the glycoprotein.

PROBLEM 10.6 Determine the single site of phosphorylation for the phosphopeptide shown in Figure 10.17. Because of the relative size of the peptide, there are 2+ and 3+ b_n ions in the spectrum, and these are annotated in Figure 10.17 at the appropriate places. Use exact mass values, because a number of calculated mass values from different segments of the peptide yield masses that differ by only a few daltons. Two mass peaks are derived from an internal sequence of residues labeled in_{17-28} and in_{17-32}.

Mass Spectral Analysis of a Phosphopeptide from the MAPKK Enzyme

Cell division is a very complex process, and one of the cellular enzyme and protein pathways (or cascades) relies on the mitogen-activated protein (or microtubule-associated protein) (MAP) series of enzymes.[28–30] These enzymes are activated and respond early in the cell division cycle; namely, in the $G_0 \rightarrow G_1$ transition phase.[31] Two important enzymes in this cycle are MAP kinase kinase

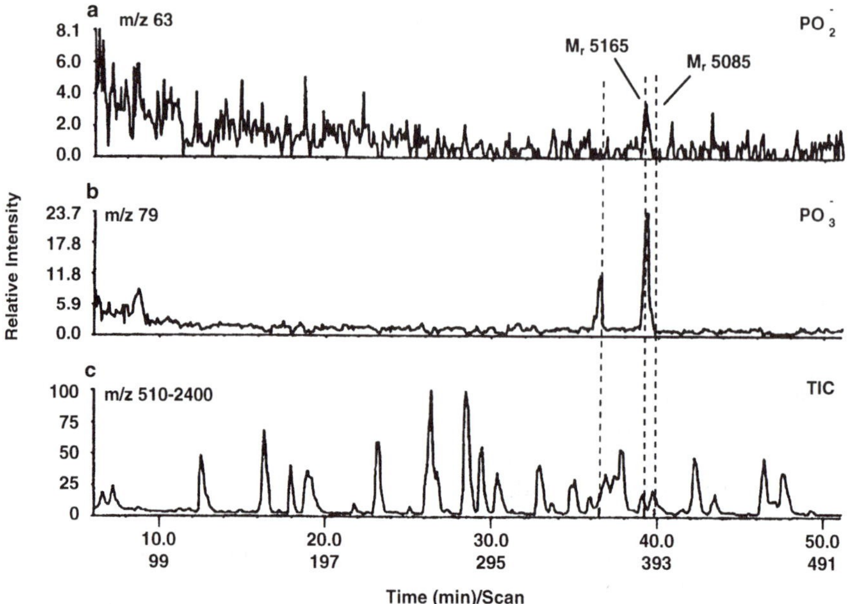

Figure 10.16 Stepped-collisional voltage (see Figure 10.6 legend) RP-HPLC-ESI mass spectral chromatogram of a tryptic digest of the deglycosylated form of the gD-2 glyco(phospho)protein. An OR of −350 V was used to produce the RIC data for (a) *m/z* 63 and (b) *m/z* 79. An OR of −115 V was used to produce the TIC in (c). A 2.1 × 250 mm C_{18} column was used, and a 5 μl/min flow from the 200 μl/min column effluent was used for ESI. (© 1997. Reproduced from reference 27 by permission of Humana Press, Totowa, New Jersey.)

(MAPKK) and MAP kinase (MAPK), and both are phosphokinases. Mos is a protein kinase which originates in germ-cell tissue and stimulates (via phosphorylation) MAPKK, which, in turn, phosphorylates MAPK.[7,13] The MAPKK enzyme was made in recombinant fashion in *E. coli* and treated with 4-vinylpyridine to reduce and alkylate the disulfide bonds.[13] More than 40 peptides were generated in a tryptic digest, and the tryptic peptide T28 was chosen for analysis. This peptide was found either as non-phosphorylated or as having one or two phosphate moieties. The sequence and ESI-product-ion mass spectrum of the peptide is shown in Figure 10.18, and the precursor is the $(M + 3H)^{3+}$ ion at 861.6 Da.[13] The sequence is composed of the L-206 to R-227 amino acid residues in the MAPKK phosphoprotein.

PROBLEM 10.7 Using Figure 10.18, determine whether the peptide has 0, 1, or 2 sites of phosphorylation. Almost all the observed ions are singly charged. There is one doubly charged "a" ion, a singly charged "a" ion, and three doubly charged b ions. Table 10.3 provides a tabulation of the ions observed in Figure 10.18. What is the identity of the 828.9-Da peak? Hint: This is a triply charged peak.

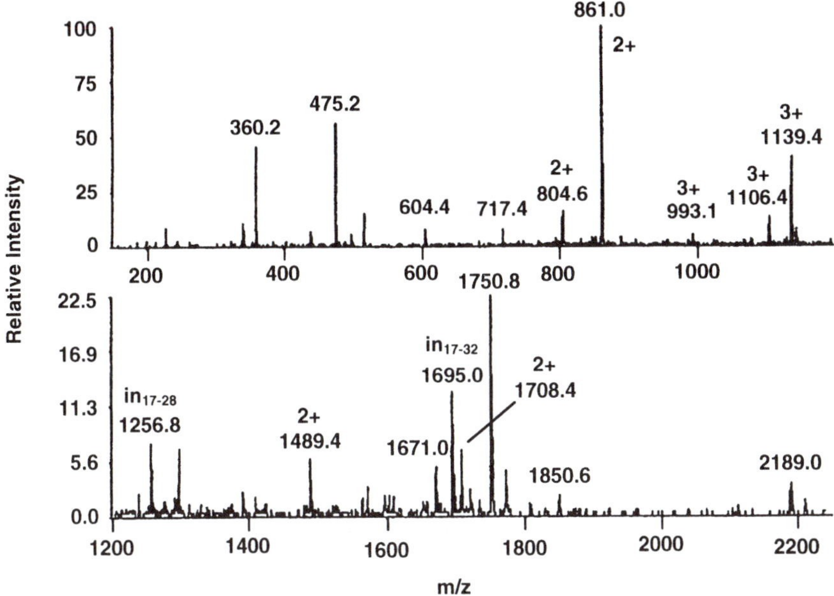

Figure 10.17 The ESI-product-ion mass spectrum of the ≃40-min elution peak in Figure 10.16; this was the C-terminal tryptic peptide of the gD-2 protein. The $(M + 4H)^{4+}$ ion at 1292 Da was the precursor ion. The sequence of the peptide is shown. (© 1996. Reproduced from reference 27 by permission of Humana Press, Totowa, New Jersey.)

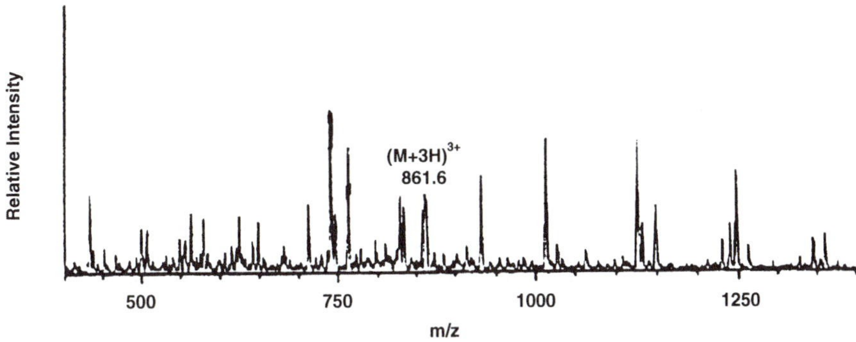

Figure 10.18. The ESI-product-ion mass spectrum of the tryptic peptide T28 from 4-vinylpyridine alkylated recombinant MAPKK phosphoenzyme. The precursor ion is the $(M + 3H)^{3+}$ ion at 861.6 Da, and the peptide sequence is given. (© 1995, American Chemical Society. Reprinted from reference 13 with permission.)

Table 10.3 Listing of observed ions in Figures 10.19 and 10.20

432.48	861.60
507.08	931.89
549.66	1013.15
563.66	1063.08
579.66	1126.31
620.23	1138.15
648.72	1149.24
712.89	1230.14
740.89	1239.46
746.71	1247.24
762.82	1262.40
828.90	1345.23
833.90	1360.40
860.81	

Mass Spectral Analysis of a Phosphopeptide from Bovine α-Casein

Bovine α-casein was investigated for its phosphate content by the characterization of tryptic peptides.[4] A gold-coated nanospray needle was used, and 1–2 μl of a tryptic digest of the α-casein was placed into the nanospray needle. The spray emitted the various peptides, and an m/z 79 precursor-ion spectrum was performed to isolate the phosphopeptides in the negative-ion mode. Tandem mass spectrometry in the positive-ion mode was then performed on the phosphopeptides for sequence information. One phosphopeptide consisted of residues D-43 to K-58, and its product-ion spectrum is presented in Figure 10.19.[4] The $(M + 2H)^{2+}$ precursor ion at 964.9 Da was used.

PROBLEM 10.8 Provide the number of phosphate residues that are found on the peptide in Figure 10.19 by analysis of the residues, and deduce the identities of the labeled peaks.

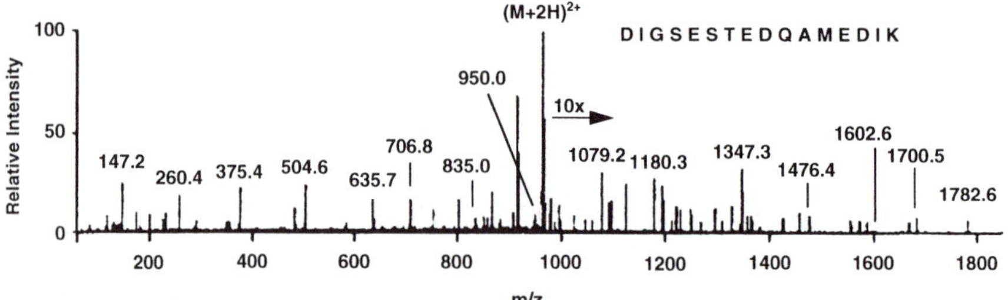

Figure 10.19 Nanospray ESI-product-ion mass spectrum of the D-43 to K-58 phosphopeptide from the tryptic digest of bovine α-casein. The precursor was the $(M + 2H)^{2+}$ species at 964.9 Da. (© 1996. Reprinted from reference 4 by permission of Academic Press, Inc., Orlando, Florida.)

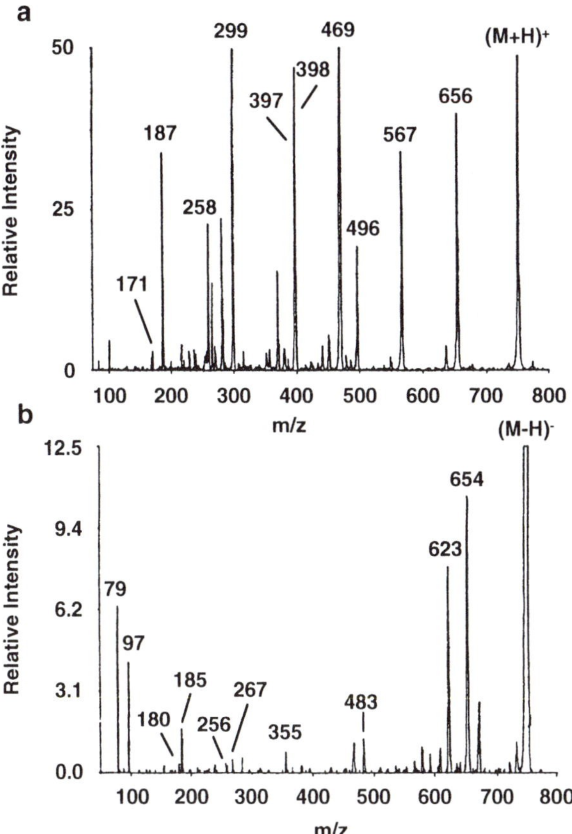

Figure 10.20 The ESI-product-ion mass spectra of a phosphopeptide in (a) the positive mode, precursor = $(M + H)^+$ at 753.7 Da, and (b) the negative mode, precursor = $(M - H)^-$ at 751.7 Da. (© 1996, American Society for Mass Spectrometry. Reprinted from reference 18 by permission of Elsevier Science, Inc.)

PROBLEM 10.9 This example allows the reader to solve the sequence of a synthetic phosphopeptide. The ESI-product-ion spectra in the positive- and negative-ion mode are given in Figures 10.20a and b, respectively, for a synthetic phosphopeptide.[18] The precursor ions used were 753.7 Da and 751.7 Da for the positive $(M + H)^+$ and negative $(M - H)^-$ ion mass spectra, respectively. Provide the sequence of the phosphopeptide using both spectra in Figure 10.20, and locate the site(s) of phosphorylation. Chapter 9 provides useful information for the solution of this problem.

REFERENCES

1. Carr, S.A.; Roberts, G.D.; Hemling, M.E. In *Mass Spectrometry of Biological Materials;* McEwen, C.N.; Larsen, B.S., Eds.; Marcel Dekker: New York, 1990; Chapter 3, pp 87–136.
2. Hunter, T. In *Methods in Enzymology: Protein Phosphorylation*; Hunter, T.; Sefton, B.M., Eds.; Academic Press: San Diego, CA, 1991; Volume 200, Chapter 1, pp 3–37.
3. Huddleston, M.J.; Annan, R.S.; Bean, M.F.; Carr, S.A. *J. Am. Soc. Mass Spectrom.* **1993**, *4*, pp 710–717.
4. Carr, S.A.; Huddleston, M.J.; Annan, R.S. *Anal. Biochem.* **1996**, *239*, pp 180–192.
5. Hubbard, M.J.; Cohen, P. *Trends Biochem. Sci.* **1993**, *18*, pp 172–177.
6. Burlingame, A.L.; Boyd, R.K.; Gaskell, S.J. *Anal. Chem.* **1996**, *68*, pp 599R–651R.
7. Mansour, S.J.; Resing, K.A.; Candi, J.M.; Hermann, A.S.; Gloor, J.W.; Herskind, K.R.; Wartmann, M.; Davis, R.J.; Ahn, N.G. *J. Biochem.* **1994**, *116*, pp 304–314.
8. Nuwaysir, L.M.; Stults, J.T. *J. Am. Soc. Mass Spectrom.* **1993**, *4*, pp 662–669.
9. Watts, J.D.; Affolter, M.; Krebs, D.L.; Wange, R.L.; Samelson, L.E.; Aebersold, R. In *Biochemical and Biotechnological Applications of Electrospray Ionization Mass Spectrometry;* Snyder, A.P., Ed.; ACS Symposium Series 619; American Chemical Society: Washington, DC, 1995; Chapter 20, pp 381–407.
10. Ding, J.; Burkhart, W.; Kassel, D.B. *Rapid Commun. Mass Spectrom.* **1994**, *8*, pp 94–98.
11. Meyer, H.E.; Hoffman-Posorske, E.; Heilmeyer, L.M.G. In *Methods in Enzymology: Protein Phosphorylation*; Hunter, T.; Sefton, B.M., Eds.; Academic Press: San Diego, CA, 1991; Volume 201, Chapter 14, pp 169–185.
12. Palczewski, K.; Buczyłko, J.; Van Hooser, P.; Carr, S.A.; Huddleston, M.J.; Crabb, J.W. *J. Biol. Chem.* **1992**, *267*, pp 18991–18998.
13. Resing, K.A.; Mansour, S.J.; Hermann, A.S.; Johnson, R.S.; Candia, J.M.; Fukasawa, K.; Vande Woude, G.F.; Ahn, N.G. *Biochemistry* **1995**, *34*, pp 2610–2620.
14. Rossomando, A.J.; Wu, J.; Michel, H.; Shabanowitz, J.; Hunt, D.F.; Weber, M.J.; Sturgill, T.W. *Proc. Natl. Acad. Sci. U.S.A.* **1992**, *89*, pp 5779–5783.
15. Buczyłko, J.; Gutmann, C.; Palczewski, K. *Proc. Natl. Acad. Sci. U.S.A.* **1991**, *88*, pp 2568–2572.
16. Huddleston, M.J.; Annan, R.S.; Bean, M.F.; Carr, S.A. In *Techniques in Protein Chemistry V*; Crabb, J.W., Ed.; Academic Press: San Diego, CA, 1994; pp 123–130.
17. Hunter, A.P.; Games, D.E. *Rapid Commun. Mass Spectrom.* **1994**, *8*, pp 559–570.
18. Busman, M.; Schey, K.L.; Oatis, J.E., Jr.; Knapp, D.R. *J. Am. Soc. Mass Spectrom.* **1996**, *7*, pp 243–249.
19. Bean, M.F.; Annan, R.S.; Hemling, M.E.; Mentzer, M.; Huddleston, M.J.; Carr, S.A. In *Techniques in Protein Chemistry VI*; Crabb, J.W., Ed.; Academic Press: San Diego, CA, 1995; pp 107–116.
20. Covey, T.; Shushan, B.; Bonner, R.; Schroder, W.; Hucho, F. In *Methods in Protein Sequence Analysis*; Jörnvall, H.; Höög, J.-O.; Gustavsson, A.-M., Eds.; Birkhäuser Verlag: Basel, Switzerland, 1991; pp 249–256.
21. Walsh, K.A.; Ericsson, L.H.; Resing, K.; Johnson, R.S. In *Methods in Protein Sequence Analysis;* Imahori, K.; Sakiyama, F., Eds.; Plenum Press: New York, 1993; pp 143–147.
22. Resing, K.A.; Johnson, R.S.; Walsh, K.A. *Biochemistry* **1993**, *32*, pp 10036–10045.
23. Gillece-Castro, B.L.; Amott, D.P.; Bier, M.E.; Land, A.P.; Stults, J.T. *Proc. 43rd ASMS Conference on Mass Spectrometry and Allied Topics* **1994**, p 302.

24. Crabb, J.W.; Johnson, C.; West, K.; Buczylko, J.; Palczewski, K.; Hou, J.; McKeehan, K.; Kan, M.; McKeehan, W.L.; Huddleston, M.J.; Carr, S.A. In *Techniques in Protein Chemistry IV*; Crabb, J.W., Ed.; Academic Press: San Diego, CA, 1993; pp 171–178.
25. Hou, J.; McKeehan, K.; Kan, M.; Carr, S.A.; Huddleston, M.J.; Crabb, J.W.; McKeehan, W.L. *Protein Sci.* **1993**, *2*, pp 86–92.
26. Ohguro, H.; Palczewski, K.; Ericsson, L.H.; Walsh, K.A.; Johnson, R.S. *Biochemistry* **1993**, *32*, pp 5718–5724.
27. Hemling, M.E.; Mentzer, M.A.; Capiau, C.; Carr, S.A. In *Mass Spectrometry in the Biological Sciences*; Burlingame, A.L.; Carr, S.A., Eds.; Humana Press: Totowa, NJ, 1996; pp 307–331.
28. Rossomando, A.J.; Payne, D.M.; Weber, M.J.; Sturgill, T.W. *Proc. Natl. Acad. Sci. U.S.A.* **1989**, *86*, pp 6940–6943.
29. Ray, L.B.; Sturgill, T.W. *Proc. Natl. Acad. Sci. U.S.A.* **1987**, *84*, pp 1502–1506.
30. Sturgill, T.W.; Ray, L.B.; Anderson, N.G.; Erickson, A.K. In *Methods in Enzymology: Protein Phosphorylation*; Hunter, T.; Sefton, B.M., Eds.; Academic Press: San Diego, CA, 1991; Volume 200, Part A, Chapter 27, pp 342–351.
31. Erickson, A.K.; Payne, D.M.; Martino, P.A.; Rossomando, A.J.; Shabanowitz, J.; Weber, M.J.; Hunt, D.F.; Sturgill, T.W. *J. Biol. Chem.* **1990**, *265*, pp 19728–19735.

ANSWERS

PROBLEM 10.1 The lower mass, 717 Da (Figure 10.3b), must have at least one phosphate. The 757-Da peak in Figure 10.3a is 40 Da higher in mass (doubly charged); therefore, it is 80 Da higher when converted to the singly charged form. This is equivalent to HPO_3; thus, the peptide in Figure 10.3a has one more phosphate than that in Figure 10.3b. Using other techniques, it was found that the 717- and 757-Da peaks have one and two phosphates, respectively. The difference between them is only 80 Da, not 98 Da. The second Ser residue retains its side-group hydroxyl in the 717-Da peptide, because it was not phosphorylated.

PROBLEM 10.2 The sequence in Figure 10.11 allows only for the Tyr at position 9 to be phosphorylated, and this is also shown in Figure 10.6c. Table 10.1 shows that a phosphotyrosine residue cannot undergo dehydration. Thus, either fragmentation to the unmodified residue ($-HPO_3$) or the unaffected, intact phosphorylated residue ($+HPO_3$) can occur as a result of CID. Scheme 10.2 provides details for arriving at the mass values. Underlined values are observed in Figure 10.11. The +80-Da (exact mass 79.98 Da) value indicates the addition of an HPO_3 to the residue mass. The phosphotyrosine internal fragment (see Chapter 9) is present at 163 + 80 − 28 (CO) = 215 Da. A number of doubly charged ions are also observed:

$$b_6^{2+} = (824.9 + 1)/2 = 412.95^{2+}$$
$$b_8^{2+} = (1025.09 + 1)/2 = 513.05^{2+}$$
$$b_9^{2+} = (1268.27 + 1)/2 = 634.64^{2+}$$

A loss of the HPO_3 moiety occurs from the $(M + 2H)^{2+}$ precursor ion as $821 - 40 = 781^{2+}$ Da. An ion 9 Da lower in mass from 781^{2+} Da appears to be

	Ac ---- R	R	L	I	E	D	A	E	
			1287.30	1174.14	1061.00	931.87	816.78	745.71	y+ 80 (79.98)
			1207.30	1094.14	981.00	851.87	736.78	665.70	y residue
b residue	198.19	354.37	467.53	580.69	709.81	824.90	895.97	1025.09	

	Y	A	A	R	G - NH_2	
	616.56					y +80
	536.59	373.41	302.33	231.25	75.07	y residue
b residue	1188.26					
b+80	1268.24					

Scheme 10.2

the $(M + 2H - HPO_3 - H_2O)^{2+}$ ion at 772^{2+} Da. This dehydration product originates from a site in the peptide other than the Tyr side group. The b series does not include the Tyr residue; however, the y series includes the Tyr-P species directly ($y_5 \cdot P$), as well as $y_8 \cdot P$ to $y_{11} \cdot P$. Note that the unmodified form of the Tyr residue (y_5) and other y fragments are observed with no dehydration (−18), and there is a $b_8 - NH_3$ ion at 1008.1 Da.

PROBLEM 10.3 Scheme 10.3 provides an analysis. The first Tyr, or Tyr-653, is the site of phosphorylation and is directly observed at the $b_7 \cdot P$ ion and $y_4 \cdot P$ ion. The $b_3 - b_6$ and y_2, y_3, $y_5 \cdot P - y_7 \cdot P$ ions are also observed. The y_8 ion is present as the doubly charged ion at 592.6 Da. No dehydration product ion is observed. The peptide has one only phosphorylation site; however, a second +80 row in Scheme 3 is provided for each b and y species. In the calculation of the y fragments, Y-654 or y_3 is observed as an unmodified residue; therefore, Y-653 (i.e., y_4), could only have either one or no phosphate moieties. In the b series, Y-653 or b_7 is observed as the $b_7 \cdot P$ ion; therefore, the possibility exists that b_8 could consist of two phosphate moieties. However, only Y-653 has a phosphate group.

PROBLEM 10.4 The top panel of Scheme 10.4 shows that the b-type ions yield only b_2–b_5, an a_2 ion, and a $b_{14} \cdot P \cdot P$ at 1568 Da. This information cannot be

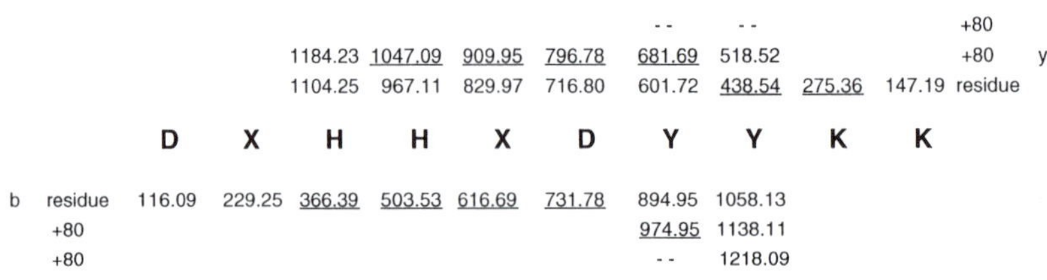

-- -- +80
1184.23 1047.09 909.95 796.78 681.69 518.52 +80 y
1104.25 967.11 829.97 716.80 601.72 438.54 275.36 147.19 residue
D X H H X D Y Y K K
b residue 116.09 229.25 366.39 503.53 616.69 731.78 894.95 1058.13
+80 974.95 1138.11
+80 -- 1218.09

Scheme 10.3

		D	V	G	A	F	X^{488}	X^{489}	V	K	G	V	A	F	E	K	
residue									877	778	650	593	494	423	276	147	y
b	residue	116	215	272	343	490											
a			187														

D	V	G	A	F	S^{488}	T^{489}		
		1402	1345	1274	1127	--	residue + 62	
		1500	1443	1372	1225	--	+80	
		1420	1363	1292	1145	1058	+80	
		1340	1283	1212	1065	978	residue	y ion
		1322	1265	1194	1047	960	-18	
		1304	1247	1176	1029	--	-18	
		y_{13}	y_{12}	y_{11}	y_{10}	y_9		

Scheme 10.4

used for the two X residues. Thus, a "y" ion analysis must be attempted. The ions y_2–y_4 and y_6–y_8 are observed in their residue states. The identities of the y_9–y_{10} ions must be obtained and this information follows. For residue 489:

$$877\ (y_8) + 101 - 18 = \underline{960}\ \text{Thr}\ (y_9^*)$$
$$877\ (y_8) + 87 - 18 = 946\ \text{Ser}$$

Only 960 Da is observed; therefore, the residue is Thr. This is used to solve for the identity of residue 488:

$$960\ (y_9^*) + 101 - 18 = 1043\ \text{Thr}$$
$$960\ (y_9^*) + 87 - 18 = \underline{1029}\ \text{Ser} \simeq 1030 = y_{10}^{**}$$

Hence, residue 488 is a serine at y_{10}^{**}. The double asterisk denotes a double dehydration.

The bottom panel of Scheme 10.4 shows the remainder of the y ion analysis, and all six mass permutation possibilities are shown. Single and double dehydrations and phosphorylation permutations from the unmodified residue mass values are calculated, and all four types with respect to the residue value are observed. Therefore, both Ser and Thr are phosphorylated. The sixth possibility in the product-ion mass spectrum is a combination of one dehydration and one phosphorylation for the y ion series. The ions $y_{10}^* \cdot P$, $y_{11}^* \cdot P$, and $y_{13}^* \cdot P$ display this combination. Note that this combination is a separate calculation with respect to the residue mass: residue $+80$ -18 = residue $+62$ Da. Also, the positions of the dehydration and phosphorylation for a given isolated $y_n^* \cdot P$ ion cannot be delineated.

PROBLEM 10.5 Scheme 5 presents the analysis of Figure 10.15, and the b ion series provides the necessary information for phosphate analysis, while only the y_2 ion is observed in the y series. The residue masses are observed for $b_3 - b_8$, while the remaining ions yield the $b_9^* - b_{12}^*$ (asterisk = dehydration)

	I287	S	H	E	X	E	S	S	S295	S	E	V	N299	
												232.25	133.12	y residue
b ion														
-18	--	183.23	320.37	449.48	562.64	691.75	778.82	865.90	952.98	1040.05	1169.17	1268.31	1382.40	
residue	114.16	201.24	338.38	467.49	580.65	709.76	796.84	883.91	970.99	1058.07	1187.18	1286.32	1400.42	
+80	--	281.22	418.36	547.47	660.63	789.74	876.82	963.89	1050.97	1138.05	1267.16	1366.30	1480.40	
		b_2	b_3	b_4	b_5	b_6	b_7	b_8	b_9	b_{10}	b_{11}	b_{12}		

Scheme 10.5

and $b_9 \cdot P - b_{12} \cdot P$ series. Therefore, Ser-295 is the residue that carries the phosphorylation. Note that the $b_9 \cdot P$ and b^*_{10} ions at 1051 Da and 1040 Da, respectively, are close in mass, as well as the $b_{11} \cdot P$ and b^*_{12} ions.

PROBLEM 10.6 Scheme 10.6 provides the information necessary in order to determine the location of the phosphate, and Figure 10.21 provides the identities of the mass spectral peaks. Very few ions are observed in the spectrum compared with the length of the peptide; however, the observed ions point to the site of phosphorylation. The b ion series does not provide information on the phos-

	I238	A	G	W	H	G	P	K	P	P	Y	T	S	T
b ion														
-18											1187.38	1288.48	1375.56	1476.67
residue	114.16	185.24	242.29	428.50	565.64	622.69	719.81	847.99	945.10	1042.22	1205.39	1306.50	1393.58	1494.68
+80											1285.37	1386.48	1473.56	1574.66

	L252	L	P	P	E	L	S	D	T	T	D	A	T	Q
-18	1589.83	1702.98	1800.10	1897.21	2026.33	2139.49	2226.57							
residue	1607.84	1721.00	1818.11	1915.23	2044.35	2157.51	2244.58	2359.67	2460.77	2561.88	2676.97	2748.05	2849.15	2977.28
+80	1687.82	1800.98	1898.09	1995.21	2007.33	2120.49	2324.56							
														b_{28}

y_{19}			y_{16}	y_{15}						y_9		
2190.15	2093.04	1963.92	1850.76	1751.63	1654.52	1525.40	1410.31	1313.19	1184.08	1068.99		+80
2110.17	2013.06	1883.94	1770.78	1671.65	1574.54	1445.42	1330.33	1233.21	1104.10	989.01	901.93	residue
2092.16	1995.04	1865.93	1752.77	1653.64	1556.52	1427.40	1312.32	1215.20	1086.08	971.00		-18
P266	E	L	V	P	E	D	P	E	D	S276	A	

-18				
residue	3074.40	3203.51	3316.67	3415.80
+80				
			b_{31}	b_{32}

							+80
830.85	717.69	604.54	475.42	360.33	263.22	134.10	residue y ion
							-18
L	L	E	D	P	E	D284	

Scheme 10.6

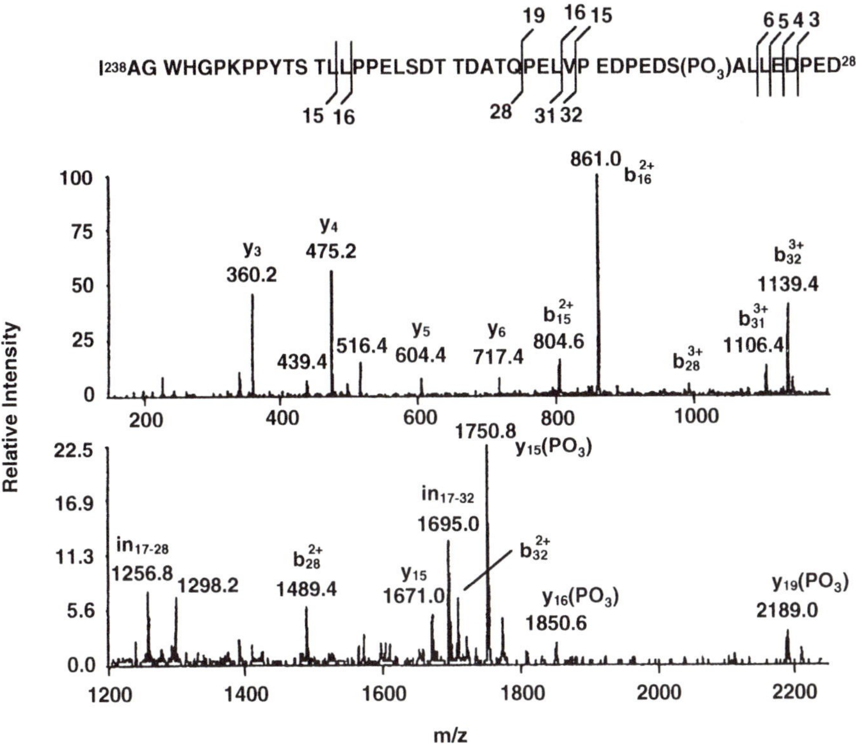

Figure 10.21 The b- and y-labeled mass spectrum and peptide sequence from Figure 10.17. (© 1996. Reproduced from reference 27 by permission of Humana Press, Totowa, New Jersey.)

phorylation site. The peptide residue series Y-248 to T-251 contains four potential phosphorylation sites, and, together with S-258, none of the dehydration, residue, or phosphorylated mass values for the residues between Y-248 and S-258 are observed. The b_{15} and b_{16} ions are observed as the b_{15}^{2+} and b_{16}^{2+} ions at 804.4 and 861 Da, respectively. The b_{28}, b_{31}, and b_{32} ions are found as 2+ and 3+ species as follows:

$$b_{28}^{2+} = (2977.28 + 1)/2 = 1489.14 \text{ Da}$$
$$b_{32}^{2+} = (3415.80 + 1)/2 = 1708.4 \text{ Da}$$
$$b_{28}^{3+} = (2977.28 + 2)/3 = 993.09 \text{ Da}$$
$$b_{31}^{3+} = (3316.67 + 2)/3 = 1106.22 \text{ Da}$$
$$b_{32}^{3+} = (3415.80 + 2)/3 = 1139.27 \text{ Da}$$

Masses at 1256.8 Da and 1695.0 Da represent the indicated internal fragment losses.

Analyzing the y series of masses, y_3–y_6 are observed and obviously have no phosphate; however, $y_{15}\cdot$P, $y_{16}\cdot$P, and $y_{19}\cdot$P are observed. The only residue between y_6 and y_{15} that can have a phosphate is Ser -276. Thus, by an indirect rationale of the y series, the site of phosphorylation has been found. An unusual ion is y_{15} at 1671 Da. This represents the $y_{15} \cdot P{-}HPO_3$ or conventional residue form. Usually, CID causes an additional dehydration reaction to occur; however, this ion is an exception.

PROBLEM 10.7 Scheme 10.7 provides the b and y ion analyses, and Figure 10.22 provides the b and y labels on the spectrum of Figure 10.18. Remember that Cys-207 has a pyridylethyl group and its residue mass is 208.28 (Table 9.12). In the b analysis, a phosphate does not appear on the Ser-212 residue, and no b ion displays either a dehydration or phosphorylation mass. Thus, the y ions are sought for phosphoryl information. The y_6 ion (Ser-222) clearly shows both a dehydration and a phosphorylation mass, and this continues up to y_{10} (Ser-218). At Ser-218, a double phosphorylation (residue mass + 160) and a dehydration/phosphorylation combination (residue + 62) is observed for y_{10} (i.e., $y_{10} \cdot P$ and $y^*_{10} \cdot P$), respectively. This trend, along with a double dehydration (y^{**}_{11}) at m/z 1149, is observed up to the y^{**}_{12} and $y^*_{12} \cdot P$ ions. Note that Thr-226 does not have a phosphorylation site, as shown in Scheme 10.7.

For m/z 828.9, careful mass calculation shows that the absent b_7 species would appear at 827.97 Da, which is close to 828.9 Da. Using the clue that m/z 828.9 is a triply charged peak, it can be related to the $(M + 3H)^{3+}$ precursor ion. The $828.9^{3+} = (M + 3H - H_3PO_4)^{3+} = 861.6 - (98/3)$. This ion represents the dehydration/phosphorylation combination of the $(M + 3H)^{3+}$ precursor ion. The doubly charged b ions are b_9^{2+}, b_{10}^{2+}, and b_{11}^{2+} at 507.08, 563.66, and 620.23 Da, respectively. The a_{10}^{2+} and a_6 ions are at 549.66 and 712.89 Da, respectively.

y ions (above I): 1360.40, 1458.39, 1378.41, 1298.43, 1280.41, 1262.40

b ions	L²⁰⁶	C	D	F	G	V	S	G	Q	L	I
-18							809.95	867.00	995.14	1108.29	1221.45
residue	114.16	322.44	437.53	584.70	641.76	740.89	827.97	885.02	1013.15	1126.31	1239.46
+80							907.95	965.00	1093.13	1206.29	1319.45
									b_9	b_{10}	b_{11}

y_{11}	y_{10}	y_9	y_8	y_7	y_6						
1247.24	1138.15	--	--	--	--	--	--	--	--	--	residue +62
1345.23	1230.14	--	--	--	--	--	--	--	--	--	+80
1265.25	1150.16	1063.08	931.89	860.81	746.71	--	512.46	413.33	356.28	--	+80
1185.27	1070.18	983.10	851.91	780.83	666.73	579.66	432.48	333.35	276.30	175.20	residue
1167.25	1052.16	965.09	833.90	762.82	648.72	--	414.47	315.34	258.29	--	-18
1149.24	1034.15	--	--	--	--	--	--	--	--	--	-18
D	S²¹⁸	M	A	N	S²²²	F	V	G	T	R²²⁷	

Scheme 10.7

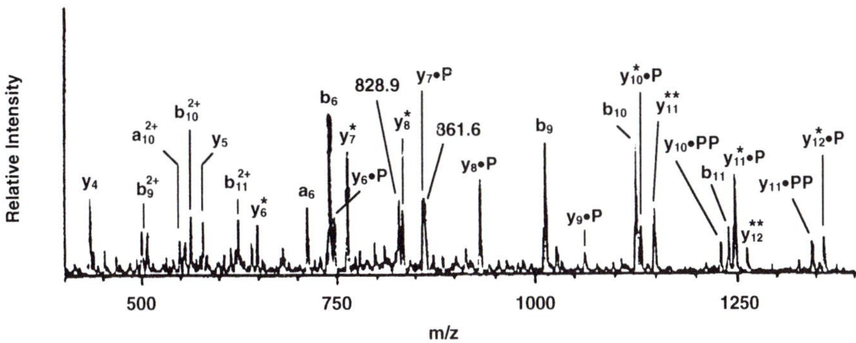

Figure 10.22 The labeled mass spectrum from Figure 10.18. (© 1995, American Chemical Society. Reprinted from reference 13 with permission.)

PROBLEM 10.8 Scheme 10.8 and Figure 10.23 supply the analyses of Figure 10.19. There is only one b ion and that is $b_{15} \cdot PP = 1782.60$ Da; however, this does not provide information as to the sites of the two phosphate groups. The information in Figure 10.19 is essentially a series of y ions. The Thr-49 residue is present as a residue mass (y_{10}), while Ser-48 has a phosphate at $y_{11} \cdot P$. This phosphate is retained in the $y_{12} \cdot P$ ion. There are no signals for y_{13}; therefore, y_{14} must be analyzed for a possible additional phosphate by Ser-46. This is observed as $y_{14} \cdot P$ and $y^*_{14} \cdot P$ ions at 1700.5 and 1602.5 Da, respectively.

PROBLEM 10.9 Scheme 10.9 shows the positive-ion mode interpretation of the mass spectral data in Figure 10.20a. The m/z 754 ion is the $(M + H)^+$ ion (Scheme 10.9), and an $(M + H - H_3PO_4)^+$ ion at m/z 656 is 98 Da lower than the $(M + H)^+$ precursor ion. Thus, there is at least one phosphate group in the peptide. Subtraction of different mass values yield a Q-V-A sequence from m/z 171 to 469 and a V-A sequence from m/z 397 to 567. It appears that m/z 171 and/or m/z 187 are terminal or near-terminal masses. Since a Ser, Thr, or Tyr was not obtained in the above analyses, there is a good probability that at least one of these residues is at or near a terminus. This line of reasoning points to the 171- and/or 187-Da ions. Assume a phosphate loss and a dehydro-residue species: tyrosine would not fit for either proposed terminal ion (171 Da or 187 Da),

	y_{14}	y_{13}	y_{12}	y_{11}	y_{10}									
residue +62	1602.55	1545.50												
+80	1700.54	1643.49												
y ions +80	1620.56	1563.51	1476.43	1347.32	1260.24									
residue	1540.58	1483.53	1396.45	1267.34	1180.26	1079.16	950.04	834.95	706.82	635.74	504.55	375.44	260.35	147.19
-18	1522.57	1465.52	1378.44	1249.32	1162.25									
-18	1504.55	1447.51												

D^{43} I G S^{46} E S^{48} T^{49} E D Q A M E D I K^{58}

Scheme 10.8

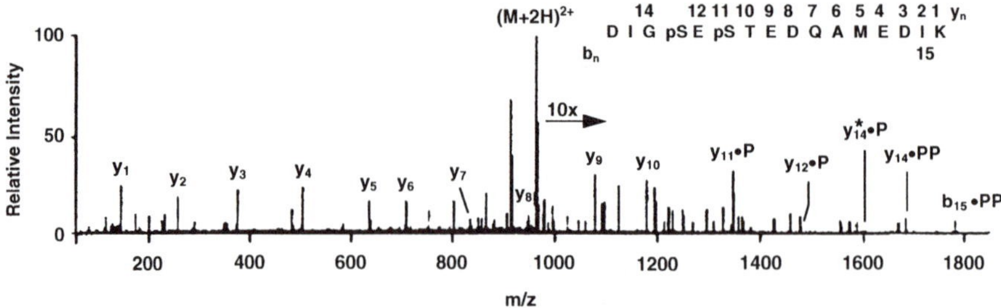

Figure 10.23 The labeled mass spectrum from Figure 10.19. (© 1996. Reprinted from reference 4 by permission of Academic Press, Inc., Orlando, Florida.)

because the 163-Da mass yields a difference of 8- and 24-Da masses, respectively, as uninterpretable. Thus, Ser and Thr must be considered. For *m/z* 171,

171 − 101 (Thr) = 70 Da = 87 (Ser) − 18 (to form dehydroalanine) + 1 (N-terminal H)

or 171 − 87 (Ser) = 84 Da = 101 (Thr) − 18 (dehydro-2-amino-butyryl) + 1

Thus, it appears that, assuming the N-terminus, a Ser and a Thr residue are present. Considering the *m/z* 187 ion:

$$187 - 101 = 86 \quad \text{and} \quad 187 - 87 = 100.$$

These calculations would not allow for both a phosphate-bearing residue and an *m/z* 187 on either the N- or C-terminus. Therefore, the N-terminus appears to be

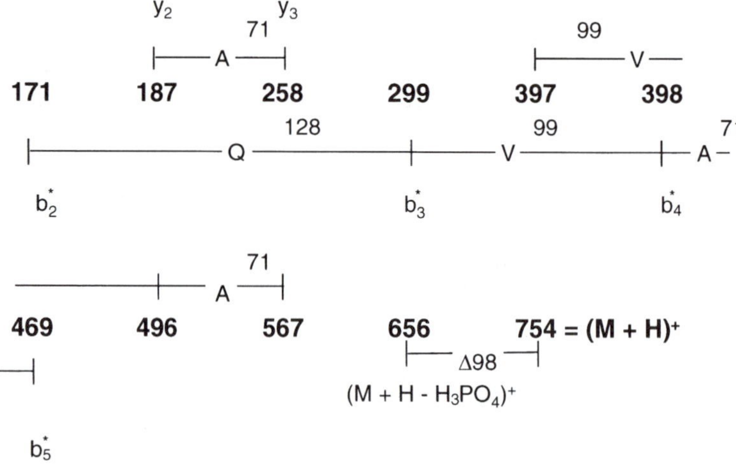

Scheme 10.9

Ser/Thr · P-Q-V-A. The b ions are marked accordingly, and the asterisk indicates the dehydration modification. This leaves *m/z* 187 as representing the C-terminus without a phosphate moiety, and an interpretation of this mass follows:

$$m/z\ 187 = 17\ \text{(OH terminal)} + 2\ \text{(ionizing proton plus N-terminal H on the peptide)} + x$$

$$x = \text{residue mass} = 168\ \text{Da}$$

Table 9.6 shows 168 Da as an Ala, Pro combination, and this indicates two residues, as (Ala, Pro) = y_2. This residue pair is connected to another Ala, as shown by *m/z* 187 and 258 (Scheme 10.9). Thus, *m/z* 258 is a y_3 ion. The *m/z* 397 and 398 cannot be doubly charged species, because their singly charged species would be of a greater mass than that of the singly charged precursor ion. So far in the analysis, no ions have been observed that retain the HPO_3 moiety, except the $(M + H)^+$. However, note the following:

$$b_3^* + 98 = 397 = b_3^* \cdot P$$

$$b_4^* + 98 = 496 = b_4^* \cdot P$$

$$b_5^* + 98 = 567 = b_5^* \cdot P$$

where P is an attached phosphate moiety (P = H_3PO_4 = 98 Da). For a sequence analysis, we have (S,T)-Q-V-A-(P,A), where either the S or T is in a dehydrated or phosphate-modified form with respect to the b and y ions.

Scheme 10.10 presents the $(M - H)^-$ = 752 Da interpretation of the peptide, and a loss of 98 Da from the precursor ion is observed at *m/z* 654. Judicious subtractions provide the A-V-Q partial sequence. Note that the N-terminal order of Thr and Ser can be obtained, where the order is T(p)-S. Threonine has the phosphate group, and b_1 and b_2 are the T(p) and T(p)-S residues,

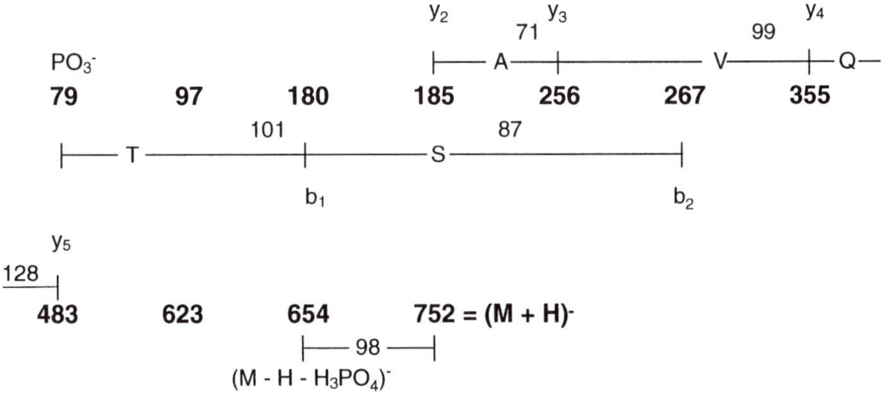

Scheme 10.10

respectively. In Scheme 10.9, it was established that y_2 consisted of a P,A combination. In Scheme 10.10, it appears that 185 Da is the mass which represents the C-terminal residue. Following the $(M + H)^+$ interpretation of m/z 187 in Scheme 10.9, the 185-Da ion in Scheme 10.10 produces the following:

$$m/z\ 185 = 17\ (\text{OH terminal}) + [\text{do not add 2 protons in negative mode}] + x$$
$$x = 168 = \text{A,P combination}$$

Neither Scheme 10.10 nor 10.11 yields a y_1 or b_6 fragment; therefore, the order of the C-terminal P,A residues cannot be determined from the spectra. The correct sequence is T(p)-S-Q-V-A-P-A, and Figures 10.24a and b present the labeled $(M + H)^+$ and $(M - H)^-$ ESI-tandem mass spectra, respectively, of the peptide.

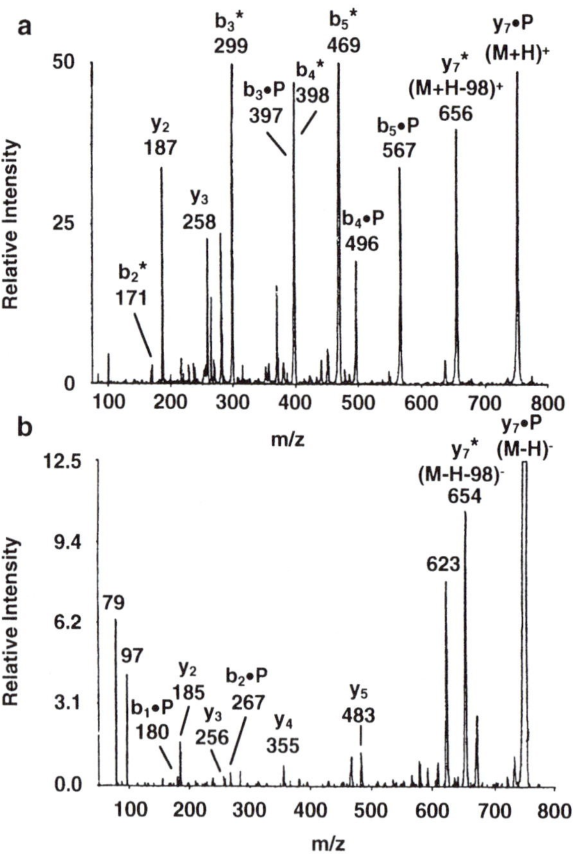

Figure 10.24 (a and b) The labeled mass spectra from Figure 10.20. (© 1996, American Society for Mass Spectrometry. Reprinted from reference 18 by permission of Elsevier Science, Inc.)

(M + H)$^+$ Analysis

	T	S	Q	V	A	P	A		
	753.71	652.61						+80	
	673.73	572.63	485.55	357.42	258.29	187.21	90.09	residue	y
	655.72	554.61						-18	ions
								-18	

	T	S	Q	V	A	P	A	
-18				380.42				
-18	84.09	171.16	299.29	398.43	469.50			b
residue	102.10	189.18	317.08	416.44	487.52			ions
+80	182.08	269.16	397.29	496.42	567.50			
+80								

(M - H)$^-$ Analysis

	T	S	Q	V	A	P	A		
								+80	
	751.71	650.61						+80	
	671.73	570.63	483.55	355.42	256.29	185.21	88.09	residue	y
	653.72	552.62						-18	ions
								-18	

	T	S	Q	V	A	P	A	
-18								
-18	82.10	169.17	297.30					b
residue	100.11	187.19	315.32					ions
+80	180.09	267.17	395.30					
+80								

Scheme 10.11

In Scheme 10.10, m/z $752 - 623 = 129 =$ E residue. Since this cannot be corroborated in the positive-ion mode spectrum (i.e., a direct loss is not observed and does not fit with m/z 171 or 187), it is dismissed as a possible residue.

Scheme 10.11 presents the positive- and negative-mode results in a tabular format in a similar fashion as those presented in previous examples, and Figure 10.24 presents the identities of the major ions.

11

Immunological Applications

Proteins and Phosphoproteins

Chapters 9 and 10 present a comprehensive overview of the role that ESI has in the elucidation of protein, enzyme, and peptide amino acid residue sequences. Modified amino acids can occur, and the investigator must remain alert to the presence of probable residue mass values which are different from those of the 20 common amino acids in an ESI-mass spectral interpretation. An experimentally determined difference between two mass spectral peptide peaks which is $\leq$1 Da of the mass of a common amino acid can provide an important clue as to the identification and position of a residue in a peptide sequence. Phosphoproteins were shown to have a significant presence in intra- and intercellular events. The mass spectral characterization of phosphoproteins takes place by monitoring the losses of 63 and 79 Da in the negative-ion mode using either up-front CID with the Carr stepping voltage method or precursor-ion analyses.

This chapter utilizes elements of and the basic knowledge of Chapters 9 and 10 and combines them for applications in the area of immunology. Immunology is driven primarily by molecular biology and biochemistry investigations; however, the analytical technique of ESI-mass spectrometry has proven a most valuable addition to this branch of science. The fundamentals of protein sequencing (Chapter 9) along with techniques in isolating a peptide amongst literally hundreds to thousands of different peptides, at picogram–femtogram (or picomole–femtomole) levels, propelled ESI-MS into the immunology field by Hunt et al. (*vide infra*) in the early 1990s. Cell-surface-displayed peptides constitute the fundamental basics of how T-cells recognize self from nonself or foreign molecules on a typical cell. The T-cells are composed of circulatory lymphocytes and a segment of the macrophage white blood cell leukocytes.[1–7] The act of recognition by a T-cell is the other important part of the equation. Phosphoproteins are prominent in this recognition scenario and mediate such major molecular process "decisions" as to whether a cell or group of cells will be lysed and destroyed. Aebersold et al. (*vide infra*) have probed the role and functions of phosphoproteins in the cell recognition process with respect to the phosphorylation phenomenon. The presentation of peptide information and the T-cell recognition process together represent a major challenge to the analytical sciences, with respect to understanding the fundamentals and details of immunology processes.

The following describes the roles of ESI-MS in the structural understanding of immunological processes, and the discussion is divided into two main areas. The first deals with the internal processes of a cell which lead to a decoration of its surface with many short peptides (of foreign and self-origin) cradled in large protein containers. The ESI-MS technique is used to sequence the short peptides for utility in deducing their protein/enzyme origin. The second section provides a glimpse into the recognition and possible destruction aspects of cells by T-cells. Cell lysis and destruction rely on a sequence of phosphorylation/dephosphorylation events within the T-cells.

Overview of Cellular Immunology

Every cell in the body, whether it is part of an organ or circulating in the blood system, presents a set of glycoprotein complexes on the surface of its outer membrane. These complexes are composed of multiple subunits, and a cleft can be found on the extracellular portion of the complex. This cleft retains a peptide between 8 and 25 amino acids in length. These peptides are residue sequences which originate from intact proteins and enzymes which are either normally found in the cell or are foreign proteinaceous matter that entered the cell. These proteins and enzymes are degraded at various sites in the cell. Some of the degradation products are 8–25 amino acid size peptides, and they are systematically bound to certain glycoproteins which are subsequently guided to the cell surface.[2,5,8–10] Leukocytes and cytotoxic T-lymphocytes (CTL) continuously survey these glycoprotein-presenting peptide structures on all cell surfaces. If these macrophages detect a peptide that is foreign to the body, complex events transpire that lead to the lysis and death of the cell(s).[1–7] It is estimated that only 200–1000 copies of a single peptide, cradled in the glycoprotein complex, are necessary per cell for activation and stimulation of the relevant surveillance macrophage CTL cells.[4,6,8,11–13] This pertains to peptide–glycoprotein complexes for many types of cells. How the macrophages determine the status of a peptide is beyond the scope of this chapter; however, a summary of the events on how the peptides reach the cell surface will be presented. The ESI-MS technique then plays an important role in determining the amino acid sequence of select peptides so as to elucidate the proteinaceous origin of immune-response events.

Cellular Display of Immunological Peptides: MHC I Peptides

There are two different types of glycoproteins that bind the small peptides mentioned above. One complex consists of a relatively large glycoprotein and a relatively small β-microglobulin protein (β2m), and this complex is termed the major histocompatibility complex I (MHC I), where histocompatibility refers to tissue acceptance.[14] The larger chain is termed the α-chain and is approximately 45 kilodaltons in mass.[5,8,9,12,15] The β2m protein is approximately 12 kDa in mass. A second type of complex, MHC II, is composed of two different glycoprotein subunits. Both MHC-type molecules require the noncovalent binding of certain peptides that are 8–25 amino acids in length to form a physically stable complex. Different biochemical pathways yield the MHC I and II–peptide complexes, and the MHC I and II units bind peptides which are 8–12 and 13–25 amino acid residues in length, respectively.[8,9,11,12,15–17]

The MHC I–peptide binding process will be considered first. The left side of Figure 11.1 presents a scheme for a typical pathway for the generation, transport, and destination of an MHC I–peptide ligand complex.[2] Self-proteins and enzymes are normally synthesized and processed within the cell and eventually need to be degraded or recycled. The 8–12 amino acid sequence can originate from the degraded self-proteins and enzymes, as well as from protein matter produced by DNA which came from a virus that infected the cell. These internally generated proteins and enzymes enter a cylindrical-shaped proteasome that comprises multiple protein units. The proteasome produces small peptides, of 8–12 amino acids in length, from the proteinaceous molecules in the cytoplasm.[12,18] These peptides are transported through the endoplasmic reticulum (ER) membrane by a protein conduit termed the transporter associated with

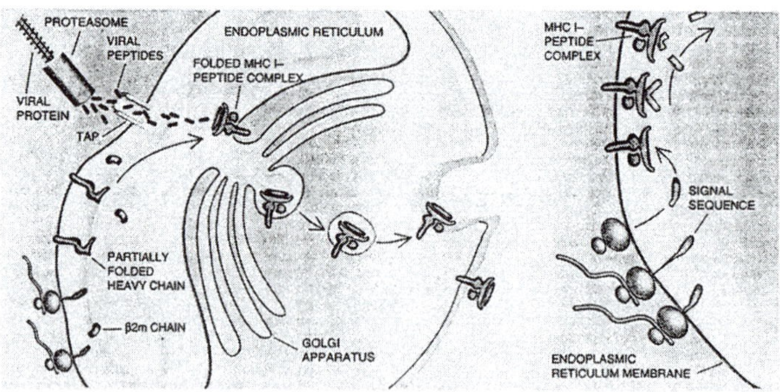

Figure 11.1 (Left side) Schematic of the cytoplasmic cellular pathway for the binding of small peptides on the MHC class-I ligand and the expression of the MHC I-bound peptide on the cell surface; (right side) alternate mechanism where the MHC I ligand binds to ER-generated peptides from the degraded leader signal sequences of peptides and proteins. (© 1994. Permission granted by Dimitry Schidlovsky Illustration, Sea Cliff, New York.)

antigen processing (TAP). The TAP[9,15,19] is composed of the TAP1 and TAP2 proteins, and they are members of a broad class of chaperonin proteins. Chaperonin proteins facilitate the folding, binding, and transportation of many proteins and enzymes in cells. There is evidence to show that the larger subunit and the β2m subunit of MHC I molecules enter the ER via the TAP assembly.[15] The MHC I–protein complex is found close to and/or in contact with the TAP conduit in order to accept a peptide molecule,[12,19,20] and the MHC I is structurally stabilized by the association with the peptide. The MHC I–peptide complex traverses the Golgi complex, which is attached to the ER (Figure 11.1). Vesicles that contain the MHC I–peptide leave the Golgi apparatus and fuse with the cell membrane. Once expressed on the surface of a cell, MHC I molecules are found as tetramer clusters.[15]

Hunt, Englehard, and coworkers, using ESI-MS,[21–23] discovered a second pathway for the binding of peptides to MHC I molecules. Using mutant cells that have a greatly diminished capacity to degrade proteins/enzymes and hence to form degraded peptides in the cytoplasm, a handful of peptides were nevertheless observed on the cell-surface MHC I complexes. These peptides did not originate from proteins and enzymes normally found in the cytoplasm. The right side of Figure 11.1 shows that the peptides originate from protein molecules in the ER. The ER proteins begin their initial synthesis as the signal sequence from normally produced proteins in the cytoplasm. That is, a signal or leader amino acid sequence is usually synthesized prior to the synthesis of the actual protein from the ribosome–mRNA complex, and the 16S and 23S ribosome subunits are shown in Figure 11.1. The protein portion of the leader-protein amino acid sequence is produced on the cytoplasm side of the ER membrane by the ribosome–mRNA complex. The leader sequence that was attached to the N-terminal end of the protein, and was synthesized by the ribosomes, is found on the interior of the ER. The leader sequence is degraded into 8–12 amino acid peptides inside the ER, and newly formed MHC I subunits can bind to the peptides.[2,23] These complexes are shuttled to the surface of the cell membrane. This alternate pathway restricts the type of peptides that can be expressed to macrophages by the MHC I complex in mutant cells. However, this pathway is also in effect in normal, wild-type cells. Thus, this contribution to the total peptide expression is relatively small given the vast amount of peptides produced in the cytoplasm–proteasome–TAP route.

In general, there are over 100 different types of MHC I-type glycoprotein molecules, each of which contains a β2m unit and a different, larger glycoprotein subunit.[5,8,14] The β2m protein in the MHC I complex always has a constant amino acid composition and sequence in humans.[8] Different MHC I complexes bind with different sets of 8–12 amino acid sequences, the latter of which originate from respective cytosol proteins and enzymes. The peptide noncovalently binds only to the larger α-subunit, yet the β2m subunit is required for proper peptide expression on the cell surface.[5,8,13,19] The MHC class-I-type molecules are found on the surface of almost all organism cells that contains a nucleus.

MHC II Peptide

The second class of MHC molecules are MHC II-type. An α- and β-glycoprotein subunit noncovalently bind together and are made in the cytoplasm. Both chains, however, span the ER membrane, where the NH_2-terminus is in the cytoplasm and the COOH-terminus is in the ER.[2] The α- and β-chains are 33–35 kDa and 25–29 kDa in mass,[5,8,14,15] respectively. Binding of peptides does not take place in the ER (Figure 11.2). Instead, a relatively large protein, the invariant chain (Ii), is bound to the MHC II complex in the ER.[2] This $\alpha\beta$–Ii complex rapidly forms trimer complexes made up of three chains of each protein.[9,24] This trimer complex moves out of the ER and into the Golgi apparatus, where carbohydrates are added onto the protein subunits (cf. Chapter 12). The invariant chain glycoprotein has a mass of 31–43 kDa, and four different forms are known.[8,9,15,24] The MHC II–Ii complexes leave the Golgi apparatus in vesicles called endosomes that exhibit lysosome characteristics, and these vesicles are termed MHC class-II compartments (MIIC).[10] In the compartment, the Ii is cleaved and degraded to a 24-amino acid length peptide, and this peptide associates with the MHC II molecule. This peptide is referred to as class II-associated invariant chain peptide (CLIP), and is composed of residues 81–104 of the invariant chain.[9] Concurrently, sections of the cell-surface membrane invaginate to form lysosome-type vesicles.[2] Extracellular protein matter that is either found on the surface or engulfed by the invagination process is subsequently degraded into peptides. The MHC II–CLIP vesicle (MIIC) fuses with the extracellular-derived peptide vesicle.[25] A protein labeled DM, which is similar in structure to the MHC II molecule, is found in a separate vesicle in the cell. This vesicle fuses with the MHC II–CLIP vesicle. The DM removes CLIP from the MHC II complex, and the latter then binds with a peptide from the surface-derived lysosome vesicle.[26] The MIIC vesicle that contains the MHC II–peptides fuses with the cell membrane so as to expose the complexes to extracellular macrophage scrutiny. The peptide is noncovalently bound by both α- and β-subunits. Class-II MHC complexes are found only in mobile white blood cells—such as

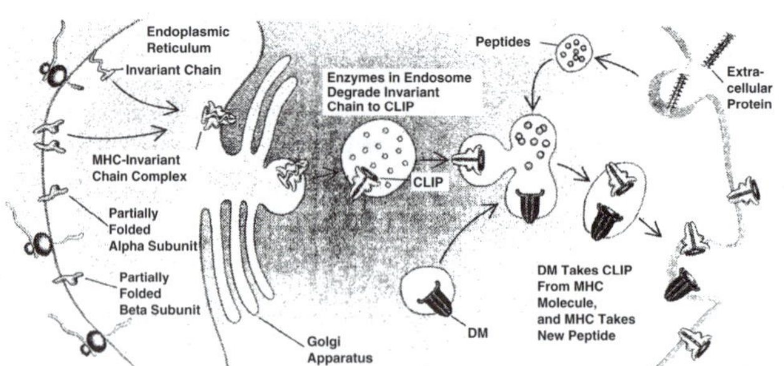

Figure 11.2 Schematic of the MHC class-II ligand pathway synthesis, MHC II binding to peptide, and cell-surface expression of the ligand–peptide complex. (© 1994. Permission granted by Dimitry Schidlovsky Illustration, Sea Cliff, New York.)

leukocytes, lymphocytes, and macrophages [antigen-presenting cells (APC)]—which ingest foreign protein.[5] There are more than 100 different types of MHC II molecules in mouse and human cells; the differences are in the amino acid sequences of the glycoprotein molecules.[5,8]

A particular individual has only about 3–6 variations (isoforms) of each MHC I and II molecule,[8,11] and each MHC isoform is able to bind to a great many different types of peptides. Terminology used to distinguish the various human MHC species from those of other organisms begins with human leukocyte (or lymphocyte) antigen (HLA), and the most prevalent MHC I HLA is HLA-2A.1.[27] The many different peptides that bind this particular class of MHC I can be characterized as having a Leu in the second amino acid position of the 8–12 amino acid sequence, and the last amino acid is uncharged and hydrophobic.[2,23,27–29] Another MHC I molecule, HLA-B27, binds a class of peptides which can be characterized as having an Arg as the second amino acid and the last residue is hydrophilic with a positively charged group. The literature documents many other class-I MHC molecules, as well as class-II MHC molecules,[5] which bind peptides that have other common features (anchor amino acid positions or motifs).[11,12,16] The HLA–DR MHC II complex consists of 19 distinct isoforms, and they all bind to the CLIP peptide. The HLA–DM complex removes CLIP such that HLA–DR can then bind to a peptide in the endosome-MIIC vesicle.[30,31] One hypothesis for the role of the invariant chain is to mitigate the binding of the many self-peptides present in the ER to the MHC II molecule.[32] Once the MHC II HLA–DR–Ii molecule reaches the endosome/lysosome MIIC vesicle, the resulting CLIP peptide becomes unstable through interaction with HLA–DM. The HLA–DR complex then has the opportunity to bind to an exogenous peptide. The peptides which bind to the various HLA molecules can vary from a small to a significant extent with respect to the identity of each amino acid position. This general variation of amino acid identity with few anchor positions is termed agretope,[29] as opposed to a specified, isolated amino acid sequence (epitope).[29,33] For mouse cells (murine), which have also been extensively studied with respect to the immune system, the various MHC species are designated as H-2 and followed by a more specific alphanumeric designation.[12] There are some structurally common MHC complexes between human and mouse cells; for example, the HLA-K^b and H-2K^b MHC I complexes, respectively, because they bind the same peptide. This is directly attributable to the fact that there are proteins and enzymes common to both organisms.

Estimates of the number of total and individual peptides that are found on cell surfaces bound to MHC I and II species have been performed. Table 11.1 presents three MHC I isoform complexes, each of which can bind to many more 8–12 amino acid peptides than are listed. Some peptides are expressed in high numbers (replicates or copies) on the surface of a cell, while most peptides are found in the low hundreds of copies per cell (or even fewer). The K^b, K^d, and D^b isoforms are three of approximately 100 different MHC I molecules known to be found on the surfaces of human cells. In Table 11.1, two different peptides derived from the same influenza virus nucleoprotein are expressed on cell sur-

Table 11.1 Representative examples of MHC I-peptide complexes in normal and infected human cells[12]

MHC I HLA species	Peptide	Peptide copies per cell	Origin of peptide	Reference(s)
K^b	RYQVYQGL	10^4	Vesicular stomatitis virus (VSV)	34
K^b	SIINFEKL	100	Ovalbumin	35
K^d	TYQRTRALV	200–500	Influenza nucleoprotein	36
K^d	KYQAVTTTL	100	Tumor cell	37, 38
K^d	SYFPEITHI	10^4	P815 tumor cells	16
D^b	ASNENMETM	200	Influenza nucleoprotein	36

faces. However, these peptides associate with different MHC I isoforms. The K^d and D^b peptides derive from the amino acid residues 147–156 and 366–374, respectively, of the intact virus nucleoprotein.[36] Usually, a cell expresses $1\text{–}2 \times 10^5$ copies of a particular MHC I isoform. Given that the average copy number of a particular peptide that is expressed on the surface of a cell is around 200 molecules,[6,12,13,27] there are at least 500–1000 different types of peptides per cell per given MHC I complex[12]. Since a human cell can have only up to six different MHC I isoforms, this yields a total of $6\text{–}12 \times 10^5$ total MHC I molecules, or 3000–6000 different peptide molecules where each peptide is at approximately the 200 copy level[12,39,40.] This neglects the few peptides where each is found in the many thousands of copies on a particular MHC I isoform.

Most cells have $10^5\text{–}10^6$ total MHC I molecules (or $6\text{–}12 \times 10^5$ as stated above). Approximately 1800 of the more than 2000 different peptides for the HLA-A2.1 and HLA-B7 MHC I isoforms[11,23,39] were found at the 0.01–0.1% levels of the total extract. This corresponds to either 10–100 or 100–1000 copies of a particular peptide, using either 10^5 or 10^6 total MHC I molecules, respectively, given only those two HLA isoforms. The remaining (approximately) 200 different peptides were found at relatively higher levels (i.e., 0.4–1.0%), and, for 10^6 total MHC I molecules, this yields 4000–10,000 copies per peptide per cell, respectively. The above values are, in general, of a similar magnitude to those of the isoforms listed in Table 11.1.

The ESI-mass spectrometric studies have shown that for an HLA-B27 MHC I isoform, each of the approximately 2000 different peptides was present at 10–1000 fmol per 10^8 cells.[23] Using Avogadro's number conversion, this yields 60–6000 copies per cell for each different peptide. Once again, these numbers are of the same relative magnitude as that for the above-mentioned HLA isoforms and other investigations.[27,41] By using more than 10^8 cells, MHC class-I peptides were found at the approximately one-copy-per-cell level.[11,27,40]

For a mouse H-$2A^d$ MHC I molecule,[42] greater than 2000 different peptides were estimated to bind to that particular isoform. Thus, for any given peptide, its amount is relatively small compared with the total amount of peptides expressed by MHC complexes on the surface of a cell.[39] Electrospray-mass spectral investigations will be presented with respect to the elucidation of a number of MHC class-I and -II peptides, along with their importance and origin.

Electrospray-MS Studies of Immunological Peptides

Mass Spectral Analysis of the MHC Complex of the Mouse Cell H-2K^b Isoform

Electrospray-mass spectrometry has been used to provide accurate mass information for the heavy chain and β2m light chain of MHC class-I complexes, as well as the peptide–MHC I unit.[43] The peptide–recombinant MHC complex of the H-2K^b isoform (mouse cells) was isolated, denatured with 8 M guanidine hydrochloride, and the protein components were separated by C_{18} RP-HPLC (Figure 11.3). The peptide elutes first, and the β2m and K^b heavy chain elute later. The peptide is from residues 52–59 of the vesicular stomatitis virus (VSV) nucleocapsid protein RGYVYQGL. Figure 11.4 provides an ESI-mass spectrum of the octapeptide where the $(M + H)^+$ and $(M + 2H)^{2+}$ ions are observed. Figure 11.5 shows the ESI-mass spectrum of the β2m light chain from the 43-min peak in Figure 11.3, and the reconstructed molecular mass is 11,818 Da. Figure 11.6 shows the ESI-mass spectrum of the K^b heavy chain from the 49-min

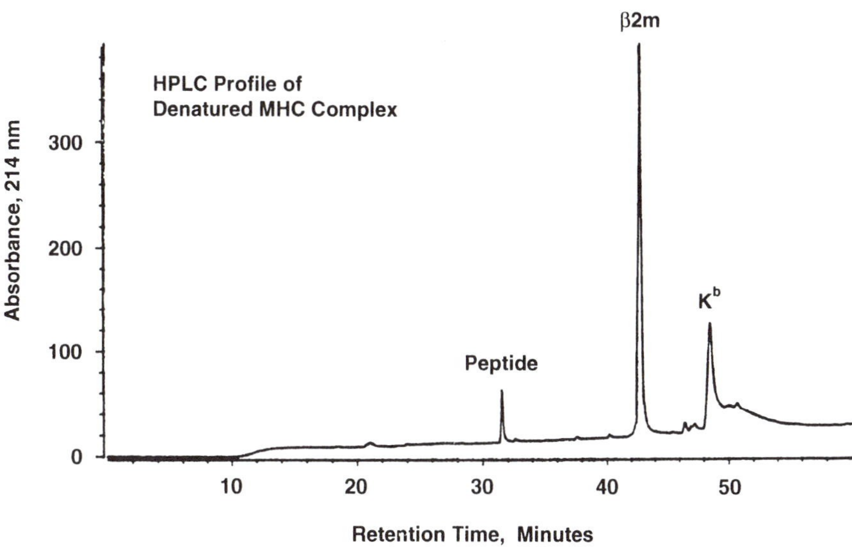

Figure 11.3 The RP-HPLC profile of the subunits of the H-2K^b MHC I and the dissociated RGYVYQGL peptide. Conditions: Vydac C_{18} column (2.1 mm × 25 cm), 0.2 ml/min and a 1%/min step in the AcCN concentration gradient. The complex was subjected to 8 M guanidine HCl denaturation for separation of the three protein components, and UV detection was at 214 nm. (© 1995. Reprinted from reference 43 by permission of Academic Press, Inc., Orlando, Florida.)

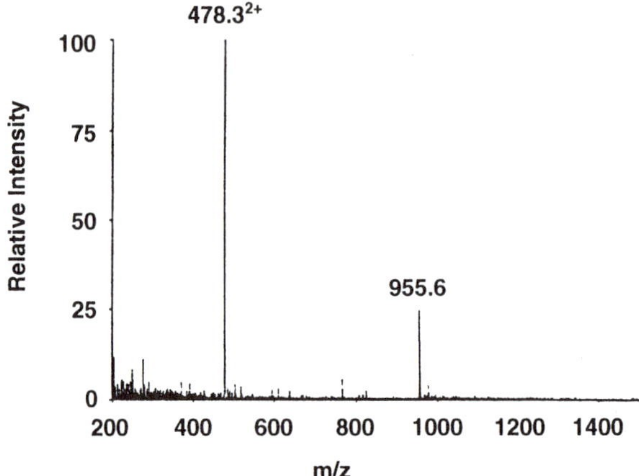

Figure 11.4 The ESI-mass spectrum of the octapeptide RGYVYQGL at 31.5 min in Figure 11.3. The HPLC eluent was collected and separately infused into the mass spectrometer. (© 1995. Reprinted from reference 43 by permission of Academic Press, Inc., Orlando, Florida.)

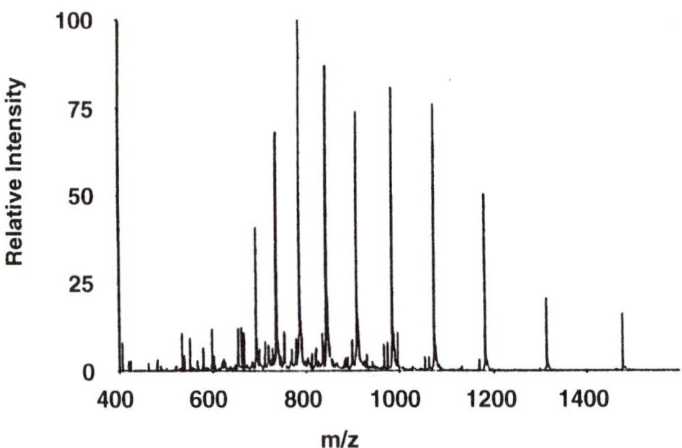

Figure 11.5 The ESI-mass spectrum of the β2m protein MHC I subunit at 42.8 min in Figure 11.3. The intact protein has a mass of 11,818 Da, and other species are observed at 11,687 Da (protein without the N-terminal Met), 11,801 (dehydration product), and 11,860 Da (N-acetylated β2m). The HPLC eluent was collected and separately infused into the mass spectrometer. (© 1995. Reprinted from reference 43 by permission of Academic Press, Inc., Orlando, Florida.)

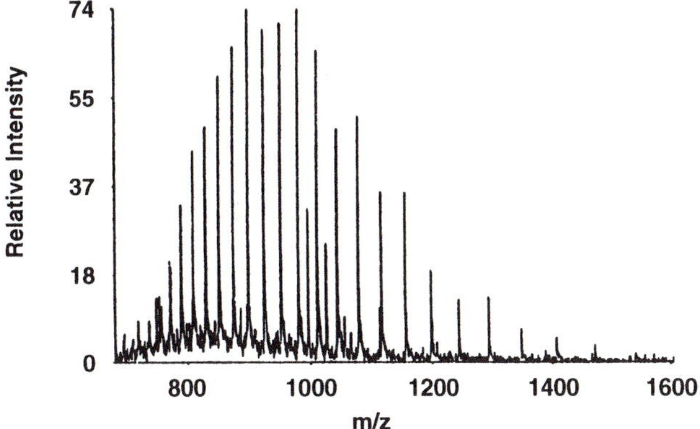

Figure 11.6 The ESI-mass spectrum of the K^b protein subunit of the MHC I complex at 49 min in Figure 11.3. Experimental masses of the K^b ligand are 32,349 Da and 32,393 Da for the intact and N-acetylated protein, respectively. The HPLC eluent was collected and separately infused into the mass spectrometer. (© 1995. Reprinted from reference 43 by permission of Academic Press, Inc., Orlando, Florida.)

peak in Figure 11.3, and the reconstructed molecular mass is 32,349 Da. This investigation used *E. coli* recombinantly derived complexes, and the peptide was synthesized. Therefore, a reasonable amount of material (nanomoles) could be obtained for the light- and heavy-chain components. The MHC I peptide extracts from cells provide amounts which are orders of magnitude lower, and some of these investigations will be analyzed.

Mass Spectral Analysis of the HLA-A2.1 MHC Class I Isoform

The HLA-A2.1 MHC class-I isoform was investigated for the general presence and identification of specific surface-displayed peptides in human B-lymphocytes by Hunt et al.[27,39] Approximately 10^9 cells were used, and the MHC I complexes from these cells were isolated and purified by immunoprecipitation. The peptides under analysis were released from the MHC I complexes by acid extraction and were separated by filtration. This same procedure was also applied to cells that do not produce either the HLA-A or the HLA-B MHC I isoform, and they served as the control. The extracts were subjected to on-line RP-HPLC-ESI-MS. Figure 11.7a presents the RP-HPLC chromatogram of the elution of the peptides released from the HLA-A2.1 isoform, and as many as 200 different peptides were observed with a signal-to-noise ratio of at least 2. Standard quantities of synthetic peptides established this signal level at approximately 30 fmol. Figure 11.7b served as a control. With the HLA-B-7 isoforms, more than 1000 peptides were observed under similar conditions as those for the A isoform (data not shown). The integrated mass spectrum of mass scans 145–155 in Figure 11.7a provided ions in the range m/z 786–1121 (data not shown).

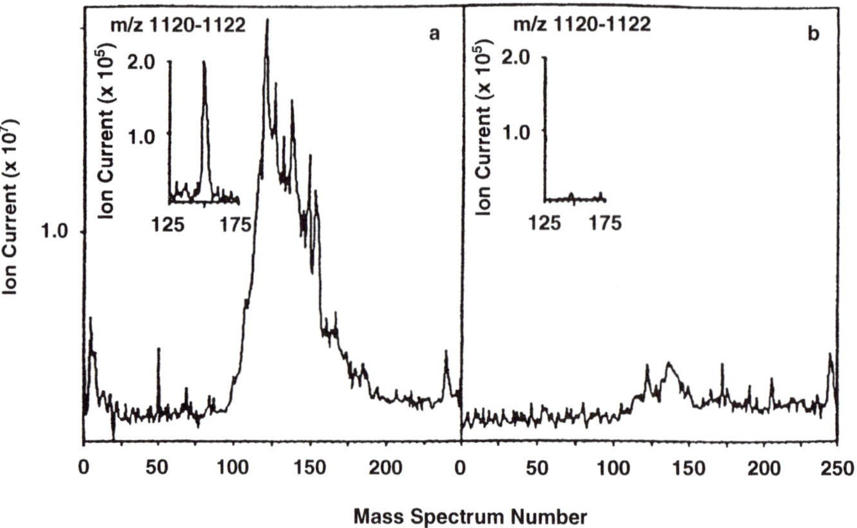

Figure 11.7 (a) The RP-HPLC-ESI-total-ion chromatogram of the extracted peptides from the MHC I HLA-A2.1 complex from human B-lymphoblastoid cells; (b) same extract as that of (a) except that the *E. coli* cells used did not produce HLA-A or HLA-B isoform MHC I complexes. (© 1992. Reprinted from reference 27 by permission of Plenum Publishing Company, New York.)

Many of the ions were very low in intensity (noise level), and relatively few ions were of significant intensity. One of the more intense ions was m/z 1121 (insets to Figure 11.7), and tandem mass spectrometry was performed on this ion (Figure 11.8). Figures 11.8a and b represent product-ion peptides of m/z 1121 in the 100–300 femtomole and 2 picomole range, respectively. A complete analysis of the monomeric peptide is also presented. Note that not all potential b and y ions are observed, but enough ions are present (underlined values) so as to deduce the peptide sequence. Some internal fragments are observed where m/z 261, 390, and 507 represent $(PY + H)^+$ or PY′, $(PYE + H)^+$ or PYE′, and $(PYEV + OH + 2H)^+$, respectively. Note that the latter ion has two additional protons and only one charge. The first couple of amino acids would be difficult to deduce from the b series; however, the y series yields a 1121 − 907 = 214-Da value for the first two residues. Table 9.6 cites this value as either the Val, Asp or Thr, Xle pair, where both Lxx and Xle mean Leu/Ile. Earlier in this chapter, it was stated that for the HLA-A2.1 isoform, the second residue in the expressed peptide was a conserved Leu residue. Thus, it is reasonable to assign the first two amino acids in the peptide as Thr-Leu in that order. The order of the first two amino acids was confirmed with the Edman degradation technique.[39] Independently, synthetic peptides that contain Leu and Ile as the second residue showed that only the Leu-containing peptide had the same HPLC retention time as that of the sample. This nonameric peptide was subjected to a database search. Briefly, this procedure takes the nine residue sequence and attempts to

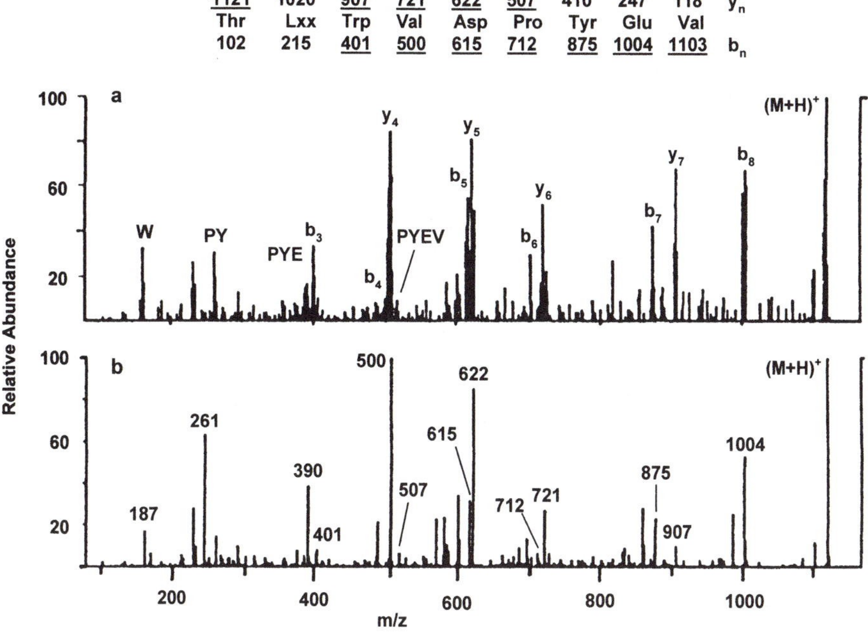

Figure 11.8 The RP-HPLC-ESI-tandem mass spectrum of *m/z* 1120–1122 (from inset in Figure 11.7a) from mass scans 145–155: (a) 100–300 fmol and (b) 2 pmol of the peptide at *m/z* 1121. (© 1992. Reprinted from reference 27 by permission of Plenum Publishing Company, New York.)

match it with a contiguous nine-residue portion of entire protein sequences in a database that contains many thousands of proteins. The peptide analyzed in Figure 11.8 was found to have the same sequence as a certain nine-residue portion of the growth factor-induced protein TIS21, whose function is largely unknown.[27,39]

Mass Spectral Analysis of the MHC I HLA-B7 Isoform

The MHC class-I HLA-B7 isoform was treated in a manner similar to that described above for HLA-A2.1,[23] except that the sequence analysis of a particular peptide required further experimental procedures. The acid extract of 2×10^8 cells that contained the peptides was subjected to RP-HPLC-MS, and the total ion chromatogram is presented in Figure 11.9. Approximately 2000 different peptides were observed in all of the extracted ESI-mass spectra, and masses spanned the range *m/z* 838–1224 (data not shown). Between 50 and 100 peptide signals were observed in each collected mass spectrum. A prominent signal at the doubly charged *m/z* 520 species, $(M + 2H)^{2+}$, was subjected to tandem mass spectrometry (Figure 11.10a).[23,44]

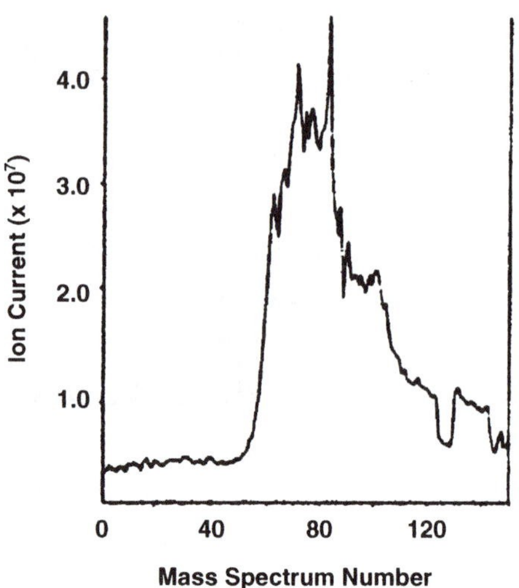

Figure 11.9 The nonapeptide extracts from the MHC I HLA-B7 isoform of JY cells were fractionated by RP-HPLC. The total ion chromatogram (TIC) is shown, where each mass scan is from 300 to 1600 Da. Mass scan rate was 1.5 s/mass spectrum. (© 1993. Reproduced from reference 23 by permission of the American Association of Immunologists.)

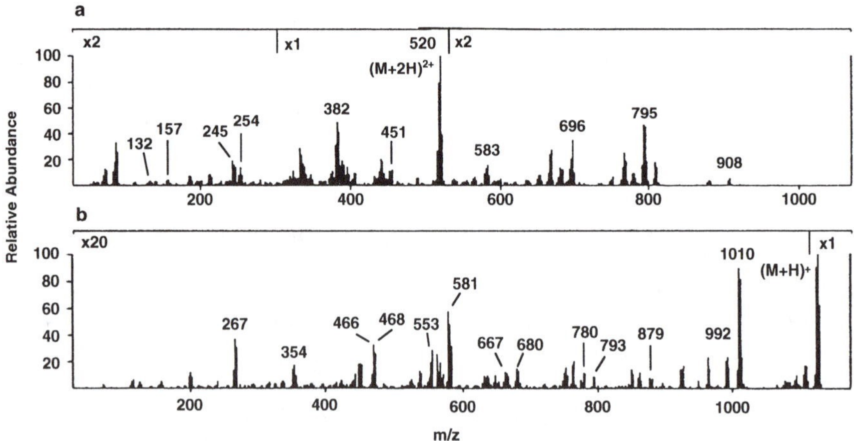

Figure 11.10 The RP-HPLC-ESI-tandem mass spectra of (a) the $(M + 2H)^{2+}$ ion at 520 Da from the TIC in Figure 11.9, and (b) the $(M + H)^{+}$ ion at m/z 1123 of the acetylated peptide in (a). (© 1993. Reproduced from reference 23 by permission of the American Association of Immunologists.)

PROBLEM 11.1 Figure 11.10a presents the ESI-tandem mass spectrum of the $(M + 2H)^{2+}$, m/z 520, from an acid extract of HLA-B7 MHC I complexes. In the analysis of this spectrum, an ambiguity arose that concerned the identity of a particular amino acid residue. The peptide was acetylated, and its ESI-tandem mass spectrum is presented in Figure 11.10b. Based on Figures 11.10a and b and Table 9.7, arrive at the sequence of the peptide. How many residues would be expected to be observed in this peptide, and how many residues undergo acetylation? Hint: In Figure 11.10b, there are six internal fragments.

Mass Spectral Analysis of Peptides Causing Graft Versus Host Disease

Investigations of the HLA-A2.1 and HLA-B7 MHC I isoforms concentrated on identification of the origin of the most intense peptides in the RP-HPLC ESI-mass spectra[23,27,39]. The present investigation searches the complex mass spectral dataspace for specific peptides that cause graft versus host disease (GVHD). Human bone marrow for transplantation must be carefully matched for compatibility of the MHC isoforms between donor and recipient. Sometimes, even this screening procedure does not preclude the occurrence of GVHD. The particular group of peptides that bind the HLA-A2.1 MHC I isoform responsible for GVHD is termed HA-2.

Affinity chromatography was used to selectively chelate HLA-A2.1 molecules from B-lymphocytes, the latter of which were known to produce and express the HA-2 peptides.[49] The MHC I–HA-2-peptide complex was then washed off and collected. The HA-2 peptides were removed from the HLA-A2.1 molecules, isolated by acid treatment, and filtered. Conventional RP-HPLC was performed and fractions were collected on a 96-well plate. Each well received an aliquot of lymphoblastoid T2 cells that had HLA-A2.1 isoforms devoid of the HA-2 group of peptides (i.e., "empty" HLA-A2.1 MHC I molecules). These cells were loaded with ^{51}Cr. Subsequently, an aliquot of cytotoxic T-lymphocytes (CTL) was added. Thus, the well(s) that contained the HA-2 peptides from the HPLC procedure (a) bound to the MHC I isoform, (b) produced cell lysis by the CTL, and (c) released radioactive ^{51}Cr.[33] Two wells displayed radioactivity, and ESI-MS produced more than 100 peptides. In order to determine which peptide(s) were responsible for CTL activity, one of the two fractions was rechromatographed by RP-HPLC. The effluent was split into an ESI-MS system and simultaneously into a 96-well plate for subsequent CTL–^{51}Cr assays.[50] The upper trace (solid line) in Figure 11.11a represents the radioactive ^{51}Cr release in fractions 50–54 (right-hand ordinate scale). Figure 11.11b shows the ESI-mass spectrum of the mass spectral TIC equivalent of the radioactive release curve in Figure 11.12a. Of the observed peptides, five were recognized by the CTL assay, because only their ion abundances tracked that of the assay response (Figure 11.11a). Figure 11.12 presents a sequence analysis of the peptide at m/z 978. Note the considerable noise found in most of the mass spectral region. Four synthetic peptides with I/L in positions 2 and 6 (Figure 11.12) were made,

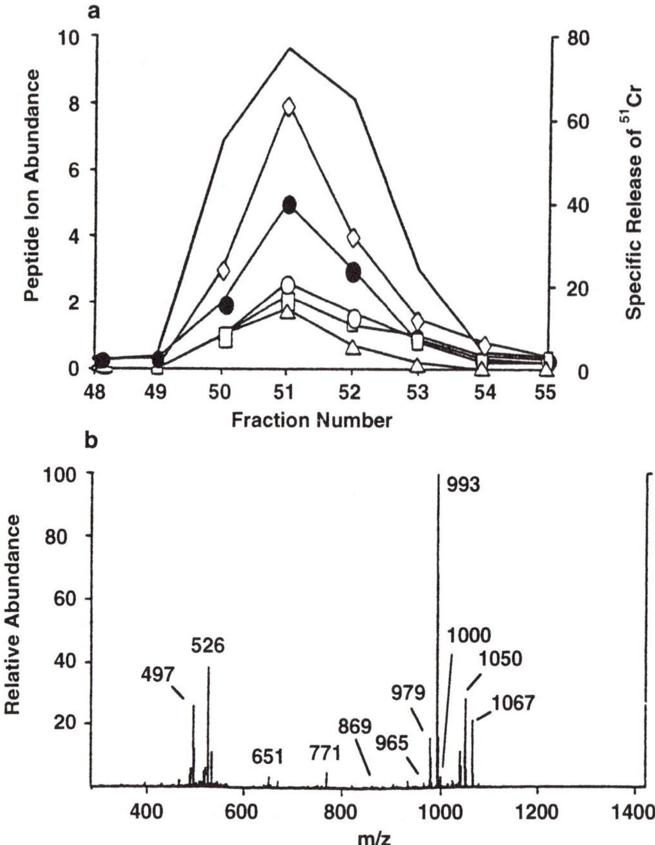

Figure 11.11 (a) Release of ^{51}Cr from human lymphoblastoid T2 cells that contain the surface HLA-A2.1 isoform when cytotoxic T-lymphocytes (CTL) are added (the upper trace and the right ordinate). Lysis and ^{51}Cr release occurs in the particular HPLC well fractions (50–54) that contain the active nonapeptides. Ion RIC traces: ◇, *m/z* 651; Δ, *m/z* 869; □, *m/z* 965; ○, *m/z* 979, and ●, *m/z* 1000. These ion values, associated with CTL recognition, were derived from the averaged mass spectrum in (b) of the mass spectral TIC equivalent of the radioactive release curve in (a). See text for details. (© 1995, American Association for the Advancement of Science. Reprinted from reference 49 with permission.)

and three of the four coeluted with the in vivo, active peptide in an HPLC-^{51}Cr assay. The peptide that contained Ile in both the 2 and 6 amino acid positions eluted at a different time. Thus, each of the three peptides was individually analyzed for CTL-mediated cell lysis. Peptide YIGEVLVSV produced a response for half-maximum cell lysis at a concentration of 40 picomolar (40 pM), while the other two isomers required more peptide (approximately 2 nanomolar concentrations for half-maximum lysis) to cause CTL-mediated ^{51}Cr release. Using known amounts of peptides in an ESI-MS format, approximately 260 HA-2

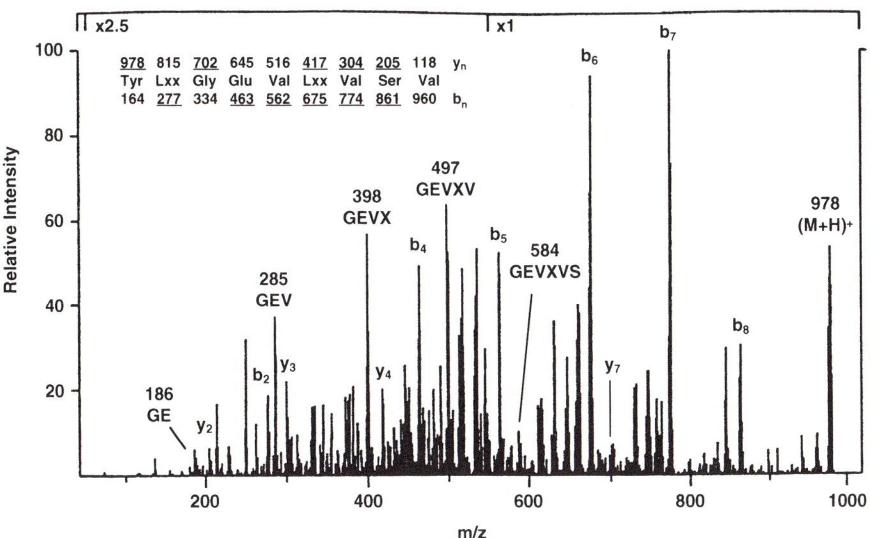

Figure 11.12 The RP-HPLC-ESI-tandem mass spectrum of *m/z* 978 (labeled *m/z* 979) in Figure 11.11b, and the deduced residue sequence. (© 1995, American Association for the Advancement of Science. Reprinted from reference 49 with permission.)

peptide HLA-A2.1 MHC I complexes are produced per cell. A database search resulted in six proteins that matched eight of nine residues, and seven proteins that corresponded to seven of nine residues in the YIGEVLVSV peptide. All the proteins are derived from nonfilamentous class-I myosins, and are involved in cellular movement and internal organelle translocation.

Mass Spectral Analysis of the MHC Class-II H-21-A^d Isoform

The MHC class-II peptide complexes can be characterized as longer in amino acid sequence than that of MHC class-I peptides. Class-II peptides essentially reflect the portion of intact proteins that are extracellular in origin. The following discussion analyzes one such peptide for sequence and origin from the mouse H-2I-A^d isoform.[42,51] The peptide fraction was isolated and subjected to RP-HPLC-UV analysis as shown in Figure 11.13a. Overall, mass spectral analysis indicated that over 2000 different peptides were present. Microcapillary RP-HPLC-ESI-MS was performed on fraction number 16 from Figure 11.13a, and the total ion chromatogram is shown in Figure 11.13b. Mass scans 55–65 were summed and displayed in Figure 11.13c. A number of intense signals are seen, yet at least 40 relatively low-intensity ions are also obtained.

PROBLEM 11.2 How many different peptides (distinguished by nominal mass value) are observed with respect to the labeled ions in Figure 11.13c.

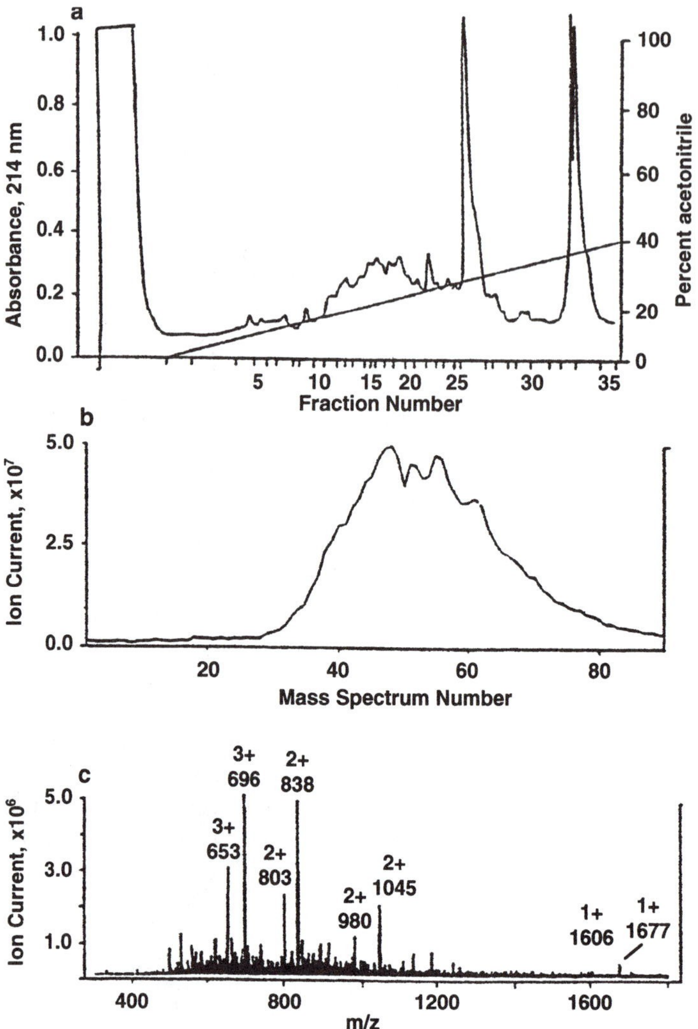

Figure 11.13 (a) The RP-HPLC-UV detection at 214 nm of the isolated peptide fraction from the MHC class-II mouse H-21-A^{d} isoform; (b) RP-HPLC-ESI-TIC of the peptides in fraction number 16 in (a); (c) averaged, reconstructed mass spectrum of mass scans 55–65 in (b). (© 1992, American Association for the Advancement of Science. Reprinted from reference 42 with permission.)

PROBLEM 11.3 The peptide at m/z 696.3^{3+} in Figure 11.13c was subjected to ESI-tandem mass spectrometry (Figure 11.14). Deduce the sequence of the peptide. The underlined masses represent doubly charged species, and only one internal fragment can be found.

The exceedingly low amounts of individual peptides (low femtomoles or less) on the MHC complexes present signal-to-noise (S/N) sensitivity concerns

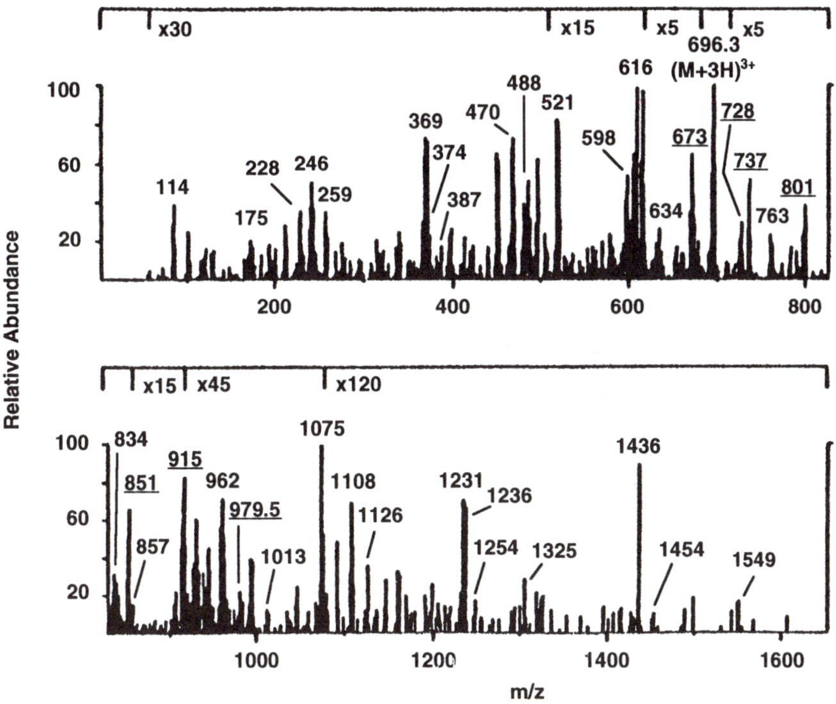

Figure 11.14 The RP-HPLC-ESI-tandem mass spectrum of the $(M + 3H)^{3+}$ ion at m/z 696.3 in Figure 11.13c. (© 1992, American Association for the Advancement of Science. Reprinted from reference 42 with permission.)

when attempts are made to identify their sequences. Relatively satisfactory analyses can be obtained on the greater abundance peptides (femtomoles–picomoles); however, many peptides of potential importance are found in the 200–1000 copies/cell range. The RP-HPLC technique is usually used in the analyses of immunogenic peptides (*vide supra*); however, capillary electrophoresis (CE) can also be considered. The RP-HPLC and CE techniques typically use 1–50 μl and 1–50 nl, respectively, of sample solution; therefore, CE has limitations with respect to sample concentration and volume applied to the CE column.[52–55] A technique was developed by Tomlinson and Naylor that uses on-line sample preconcentration and interfaces to CE in order to characterize dilute solutions of immunological peptides. The entire method relies on RP-HPLC, CE, and isotachophoresis (ITP) with mass spectrometry detection, and these techniques are reviewed in Chapter 2. The method is outlined along with an analysis of an MHC peptide. Mouse cells were lysed with the CHAPS zwitterionic detergent, and the EL-4 peptide–MHC I H-2-K^b complexes were concentrated by immunoaffinity procedures that utilize sepharose columns. The choice of detergent is very important, since it can be carried over and deleteriously affect the separation and recovery performance of the peptides in the CE system.[56] The peptides were isolated by acetic acid release from the MHC species. The peptides were then subjected to C_{18} RP-HPLC with a trifluoroacetic acid-acetonitrile gradient in order to fractionate the material. The acetonitrile was evaporated, leaving 40–70 μl of peptide solution, and between 5 and 50 peptides could be found in the HPLC fractions.[3] The HPLC fractions that contained peptides were diluted to approximately 150 μl.[56] A schematic of the experimental apparatus highlighting the CE system is shown in Figure 11.15. The concentration aspects of the method take place in the adsorptive phase (styrene divinylbenzene) membrane (inset of Figure 11.15), and the technique is called membrane preconcentration-capillary electrophoresis or mPC-CE.[52–54] Fifty microliters of the 150-μl peptide fraction is pressure-applied to the membrane in an off-line fashion, and the peptides adhere to the membrane. The CE separation buffer (2 mM NH_4OAc, 1% HOAc, pH 2.9) is used to wash off any unretained material and salts. The next steps are the elution, concentration, and characterization of the peptide molecular mass by mass spectrometry. Transient isotachophoresis takes place (tITP) when the following protocol is used, and the underlying principles are presented in Chapter 2. A leading stacking buffer (LSB) of 0.1–5% aqueous NH_4OH is introduced to the membrane, followed by an 80% aqueous MeOH solution which elutes the peptides from the membrane. A trailing stacking buffer

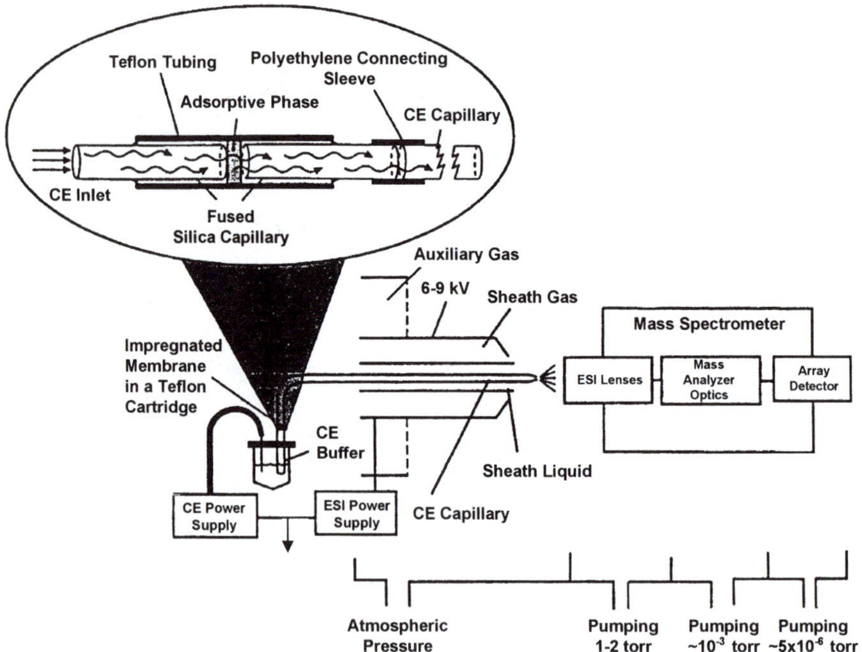

Figure 11.15 Diagram of the membrane preconcentration-capillary electrophoresis (mPC-CE) assembly interfaced to an ESI-mass spectrometry system. (© 1996, American Association for Mass Spectrometry. Reprinted from reference 54 by permission of Elsevier Science, Inc.)

(TSB) follows next and this consists of 1% HOAc and 2 mM NH_4OAc, pH 2.9.[52,55] The CE separation buffer consists of the LSB and TSB, or 2 mM NH_4OAc and 1% HOAc, pH 2.9, and this is used to develop the three solvent zones of the tITP.[52,54] The tITP of the peptide occurs between the LSB and TSB and is shown in Figure 11.16.[52,54] The LSB and TSB converge upon application of the CE electric field and concentrate (focus or stack) the peptides into narrow bands. An ESI sheath flow was used (isopropanol:H_2O:HOAc at 60:40:1 v/v) at a flow rate of 2–3 μl/min.

For tandem mass spectrometry peptide sequencing purposes, the remaining 100 μl of the 150-μl peptide fraction is subjected to the mPC-tITP-CE procedure. Attomoles–femtomoles of various peptide mixtures were applied successfully in this technique.[55,57] Pressure injection of the peptide solution allows a more rapid loading of peptides onto the membrane. In addition, off-line loading allows a greater solution volume (60–100 μl), and hence more peptide, to be concentrated, as opposed to performing a relatively large-volume on-line sample injection onto the CE column. Many other parameters—such as cell lysis surfactant; coated versus uncoated capillary columns; CID collision-cell parameters; capillary and ESI needle i.d.; and ratio and concentration of the LSB, peptide elution solvent, and TSB—affect the sensitivity, clarity, resolution, and peak broadening of the tandem mass spectra of peptides.[52,54,56,57] A modified form of the mPC concept was developed by Aebersold et al., whereby the preconcentration membrane was replaced with C-18 beads (5-μm beads with 300 Å pores).[58] The beads were held between two polytetrafluoroethylene (PTFE) membrane disks and occupied a length of 1 mm in a 50-μm i.d. capillary.

PROBLEM 11.4 Using the HPLC-mPC-tITP-CE-ESI-MS technique, a peptide was isolated from the mouse H-2-K^b MHC I complex. The $(M + 2H)^{2+}$ ion at 503.6 Da (504 Da nominal mass) was fragmented by tandem mass spectrometry, and its mass spectrum is shown in Figure 11.17.[56] Deduce the identity of

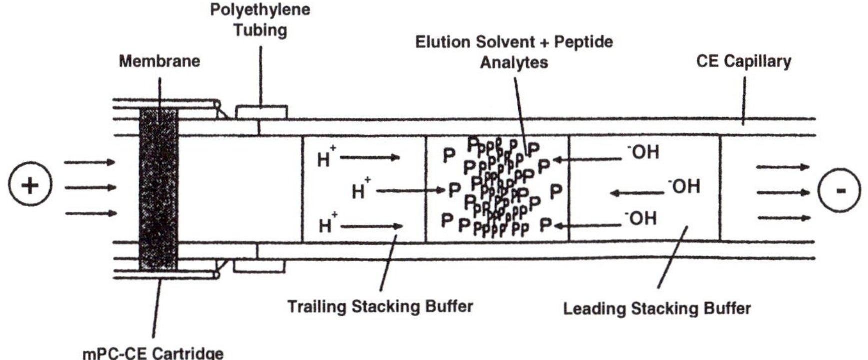

Figure 11.16 Details of the mPC-CE when used in the transient isotachophoresis mode (tITP). P, peptides. (© 1996, American Society for Mass Spectrometry. Reprinted from reference 54 by permission of Elsevier Science, Inc.)

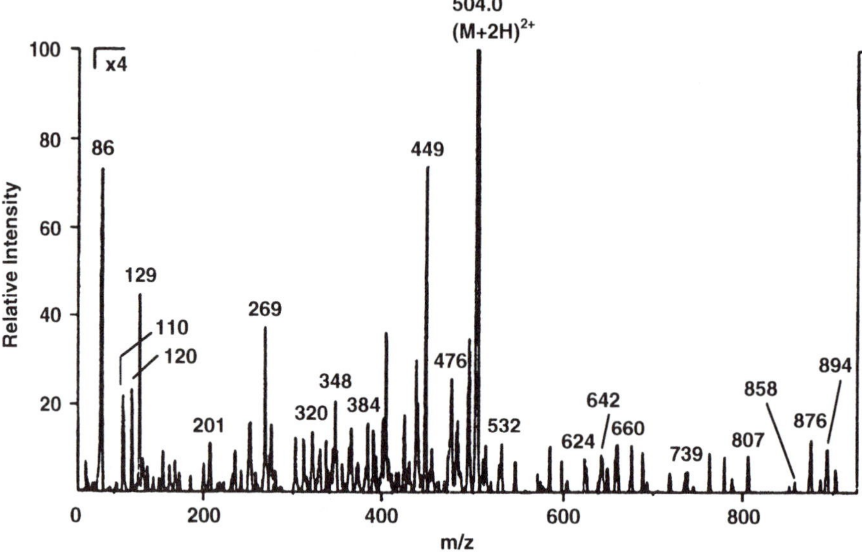

Figure 11.17 The mPC-tITP-CE-ESI-tandem mass spectrum of a peptide isolated from the mouse H-2-K^b MHC I complex. The $(M + 2H)^{2+}$ precursor ion is at 503.6 Da (nominal mass of 504.0 Da). (© 1996. Reprinted from reference 56 by kind permission of Elsevier Science-NL, Sara Burgerhartstraat 25, 1055 KV Amsterdam, The Netherlands.)

the peptide. Mass values are rounded to the nearest whole number. A number of immonium ions are present.

T Cell Component of the Immune System

Figure 11.18 presents a schematic of the T-lymphocyte cell[59] and the antigen (or peptide)-presenting cell (APC) with an attached MHC complex.[14,59–65] The peptide to be interrogated by the immune system T-cell is shown as a small oval held by the MHC complex[66] and the α- and β-subunit proteins of the T-cell receptor complex (TCR). Note that the CD4/CD8 protein is another receptor, or coreceptor; however, it is not directly involved with the peptide in question. The protein is labeled CD8 when an MHC I is present and CD4 when an MHC II complex is being interrogated. The TCR is composed of the α-, β-, CD3 protein, and phosphoprotein subunits.[66] The eta (ζ) phosphoprotein has three pairs of phosphate sites, while the other CD3 subunits have only one pair of phosphate sites. The amino acid residue regions that contain the phosphate sites (shaded portions in Figure 11.18) are referred to as immunoreceptor tyrosine-based activation motifs or ITAMs. Select sections of the α- and β-subunits interact directly with the peptide presented by MHC complexes. On the receptor site ends of the α- and β-subunits, five distinct regions can be described as the Vα, Jα, Vβ, Dβ, and Jβ variable regions.[63,67,68]

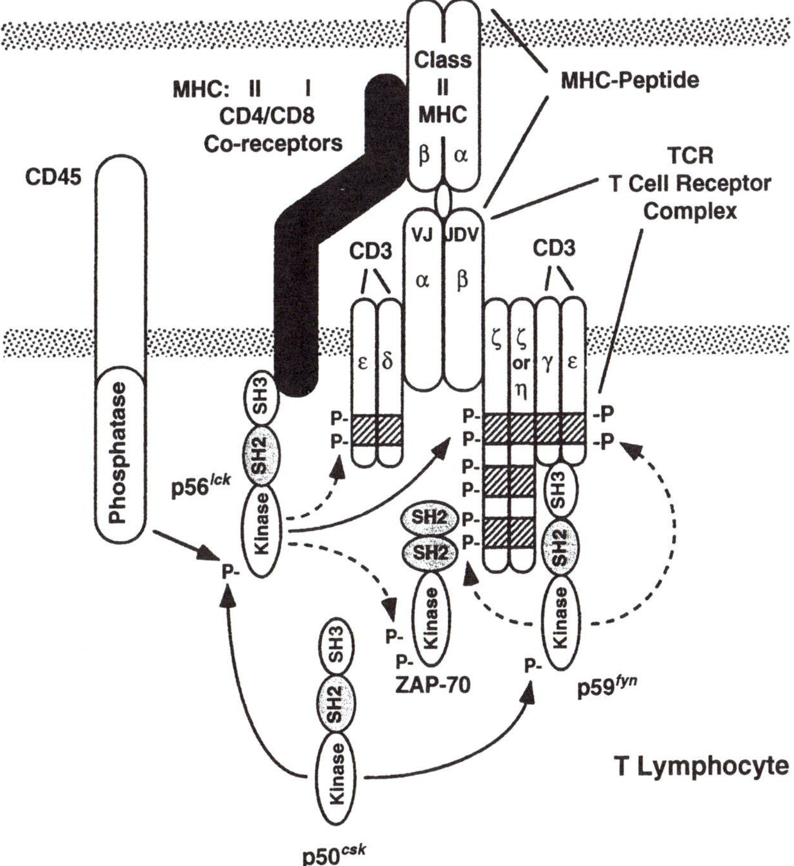

Figure 11.18 Schematic of the basic interaction between the MHC-peptide on an antigen-presenting cell (APC) and a T-cell's phosphoprotein recognition, including the APC destruction signal transduction system. (© 1995, American Chemical Society. Reprinted from reference 59 with permission.)

There are a number of important kinases in the T-cell that activate certain structures in the central TCR complex. The majority of the phosphoproteins in the T-cell are protein tyrosine kinases (PTKs).[60–62] Thus, all phosphate sites in Figure 11.18, including the four kinases, are on tyrosine residues. Two families of phosphoproteins in the PTKs are the src and syk. The src family includes $p56^{lck}$ and $p59^{fyn}$, and the syk family includes ZAP-70. These proteins act on different phosphate sites of the TCR complex (Figure 11.18),[61,62,69,70] and they regulate reversible phosphorylation processes on the Tyr residues.[71] In addition, the src members are themselves phosphorylated in a reversible fashion, and this is effected by the PTK $p50^{csk}$ kinase and the transmembrane CD45 phosphatase

(Figure 11.18). Thus, upon an optimal TCR α- and β-subunit interaction with an MHC peptide, a series of phosphorylation and dephosphorylation events occur that lead to cell multiplication. Subsequently, cytokine molecules, which include interleukin (IL),[69] tumor necrosis factor (TNF-α), IFN-γ,[72] and chemotactic cytokines, are released by T-cells, and they cause inflammation and tissue destruction.[62,73,74] Thus, it is important to be able to assess the various actions of the PTK and TCR species—which produce the initial intracellular chain of events in a T-cell, starting from MHC-antigen peptide recognition—because of the destructive consequences. Molecules of the kinase–TCR system in a T-cell were analyzed for the structural elucidation of phosphorylation sites, and some of these investigations are presented below.

Mass Spectral Analysis of a Phosphopeptide from ZAP-70

The phosphorylation sites of ZAP-70 were investigated by RP-HPLC and immobilized metal-ion affinity chromatography (IMAC) connected on-line with RP-HPLC.[59,71,75,76] Because of the extremely low concentrations of the phosphoproteins in Figure 11.21, both ZAP-70 and p56^{lck} phosphoproteins were expressed in a recombinant fashion, and biochemical reactions were performed in vitro. The ZAP-70 protein is a phosphoprotein that consists of 593 amino acids, and Figure 11.18 shows that the p56^{lck} kinase phosphorylates ZAP-70 on a number of sites throughout the sequence.

The ZAP-70 protein was immunoprecipitated from the cell expression system and was allowed to react with p56^{lck} on a spot on a sodium dodecyl sulfate-polyacrylamide gel. Radioactive ^{32}P from ATP was also added to the gel in order to radioactively label the phosphorylation sites. The reaction was terminated, and two proteins were separated by electrophoresis on the gel. A small amount of radioactivity was observed for the ZAP-70 protein when p56^{lck} was not present on the gel, because a low degree of autophosphorylation occurred during the expression of ZAP-70 in the baculovirus cell expression system. An intense radioactive spot occurred on the gel when p56^{lck} was added. The ZAP-70 radioactive spot on the gel was digested with trypsin, the spot was isolated, and the peptides were eluted and spotted onto a thin-layer chromatography (TLC) plate. The tandem of electrophoresis and ascending TLC in a solvent system resulted in a two-dimensional array of peptides and phosphopeptides, and the latter were observed by autoradiography.[59,71,75] A phosphopeptide was subsequently isolated from each spot, and each was separately subjected to RP-HPLC-ESI-mass spectral analysis.

The RP-HPLC-ESI-mass spectral analysis[59,71] of one phosphopeptide is shown in Figure 11.19. The dashed line indicates the acetonitrile elution concentration gradient of the HPLC system. The minor peak near 20 min displays an ESI-mass spectrum with an $(M + H)^+ = 1247.0$ Da and $(M + 2H)^{2+} = 624.5$ Da. The broad elution profile between 20 and 35 min represents the contaminants that remain from the excision of the phosphopeptide TLC spot. This experiment used approximately 16.7 pmol of phosphopeptide.

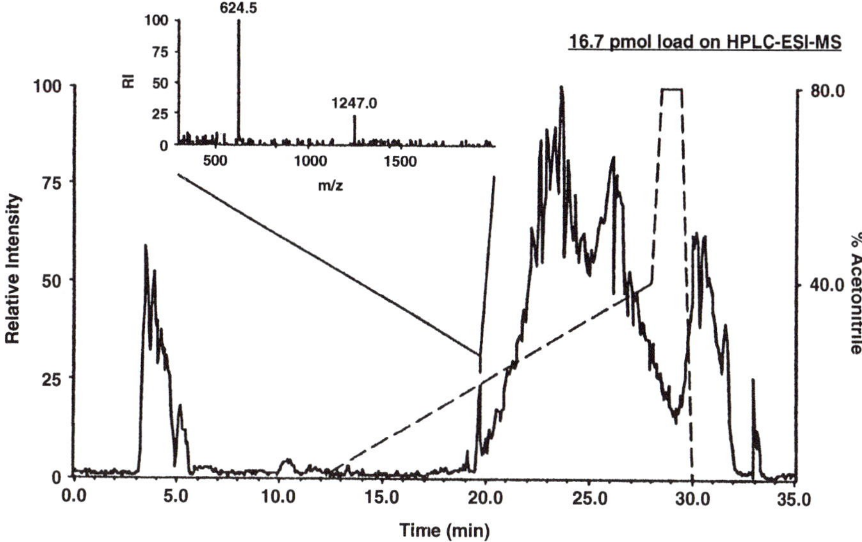

Figure 11.19 The RP-HPLC-ESI-TIC of the elution of a single tryptic phosphopeptide from the ZAP-70 phosphoprotein. The phosphopeptide is indicated at approximately 20 min, and its ESI-mass spectrum is shown. The broad "hump" represents TLC plate contaminants. See text for details. (© 1994, American Society for Biochemistry and Molecular Biology, Bethesda, MD. Reproduced from reference 71 with permission.)

This phosphorylated tryptic peptide mass corresponds with the amino acid residue sequence $K^{176}LY^{178}SGAQTDGK^{186}$ of the ZAP-70 protein. Here, Y^{178} is the site of phosphorylation, and the Ser and Thr residues are not phosphorylated.

Chapter 10 presents information on IMAC-ESI-MS technology, and the IMAC-HPLC-ESI-MS serial system was used to probe the same K-176 to K-186 phosphopeptide shown in Figure 11.19. Figure 11.20 provides this information[59,71] where the IMAC column bound approximately 5 pmol of the phosphopeptide by ferric ion interaction and allowed the other components to elute through. The phosphopeptide subsequently was eluted from the IMAC column and onto the RP-HPLC column, where conventional HPLC chromatography occurred. Note the significant clean-up of the signal because of the added dimension (IMAC) of separation.

With this same column technology, two additional phosphopeptides on ZAP-70 were found that were phosphorylated by $p56^{lck}$. These consisted of residues 64-Y^{69}-75 and 485-Y^{492}-Y^{493}-496, where the phosphorylated tyrosine residues are indicated. The tyrosine 69 and 178 phosphorylation sites occur in the SH2 domain of ZAP-70. The SH2 domain is a conserved amino acid sequence in signal-transduction proteins, and these regions bind to other proteins that contain phosphotyrosine moieties (Figure 11.21). The Y^{492} and Y^{493} residues are found in the kinase region of ZAP-70. Autophosphorylation (i.e.,

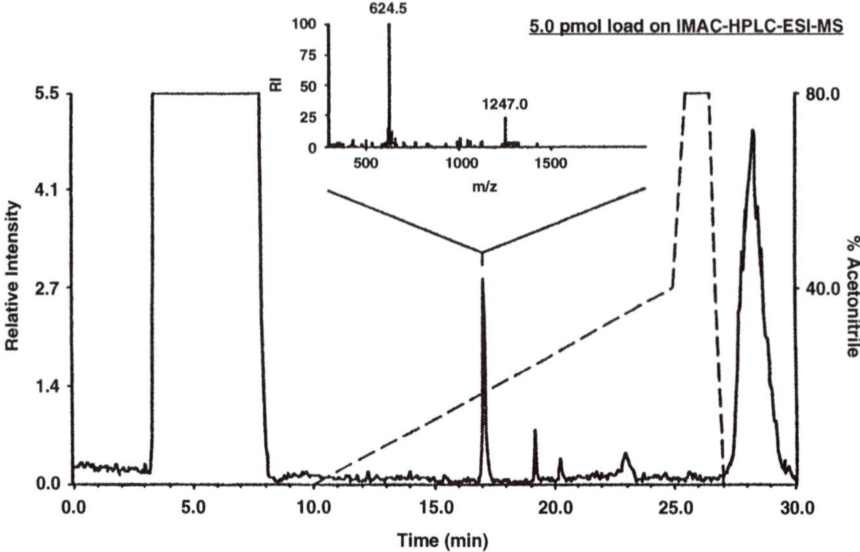

Figure 11.20 The IMAC-HPLC-ESI-TIC of the same phosphopeptide preparation as in Figure 11.19. Note the increased clarity of the phosphopeptide analyte and its ESI-mass spectrum is provided. (© 1994, American Society for Biochemistry and Molecular Biology, Bethesda, MD. Reproduced from reference 71 with permission.)

ZAP-70 catalyzing its own phosphorylation), occurs on tyrosine residues that originate between the SH2 regions (125-Y^{126}-128) and the SH2–kinase region (284-Y^{292}-298 and 283-Y^{292}-298).

Mass Spectral Analysis of a Phosphopeptide from CD3-ζ

Figure 11.18 shows that the $p56^{lck}$ kinase also provides phosphorylation capabilities to the CD3-ζ phosphoprotein subunit of the TCR complex. In vitro phosphorylation was performed by recombinantly expressed $p56^{lck}$ on a synthetically made CD3-ζ subunit, because these native phosphoproteins are found in concentrations too low to be isolated in a relatively pure state. Handling procedures performed for this phosphorylation reaction pair were similar to those for ZAP-70 and $p56^{lck}$.

A synthetic peptide that corresponded to the 52–164 residue sequence of the CD3-ζ subunit was phosphorylated with $p56^{lck}$ in the presence of ^{32}P in a vial.[75] The reaction was terminated, and the phosphorylated CD3-ζ subunit was purified over a C_4 column. Tryptic peptides were produced, and they were separated in a two-dimensional TLC preparation similar to that used for ZAP-70 (see section "Mass Spectral Analysis of a phosphopeptide from ZAP-70"). The phosphopeptide portion of the tryptic peptides was observed by autoradiography. A total of 10 phosphopeptides (TLC spots) were observed, and each was subjected

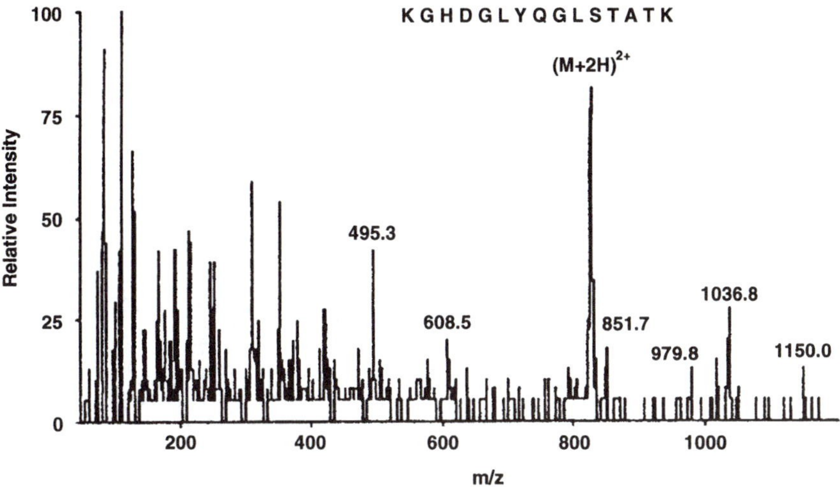

Figure 11.21 The RP-HPLC-ESI-product-ion mass spectrum of the $(M + 2H)^{2+}$ ion at 828.7 Da of a tryptic phosphopeptide from the CD3-ζ subunit. See text for details. (© 1994. Reprinted from reference 75 by permission of Academic Press, Inc., Orlando, Florida.)

to a separate RP-HPLC-ESI-MS analysis. Product-ion mass spectrometry was performed in order to deduce the sites of phosphorylation for each peptide. Two of these phosphopeptides are presented for mass spectral interpretation.

PROBLEM 11.5 The peptide from the residues K-136 to K-150 provided a product-ion mass spectrum in Figure 11.21, and the sequence is provided in the figure. Deduce the site of phosphorylation and provide the identity of each peak in the spectrum. The $(M + 2H)^{2+}$ precursor ion at 828.7 Da was used to effect CID.

PROBLEM 11.6 Another tryptic phosphopeptide recovered from the TLC plate from the CD3-ζ phosphoprotein underwent RP-HPLC-ESI-tandem mass spectral analysis (Figure 11.22). The precursor ion was the $(M + 2H)^{2+}$ species at 695.0 Da. Provide the complete sequence and site of phosphorylation of the phosphopeptide.

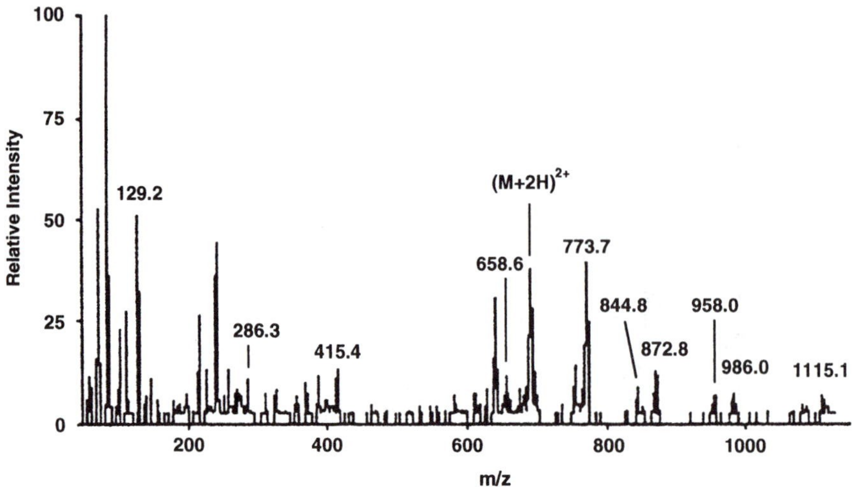

Figure 11.22 The RP-HPLC-ESI-product-ion mass spectrum of the $(M + 2H)^{2+}$ ion at 695.0 Da of a tryptic phosphopeptide from the CD3-ζ subunit. See text for details. (© 1994. Reprinted from reference 75 by permission of Academic Press, Inc., Orlando, Florida.)

REFERENCES

1. Cole, B.C. *ASM News* **1996**, *62*, pp 471–475.
2. Engelhard, V.H. *Sci. Am.* **1994**, *271*, pp 54–61.
3. Cox, A.L.; Skipper, J.; Chen, Y.; Henderson, R.A.; Darrow, T.L.; Shabanowitz, J.; Engelhard, V.H.; Hunt, D.F.; Slingluff, C.L., Jr. *Science* **1994**, *264*, pp 716–719.
4. Demotz, S.; Grey, H.M.; Sette, A. *Science* **1990**, *249*, pp 1028–1030.
5. Tizard, I.R., Ed.; *Immunology, An Introduction*; 3rd Edition; Saunders College Publishing: Orlando, FL, 1992.
6. Christinck, E.R.; Luscher, M.A.; Barber, B.H.; Williams, D.B. *Nature* **1991**, *352*, pp 67–70.
7. Dustin, M.L.; Springer, T.A. *Annu. Rev. Immunol.* **1991**, *9*, pp 27–66.
8. Brodsky, F.M.; Guagliardi, L.E. *Annu. Rev. Immunol.* **1991**, *9*, pp 707–744.
9. Cresswell, P. *Annu. Rev. Immunol.* **1994**, *12*, pp 259–293.
10. Peters, P.J.; Neefjes, J.J.; Oorschot, V.; Ploegh, H.L.; Geuze, H.J. *Nature* **1991**, *349*, pp 669–676.
11. Engelhard, V.H. *Annu. Rev. Immunol.* **1994**, *12*, pp 181–207.
12. Rammensee, H.G.; Falk, K.; Rötzschke, O. *Annu. Rev. Immunol.* **1993**, *11*, pp 213–244.
13. Vitiello, A.; Potter, T.A.; Sherman, L.A. *Science* **1990**, *250*, pp 1423–1426.
14. Kostyu, D.D.; Amos, D.B. In *The Metabolic Basis of Inherited Disease*; 6th Edition; Scriver, C.R.; Beaudet, A.L.; Sly, W.S.; Valle, D., Eds.; McGraw-Hill Information Services Company: New York, 1989; Chapter 4, pp 225–249.
15. Germain, R.N.; Margulies, D.H. *Annu. Rev. Immunol.* **1993**, *11*, pp 403–450.
16. Falk, K.; Rötzschke, O.; Stevanović, S.; Jung, G.; Rammensee, H.G. *Nature* **1991**, *351*, pp 290–296.

17. Gulden, P.H.; Hackett, M.; Addona, T.A.; Guo, L.; Walker, C.B.; Sherman, N.E.; Shabanowitz, J.; Hewlett, E.L.; Hunt, D.F. In *Mass Spectrometry in the Biological Sciences*; Burlingame, A.L.; Carr, S.A., Eds.; Humana Press: Totowa, NJ, 1996; pp 281–305.
18. Goldberg, A.L. *Science* **1995**, *268*, pp 522–523.
19. Androlewicz, M.J.; Ortmann, B.; Van Endert, P.M.; Spies, T.; Creswell, P. *Proc. Natl. Acad. Sci. U.S.A.* **1994**, *91*, pp 12716–12720.
20. Grandea, A.G., III; Androlewicz, M.J.; Athwal, R.S.; Geraghty, D.E.; Spies, T. *Science* **1995**, *270*, pp 105–108.
21. Henderson, R.A.; Michel, H.; Sakaguchi, K.; Shabanowitz, J.; Appella, E.; Hunt, D.F.; Engelhard, V.H. *Science* **1992**, *255*, pp 1264–1266.
22. Henderson, R.A.; Cox, A.L.; Sakaguchi, K.; Appella, E.; Shabanowitz, J.; Hunt, D.F.; Engelhard, V.H. *Proc. Natl. Acad. Sci. U.S.A.* **1993**, *90*, pp 10275–10279.
23. Huczko, E.L.; Bodnar, W.M.; Benjamin, D.; Sakaguchi, K.; Zhu, N.Z.; Shabanowitz, J.; Henderson, R.A.; Appella, E.; Hunt, D.F.; Engelhard, V.H. *J. Immunol.* **1993**, *151*, pp 2572–2587.
24. Roche, P.A.; Marks, M.S.; Cresswell, P. *Nature* **1991**, *354*, pp 392–394.
25. Rudensky, A.Y.; Marić, M.; Eastman, S.; Shoemaker, L.; DeRoos, P.C.; Blum, J.S. *Immunity* **1994**, *1*, pp 585–594.
26. Sanderson, F.; Kleijmeer, M.J.; Kelly, A.; Verwoerd, D.; Tulp, A.; Neefjes, J.; Geuze, H.J.; Trowsdale, J. *Science* **1994**, *266*, pp 1566–1569.
27. Hunt, D.F.; Henderson, R.A.; Shabanowitz, J.; Sakaguchi, K.; Michel, H.; Sevilir, N.; Cox, A.L.; Appella, E.; Engelhard, V.H. *Science* **1992**, *255*, pp 1261–1263.
28. Kubo, R.T.; Sett, A.; Grey, H.M.; Appella, E.; Sakaguchi, K.; Zhu, N.Z.; Arnott, D.; Sherman, N.; Shabanowitz, J.; Michel, H.; Bodnar, W.M.; Davis, T.A.; Hunt, D.F. *J. Immunol.* **1994**, *152*, pp 3913–3924.
29. Resing, K.A.; Johnson, R.S.; Walsh, K.A. *Biochemistry* **1993**, *32*, pp 10036–10045.
30. Roche, P.A. *Science* **1996**, *274*, pp 526–527.
31. Weber, D.A.; Evavold, B.D.; Jensen, P.E. *Science* **1996**, *274*, pp 618–620.
32. McCormack, A.L.; Monji, T.; Pious, D.; Yates, J.R. *Proc. 42nd ASMS Conference on Mass Spectrometry and Allied Topics* **1994**, p 645.
33. Slingluff, C.L., Jr.; Cox, A.L.; Henderson, R.A.; Hunt, D.F.; Engelhard, V.H. *J. Immunol.* **1993**, *150*, pp 2955–2963.
34. Van Bleek, G.M.; Nathenson, S.G. *Nature* **1990**, *348*, pp 213–216.
35. Rötzschke, O.; Falk, K.; Stevanović, S.; Jung, G.; Walden, P.; Rammensee, H.G. *Eur. J. Immunol.* **1991**, *21*, pp 2891–2894.
36. Falk, K.; Rötzschke, O.; Deres, K.; Metzger, J.; Jung, G.; Rammensee, H.G. *J. Exp. Med.* **1991**, *174*, pp 425–434.
37. Wallny, H.J.; Deres, K.; Faath, S.; Jung, G.; Van Pel, A.; Boon, T.; Rammensee, H.G. *Int. Immunol.* **1992**, *4*, pp 1085–1090.
38. Rötzschke, O.; Falk, K.; Stevanovic, S.; Jung, G.; Walden, P.; Rammensee, H.G. *Eur. J. Immunol.* **1991**, *21*, pp 2891–2894.
39. Hunt, D.F.; Shabanowitz, J.; Michel, H.; Cox, A.L.; Dickinson, T.; Davis, T.; Bodnar, W.; Henderson, R.A.; Sevilir, N.; Engelhard, V.H.; Sakaguchi, K.; Appella, E.; Grey, H.M.; Sette, A. In *Methods in Protein Sequence Analysis*; Imahori, K.; Sakiyama, F., Eds.; Plenum Press: New York, 1993; pp 127–133.
40. Bevan, M.J.; Hogquist, K.A.; Jameson, S.C. *Science* **1994**, *264*, pp 796–797.
41. DiBrino, M.; Parker, K.C.; Shiloach, J.; Knierman, M.; Lukszo, J.; Turner, R.V.; Biddison, W.E.; Coligan, J.E. *Proc. Natl. Acad. Sci. U.S.A.* **1993**, *90*, pp 1508–1512.

42. Hunt, D.F.; Michel, H.; Dickinson, T.A.; Shabanowitz, J.; Cox, A.L.; Sakaguchi, K.; Appella, E.; Grey, H.M.; Sette, A. *Science* **1992**, *256*, pp 1817–1820.
43. Papadopoulos, N.J.; Sacchettini, J.C.; Nathenson, S.G.; Angeletti, R.H. In *Techniques in Protein Chemistry VI*; Crabb, J.W., Ed.; Academic Press: San Diego, CA, 1995; pp 375–383.
44. Arnott, D.; Shabanowitz, J.; Hunt, D.F. *Clin. Chem.* **1993**, *39*, pp 2005–2010.
45. Thorne, G.C.; Ballard, K.D.; Gaskell, S.J. *J. Am. Soc. Mass Spectrom.* **1990**, *1*, pp 249–257.
46. Thorne, G.C.; Gaskell, S.J. *Rapid Commun. Mass Spectrom.* **1989**, *3*, pp 217–221.
47. Cox, K.A.; Gaskell, S.J.; Morris, M.; Whiting, A. *J. Am. Soc. Mass Spectrom.* **1996**, *7*, pp 522–531.
48. Dikler, S.; Kelly, J.W.; Russell, D.H. *J. Mass Spectrom.* **1997**, *32*, pp 1337–1349.
49. den Haan, J.M.M.; Sherman, N.E.; Blokland, E.; Huczko, E.; Koning, F.; Drijfhout, J.W.; Skipper, J.; Shabanowitz, J.; Hunt, D.F.; Engelhard, V.H.; Goulmy, E. *Science* **1995**, *268*, pp 1476–1480.
50. Bodnar, W.M.; Hunt, D.F.; Cox, A.L.; Davis, T.A.; Michel, H.; Shabanowitz, J.; Henderson, R.A.; Slingluff, C.; Engelhard, V.H. *Proc. 41st ASMS Conference on Mass Spectrometry and Allied Topics*, **1993**, pp 1076a–1076b.
51. Sette, A.; Demars, R.; Grey, H.M.; Oseroff, C.; Southwood, S.; Appella, E.; Kubo, R.T.; Hunt, D.F. *Chem. Immunol.* **1993**, *57*, pp 152–165.
52. Tomlinson, A.J.; Guzman, N.A.; Naylor, S. *J. Cap. Elec.* **1995**, *6*, pp 247–266.
53. Tomlinson, A.J.; Naylor, S. *J. Cap. Elec.* **1995**, *5*, pp 225–233.
54. Tomlinson, A.J.; Benson, L.M.; Jameson, S.; Johnson, D.H.; Naylor, S. *J. Am. Soc. Mass Spectrom.* **1996**, *8*, pp 15–24.
55. Tomlinson, A.J.; Benson, L.M.; Braddock, W.D.; Oda, R.P.; Naylor, S. *J. High Resolut. Chromatogr.* **1995**, *18*, pp 381–383.
56. Tomlinson, A.J.; Jameson, S.; Naylor, S. *J. Chromatogr., A* **1996**, *744*, pp 273–278.
57. Tomlinson, A.J.; Naylor, S. *J. Liq. Chromatogr.* **1995**, *18*, pp 3591–3615.
58. Figeys, D.; Ducret, A.; Aebersold, R. *J. Chromatogr., A* **1997**, *763*, pp 295–306.
59. Watts, J.D.; Affolter, M.; Krebs, D.L.; Wange, R.L.; Samelson, L.E.; Aerbersold, R. In *Biochemical and Biotechnological Applications of Electrospray Ionization Mass Spectrometry;* Snyder, A.P., Ed.; ACS Symposium Series 619; American Chemical Society: Washington, DC, 1995; Chapter 20, pp 381–407.
60. Peri, K.G.; Veillette, A. *Chem. Immunol.* **1994**, *59*, pp 19–39.
61. Isakov, N.; Wange, R.L.; Samelson, L.E. *J. Leukocyte Biol.* **1994**, *55*, pp 265–271.
62. Perlmutter, R.M.; Levin, S.D.; Appleby, M.W.; Anderson, S.J.; Alberola-Ila, J. *Annu. Rev. Immunol.* **1993**, *11*, pp 451–499.
63. Garcia, K.C.; Degano, M.; Stanfield, R.L.; Brunmark, A.; Jackson, M.R.; Peterson, P.A.; Teyton, L.; Wilson, I.A. *Science* **1996**, *274*, pp 209–219.
64. Garboczi, D.N.; Ghosh, P.; Utz, U.; Fan, Q.R.; Biddison, W.E.; Wiley, D.C. *Nature* **1996**, *384*, pp 134–141.
65. Banchereau, J.; Bazan, F.; Blanchard, D.; Brière, F.; Galizzi, J.P.; van Kooten, C.; Liu, Y.J.; Rousset, F.; Saeland, S. *Annu. Rev. Immunol.* **1994**, *12*, pp 881–922.
66. Marx, J. *Science* **1995**, *267*, pp 459–460.
67. Murray, D.L.; Ohlendorf, D.H.; Schlievert, P.M. *ASM News* **1995**, *61*, pp 229–235.
68. Herman, A.; Kappler, J.W.; Marrack, P.; Pullen, A.M. *Annu. Rev. Immunol.* **1991**, *9*, pp 745–772.
69. Minami, Y.; Kono, T.; Miyazaki, T.; Taniguchi, T. *Annu. Rev. Immunol.* **1993**, *11*, pp 245–267.

70. Wange, R.L.; Isakov, N.; Burke, T.R.; Otaka, A.; Roller, P.P.; Watts, J.D.; Aebersold, R.; Samelson, L.E. *J. Biol. Chem.* **1995**, *270*, pp 944–948.
71. Watts, J.D.; Affolter, M.; Krebs, D.L.; Wange, R.L.; Samelson, L.E.; Aebersold, R. *J. Biol. Chem.* **1994**, *269*, pp 29520–29529.
72. Farrar, M.A.; Schreiber, R.D. *Annu. Rev. Immunol.* **1993**, *11*, pp 571–611.
73. McFarland, H.F. *Science* **1996**, *274*, pp 2037–2038.
74. Miyajima, A.; Kitamura, T.; Harada, N.; Yokota, T.; Arai, K. *Annu. Rev. Immunol.* **1992**, *10*, pp 295–331.
75. Affolter, M.; Watts, J.D.; Krebs, D.L.; Aebersold, R. *Anal. Biochem.* **1994**, *223*, pp 74–81.
76. Tomer, K.B.; Moseley, M.A.; Deterding, L.J.; Parker, C.E. *Mass Spectrom. Rev.* **1994**, *13*, pp 431–457.

ANSWERS

PROBLEM 11.1 Prior to data analysis, the HLA-B7 can be considered to bind peptides of a similar length as those of HLA-A2.1. Therefore, a sequence of nine residues is a reasonable assumption. Figure 11.23 provides the mass spectral details regarding the native peptide and the acetylated form. In Figure 11.23a, virtually all the information is contained in the b ion series. The identity of the ninth residue can be determined by subtraction of the b_8 value from that of the M value. This yields $1038 - 908 = 130$ Da, where $1038 = \text{M}$. After subtracting the C-terminal hydroxyl value of 17 Da, the residue mass of 113 Da (Lxx) remains. The M value of 1038 Da $= 520(2) - 2$. The alert reader will note that b_4 is really a $b_4 - H_2O$ ion. Only y_1 and y_2 are observed in the y series, and the y_1 value of 132 Da confirms the Lxx identity of the C-terminal residue.

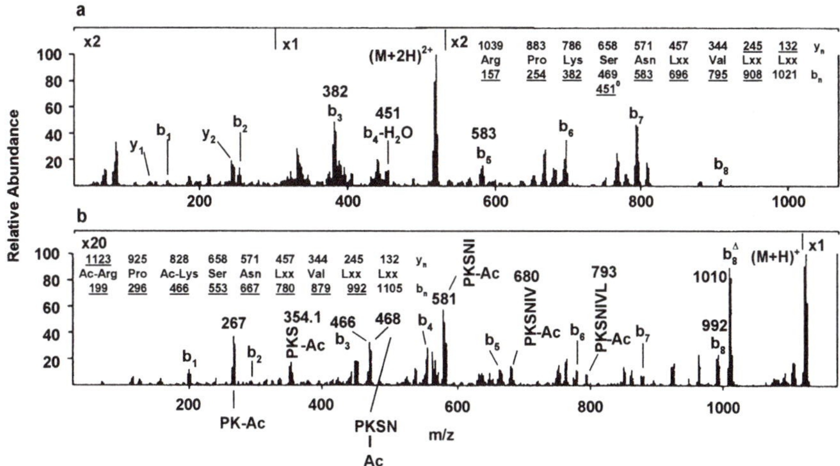

Figure 11.23 (a and b) Same as Figure 11.10, including the proper ion residue sequence labels and peptide sequence. Observed masses are underlined. (© 1993. Reproduced from reference 23 by permission of the American Association of Immunologists.)

In addition to the identity ambiguities of the three-Lxx residue, the third residue has a nominal mass of 128 Da in Figures 11.10a and 11.23a. This can be due to either Lys or Gln. Acetylation of the peptide shows that the b_3 value of 382 Da (Figure 11.10a) shifts to 466 Da (Figure 11.23b) upon acetylation. This produces a difference of 466 – 382 = 84 Da and can be attributed to the acetylation sites at the N-terminal and at the third residue. Thus, the third residue must be Lys. Once again, the b series dominates the ESI-tandem mass spectrum. Six internal fragments are observed, and they all have proline at the N-terminus, with the acetylated lysine residue at the second residue position.

In Figure 11.23b, an intense peak is observed at m/z 1010 (b_8^{Δ}), and this is 18 Da *higher* than the b_8 ion peak at m/z 992. This phenomenon has been observed for angiotensins II and III[45–47] under FAB mass spectrometry conditions and for bradykinin and splenopentin under matrix-assisted laser desorption/ionization (MALDI)-post-source decay time-of-flight (TOF)-MS.[48] Apparently, the C-terminal hydroxyl moiety of the peptide migrates to the carboxyl C-terminal portion of the b_n ion, and this was supported by ^{18}O-labeling experiments. The b_n + OH neutral species also adds an additional proton to form the $(b'_n + OH)^+$ or $(b_n + OH + H)^+$ charged species. This can be confusing, because the resulting ion structure appears to be of the y_n ion type, except that the $(b_n + OH + H)^+$ ion does not include the true C-terminal residue of the original peptide.

The peptide in question was subsequently identified in a database search as originating from the B-lymphocyte cell surface CD20 antigenic marker.[23] This peptide belongs to a class of peptides that bind to the HLA-B7 MHC I isoform, and can it be characterized as having a proline and Lxx residue in positions 2 and 9, respectively.

PROBLEM 11.2 The following groups of ions show mathematical relationships based on the $(M + 3H)^{3+}$, $(M + 2H)^{2+}$, and $(M + H)^+$ conversions: 653^{3+}, 980^{2+}; 696^{3+}, 1045^{2+}; 803^{2+}, 1606^+; 838^{2+}, 1677^+. Thus, four different peptides can be readily discerned.

PROBLEM 11.3 In preparation for the sequence analysis, the reader should have noted the many b_n-18 and y_n-18 ions that are due to dehydration gas-phase reactions of the b and y sequence ions. Figure 11.24 provides labels for the ions, as well as the peptide sequence. The m/z 1549 ion represents the b_{13}-H_2O ion, and the top row of six ions represent the doubly charged y ions. Here, y_1–y_{10} are observed as singly charged ions, and y_{11}–y_{16} are present as doubly charged species. All residues can be accounted for by successive mass differences in the b and y series. The identity of the first residue can be obtained in two ways. First, y_{16}^{2+} can be subtracted from the $(M + H)^+$ ion by the following:

$$(M + 3H)3+ = 696.3\text{Da}$$

$$(696.3 \times 3) - 2H = 2086.9 \simeq 2087\text{Da} = (M + H)^+$$

$$y_{16}^{2+} = 979.5; (979.5 \times 2) - H = 1958\text{Da}$$

Thus, 2087 – 1958 = 129 Da = Glu residue.

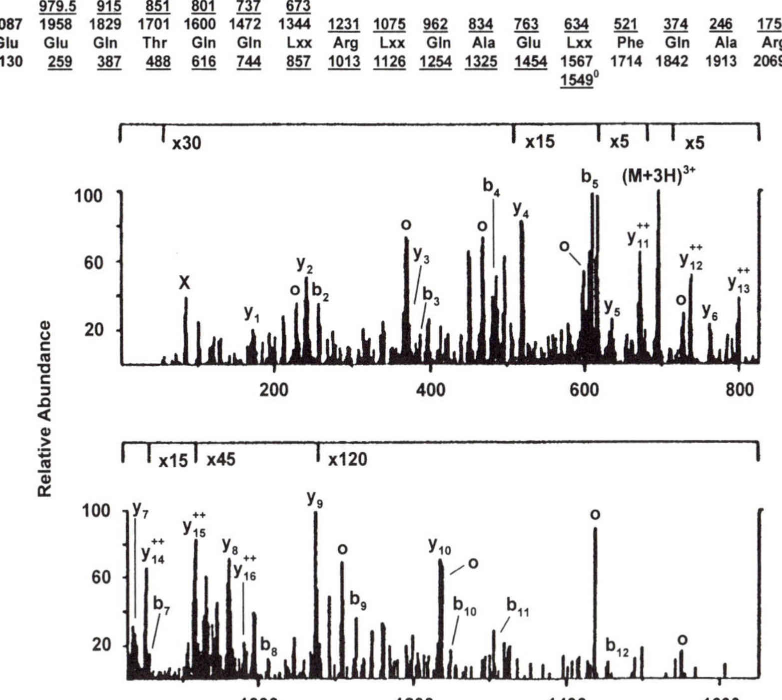

Figure 11.24 The labeled mass spectrum from Figure 11.14, including the labeled sequence of the peptide. Peaks labeled with a (○) indicate a water loss from a "b" or "y" ion. (© 1992, American Association for the Advancement of Science. Reprinted from reference 42 with permission.)

The second method relies on the b ion series where $b_2 = 259$ Da. Table 9.6 shows that, of the three possibilities, 258 Da is that of a Glu, Glu residue pair. Remember that the N-terminal hydrogen must be subtracted from 259 Da to yield the residue mass of 258 Da. Here, b_2 could not be the Pro-CMCys sequence, and it also cannot be the tetrapeptide. The latter is ruled out because the $y_{16}^{2+} - y_{15}^{2+}$ differential yields 64.5^{2+} Da, or a 129 Da Glu residue. The peptide consists of 17 amino acid residues. The peptide was identified, by database search, as mouse apolipoprotein-E or Apo-E, which is secreted into the extracellular fluid. The 17 amino acids constitute residues 236–252 in Apo-E, and this results in Lxx positions 7, 9, and 13 as Ile, Leu, and Ile, respectively.

PROBLEM 11.4 Scheme 11.1 provides the information necessary to arrive at the structure of the peptide, and Figure 11.25 provides the labeled mass spectrum. The asterisk represents a dehydration. Note that b_1, y_1, and b_8 are not observed; however, the first residue can be obtained as follows: y_7 establishes the second residue as Ser, and the average mass of the peptide is present as the $(M + 2H)^{2+}$

	I	S	F	K	F	D	H	L	
				641.74					y_n^*
	1007.17	894.02	806.94	659.76	531.59	384.41	269.32	132.18	y_n
b_n	114.17	201.24	348.42	476.59	623.77	738.86	876.00	989.16	
a_n			320.42	448.59					
b*							857.98		

Scheme 11.1

ion. Thus, the $(M + H)^+$ would be $(504.0^{2+} \times 2) - 1 = (M + H)^+ = 1007.0$ Da, $1007.0 - 894 = 113.0$ Da, and this is equal to an Xle residue.

The identity of the last residue follows a similar logic. The b_7 ion is established as a His residue, and the b_8 value would be $1007 - 1 - 17 = 989$ Da. Therefore, the last residue's mass would be $989 - 876 = 113$ Da, or an Xle residue.

The fourth residue has a mass of 128 Da and could be either a Lys or Gln. The low-mass immonium ions provide a clue as to the identity of the residue in the form of m/z 129. Table 9.13 relates the fact that m/z 129 for Gln is usually of a low intensity, while that of Lys is assumed to be of moderate intensity. The approximately 50% signal intensity of m/z 129 appears to be that of Lys. Scheme 11.1 provides the correct I/L values of the first and last residues, and Lys is the fourth residue.

PROBLEM 11.5 Scheme 11.2 provides the analysis and the observed peaks are underlined. The amino acid site of phosphorylation should be immediately recognized as Tyr-142, because this chapter has stated the fundamental role of

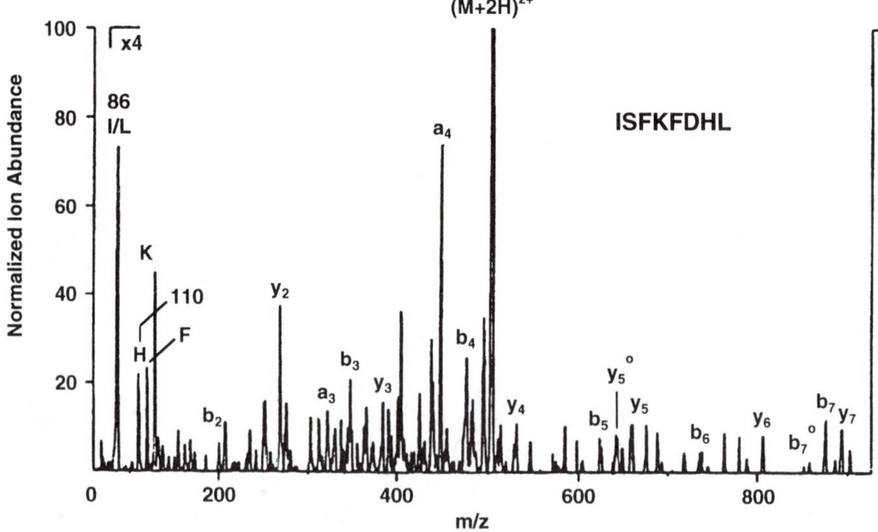

Figure 11.25 Labeled mass spectrum from Figure 11.17 together with the sequence of the octapeptide analyte. (© 1996. Reprinted from reference 56 by kind permission of Elsevier Science-NL, Sara Burgerhartstraat 25, 1055 KV Amsterdam, The Netherlands.)

	K136	G	H	D	G	L	Y142	Q	G	L	S	T	A	T	K150
-18															
b residue	129.17	186.05	323.19	438.28	495.33	608.49	771.67								
+80							851.65	979.78	1036.83	1150.0					
					b5	b6	b7•p	b8•p	b9• p	b10•p					

Scheme 11.2

tyrosine in the phosphorylation processes in T-cells. Note that no dehydration ions occur (Table 10.1), and four phosphorylated b ions are observed. The y series of ions was not observed. Only a few b ions are observed; however, they bracket the site of phosphorylation.

PROBLEM 11.6 Scheme 11.3 provides the sequence of the phosphopeptide, residue numbers, and identities of the mass spectral peaks. There are five b and two a peaks that are phosphorylated, and all ions listed in Scheme 11.3 are observed in Figure 11.22. No y ions are observed. There is either a "b" or an "a" ion present in the spectrum for the first eight residues. The phosphate occurs on the fourth residue, Y^{83}, and is observed as the $b_4 \cdot P$ ion. The $(M + 2H)^{2+}$ species establishes the mass of 1388.0 Da for the phosphopeptide, and the difference in mass between the $b_8 \cdot P$ fragment and M_r value, in residue mass, is

$$1388.0 - 1115.07 = 272.93$$
$$272.93 - 17 \text{ (C-terminal OH)} = 255.93 \simeq 256\text{Da}$$

Table 9.6 shows that seven different residue combinations can fit the residue mass of 256 Da. Since the peptide is a tryptic peptide, the C-terminal residue must be either a Lys or an Arg residue. Therefore, some combinations can be eliminated, and this leaves the following three possibilities: Lys, Lys; Gln, Lys; Ala, Gly, Lys. With no y ions to distinguish between the nominal isobaric Lys, Lys; Gln, Lys combinations and Ala, Gly, Lys species, we are left with these three possibilities. The correct sequence is Lys, Lys and it is presented in Figure 11.22 and Scheme 11.3. The amino acid residues for this phosphopeptide span the R-80 to K-89 region of the CD3-ζ TCR subunit.

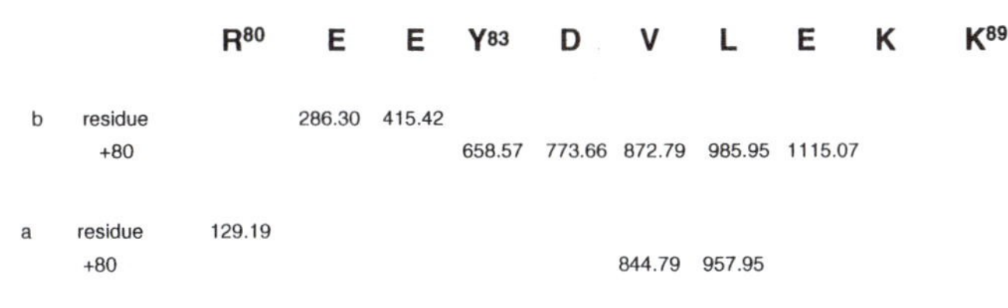

R80 E E Y83 D V L E K K89
b residue 286.30 415.42
+80 658.57 773.66 872.79 985.95 1115.07
a residue 129.19
+80 844.79 957.95

Scheme 11.3

12

Interpretation of Glycoprotein Mass Spectra

One of the important, if not the most important, posttranslational modifications (PTM) to a protein or peptide is the addition of carbohydrate/polysaccharide moieties.

In 1996, Gahmberg et al.[1] investigated the SWISS-PROT database and found that, of the 1823 mammalian proteins that had a modification to the protein backbone, 91.7% were listed as a glycoprotein. Glycoproteins essentially are proteins or peptides that are posttranslationally modified with the addition of one or more oligosaccharide entities. An oligosaccharide or polysaccharide is a biological polymer of (mono) saccharide (carbohydrate) residues. The majority of proteins of the mammalian cell surface and internal endoplasmic reticulum are glycoproteins.[1]

Glycoproteins find themselves in many different functional roles, and excellent reviews can be found on this topic by Varki[2] and Paulson.[3] Despite their importance in cell-surface recognition markers for microorganisms, antibodies, control of protein turnover, intercellular recognition events, enzymes, structural integrity of enzymes and proteins, and as a toxic component of certain glycoproteins, it was recently stated that ". . . the functions of DNA and proteins are generally known . . . [while] . . . it is much less clear what carbohydrates do . . .".[2,4,5] Evidence for this seemingly unusual claim is clearly outlined by Varki:

1. The function of the carbohydrate moiety on a glycopeptide is very difficult to predict.
2. Uncommon carbohydrate residues, sequences, or modifications of residues usually dictate the function of an entire moiety.
3. Different roles and functions can be found for the same polysaccharide sequence when it is attached to different protein/peptide

hosts (glycoconjugates) and is at different sites in the polypeptide host.

These last two concepts can be found in this chapter in the analyses of the structure of various oligosaccharide moieties on different proteinaceous hosts.

Carbohydrate Structure

Table 12.1 presents structures of the most commonly found carbohydrate residues in the polysaccharide portion of glycoproteins. The integrity of a saccharide residue is analogous to that of an amino acid residue (Chapter 9). The addition of a water molecule provides the intact, neutral monomer sugar unit for all residues in Table 12.1. Thus, a hydrogen would be added to the singly bonded oxygen in each structure in Table 12.1 (left-most oxygen in each structure), and the hydroxyl group would be added to the right-most carbon that has only three bonds. The permethylated structures in Table 12.1 will be addressed in a subsequent section of this chapter. At first glance, there is a great deal of similarity between the various sugar residues. They have six-membered rings with an internal ether oxygen heteroatom. Alcoholic functionalities abound, and a few residues have an N-acetyl or carboxylic acid group in place of an alcohol group.

Table 12.1 will be an important reference for deducing the structures and masses of various polysaccharide elements in the glycoprotein examples in this chapter. Except for fucose, all sugars are in the D mirror-image form. Look carefully at α-L-fucose and β-D-galactose in Table 12.1. Except for the first carbon in the ring (carbon-1 or C-1) and the C-6 moiety, they are mirror images of each other. The particular configuration of the C-6 moiety—namely, whether C-6 is "above" or "below" the 'plane' of the ring—dictates the L- or D- form. This is strictly a mirror image or stereoisomeric property. As it turns out, essentially all sugar monomers in glycoproteins are in the D mirror-image form except for fucose. The mirror image of a D-sugar configuration yields an L-form, and, neglecting the configuration of functionalities about C-1, L-fucose can become identical to L-galactose (mirror image of D-galactose) with the addition of an oxygen atom on C-6. The configuration and position of the hydroxyl groups on a sugar residue (intraresidue orientation of the particular hydroxyl moieties) dictate the identity of the sugar, and they are fixed for a particular sugar. The exception is the hydroxyl on C-1. For a given sugar, the C-1 hydroxyl group (or the residue bond orientation on C-1) can be above or below the plane of the ring. An example is given in Table 12.1 for D-mannose. If the C-1 hydroxyl or residue bond is "below" the plane of the ring, the bond is designated α; if it is "above" the ring, it is designated β. Note that for fucose and all L-forms, this is opposite in designation; that is, the α-bond is "above" the plane of the ring. The α- or β-stereochemical position of the C-1 hydroxyl is termed an anomeric property; that is, α-D-mannose and β-D-mannose are anomers.

Table 12.1 Saccharide residues

Underivatized residue	Abbreviation	Underivatized structure	Residue formula	M_r	Permethylated structure	Additional methyl groups	Residue formula	M_r
Pentose								
β-D-Xylose	Xyl	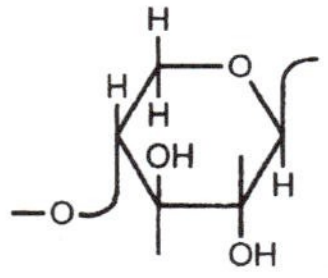	$C_5H_8O_4$	132.116	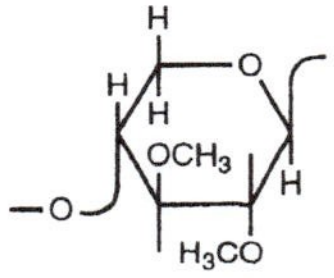	2	$C_7H_{12}O_4$	160.170
Deoxyhexose								
α-L-Fucose 6-Deoxy-L-Galactose	Fuc	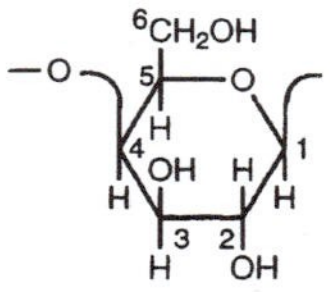	$C_6H_{10}O_4$	146.143	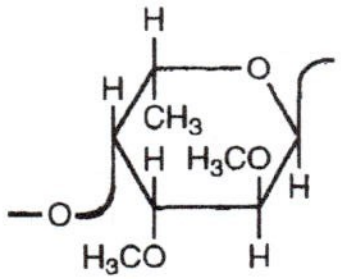	2	$C_8H_{14}O_4$	174.197
Hexose	Hex							
β-D-Galactose	Gal	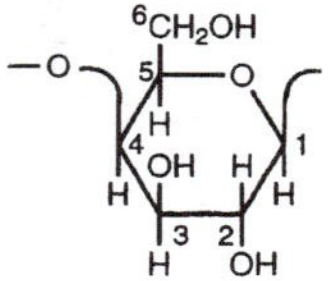	$C_6H_{10}O_5$	162.142	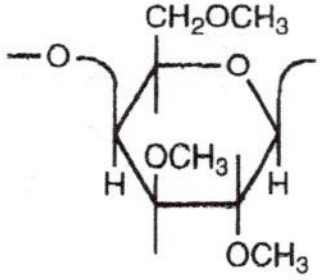	3	$C_9H_{16}O_5$	204.223

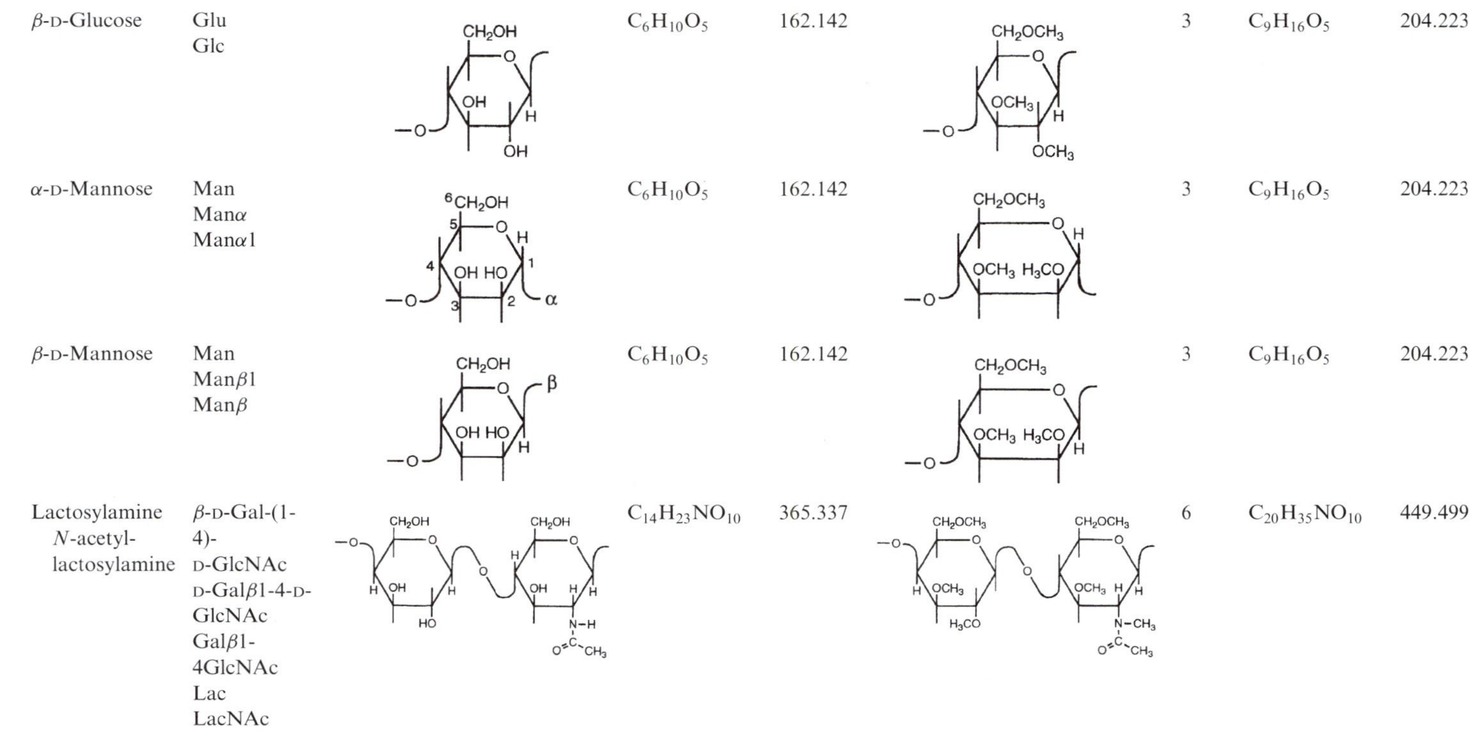

β-D-Glucose	Glu Glc		$C_6H_{10}O_5$	162.142		3	$C_9H_{16}O_5$	204.223
α-D-Mannose	Man Manα Manα1		$C_6H_{10}O_5$	162.142		3	$C_9H_{16}O_5$	204.223
β-D-Mannose	Man Manβ1 Manβ		$C_6H_{10}O_5$	162.142		3	$C_9H_{16}O_5$	204.223
Lactosylamine *N*-acetyl- lactosylamine	β-D-Gal-(1- 4)- D-GlcNAc D-Galβ1-4-D- GlcNAc Galβ1- 4GlcNAc Lac LacNAc		$C_{14}H_{23}NO_{10}$	365.337		6	$C_{20}H_{35}NO_{10}$	449.499

(*continued*)

Table 12.1 (*continued*)

Underivatized residue	Abbreviation	Underivatized structure	Residue formula	M_r	Permethylated structure	Additional methyl groups	Residue formula	M_r
Diacetyl-chitobiose	D-GlcNAcβ1-4-D-GlcNAc	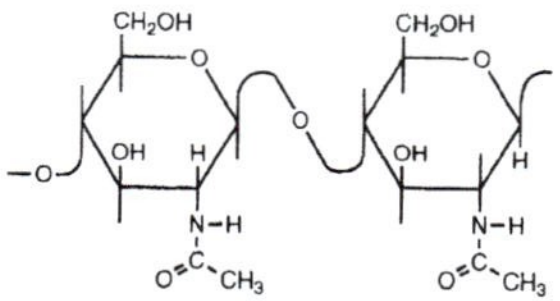	$C_{16}H_{26}N_2O_{10}$	406.39	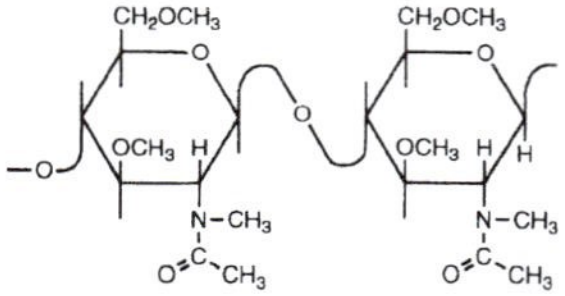	6	$C_{22}H_{38}N_2O_{10}$	490.552
Hexuronic acid β-D-Glucuronic acid	GlcA GlcU	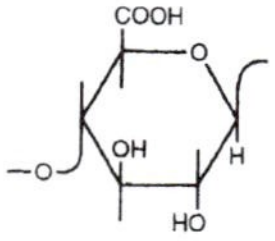	$C_6H_8O_6$	176.126	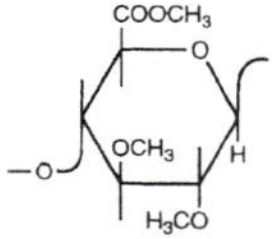	3	$C_9H_{14}O_6$	218.207
2-Acetamido-2-deoxy-glucuronic acid 2-*N*-acetyl-glucuronic acid	GlcUNAc	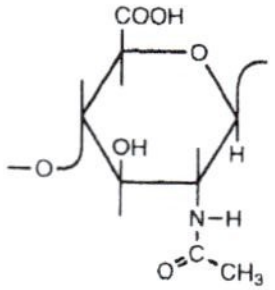	$C_8H_{11}NO_6$	217.18	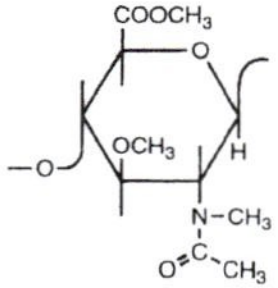	3	$C_{11}H_{17}NO_6$	259.261
N-*acetyl-hexosamine*	HexNAc							
N-acetyl-α-D-glucosamine	GlcNAc	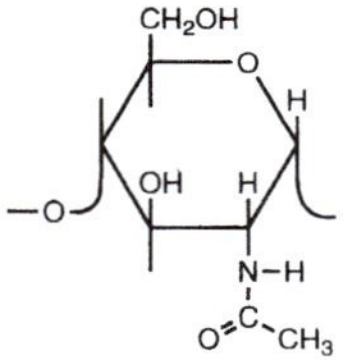	$C_8H_{13}NO_5$	203.195	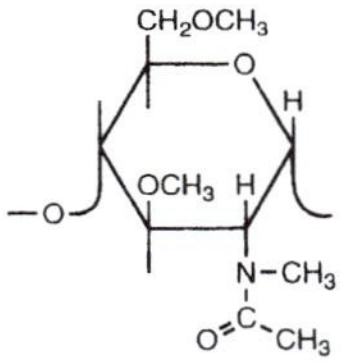	3	$C_{11}H_{19}NO_5$	245.276

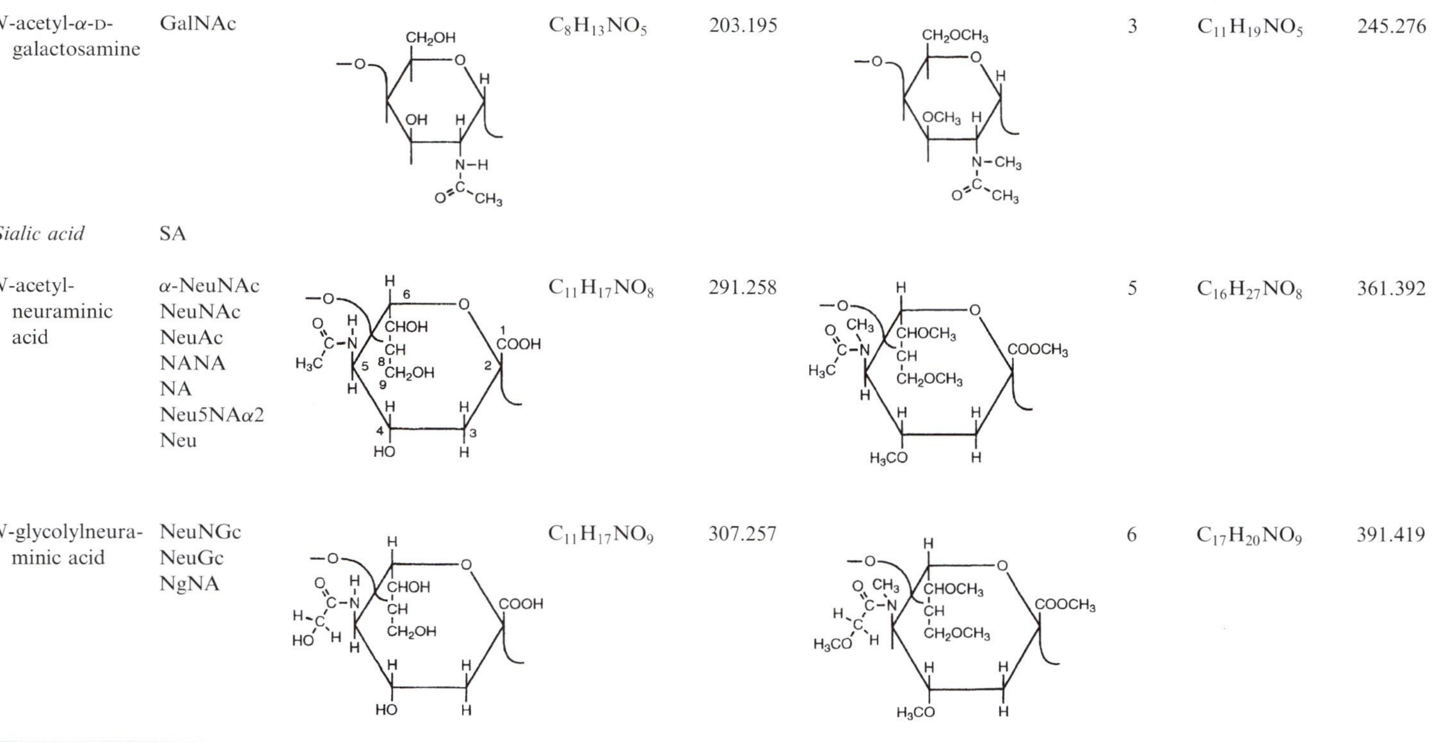

N-acetyl-α-D-galactosamine	GalNAc		$C_8H_{13}NO_5$	203.195		3	$C_{11}H_{19}NO_5$	245.276
Sialic acid	SA							
N-acetyl-neuraminic acid	α-NeuNAc NeuNAc NeuAc NANA NA Neu5NAα2 Neu		$C_{11}H_{17}NO_8$	291.258		5	$C_{16}H_{27}NO_8$	361.392
N-glycolylneuraminic acid	NeuNGc NeuGc NgNA		$C_{11}H_{17}NO_9$	307.257		6	$C_{17}H_{20}NO_9$	391.419

The "above" and "below" designations are not entirely accurate, because the sugar ring cannot exist in a planar form. Instead, the ring exists in a boat, twisted, or chair-type molecular conformation. These concepts are beyond the scope of this chapter, and further information can be found in any recent organic or biochemical textbook.

Even more important, by now the reader is probably asking the question of how it is possible that a mass spectrometer can produce data for differentiation purposes with respect to the sugar isomers in Table 12.1. The biochemical literature quite convincingly answers this question, at least twofold: chemical modifications to the polysaccharide, and use of specific, stereochemically sensitive glycosidase enzymes.[6] Two other criteria are routinely used: faith in the probability of occurrence of certain carbohydrate species based on established work, and independent techniques—such as nuclear magnetic reconance (NMR) technology.[6]

Enzyme and Chemical Reactions on Carbohydrates

No matter how short or long a polypeptide is, there is usually a terminal amine (i.e., free hydrogen end) and a terminal hydroxyl. Exceptions are cyclic peptide and modified terminal groups (Chapter 9). In an analogous fashion, carbohydrate polysaccharides have two types of terminal sugar(s) that have only one bond to the rest of the chain. One terminal sugar is 1 Da (hydrogen addition) higher in mass than its residue form. The other end of the polysaccharide is connected to the protein, and the mass of this terminal monomer unit is equal to its residue mass. Fortunately, polysaccharides, with few exceptions, attach only to the asparagine (Asn), serine (Ser), or threonine (Thr) amino acid residues, as opposed to most or all of the 20 residues (Chapter 9). However, a polypeptide can have multiple numbers of these three residues, and, usually, only a portion of these residues are glycosylated. The polysaccharide identity can have a wide arrray of permutations. In many cases, it is desirable to remove the carbohydrate chain from the protein portion for subsequent analysis. Regardless of the method, hydrolysis takes place—where the hydrogen adds to the amino acid residue side-chain R-group and the hydroxyl adds to the terminal sugar.

Tables 12.2 and 12.3 present chemical methods of removing the polysaccharide from a glycoprotein with respect to Ser/Thr and Asn, respectively. Either acid or base hydrolysis provides a release mechanism for the glycoprotein conjugate attached to the Asn amino acid residue (Table 12.3), while base hydrolysis is used for the Ser/Thr residue (Table 12.2). The first reaction in Table 12.4 provides an important point in an appreciation of the products in Tables 12.2 and 12.3. On the reactant side in Tables 12.2 and 12.3, the polysaccharides are bound to the protein, and when they are released, a hydroxyl or aldehyde is formed on C-1. The important point is that a free hydroxyl occurs when the ring remains intact, and an aldehyde group forms when the ring is cleaved. The first reaction in Table 12.4 shows that these two species can be found in equilibrium. Either sugar form is designated as a reducing species, because (a) the aldehyde group, which is the actual reducing species, can be converted to an alcohol, and

(b) the intact ring form is a potential reducing species because it can open up to an aldehyde, which can be directly reduced. Both forms are also known as aldohexoses; namely, they are capable of forming an aldehyde moiety on C-1. When the aldehyde is converted to an alcohol group, all six carbons have an alcohol group or saturated oxygen species, and the molecule is necessarily in a linear form. This is called an alditol (Table 12.4, reaction 4) of the aldohexose (aldose). If an N-acetyl group is present on the linear or open-ring alditol, the latter is called an aldaminitol (Table 12.4, reaction 5; Tables 12.2 and 12.3, reaction 3).

In chemical reaction nomenclature, an oxidizing species can be reduced and a reducing species can be oxidized. This is a consequence of the gain or loss of electrons (Lewis acid–base theory). Put another way, an oxidizing species gains electrons (gets reduced) and a reducing species loses electrons (gets oxidized). Biochemical nomenclature refers to the C-1 aldehyde moiety on a sugar monomer as the reducing species. Benedict's solution contains water-soluble Cu^{2+}. The presence of an aldehyde causes Cu^{2+} (aq) to precipitate as Cu_2O (monovalent copper), and the aldehyde reducing sugar is oxidized (loses two electrons to form an acid).[26,31] However, when an aldehyde is converted into an alcohol, the aldehyde actually gains two electrons when placed into the context of an oxidation–reduction chemical equation. Now, the aldehyde moiety is noted as an oxidizing species, because it gains two electrons (gets reduced) upon conversion to an alcohol.[25,32]

Further important information can be noted in Tables 12.2–12.4. In Table 12.2, Ser and Thr undergo a β-elimination reaction in the base-catalyzed release of the polysaccharide chain, and note that this α, β nomenclature is a completely different issue with respect to the α, β position of the C-1 carbon on the sugar unit. The α and β here refer to the α- and β-carbon atoms in the Ser/Thr residues. Thus, Ser and Thr each are reduced in mass by 2 Da. For Asn in Table 12.3, the acid treatment yields aspartic acid (Asp), while base treatment yields the original Asn residue. There is a difference of 1 Da in mass for the acid treatment, and this difference can be used as an indicator in structural analyses. Note that for the base-catalyzed reactions in both Tables 12.2 and 12.3, deleterious effects—such as peeling (reverse aldol condensation)—can take place as noted in the footnote to Tables 12.2 and 12.3.[7,11,20] The terminal sugar can internally rearrange via a β-elimination as shown in reaction 2, Table 12.4, and can continue to destroy itself (Table 12.4, reaction 3) if the third carbon in the terminal sugar is linked to the rest of the polysaccharide instead of the fourth carbon as shown in Tables 12.2 and 12.3. This can be prevented by having $NaBH_4$ present during base-catalyzed hydrolysis (reaction 3 in Tables 12.2 and 12.3). The borohydride reagent reduces the aldehyde "reducing species" to an alcohol, and this latter moiety is stable to the destructive effects of hydroxide base. Note the action of borohydride and borodeuterohydride on the aldehyde moiety in reactions 4, 5, and 6, respectively, in Table 12.4. An alditol derivative is formed from *N*-acetylglucosamine, and this species can be labeled by several names, including aldaminitol, *N*-acetylalditol, glucosaminitol, *N*-acetylglucosaminitol, and GlcNAc-1-ol. The use of base to remove a polysaccharide moiety

Table 12.2 Chemical methods of cleaving Ser- or Thr-linked carbohydrate residues in glycoproteins

Method	Mechanism	References
1. Base-catalyzed β-elimination,[a] Ser-linked	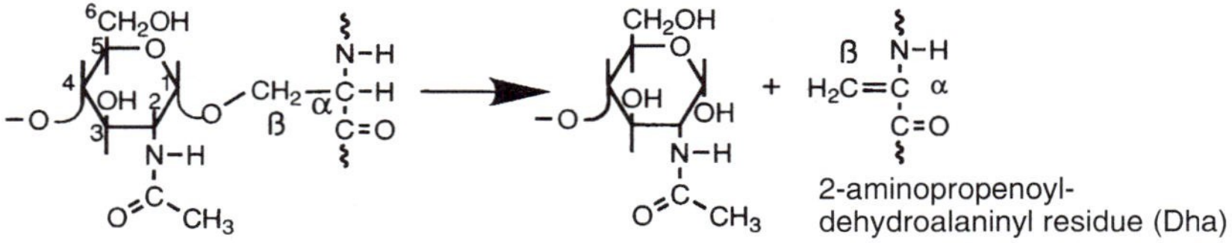 2-aminopropenoyl-dehydroalaninyl residue (Dha)	7–10
2. Base-catalyzed β-elimination,[a] Thr-linked	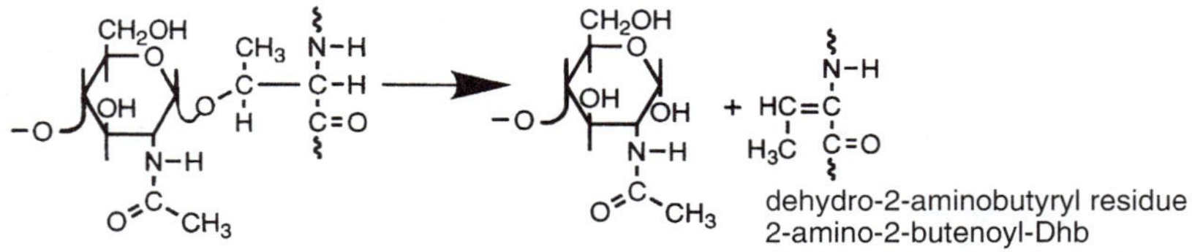 dehydro-2-aminobutyryl residue 2-amino-2-butenoyl-Dhb	7–10
3. Mild NaOH, $NaBH_4$[b]	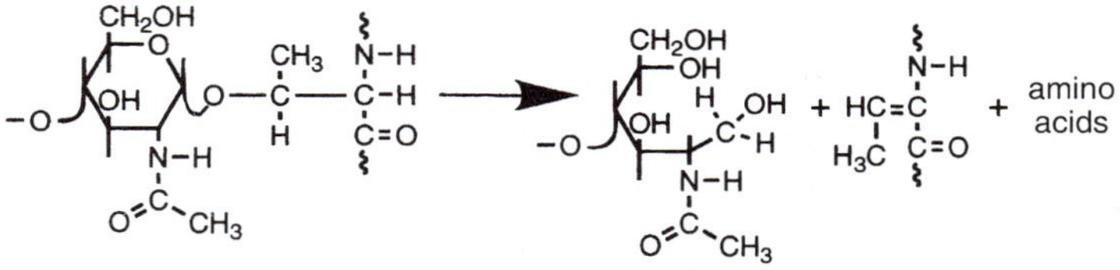 + amino acids	7, 11–19

Method	Reaction	References
4. Mild hydrazinolysis, 65°C	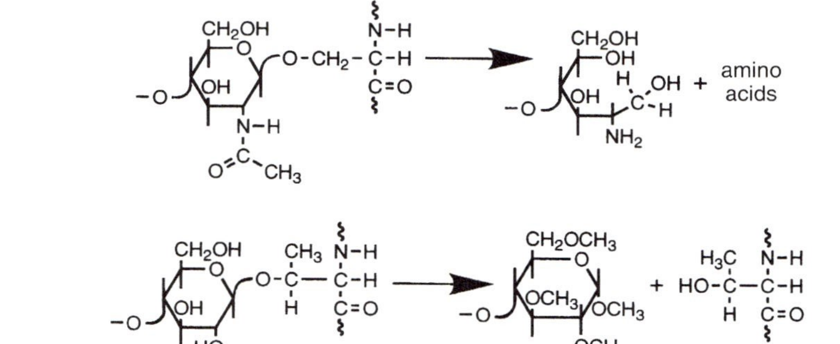	13–20
5. Direct base methylation NaOH/ Me_2SO or CH_3I		10, 21

[a] β-elimination refers to the β-carbon on the amino acid residue. A 3-substituted residue is susceptible to peeling (see text and Table 12.4).

[b] Reductive β-elimination; the borohydride stabilizes a 3-substituted residue from peeling.

Table 12.3 Chemical methods of cleaving Asn-linked carbohydrate residues in glycoproteins

Method	Mechanism	References
1. Acidic hydrolysis	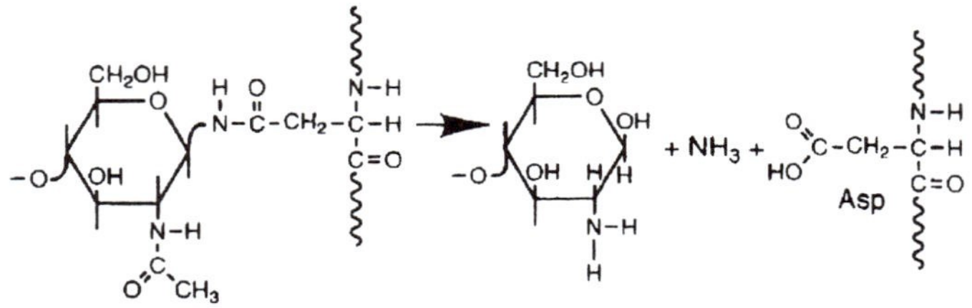 → + NH_3 + Asp	7, 20, 22–24
2. Base hydrolysis[a] NaOH oxidation	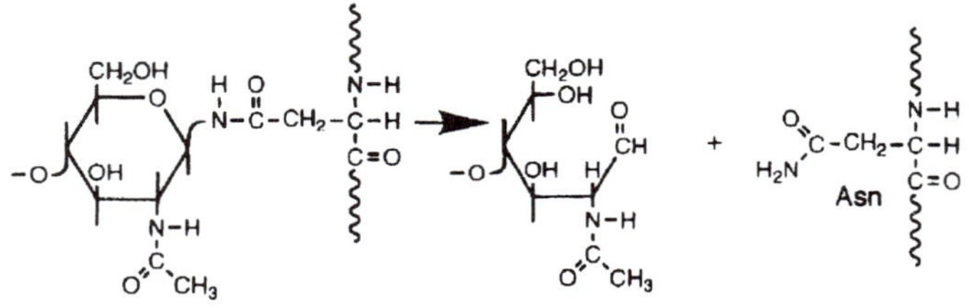 → + Asn	7

3. Mild NaOH, $NaBH_4$[b]	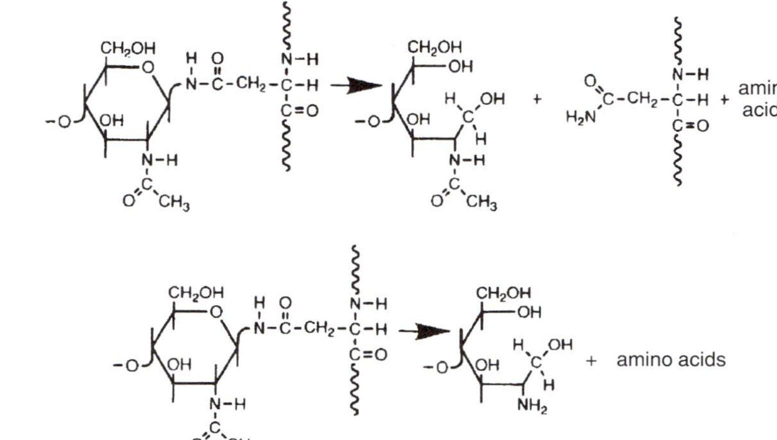	7, 11, 13–18
4. Hydrazinolysis, 95°C		13–15, 17, 18, 20

[a]A 3-substituted sugar is susceptible to peeling (see text and Table 12.4).

[b]Reductive amination; the borohydride stabilizes a 3-substituted residue from peeling.

Table 12.4 Chemical reactions with carbohydrate residues

Method	Mechanism	References
1. Mild aqueous H^+ or OH^- hydrolysis	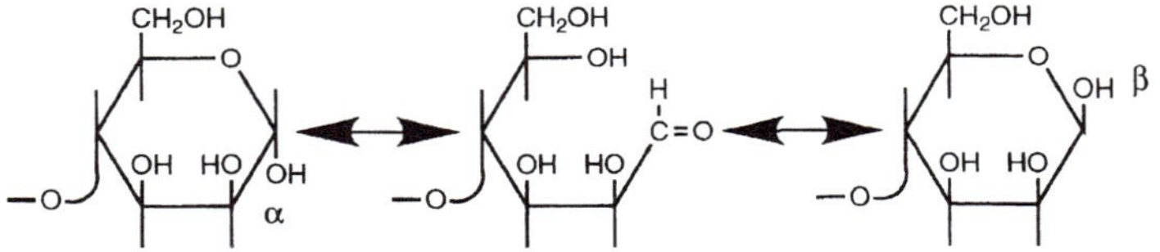 α ⟷ β	25, 26
2. Excess NaOH	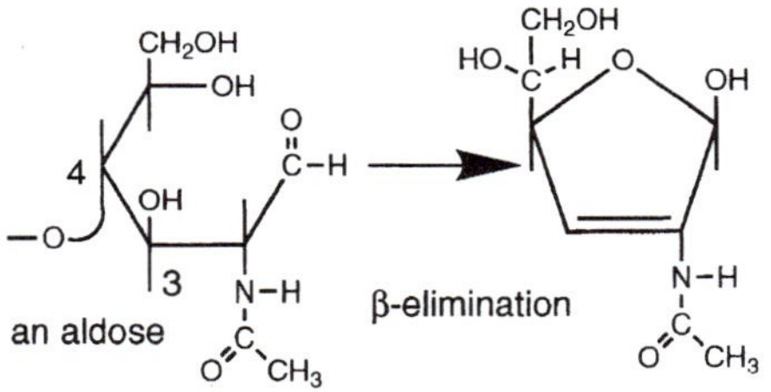 an aldose → β-elimination	7
3. Excess NaOH if 3-linked	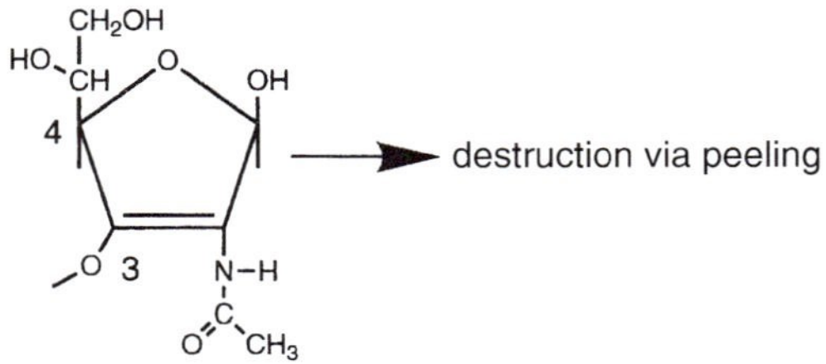 → destruction via peeling	7

4. $NaBH_4$ reduction NaOH	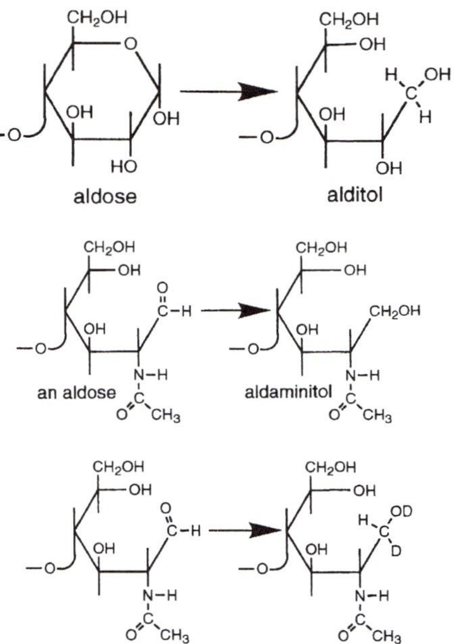	27
5. $NaBH_4$ NaOH		7, 28
6. $NaBD_4$ reduction		17

(continued)

Table 12.4 *(continued)*

Method	Mechanism	References
7. re-N-acetylation	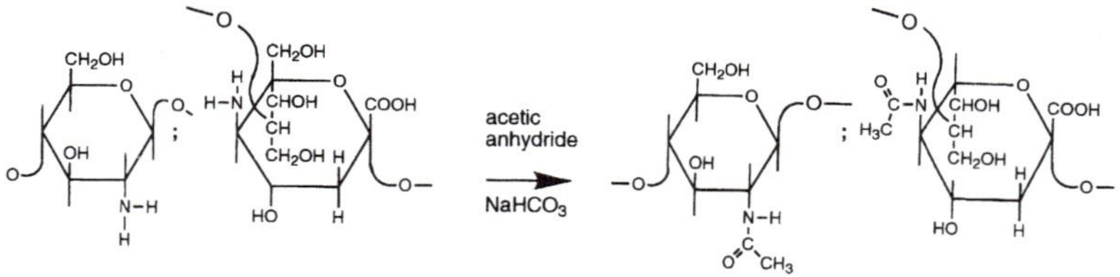 acetic anhydride, $NaHCO_3$	29
8. Permethylation CH_3I or $H_3C\text{-}S\text{-}CH_2 \cdot Na^+$ methylsulfinyl carbanion	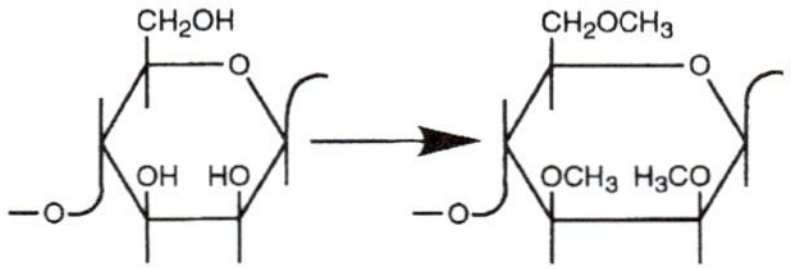	10, 13, 21, 27, 30

from a protein causes β-elimination and amination to occur on the Ser/Thr and Asn residues, respectively. Thus, coupled with $NaBH_4$, reductive β-elimination and reductive amination characterize these cleavage reactions. Note that the carbohydrate portion will increase in mass by 17 Da because of the addition of a hydroxyl group. A drawback of both the reductive cleavage reactions is the propensity to hydrolyze the polypeptide chain. These reactions may not be desirable if the sequence of the polypeptide needs to be determined by separate methods (Chapter 9).

Another method of cleaving the carbohydrate polymer from a peptide chain is hydrazinolysis (reaction 4 in Tables 12.2 and 12.3). Cleavage from Ser/Thr or Asn produces a stable alditol sugar moiety, which terminates the polysaccharide chain, and this produces an increase of 17 Da to the carbohydrate. Hydrazinolysis is deleterious in that it destroys the polypeptide chain by hydrolyzing the residue peptide bonds. Also, hydrazinolysis and the acid cleavage of Asn-linked carbohydrates (Table 12.3, reaction 1) cause de-N-acetylation of all amine-based sugar residues. These residues include the *N*-acetyl-glucosamine, -galactosamine, -glucuronic acid, and sialic acid residues cited in Table 12.1, and the protein residue regains the amine functionality. Thus, after these carbohydrate–polypeptide cleavage reactions take place, a re-N-acetylation reaction is performed, as shown in reaction 7, Table 12.4.

Enzymatic Removal of Polysaccharides

There are two classes of enzymes that remove oligosaccharide polymers from polypeptide backbones. The first class is the endoglycosidases. Table 12.5 shows three types, and, as the name implies, these enzymes cleave bonds from internal carbohydrate residues as opposed to that of terminal residues.

The endo-α-*N*-acetylgalactosaminidase, or *O*-glycanase, releases the polysaccharide from the O-linked Ser/Thr residues. Enzyme reactions produce "clean" predictable products as opposed to the potential deleterious products with chemical reactions. Unfortunately, with *O*-glycanases, not all O-linked glycoproteins are cleaved. Thus, a combination of enzymes and chemical methods are needed because a universal *O*-glycanase has not yet been found,[19,20,42] and Maley et al.[35] provide an extensive list of glycosidases.

Endo-β-*N*-acetylglucosaminidase, or Endo-F, cleaves the glycosidic bond between the first two sugar residues attached to the peptide, and, more specifically, this occurs for all carbohydrates attached to an Asn residue (Table 12.5, reaction 2). The Asn residue remains connected to the terminal sugar residue (which is essentially always a β-*N*-acetylglucosamine residue—see section "Asn-Linked Polysaccharide ESI-Mass Spectra"), and the penultimate sugar residue (which is essentially a β-*N*-acetylglucosamine residue) is the terminal reducing sugar for the released polysaccharide.

Reaction 3 in Table 12.5 lists a glycosidase called *N*-glycanase or PNGase F, which cleaves all N-linked (Asn–carbohydrate) glycoproteins. This enzyme leaves the entire carbohydrate portion intact and converts the Asn residue to an Asp residue.

Table 12.5 Endoglycosidase enzymes used to cleave internal amino acid residue – carbohydrate bonds in glycoproteins

Enzyme	Site of Action	Reference(s)
1. Endo-α-*N*-acetyl-galactosaminidase[a]; *O*-glycanase from *Diplococcus pneumoniae*	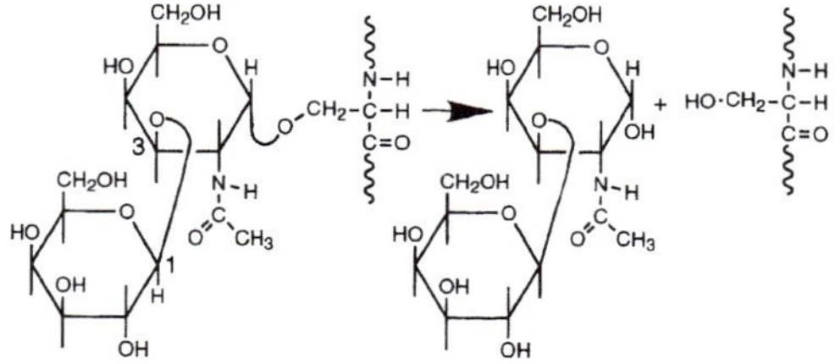 Galβ1-3GalNAcα1-Ser/Thr	33–35
2. Endo-β-*N*-acetylglucosaminidase (Endo-F) from *Flavobacterium meningosepticum* or endoglycosidase H (Endo-H) from *Streptomyces plicatus*	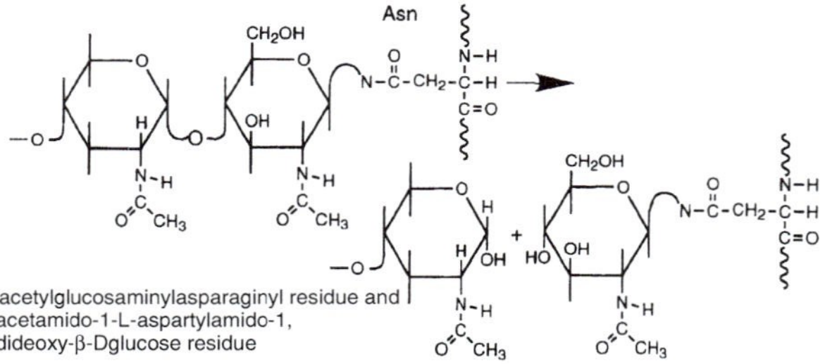 N-acetylglucosaminylasparaginyl residue and 2-acetamido-1-L-aspartylamido-1, 2-dideoxy-β-Dglucose residue	7, 35–39

3. Peptide: *N*-glycosidase F; *N*-glycanase; PNGase F; peptide-N^4-(*N*-acetyl-β-glucosaminyl) asparagine amidase from *Flavobacterium meningosepticum*	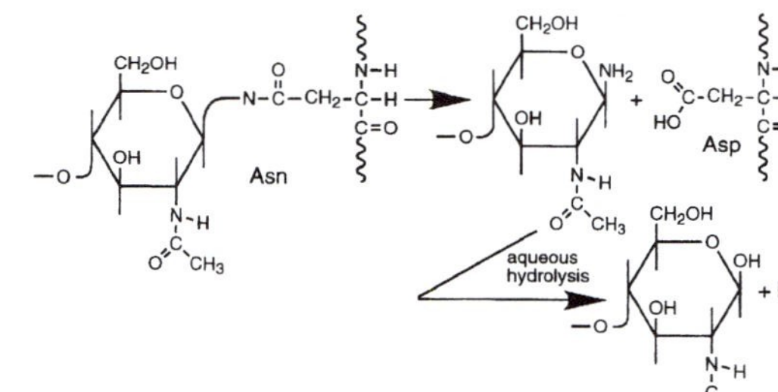	35, 38, 40, 41 37

[a]The *O*-glycanase from *Streptomyces plicatus* releases a broader distribution of oligosaccharides than that of *D. pneumoniae*, the latter of which releases only a Galβ1-3GalNAc moiety.

One drawback that is common to all three enzymes is that the protein portion of the peptide may be folded such that some of the carbohydrate species are buried and, hence, inaccessible to enzyme action. Thus, it usually is prudent to denature the glycoprotein and reduce any disulfide bonds prior to the addition of glycosidases.

Table 12.6 provides a listing of exoglycosidases, and they cleave only respective terminal (nonreducing end) sugar residues. These series of glycosidases can operate on polysaccharides that are either N- or O-linked and can be very specific. For example, the α-mannosidase derived from jack-bean meal can cleave a terminal α-mannose residue if it is connected from its C-1 position to

Table 12.6 Exoglycosidase enzymes used to selectivity cleave terminal carbohydrate residues

Enzyme	Representative sites of action	Reference(s)
1. α-Mannosidase from jack-bean meal	Cleaves the α-C_1 reducing hydroxyl bond in Manα1–2 hexose, Manα1–3 hexose, Manα1–6 hexose, Manα1–2, 3, or 6 Man	33, 34, 36
2. α-Mannosidase I from *Aspergillus saitoi*	Manα1–2 Man	33
3. β-Mannosidase		28
from rat epididymis *Helix pomatia*	Cleaves the C_1 glycosidic bond on a β-mannose residue Manβ1–4GlcNAc	43
4. β-galactosidase		
from coffee bean	Galα1–3, 4, or 6 Hex	33, 36
from figs.	Terminal Galα1-	44
5. β-galactosidase		
from jack-bean meal	Galβ1–3, 4, or 6 Hex	33, 36
from *Diplococcus pneumoniae* and *Streptococcus pneumoniae*	Galβ1–4GlcNAc; Galβ1–4GalNAc	34, 43, 45
6. α-*N*-acetylhexosaminidase from pig liver	GlcNAcα1–2, 3, 4, 6 Hex; GalNAcα1–2, 3, 4, 6 Hex	46
7. β-*N*-acetylhexosaminidase		
from jack-bean meal	GlcNAcβ1–2, 3, 4, 6 Hex; GalNAcβ1–2, 3, 4, 6 Hex	33, 36
from *D. pneumoniae*	GlcNAcβ1–3, 6 Gal; GlcNAcβ1–2, 4 Man	33, 47
from chicken liver	GlcNAcβ1–3, 4 Hex	43
8. α-Fucosidase		
from *Charonia lampas*	Fucα1–2 Gal; Fucα1–4, 6 GlcNAc	33
from almond	Fucα1–3, 4 GlcNAc	
9. α-Fucosidase, bovine epididymis	Fucα1–6 GlcNAc; Fucα1–2, 3, 4 GlcNAc	47
10. α-*N*-acetylneuraminidase		
from *Arthrobacter urefaciens* and *Vibrio cholerae*	NeuNAcα2–3, 6Gal; NeuNAcα2–6, 8NeuNAc; NeuNAcα2–6GalNAc; NeuNAcα2–6GlcNAc (slow reaction)	33, 34, 43, 48
from Newcastle disease virus	NeuNAcα2–3 Gal; NeuNAcα2–8NeuNAc	33, 43, 49
11. α-glucosidase from porcine liver	Glcα1–2, 3 Glc; Glcα1–3 Man	33

the 2, or 3, or 6 position on the adjacent residue, the latter being the second-last residue from the reducing end of the polysaccharide. Remember, the reducing end of the polysaccharide is connected to the polypeptide backbone; therefore, that makes a terminal sugar a nonreducing residue. An analogy can be made between the amine/carboxyl-terminal moieties in proteins and the nonreducing end/reducing end of a polysaccharide. Thus, Manα1–2, 3, 6 Man means the terminal mannose is in the α-configuration on C-1, and α-mannosidase will cleave the glycosidic bond between the Manα1 connected with either the 2, 3, or 6 position of the adjacent mannose residue. The α-mannosidase I from *A. saitoi* is even more specific in that it will only cleave the glycosidic bond between the C-1 of a terminal α-mannose and the 2 position of the adjacent mannose residue.

The α-galactosidase enzyme from coffee beans (entry 4 in Table 12.6) will cleave the glycosidic bond between a terminal α-galactose residue and the 3, or 4, or 6 position of a neighboring hexose residue. Thus, this enzyme is less specific than α-mannosidase.

In deducing the structure and sequence of a polysaccharide, one method is to judiciously use the exoglycosidases in the correct sequence so as to determine not only the sugar identity, but also the anomeric configuration and information as to the linkage position on both sugar residues. It has been noted in the literature that care must be taken regarding the purity of the exoglycosidase preparation, because relatively high levels of contaminating exoglycosidases can be present.[50]

Mass Spectrometry Fragmentation of Carbohydrates

The proton affinity of the ether, hemiacetal, and acetal functional groups of polysaccharides are significantly lower than the primary and secondary amine groups on the lysine, arginine, and histidine amino acid residues of proteins.[51,52] This can have significant consequences, especially for quadrupole instruments. Significant amounts of additional mass (carbohydrate) to that of polypeptides may cause the mass-to-charge ratio to be higher than the quadrupole mass range, despite the multiple-charging phenomenon (Chapter 4). Underivatized polysaccharides have relatively low ionization efficiency, and their sensitivity is poorer in a given mass range compared with that of a polypeptide. Yet the glycopeptide or separated polysaccharide can be protonated to yield mass spectra with reasonable signal-to-noise ratio.

Just as a peptide can fragment into a, b, c, x, y, and z fragments, polysaccharides also can cleave into predictable fragments, including the phenomenon of proton transfer to certain fragments. Scheme 12.1 shows a Galβ1-Manβ1-Manβ1 trisaccharide attached to a Ser residue.[53] The possible fragments are outlined, and the nomenclature of the Y, Z, B, and C ions is similar to that of proteins, except that here we will use capital letters to denote the ring and glycosidic bond cleavages. Scheme 12.2 provides details as to the mechanism of fragmentation for the A ion in the positive mode and the B and C ions in the

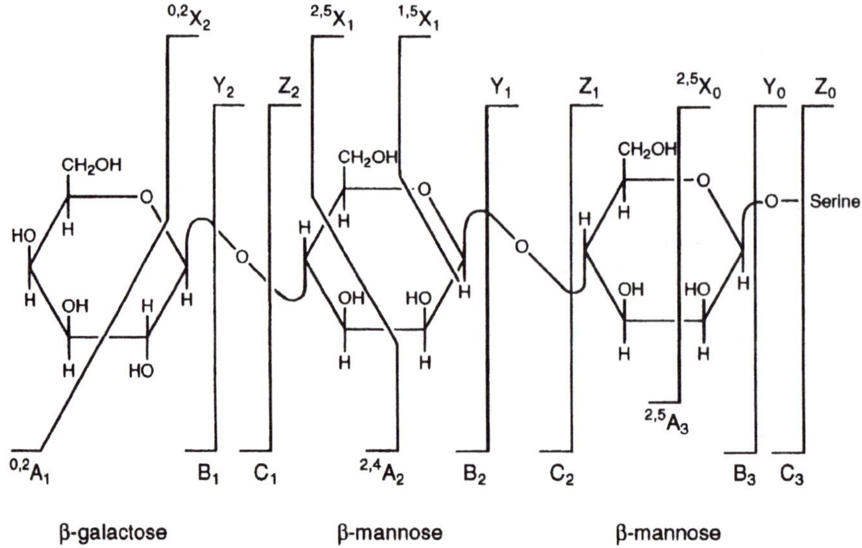

Scheme 12.1 Mass spectral fragmentation scheme of a hexose oligosaccharide attached to a serine amino acid.

positive- and negative-ion modes.[18,53–57] The B and C ions include the non-reducing end of the molecule in an analogous fashion to the b and c ions, respectively, retaining the terminal amine residue in a protein. The A ion for carbohydrates is essentially analogous to the polypeptide d and v ions. Conditions in the mass spectrometer that can cause this fragmentation include ESI-CID-MS (Chapter 1) and ESI-tandem MS. Scheme 12.3 provides fragmentation details of the X ion in the positive mode and the Y and Z ions in both modes of polarity.[18,53,57] These ions retain the reducing-end sugar unit, and the Y and Z ions can be considered as analogues of the protein-residue y and z ions, the latter of which include the terminal-carboxyl amino acid residue in protein fragmentation. Note that in both polarity modes, for a given fragmentation mechanism, the B and Y ions are complementary to each other (Schemes 12.2 and 12.3) and the C and Z ions are complementary. The major difference in each complementary pair is not in the structure of the neutral and ionized products, but rather concerns which of the products carries the charge. Table 12.7 provides a synopsis of how to calculate the mass of a particular fragment in the positive- and negative-ion modes.[53] Table 12.7 essentially summarizes Schemes 12.2 and 12.3. Usually, the B and Y ions dominate in an ESI-mass spectrum. The polysaccharide fragmentation schemes do not produce ions that are analogous to the a and x protein-fragment ions.

Unlike the analogous d, v, and w protein fragments (Chapter 9), the A and X partial ring-containing polysaccharide-fragment ions can be observed in quadrupole ESI-mass spectra, albeit at relatively low intensities.

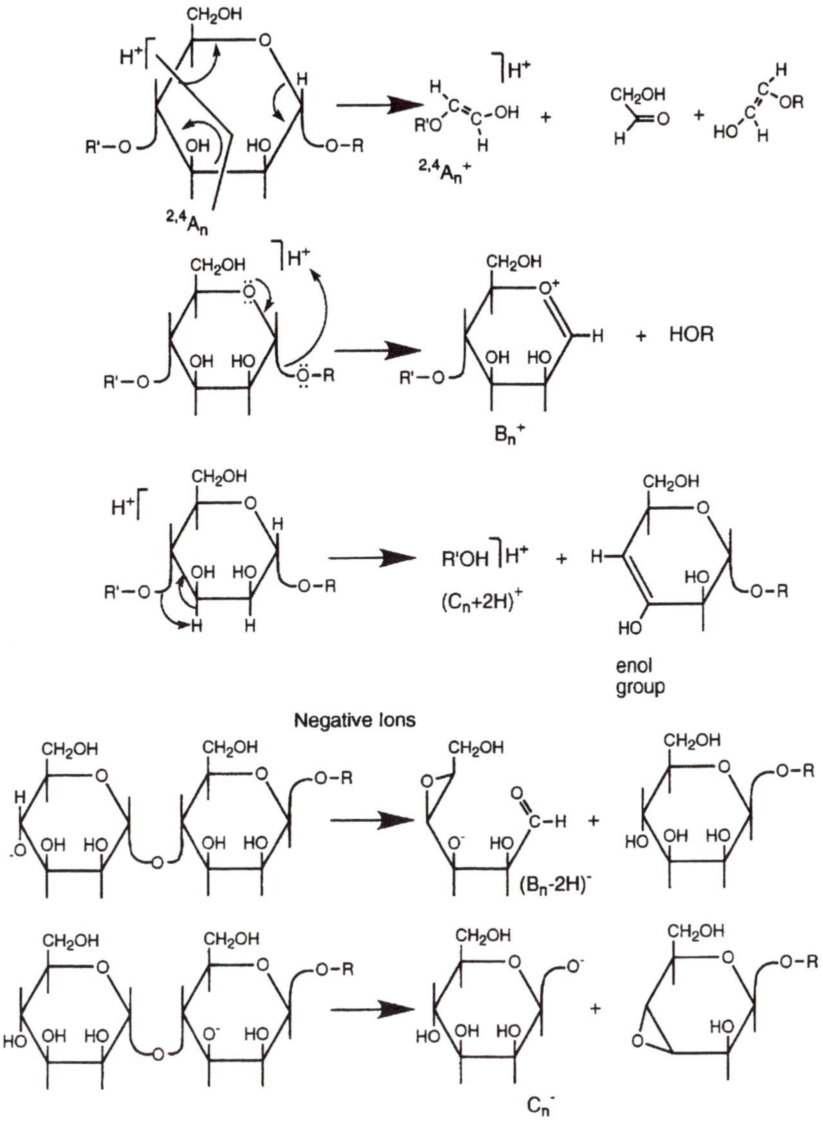

Scheme 12.2 Mass spectral fragmentation of hexose A, B, and C ions.

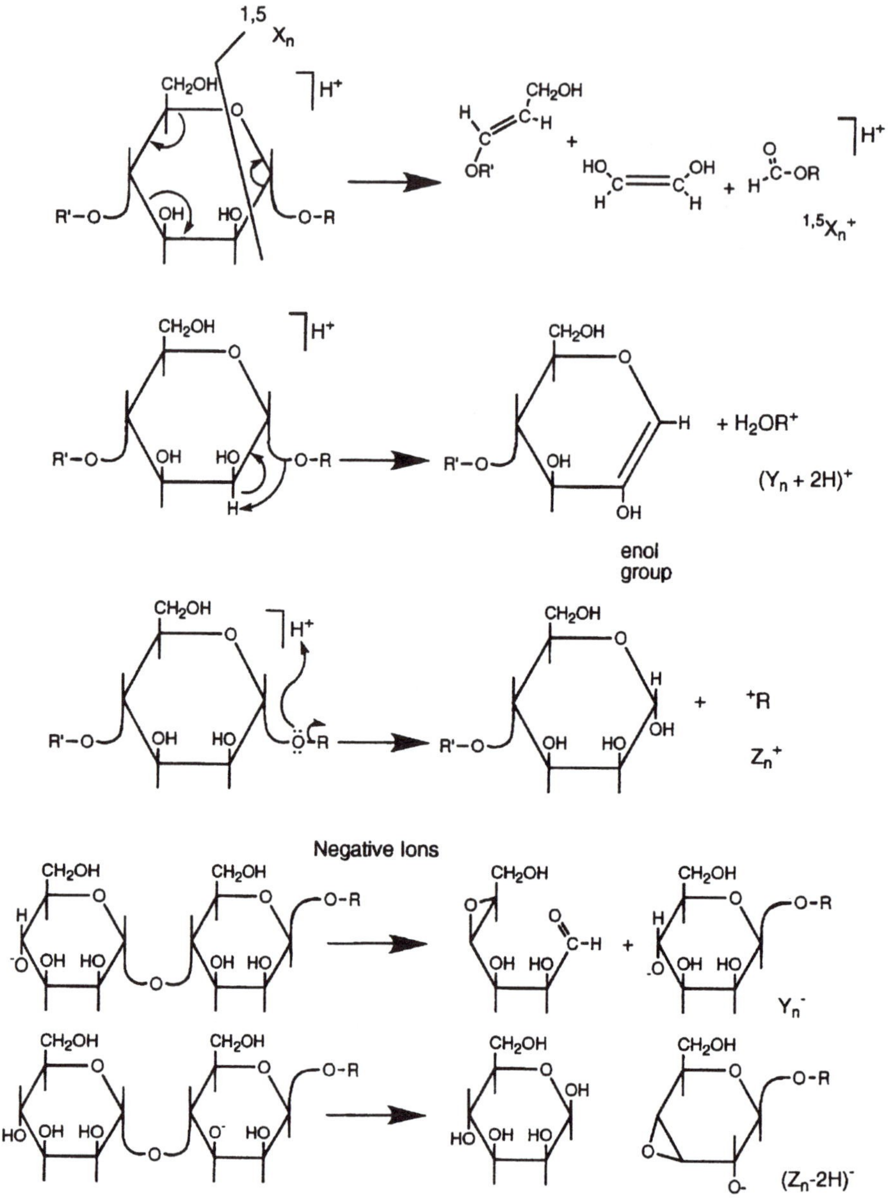

Negative Ions

Scheme 12.3 Mass spectral fragmentation of hexose X, Y, and Z ions.

Table 12.7 Hydrogen addition/deletion shorthand nomenclature for oligosaccharide ions

Positive-ion mode	Negative-ion mode
Nonreducing-end ions	
A_n^+ = (ring fragment + H)$^+$	$A_n^- = (A_n - H)^-$
$B_n^+ = B_n^+$	$B_n^- = (B_n - 2H)^-$
$C_n^+ = (C_n + 2H)^+$	$C_n^- = C_n^-$
Reducing-end ions	
X_n^+ = (ring fragment + H)$^+$	$X_n^- = (X_n - H)^-$
$Y_n^+ = (Y_n + 2H)^+$	$Y_n^- = Y_n^-$
$Z_n^+ = Y_n^+ - H_2O$	$Z_n^- = (Z_n - 2H)^-$

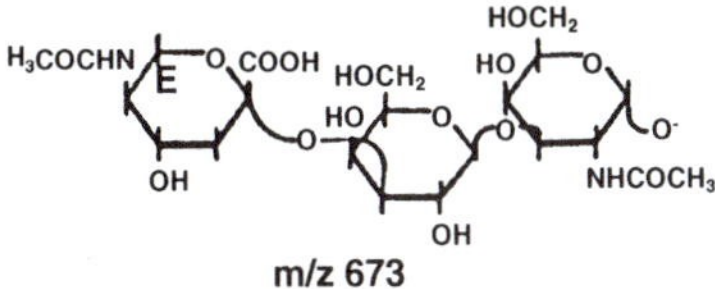

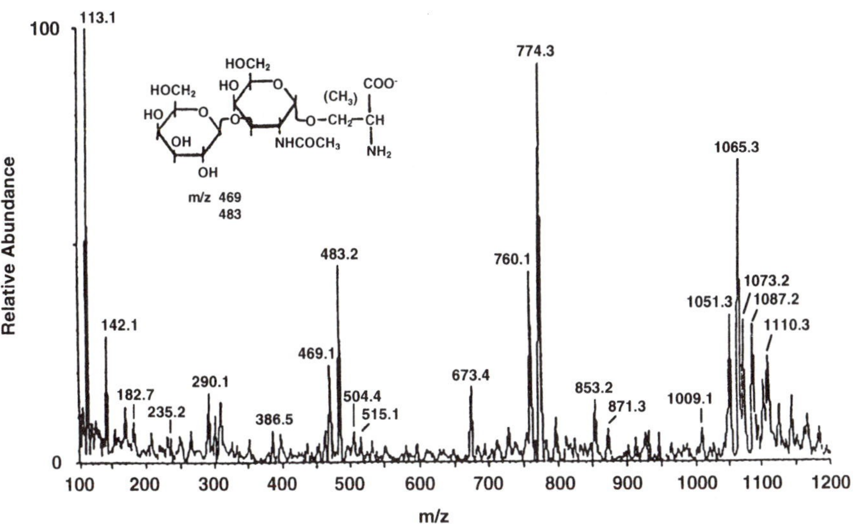

Figure 12.1 Negative-ion ESI-mass spectrum of O-linked tetrasaccharide monopeptides (see text), where the mixture contains Ser and Thr glycomonopeptides. (© 1994. Reprinted from reference 58 by permission of Marcel Dekker, Inc.)

Ser/Thr-Linked Polysaccharide ESI-Mass Spectra

Negative Ion Mass Spectral Analysis of Two Tetrasaccharides

The first example comes from two tetrasaccharide monopeptides: a tetrasaccharide–Ser and a tetrasaccharide–Thr.[58] Both are normally excreted in the urine in free form as catabolic products. The structure is as follows:

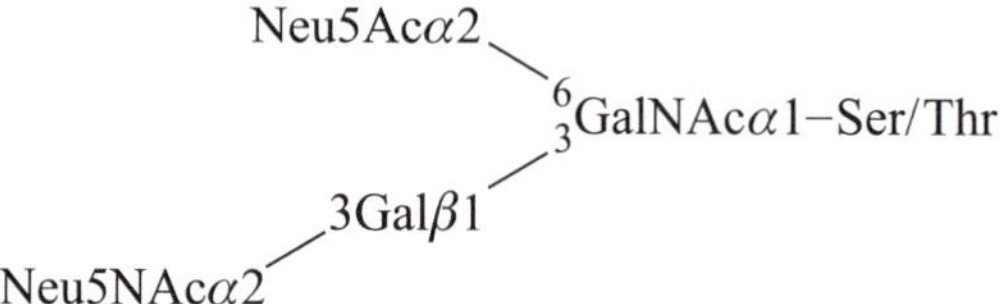

Note that two of the three glycosidic bonds and the GalNAc–Ser/Thr bond involve the α-anomer configuration. In analyzing the carbohydrate structure, it is usually best, from a "reference" point of view, to draw the structure and denote the neutral mass of each component. Thus, we have the following:

```
        H–Neu5NAc
          292.26      GalNAc–Ser
                  Gal   202.19  104.09
  H–Neu5NAc      162.14
     292.26
                                –Thr
                                118.12
```

Referring to Table 12.1, the sialic acid Neu5NAc residues each have a mass increase of 1 Da (one hydrogen). The Gal residue has the normal two-bond linkage residue. Therefore, GalNAc must be reduced by 1 Da (one hydrogen), because it displays a third bond. The Ser/Thr residue mass values must be reduced by 1 Da (one hydrogen) since attachment is by a dehydration reaction that consists of the loss of the hydroxyl moiety on GalNAc and the loss of a hydrogen from the hydroxyl side group of the amino acids. Because they are attached to the tetrasaccharide by their hydroxyl side group, 18 Da must be added to the mass of the amino acids. Thus, their amino and carboxyl functionalities remain intact. Figure 12.1 presents the negative-ion mode ESI-tandem mass spectrum of a mixture of the two species at an elevated skimmer–cone voltage (Chapter 1) to induce fragmentation.[58] The neutral mass of the tetrasaccharide–Ser and tetrasaccharide–Thr is 1052.94 Da and 1066.94 Da, respectively. The $(M - H)^-$ species have masses of 1051.94 and 1065.94, respectively, and these values can be observed in Figure 12.1. The difference between the calculated and observed values for both species is 0.64 Da, and the loss of a proton occurs on a sugar hydroxyl group. Note the sodium adducts for both species as follows:

Tetrasaccharide – Ser: $(M - 2H + Na)^- = 1052.94 - 2 + 23 = 1073.94 \simeq 1073.2$ Da (observed)

Tetrasaccharide–Thr: $(M - 2H + Na)^- = 1066.94 - 2 + 23 = 1087.94 \simeq 1087.2$ Da (observed)

PROBLEM 12.1 What is the identity of the *m/z* 1110.3 ion in Figure 12.1?

A loss of a Neu5NAc residue yields the following intense ions:

$$\text{Ser:}\quad 1051.94\ \text{Da} - 291.26 = 760.68 \simeq 760.1\ \text{Da}$$
$$\text{Thr:}\quad 1065.94 - 291.26 = 774.68 \simeq 774.3\ \text{Da}$$

Note that the loss of either Neu5NAc residue could produce these masses for each tetrasaccharide species. The Neu5NAc residue is lost as a 291.26-Da fragment in the negative-ion mode, and the remainder of the molecule is found as a Y ion (Scheme 12.3 and Table 12.7). The negative charge is retained on the glycosidic bond oxygen on either the C-3 of Gal or the C-6 of GalNAc, depending on which sialic acid residue was lost.

A loss of two Neu5NAc groups yields the following:

$$\text{Ser:}\quad 1051.94 - 290.26 - 292.26 = 469.42 \simeq 469.1\ \text{Da}$$
$$\text{Thr:}\quad 1065.94 - 290.26 - 292.26 = 483.42 \simeq 483.2\ \text{Da}$$

One of the Neu5NAc residues leaves as a B ion (Scheme 12.2) and transfers one of its protons to the Galβ1 or GalNAcα1 glycosidic bond oxygen (where the Neu5NAc sialic acid residue was attached). The other Neu5NAc residue is a neutral leaving group from a Y-type cleavage (negative-ion mode—Scheme 12.3 and Table 12.7) and the glycosidic bond oxygen imparts a net negative charge on the molecule.

A loss of the amino acid and an Neu5NAc residue yields the following:

$$\begin{aligned}\text{Ser:}\quad 1051.94 - 290.26 - 104.09 &= 657.59\\ 657.59 + 16 &= 673.59\ \text{Da}\\ \text{Thr:}\quad 1065.94 - 290.26 - 118.12 &= 657.56\\ 657.56 + 16 &= 673.56\ \text{Da}\end{aligned}$$

The amino acid side-chain oxygen remains with the oligosaccharide; thus, 16 Da must be added to the ion. This free oxygen retains the negative charge and is not neutralized with a hydrogen from the amino acid leaving group. The sialic acid leaving group transfers one of its hydrogen atoms to the glycosidic bond oxygen (where it was attached) and does not produce a negative charge. The sialic acid Neu5NAc residue leaves as a $(B - 2H)^-$ ion.

Positive Ion Mass Spectral Analysis of Two Tetrasaccharides

Using the same tetrasaccharide–Ser/Thr molecules, a positive-ion ESI-product-ion mass spectrum is shown in Figure 12.2,[58] and an analysis follows:

$$\text{Ser:}\quad (M + H)^+ = 1053.94\ \text{Da};\ (M + Na)^+ = 1075.94\ \text{Da}$$
$$\begin{aligned}\text{Thr:}\quad (M + H)^+ &= 1067.94\ \text{Da} \simeq 1067.3\ \text{Da}\\ (M + Na)^+ &= 1089.94\ \text{Da} \simeq 1089.4\ \text{Da}\end{aligned}$$

PROBLEM 12.2 Explain the origin of the m/z 784.1 and 798.3 ions in Figure 12.2; what type of ions (Table 12.7) are they?

PROBLEM 12.3 Explain the origin of the m/z 600 and 614.3 ions in Figure 12.2; what type of ions (Table 12.7) are they? Pay particular attention to the residue loss in terms of the mass of the leaving group at each cleavage stage.

The m/z 309.2 and 323.3 signals are represented by the GalNAcα1–Ser/Thr species and arise by the expression $(M + H)^+ - 2$Neu5NAc-Gal as follows:

$$\text{Ser:}\quad 1052.94 - 2(292.26) + 2H - 163.14 + 1 + H^+ = 309.28 \text{ Da}$$

$$\text{Thr:}\quad 1066.94 - 2(292.26) + 2H - 163.14 + 1 + H^+ = 323.28 \text{ Da}$$

Alternatively:

$$\text{Ser:}\quad \text{GalNAc–Ser} = 204.19 + 104.09 + H^+ = 309.28 \text{ Da}$$

$$\text{Thr:}\quad \text{GalNAc–Thr} = 204.19 + 118.12 + H^+ = 323.31 \text{ Da}$$

These are Y-type ions. The m/z 204 ion is a GalNAc residue, and its origin is of the B type (Scheme 12.2). Both side-chain sugar groups (or antennae) transfer a hydrogen to the 3 and 6 hydroxyl groups on GalNAc, while the amino acid retains the R-group hydroxyl oxygen atom. The m/z 292 ion is characteristic of Neu5NAc and this is a B-type ion. The m/z 274 ion is characteristic of Neu5NAc with the loss of water.

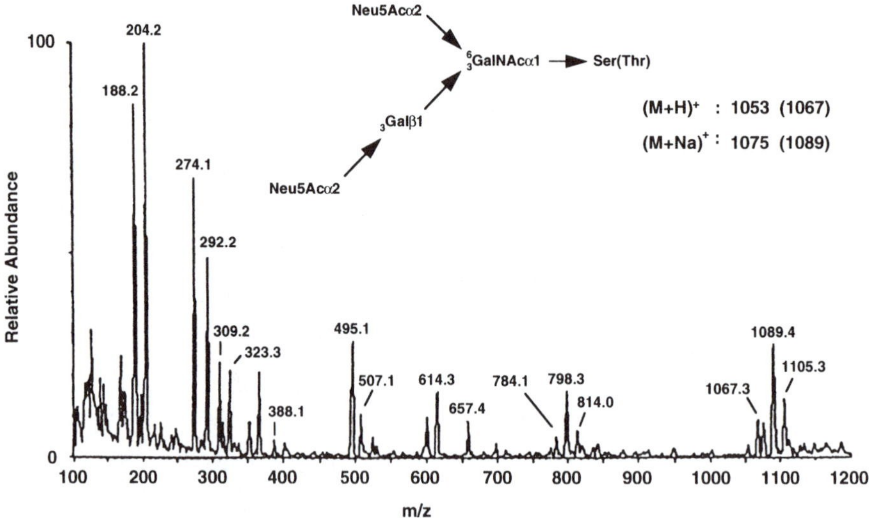

Figure 12.2 Positive-ion ESI-mass spectrum of O-linked tetrasaccharide monopeptides (see text), where the mixture contains Ser and Thr glycomonopeptides. (© 1994. Reprinted from reference 58 by permission of Marcel Dekker, Inc.)

PROBLEM 12.4 What combination of sugars could yield the m/z 495.1 ion in Figure 12.2, and what type of ion is it?

Mass Spectral Analysis of a Synthetic Oligosaccharide

The following O-linked oligosaccharide was synthetically made and analyzed by ESI-MS and ESI-tandem MS conditions. The species below represents a T_n antigen immunological marker, which is associated with breast tumors.[59]

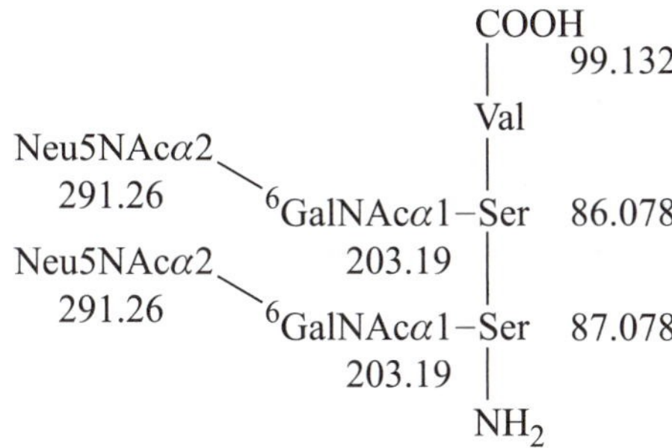

The mass of each *residue* is given. Note that the terminal Ser mass is equal to its residue mass, because the presence of the second amine proton and absence of the side-group hydroxyl proton mathematically cancel each other.

The negative-ion mode ESI-up-front CID mass spectrum of this glyconjugate is shown in Figure 12.3. The $(M - H)^-$ and $(M - 2H)^{2-}$ peaks are present at m/z 1279.2 and 638.9 (639.1 Da, calculated), respectively.[59]

PROBLEM 12.5 Derive the origin of m/z 493.4 in Figure 12.3; what type of ion is it? The m/z 290 ion is a B_n ion of the sialic acid as 292.26 – 2H = 290.26 Da.

PROBLEM 12.6 Derive the origin of m/z 987.8 and 696.5 in Figure 12.3.

Figure 12.4 presents the positive-ion mode ESI-CID mass spectrum of the di-O-linked polysaccharide[59] and an analysis follows: the $(M + H)^+$ and $(M + Na)^+$ peaks observed at 1280.5 Da and 1302.4 Da, respectively, are close to the calculated values of 1281.2 Da and 1303.2 Da, respectively.

PROBLEM 12.7 What is the structure of the m/z 1325.0 peak?

In Figure 12.4, we see the GalNAc (or GNAc) B-type ion at m/z 204.0, and the B-type ion of Neu5NAc at m/z 292.2 along with its dehydrated species at m/z 274.2. Also present is the B ion of Neu5NAcα2–6GalNAc at m/z 495.2. Small mass differences can be noted for the same sugar loss in the positive- and negative-ion modes:

$(M - H - \text{Neu5NAc})^- = 987.9$ Da $\quad (M - H - 2\text{Neu5NAc})^- = 696.50$ Da
$(M + H - \text{Neu5NAc})^+ = 989.9$ Da $\quad (M + H - 2\text{Neu5NAc})^+ = 696.68$ Da

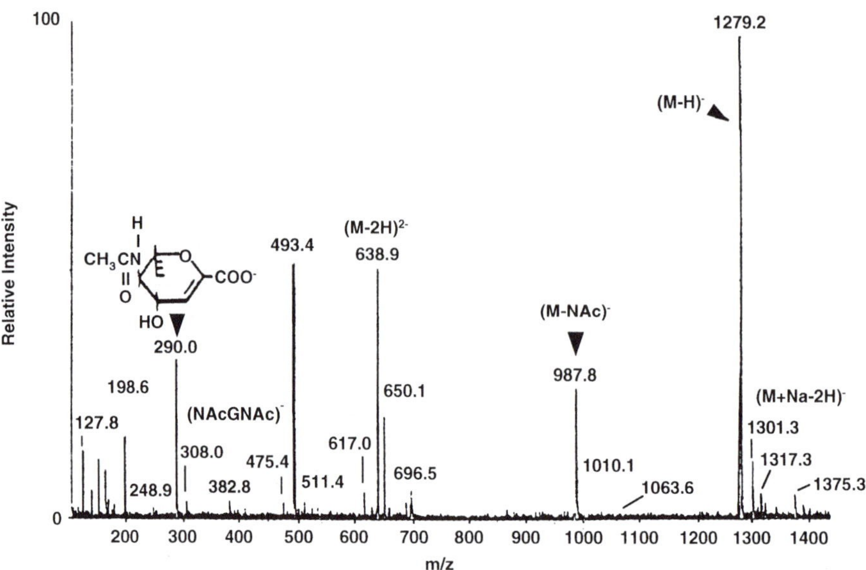

Figure 12.3 Negative-ion ESI-mass spectrum of a tripeptide (see text) with an O-linked disaccharide on each of two Ser residues. (© 1994. Reproduced from reference 59 by permission of John Wiley & Sons, Limited.)

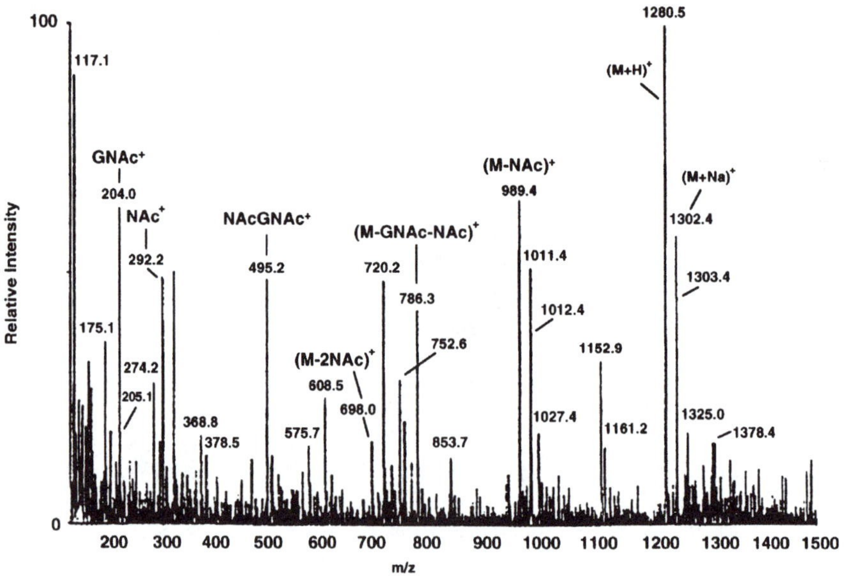

Figure 12.4 Positive-ion ESI-up-front CID mass spectrum of a tripeptide (see text) with two O-linked disaccharides on each of two Ser residues. (© 1994. Reproduced from reference 59 by permission of John Wiley & Sons, Limited.)

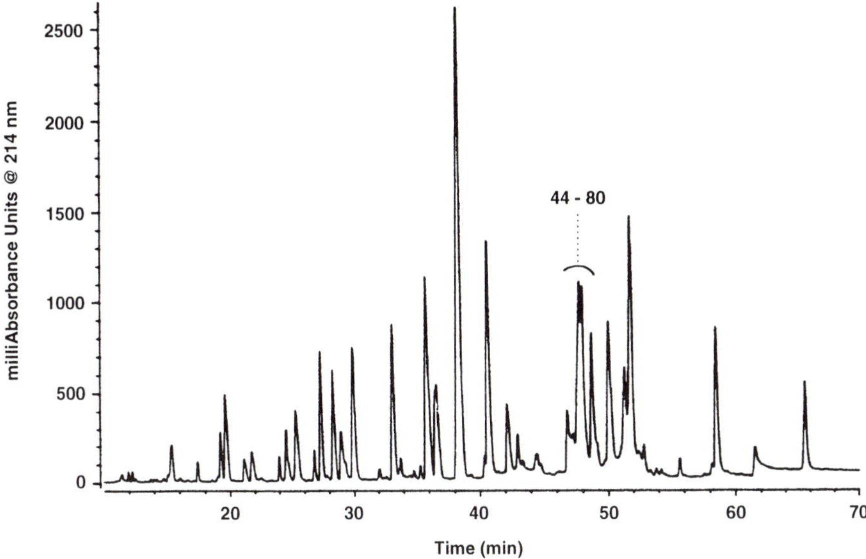

Figure 12.5 The RP-HPLC chromatogram of the carboxymethylated trypsin digest of coagulation human factor IX glycoprotein. The amino acid 44–80 tryptic glycopeptide (O-linked tetrasaccharide) is highlighted. Detection was at 214 nm. Column conditions: Vydac C_{18} column (4.6 × 250 mm) equilibrated with 0.1% TFA solution at 30°C. An increasing gradient of 0.1% TFA in AcCN was used that resulted in a 60% v/v of AcCN after 90 min of chromatography. (© 1993, American Chemical Society. Reprinted from reference 60 with permission.)

These ions involve the gain or loss of a proton from the neutral mass of the parent molecule.

Mass Spectral Analysis of a Glycopeptide of Coagulation Human Factor IX

Coagulation human factor IX (Christmas factor) is a 57-kDa glycoprotein. When calcium ions are present, factor IX binds to the phospholipid region on the surface of endothelial cells. Upon activation, a peptide portion is released, and this and the rest of the factor IX glycoproteins are but a few of the many components in the coagulation of blood.

The glycoprotein was first treated to cleave all −S−S− bonds by carboxymethylation (Tables 9.7 and 9.13). Figure 12.5 shows a reversed-phase high-pressure liquid chromatography (RP-HPLC) separation of the carboxymethylated tryptic digest of the glycoprotein.[60] A glycopeptide that includes amino acid residues 44–80 is highlighted. Prior investigations[60] concluded that this peptide contained carbohydrate and it was thus isolated and further digested with thermolysin (Table 9.4). An RP-HPLC of the resulting fragments is

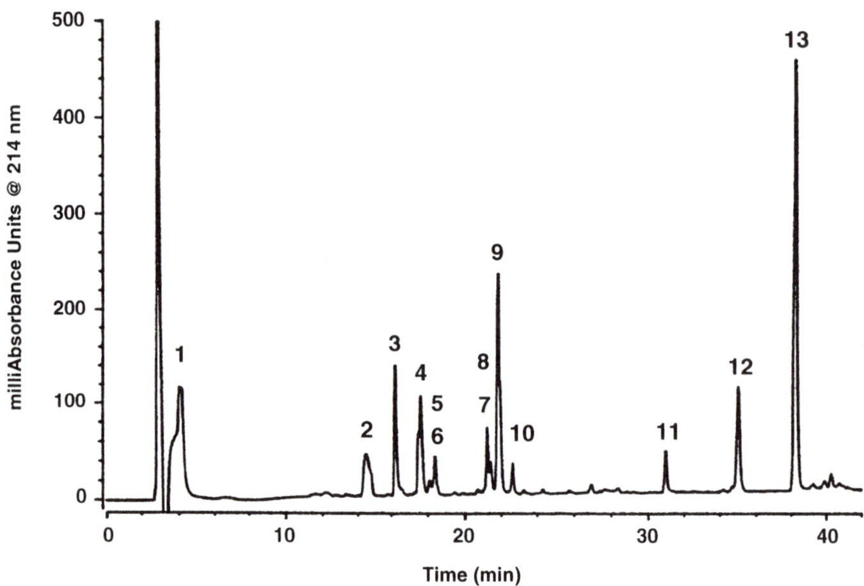

Figure 12.6 The amino acid 44–80 glycopeptide (Figure 12.5) was treated with the thermolysin enzyme, and an RP-HPLC chromatogram of the thermolytic digest is shown. Chromatographic conditions were the same as those in Figure 12.5. (© 1993, American Chemical Society. Reprinted from reference 60 with permission.)

shown in Figure 12.6.[60] Peak 4 in Figure 12.6 is attributed to the following glycopeptide:

$$\begin{array}{l} H_2N\text{—Leu–Asn–Gly–Gly–Ser–CMCys–Lys–Asp–Asp—COOH} \\ \qquad\quad 57 \qquad\qquad\qquad\qquad | \qquad\qquad\qquad\qquad\quad 65 \\ \qquad\qquad\qquad\qquad\qquad\quad O \\ \qquad\qquad\qquad\qquad\qquad\quad | \\ \text{Neu5NAc}\alpha\text{2–6Gal}\beta\text{1–4GlcNAc}\beta\text{1–3Fuc}\alpha\text{1} \end{array}$$

PROBLEM 12.8 Assuming the peptide portion remains intact, corroborate the sequence of carbohydrate residues in the ESI-CID-mass spectrum in Figure 12.7. At least one doubly charged ion can be found in the spectrum. Can all four sugar residues be placed in the above-indicated order with just the information given in Figure 12.7?

Mass Spectral Analysis of a Glycopeptide from Human Platelet Derived Growth Factor β-Chain

A tryptic glycopeptide was analyzed from a mixture of human platelet-derived growth factor β-chain glycoproteins. This protein was made in a recombinant

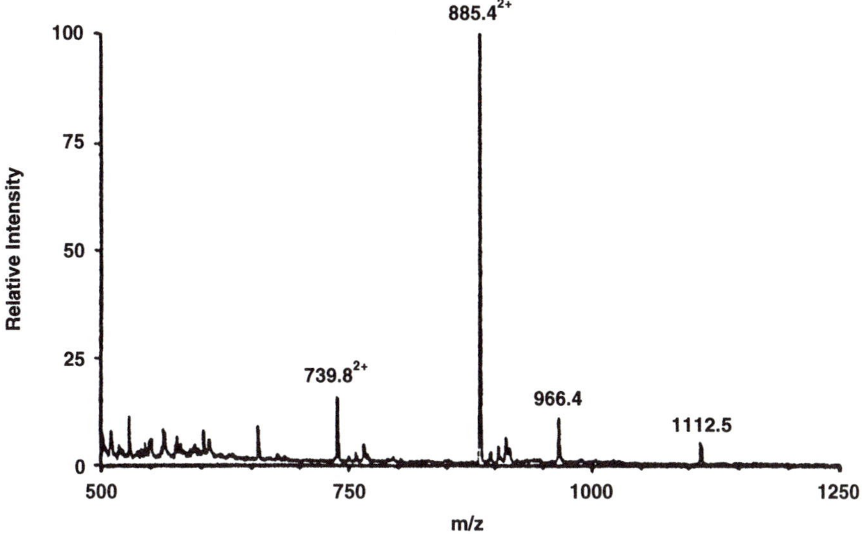

Figure 12.7 The ESI-mass spectrum of the peak 4 glycopeptide in Figure 12.6 (see text for details). (© 1993, American Chemical Society. Reprinted from reference 60 with permission.)

fashion from yeast cells. The glycopeptide is the 86–98 sequence of amino acids from the intact protein, and the sequence is as follows:

H_2N-Lys-Ala-Thr-Val-Thr-Leu-Glu-Asp-His-Leu-Ala-CMCys-Lys-COOH

Figure 12.8 presents the ESI-tandem mass spectrum of the $(M + H)^+$ ion of the glycopeptide at 1649.8 Da.[61] During CID, a portion of the glycopeptide was deglycosylated. In Figure 12.8, there are five peptide fragments labeled with an asterisk. These fragments represent the peptide portion without the carbohydrate residues.

PROBLEM 12.9 With the above information, deduce what type of polysaccharide (e.g., polyfucose, polyhexose, polyhexosamine) is present, and at what amino acid residue is it attached? The indicated VT internal fragment has a nominal mass of 200 Da (Table 9.6). What peptide fragment (i.e., a, b, c, x, y, z) also has a nominal mass of 200 Da? This label does not appear in Figure 12.8.

Mass Spectral Analysis of a Glycopeptide from LCAT

Human lecithin:cholesterol acyltransferase (LCAT) is an enzyme composed of 416 amino acid residues, and it promotes the transfer of the 2-position acyl moiety from phosphocholine to blood plasma cholesterol. The enzyme has at least seven different tryptic peptides that contain polysaccharide chains. One of these peptides will be investigated.[62]

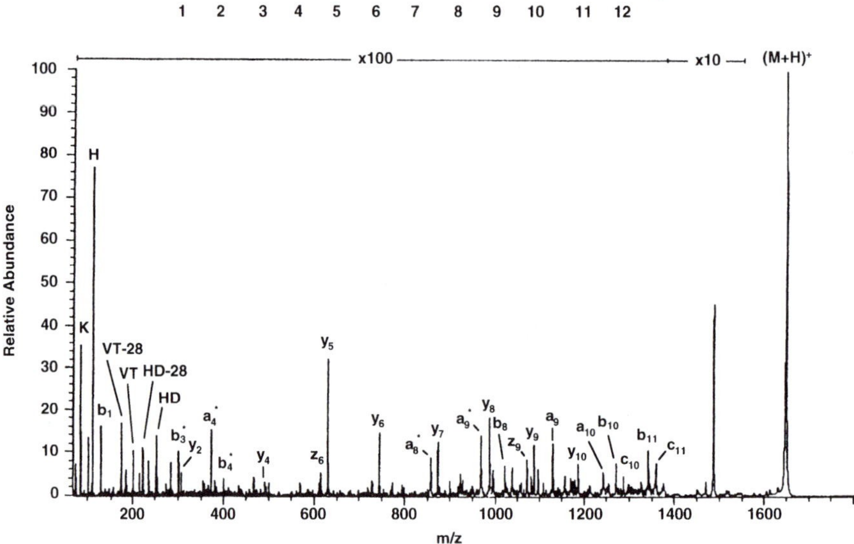

Figure 12.8 High-energy ESI-tandem mass spectrum of a tryptic O-linked glycopeptide of the β-chain of human platelet-derived growth factor (see text for details). (© 1993. Reprinted from reference 61 by permission of Academic Press, Inc., Orlando, Florida.)

The peptide in question is the carboxyl end of the glycoprotein at residues 400–416: QGPPASPTASPEPPPPE.

There are three potential sites for O-glycosylation, and Ser-405 is not glycosylated. Figure 12.9a shows a spectrum of the intact glycopeptide where the doubly and triply charged peaks are labeled A2 and A3, respectively.[62] Separate saccharide analyses of similar human glycoproteins[63,64] showed that the LCAT terminal glycopeptide may contain a total of two sialic acid Neu5NAc, two Gal, and two GalNAc residues. The intact glycopeptide was then treated with a glycosidase and then with a second, more specific glycosidase. The resulting product displayed the ESI-mass spectrum shown in Figure 12.9b, where the 1+ and 2+ species are labeled as A and A2, respectively.

PROBLEM 12.10 With the above-given information—first, provide a *corrected* amino acid sequence from that given above. Table 9.13 provides useful information. Then, deduce which glycosidases were used and the structure and sequence of the glycopeptide.

Mass Spectral Analysis of a Glycopeptide from Erythropoietin

The glycoprotein erythropoietin (EPO) is a hormone involved in the regulation of the level of red blood cells. It is found in the kidney and has four different

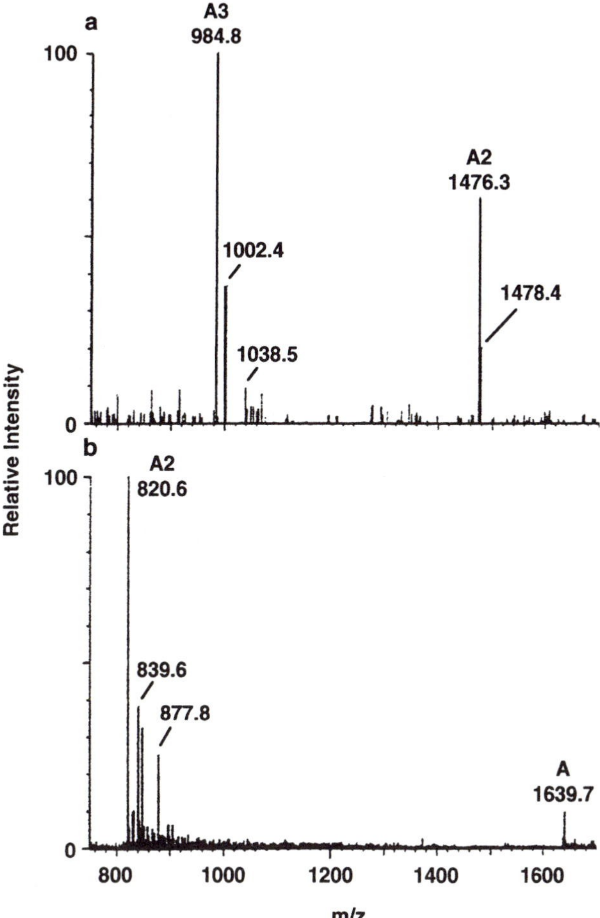

Figure 12.9 The ESI-mass spectra of the C-terminal tryptic glycopeptide of LCAT. (a) Mass spectrum of intact glycopeptide and (b) mass spectrum after treatment with two separate glycosidases. See text for details. (© 1995. Reprinted from reference 72 by permission of Cambridge University Press.)

amino acid sites that have attached carbohydrate residues. One of the sites is O-linked to a Ser-126 residue. The previous examples presented either standards or digests of oligosaccharides from their respective glycopeptides where only one structure was found for a given Ser or Thr residue. Sequence information could be obtained in some cases under "up-front" CID conditions. However, in the standard ESI-mass spectrum, essentially only one motif of a polysaccharide was found for a given Ser/Thr residue.

Suppose an isolated glycopeptide from a glycoprotein yielded a few to many distinct masses in a conventional ESI-mass spectrum, where the mass differences could be attributed to known sugar residues. Barring an uncommon glycosidic bond lability under the gentle ESI conditions, this would be a normal and expected phenomenon in glycoconjugate analytical biochemistry. Cells routinely produce copies of glycoproteins with identical polypeptide sequences and vary in carbohydrate residue lengths when attached to all three amino acids (Ser, Thr, and Asn). In most cases, a successive series of masses can be observed, each of which has an increase of one additional sugar unit. Thus, what appears to be a CID spectrum of one glycopeptide molecule is usually a conventional mass spectrum that consists of separate, distinct saccharide–polypeptide molecular entities.

Sometimes, a glycoprotein (or protein, in general) is desired in significant amounts; however, the native cellular source usually produces very little of the substances. The DNA responsible for production of the protein is spliced from the native genomic material and inserted into the genome of another cell that has the capacity to make significant amounts of the desired product. Producing the product in this fashion is called recombinant technology,[65] and the substance, EPO here, is labeled recombinant EPO or r-EPO. For proteins, this is a somewhat harmonious arrangement when no posttranslational modifications are needed. Modifications such as polysaccharide addition are much more difficult to engineer in advance. Thus, the investigator is at the mercy of the cell that produces the r-glycoprotein, because the surrogate cell may not have the proper inherent genetic information to duplicate the modifications with respect to that of the native substance. Since the native state of glycoproteins has just been described as being heterogeneous in nature, duplication would be very fortuitous in a surrogate cell. This takes into consideration the different glycoprotein molecules and their amounts. Terminology will be introduced in the "Asn-Linked Polysaccharide ESI-Mass Spectra" section which describes this phenomenon, as it is not required at present.

The disulfide bonds of r-EPO were reduced, and the free thiol groups were pyridylethylated. The glycoprotein was divided into two portions and digested with Lys-C and trypsin (Chapter 9). An RP-HPLC chromatogram of the Lys-C digest is shown in Figure 12.10.[66] The glycopeptide in peak number 5 was found to contain an O-linked Ser residue. The glycopeptide can be characterized as Glu-117 to Arg-140 and the sequence of the glycopeptide follows:

H_2N-EAISPPDAASAAPLRTITADTFRK-COOH
117 120 125 130 140

In a separate series of experiments, trypsin digestion was accomplished, and a tryptic peptide from Glu-117 to Arg-131 was isolated. Analysis of this peptide showed that the site of glycosylation was in this smaller peptide. Thus, the polysaccharide could be located either on Ser-120 or Ser-126. Figure 12.11 presents an ESI-mass spectrum of the tryptic Glu-117 to Arg-131 glycopeptide.[66]

PROBLEM 12.11 Identify the neutral masses for each labeled peak in Figure 12.11. As an aid, group all masses and order them in sequence. Then, subtract a lower mass from successively higher masses until an integral, single sugar residue mass value is *not* obtained. Treat the letter and mass values as equations,

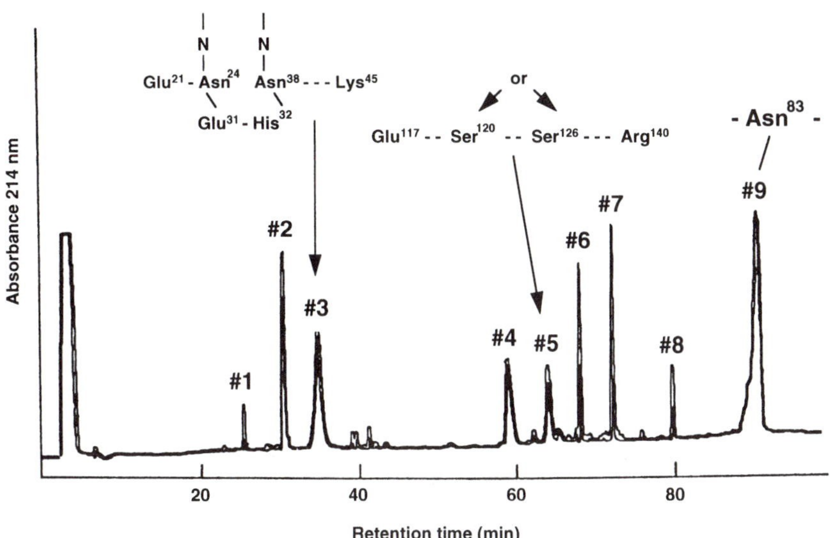

Figure 12.10 The RP-HPLC chromatogram of a Lys-C enzymatic digest of the pyridylethylated glycoprotein recombinant-erythropoietin. Chromatography conditions: HiPore RP-318 column (4.6 × 250 mm) with a 0.75 ml/min flow rate. Solvent A = 0.1% aq TFA and solvent B = 0.09% TFA in 95% AcCN with a 10-min hold. Development was effected up to 20% B with a 20-min hold and then a 50-min gradient to 60% B. Absorbance was at 214 nm. (© 1994. Reprinted from reference 66 by permission of Academic Press, Inc., Orlando, Florida.)

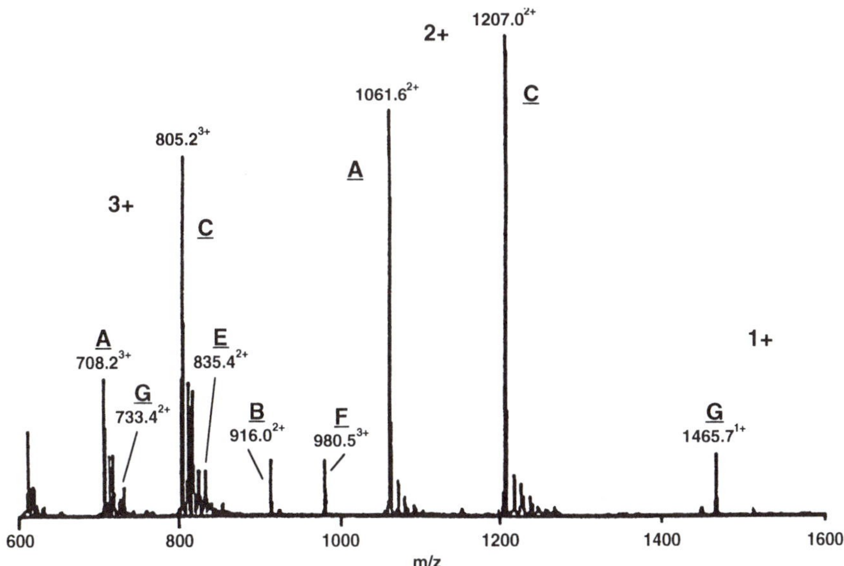

Figure 12.11 The ESI-mass spectrum of a mixture of O-linked glycopeptides from r-erythropoietin. The glycopeptides have the common tryptic peptide Glu-117 to Arg-131 core. See text for details. (© 1994. Reprinted from reference 66 by permission of Academic Press, Inc., Orlando, Florida.)

and use these derived equations as simultaneous equations to produce possible carbohydrate sequences. Use the sequence and structural motifs of the previous examples in this chapter as guides to eliminate possible isomeric structures (sequences). From this given information only, can the position of the polysaccharide on the polypeptide be determined?

Mass Spectral Analysis of an Oligosaccharide from a Bacterial Enzyme

This section presents an example of a recombinant glycoprotein where a number of endo-β-*N*-acetylglucosaminidases produced by *Flavobacterium meningosepticum* were cloned and produced by the *E. coli* organism. However, upon conventional gel analysis, it was found that the cloned glycoproteins were slightly lower in mass than that of the native glycoprotein in *F. meningosepticum*. The difference was found in the polysaccharide portion of the glycoprotein, and a number of uncommon sugar units were found in the polysaccharide.[9] Trypsin digestion produced a tryptic peptide that contained the polysaccharide, and the sequence of the polypeptide portion is SILDS*TK, where the polysaccharide is connected to the S* serine residue. The M_r of the polypeptide portion is 761.86 Da, which does not include the hydrogen on the hydroxyl side group of the S* serine residue. Figure 12.12 provides details of the structure, and the abbreviations represent the following (Table 12.1): 2-*O*-Me-Man, 2-*O*-methyl mannose (nonreducing end); GlcUNAc, 2-*N*-acetylglucuronic acid or 2-acetamido-2-deoxyglucuronic acid; GlcU, glucuronic acid; Glu, glucose; 2-O-Me-GlcU, 2-*O*-methylglucuronic acid; Man, mannose (reducing end); and 2-O-Me-Rha, 2-*O*-

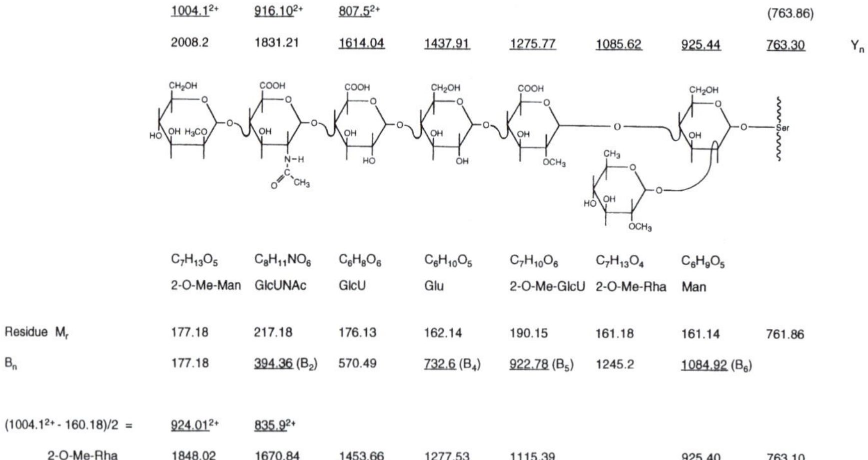

Figure 12.12 Structural details of an O-linked heptasaccharide along with ESI-mass spectral m/z values.

methylrhamnose. Summation of the seven sugar residue masses yields a value of 1245.1 Da, and added to that of the polypeptide this yields an M_r of 2006.96 Da. Note that the anomeric oxygen position on the C-1 atoms of all the sugar residues cannot be determined by mass spectrometry methods; thus, an "above or below" the ring annotation is not provided. The tryptic glycopeptide was analyzed, and then the polysaccharide portion was released from the polypeptide by reductive β-elimination to form the alditol on the reducing end (Table 12.4, equation 4). This converted the mannose reducing end to a mannitol residue. The absence of mannose and the presence of mannitol gave evidence that Man was the residue connected to the polypeptide chain.

The tryptic polypeptide produced the doubly and triply charged ions at m/z 1004.1 and 669.7 (data not presented) that yield a neutral mass of 2006.2 Da, and this is close to the calculated value of 2006.96 Da.[67] A tandem mass spectrum of the doubly charged ion of the polypeptide (m/z 1004.1^{2+}) yielded the product-ion spectrum shown in Figure 12.13.[67] There are four additional doubly charged ions (neutral mass) in the product-ion spectrum and they are m/z 807.4^{2+} (1612.8), 835.7^{2+} (1669.4), 916.0^{2+} (1830), and 925.1^{2+} (1848.1). An analysis of this spectrum is given in Figure 12.12, and the underlined values indicate the observed ions. Note that an ion is observed in Figure 12.12 for each Y_n species, including three doubly charged ions.

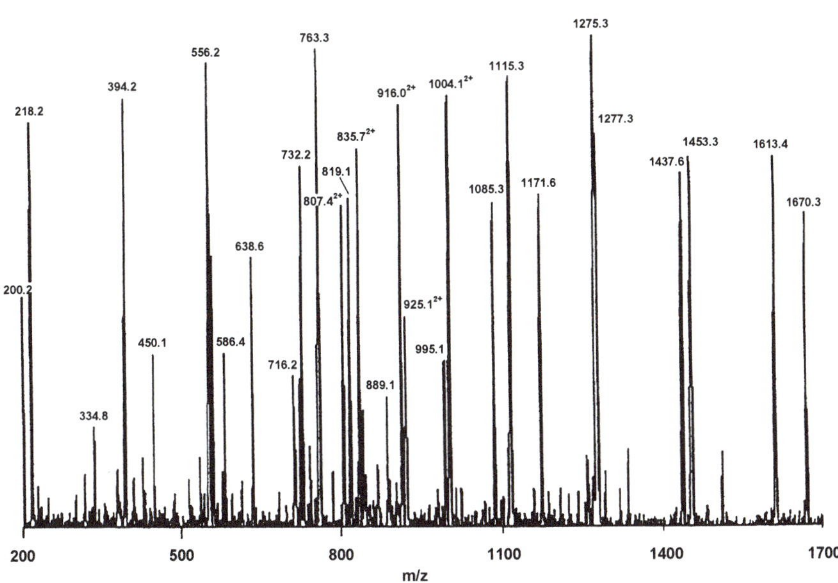

Figure 12.13 The ESI-tandem mass spectrum of a glycopeptide from *F. meningosepticum* P40; $(M + 2H)^{2+}$ = 1004.1 Da was the precursor ion. (© 1995. Reproduced from reference 67 by permisson of the American Society for Biochemistry and Molecular Biology, Bethesda, MD.)

The polypeptide ion mass is deduced by adding a proton to the Ser hyroxyl side group and an additional ionizing proton. These protons add 2 Da to the 761.86 Da polypeptide mass to yield 763.86 Da, and this is close to the observed value of 763.30 Da. The next ion, *m/z* 925.44, is calculated by the addition of the mannose residue mass of 161.14 Da and 1 Da (hydrogen) from the 2-O-Me-Rha moiety to *m/z* 763.30. The *m/z* 1085.62 ion results from *m/z* 925.44 + 160.18 (2-O-Me-Rha). The rest of the Y_n ions are calculated by adding respective residue masses. Not all of the B_n ions are observed, and the series at the bottom of Figure 12.12 is characterized by an initial loss of 2-O-Me-Rha from the $(M + 2H)^{2+}$ ion. Further ions are produced by the cleavage of residues, starting from the nonreducing end. Note that the lower series (absence of 2-O-Me-Rha) and the Y_n series in Figure 12.12 are correlated by the 2-O-Me-Rha residue, for example:

Y_n ion	1831.21 (916.10^{2+})	1614.04	1437.91	1275.77
2-O-Me-RHA	–160.18	–160.18	–160.18	–160.18
Lower ion series in Figure 12.12	1671.03	1453.86	1277.73	1115.59

This observation firmly places a deoxyhexose (2-O-Me-rhamose) residue as a branching residue on the reducing-end hexose moiety, because the residues from 2-O-Me-Man to 2-O-Me-GlcU can be found as an ion with or without the terminal 2-O-Me-Rha residue. Note the relatively close agreement in mass between the experimental values in Figure 12.13 and the calculated values in Figure 12.12. The Y_n calculated ion of 763.86 Da is observed at 763.30 Da, and the calculated doubly charged ion of 924^{2+} Da is observed at 925.1^{2+} Da. Two other ions in Figure 12.13 are *m/z* 218.2 and 556.2, which are internal fragments and can be identified as GlcU and GlcUNA-GlcU-Glc, respectively. If the investigator did not know a priori the identity of each residue, the information in Figure 12.13 would not be enough to *identify* any of the individual species. The sequence of hexose sugar units of respective masses can be obtained with confidence, and educated guesses could be made as to the identity of each residue. However, on which hydroxyl group do the methyl groups reside for three of the residues? Which position is the acid located on the three acidic residues? How do we even know that there are acid or methyl groups on the molecule just from the ion mass value? Further investigations, utilizing chemical treatment and derivatization methods, are required in order to address these issues. The identities of the individual residues were assessed by select derivatization techniques and a comparison with standards by use of gas chromatography (GC)-MS. With the GC-MS composition analysis and sequential M_r values, the basic information concerning residue identities and a sequence can be attained.[9,67] Isolation of the polysaccharide from the polypeptide was accomplished by two methods. One method produced an alditol with complete methyl derivatization of all acidic protons on the sugar residues (Table 12.2, reaction 3 and Table 12.4, reaction 8). The other method used direct base methylation to yield an intact reducing end ring (Table 12.2, reaction 5 or a combination of Table 12.2, reaction 1 and Table 12.4, reaction 8) with complete methyl derivatization. The derivatization with CH_3I substitutes all acidic protons (including hydroxyl, amine, and acid protons) with a methyl group.

PROBLEM 12.12 Draw the complete structure of the permethylated alditol and aldose of the heptasaccharide shown in Figure 12.12. Calculate the M_r and singly charged m/z value for each. What is the difference in daltons between the two molecules? The permethylated species was ionized by addition of sodium ions. Sodium ions were added to increase the efficiency of ionization because of the inherently low proton affinity that carbohydrates, in general, possess.[52,68]

During base-catalyzed release of the glycan from the polypeptide (Table 12.2, reaction 3), and permethylation, cleavage of the glycosidic bonds occurs between the oxygen and C-4 for all three carboxyl-containing residues. Because the C-5 hydrogen is labile when it is next to a carboxyl or ester (i.e., an α-hydrogen), the carboxyl causes a β-elimination of the residue on C-4. This occurred at all three ester-containing residues from the compound shown in Figure 12.46 (Answers section, page 453), because the glycosidic linkages are all at C-4. Thus, in addition to the intact permethylated heptasaccharide derivative in the reaction vessel, a number of by-products were formed.[67]

PROBLEM 12.13 List, by residue, the various reaction by-products that can be formed by the β-elimination pathway of linked 4-*O*-hexuronyl residues for the glycan portion in Figure 12.46. How many fragments originate from the reducing and nonreducing ends?

Upon ESI-MS, all by-products were observed, along with either the aldose or alditol (data not presented). The appearance of these fragment species provided evidence that the residues have a 1–4 linkage. There are several reasons as to why the entire polysaccharide was derivatized as opposed to just being treated with basic solution. A primary reason was to increase the sensitivity of detection so as to provide for a satisfactory signal-to-noise ratio. A second reason is that permethylation increases the overall hydrophobicity of the molecule, and this aids in the passage of the biomolecules through an RP-HPLC column. This concept will be dealt with further in the next section, "Asn-Linked Polysaccharide ESI-Mass Spectra". A third reason for permethylation is that differential permethylation can yield further clues for structural analyses. The following four samples were made:

1. CH_3-derivatized alditol,
2. CH_3-derivatized aldose,
3. CD_3-derivatized aldose or deuteriomethylated (CD_3) alditol,
4. exchange of the ester CD_3 groups, using the product from (3), with a CH_3 remethylation reaction.

The CD_3-derivatized alcohol groups do not exchange with the CH_3 remethylation reaction. Reactions (1) and (3) account for the number of added methyl groups, and reaction (3) highlights inherent (already present) CH_3 groups versus the added CD_3 methylation sites. Reaction (4) exposes the number of ester sites. These differences were observed in ESI-mass spectra (data not presented) of the intact and fragment entities of the heptasaccharide.[67]

Asn-Linked Polysaccharide ESI-Mass Spectra

The section "Mass Spectral Analysis of a Glycopeptide from Erythropoietin" has already presented information on basic recombinent (r) technology, and, in particular, states that an r-glycoprotein may not have the same glycan portion as that of the native glycoprotein. There is another aspect that serves to make the structural elucidation of a glycoprotein even more interesting. Cellular machinery rarely makes a glycoprotein molecule that is a particular and distinct glycoprotein. Cells usually manufacture different glycoprotein molecular species with the same core protein. The cause of this is the variability of the structure of the glycan portion, where the glycans are connected to the same amino acid residue to form a series of different glycoprotein molecules. These different species of glycoprotein, which differ by either the amount (number and mass) or type (isomers) of glycosylation on a specific amino acid site, exist as a mixture manufactured by the cell. Each individual molecular species can be labeled a glycoform, and the number of glycoforms characterize the degree of microheterogeneity at a specific amino acid site. Some glycoforms differ by type—namely, isomers—and are labeled as glycomers within a glycoform distribution.[51] The more different kinds of polysaccharide chains that are manufactured for a specific amino acid residue for a certain protein, the greater is the microheterogeneity, or site-specific (amino acid residue) heterogeneity. The glycoform distribution at a certain glycoprotein amino acid residue is also known as the glycoprotein heterogeneity.[34,51,66,69,70] The glycomer distribution includes isomeric residue(s), linkage, and branching features.[51]

The term heterogeneity is usually reserved for the existence of N-(Asn)- and O-(Ser/Thr)-linked carbohydrates on a particular glycoprotein.

The O-linked carbohydrate motif was discussed in the Section "Ser/Thr-Linked Polysaccharide ESI-Mass Spectra". In most cases, the amount of saccharide residues in an O-linked species is relatively low. The chain is usually composed of only a few carbohydrate residues,[5,71] even though there are exceptions.[3,5,20] These O-linked polysaccharides can display fairly consistent types of structures and sequences, and they can display quite variable and unique sequences.

This is in contrast to the N-linked oligosaccharides. These have three different types of structures or motifs, and their variablity resides on the terminal nonreducing end. All three have a common "core"; namely, the residues at and near the reducing end of the oligosaccharide. These structures are predictable to a large extent, with only minor variations. The three motifs are labeled high-mannose, hybrid, and complex. The hybrid has elements of both the high-mannose and complex polysaccharides. These three types of N-linked oligosaccharides are termed glycotypes, and they too have glycoforms and glycomers.

High Mannose Glycotype

The structure of a high-mannose form is presented in Scheme 12.4, along with abbreviated forms of the structure. Note that there are nine Man and two GlcNAc residues, and this is essentially the largest form of the high-mannose glycotype that has been found. Other high-mannose glycoforms are characterized by fewer Man residues at various branching points. However, these lower mass glycoforms can all be generated by the basic architecture of the $Man_9GlcNAc_2$ structure shown in Scheme 12.4. For example, linear sequences of nine, eight, or seven Man residues with branching Man residues have not been observed. Thus, four Man residues is the maximum in a linear arrangement for the high-mannose glycotype. Likewise, with $Man_8GlcNAc_2$, $Man_7GlcNAc_2$, and so on, only certain structures are possible. There is only one $Man_9GlcNAc_2$ glycoform and three $Man_8GlcNAc_2$ glycoforms, and these and other permutations are shown Figure 12.14.[68,69] Note that the Man–Man linkages are all of the α-anomer type, and the Man–GlcNAc and GlcNAc–GlcNAc linkages are of the β-configuration. Table 12.8a presents the mass of the high-mannose glycoforms, as well as the permethylated species. The hybrid glycotype is listed in Table

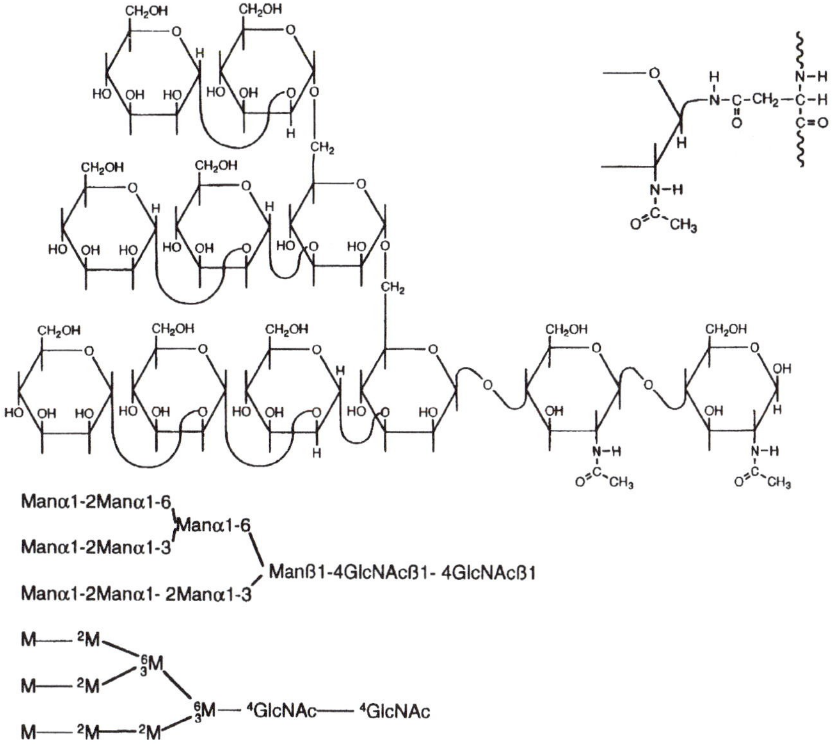

Manα1-2Manα1-6
Manα1-6
Manα1-2Manα1-3
Manβ1-4GlcNAcβ1- 4GlcNAcβ1
Manα1-2Manα1- 2Manα1-3

M—— 2M
6/3M
M—— 2M
6/3M— 4GlcNAc—— 4GlcNAc
M—— 2M——2M

Scheme 12.4 Asparagine-linked oligosaccharides: $Man_9GlcNAc_2$ high-mannose glycotype.

Man$_9$

9-26-ABC

Man$_8$

8-24-BC

8-24-AC

8-24-AB

Man$_7$

7-22-A

7-22-B

7-22-C

7-18-AB

7-21-AC

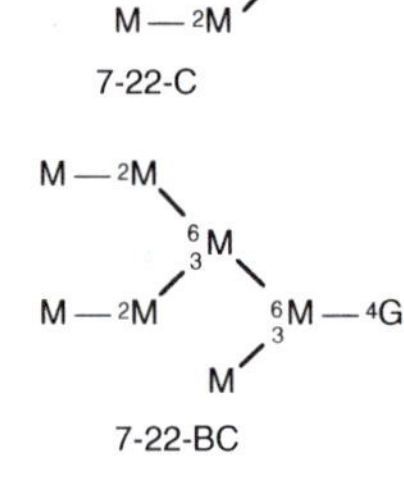

7-22-BC

Man$_6$

6-16-A

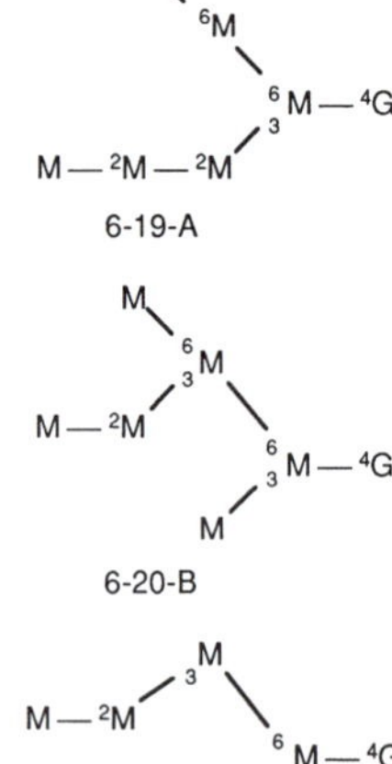

6-19-A

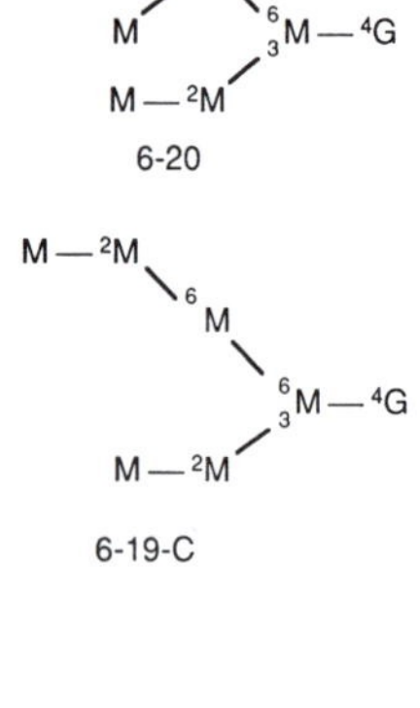

6-20

6-20-C

6-20-B

6-19-C

6-16-B

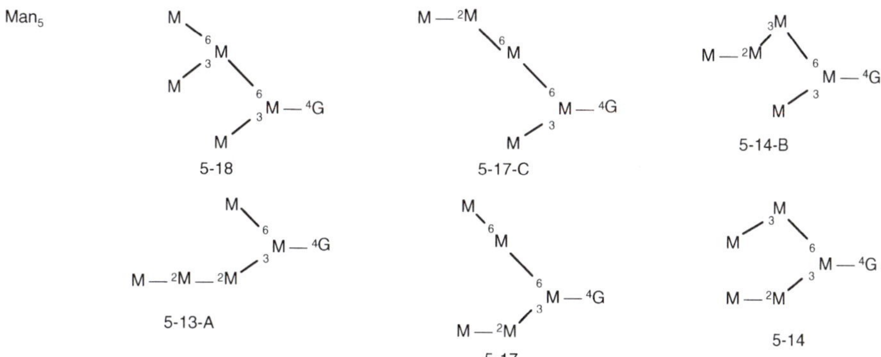

Figure 12.14 (facing page and above) The 23 N-linked high-mannose glycans. (© 1994, John Wiley & Sons, Limited. Reproduced from reference 69 with permission.)

Table 12.8a Average mass values of oligosaccharides found in the high-mannose glycotype

Carbohydrate Man_n	M_r (Da)[a] of Man_n plus GlcNAc	or $GlcNAc_2$	Permethylated M_r (Da)[b] of Man_n plus GlcNAc	or $GlcNAc_2$	Permethylated Man_n $GlcNAc_2$ $(M_r + 2Na)^{2+}/2$
Man_3	707.622	910.817	904.013	1149.289	597.644
Man_4	869.862	1073.057	1108.236	1353.512	699.756
Man_5	1032.002	1235.197	1312.46	1557.736	801.868
Man_6	1194.042	1397.237	1516.68	1761.958	903.979
Man_7	1356.182	1559.377	1720.90	1966.181	1006.090
Man_8	1518.312	1721.507	1925.128	2170.404	1108.202
Man_9	1680.462	1883.657	2129.351	2374.627	1210.313

[a]Mass values include the H– nonreducing and –OH terminal reducing end moiety.

[b]Mass values include the CH_3– on the nonreducing end and the $-OCH_3$ terminal methoxy moiety on the reducing end.

Table 12.8b Average mass values of oligosaccharides found in the hybrid glycotype

Carbohydrate	Composition	M_r (Da)[a]	Permethylated M_r (Da)[b]	$(M_r + 2Na)^{2+}/2$
Hybrid	$Man_5GlcNAc_3$	1438.297	1803.012	924.060
Bisecting hybrid	$Man_5GlcNAc_4$	1641.507	2048.288	1047.144
Bisecting hybrid	$Hex_6GlcNAc_4$	1803.620	2252.512	1149.256
Bisecting hybrid	$Hex_5GlcNAc_5$	1844.702	2293.563	1169.782
Bisecting hybrid	$Hex_6GlcNAc_5$	2006.815	2497.786	1271.893

[a]Mass values include the H– nonreducing and –OH terminal reducing end moiety.

[b]Mass values include the CH_3– on the nonreducing end and the $-OCH_3$ terminal methoxy moiety on the reducing end.

12.8b, and Scheme 12.5 presents a typical hybrid glycotype. The upper branch or antenna, which consists of three Man residues and is connected by a 1–6 linkage to the mannose trisubstituted residue, is of the high-mannose type. The lower branch, with a Man1–3Man linkage, is of the complex glycotype. The latter will be treated separately in this chapter. This overall structure can be labeled as biantennary, where each antenna is composed of elements of two different glycotypes. Note that in Schemes 12.4 and 12.5, both the high-mannose and the hybrid glycotypes share a common $(Man)_2$ManGlcNAcGlcNAc core polysaccharide structure.

Mass Spectral Analysis of the Glycotypes of Ovalbumin

The protein ovalbumin has only one N-glycosylation site, at Asn-292,[72] and it is of the high-mannose and hybrid glycotypes. Thus, the cell manufactures two types of glycotypes that are connected to the same Asn residue in ovalbumin. The ESI-mass spectrum of ovalbumin[73] is presented in Figure 12.15. The inset shows details of the $(M + 25H)^{25+}$ and $(M + 24H)^{24+}$ protonated forms of the glycoprotein. Note that the inset shows the glycoform distribution for the $(M + 25H)^{25+}$ ion, as well as the actual m/z values and deconvoluted M_r values for each glycoform in the $(M + 25H)^{25+}$ envelope.

PROBLEM 12.14 With the knowledge of only one N-linked glycosylation site for ovalbumin, how can the different peaks in the $(M + 25H)^{25+}$ envelope be labeled as glycoforms?

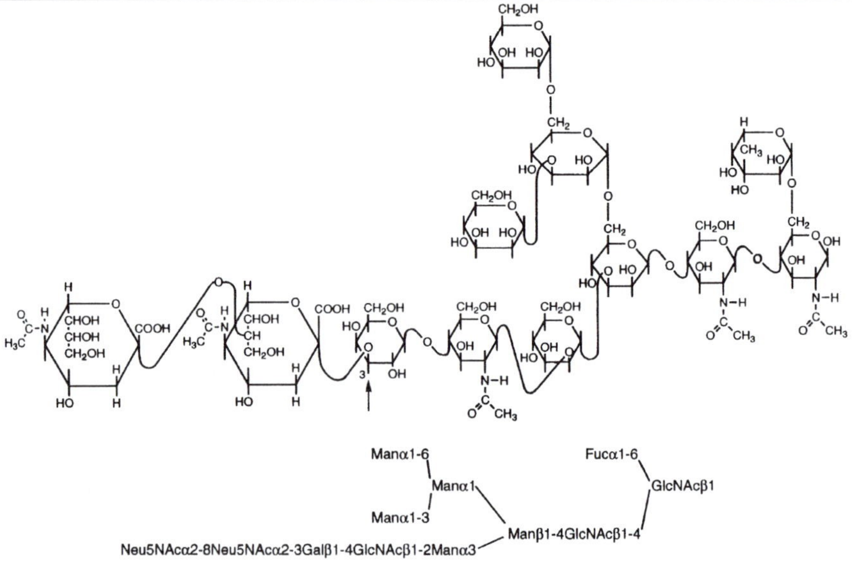

Manα1-6
Manα1
Manα1-3
Fucα1-6
GlcNAcβ1
Manβ1-4GlcNAcβ1-4
Neu5NAcα2-8Neu5NAcα2-3Galβ1-4GlcNAcβ1-2Manα3

Scheme 12.5 Asparagine-linked oligosaccharides: hybrid fucosylated, disialylated glycotype.

The average mass or M_r of the ovalbumin polypeptide is known as 42,750 Da from DNA analyses. There are posttranslational modifications that add an additional 119 Da, and these include phosphorylation of a Ser residue, N-terminus acetylation, and a disulfide bond. This increases the M_r to 42,869 Da, excluding the mass of the glycan portion.

PROBLEM 12.15 When ovalbumin is treated with endoglycosidase H, the resulting species produces the ESI-mass spectrum in Figure 12.16.[73] What is the M_r value? Describe both ways that the M_r value can be deduced. Schemes 12.4 and 12.5 and Table 12.5 will help to solve this question.

From Problem 12.15, the observed average mass of the GlcNAc–polypeptide is 43,073 Da, and subtraction of the GlcNAc residue yields the polypeptide M_r of 42,870 Da. It has been shown that high-mannose and hybrid glycoforms are found in ovalbumin; therefore, this must be reflected in Figures 12.15 and 12.16. The ovalbumin hybrid species do not contain the Neu5NAc or fucose residues as are shown in Scheme 12.5. Subtraction of the polypeptide mass from each of the deconvoluted mass values in Figure 12.15, along with either a 17-Da

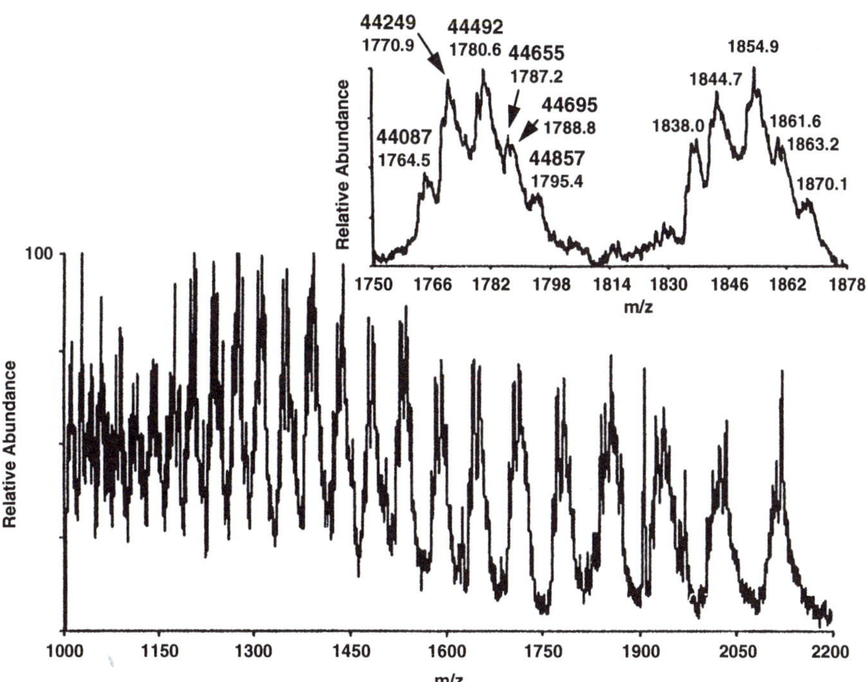

Figure 12.15 Positive-ion ESI-mass spectrum of ovalbumin. Inset shows the heterogeneity details of the $(M + 24H)^{24+}$ and $(M + 25H)^{25+}$ ions. (© 1992, American Chemical Society. Reprinted from reference 73 with permission.)

(closed-ring aldose) or 18-Da (open-ring aldose) addition to the terminal reducing GlcNAc residue, yields mass values of the two glycotype glycoforms.

PROBLEM 12.16 Calculate the glycan mass values from the above information for ovalbumin (Figures 12.15 and 12.16), and provide an identity for each glycan mass value with respect to glycoform and glycotype.

Problem 12.16 makes use of a rather low-mass spectrometric resolution analysis, because there could be other glycoforms hidden in the peaks in Figure 12.15. *N*-glycanase digestion (Table 12.5) of ovalbumin, along with chromatographic separation techniques, can yield a mixture of the glycoforms of both glycotypes, without the large polypeptide portion of ovalbumin. A direct infusion of the complete glycoform mixture[73] yielded the ESI-mass spectrum shown in Figure 12.17. Even though it is heterogeneous in nature, it is much easier to analyze the carbohydrates when removed from the large polypeptide than of those in Figure 12.15. The two most intense ions, adducted with Na^+, contain the following polysaccharides: 1235 + 23 = 1258 Da = G_2h_5 and 1397 + 23 = 1420 Da = G_2h_6. These two polysaccharide species represent the high-mannose glycotype, and many other ions are easily observed compared with those of Figure 12.15. These two species were found in Figure 12.15

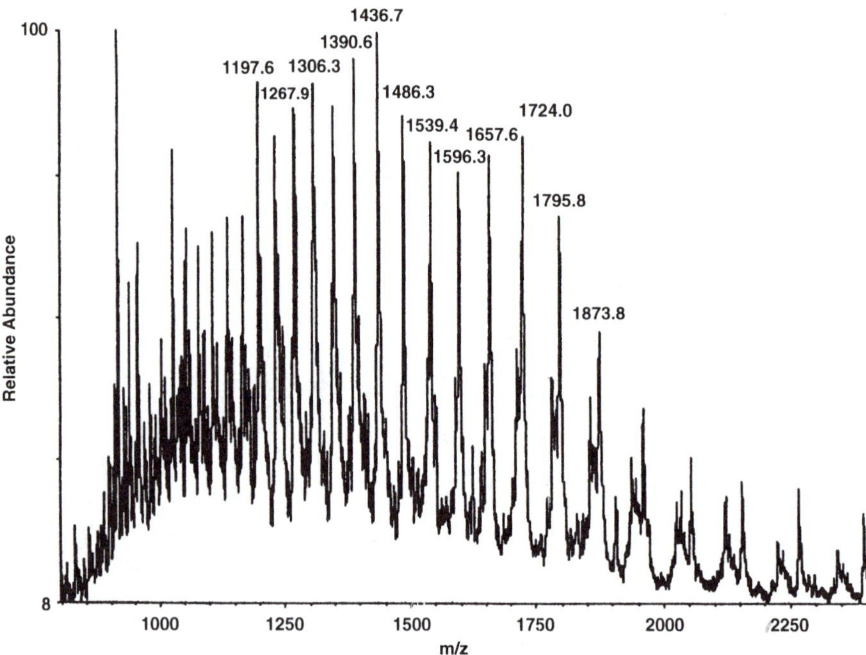

Figure 12.16 Positive-ion ESI-mass spectrum of ovalbumin after treatment with endoglycosidase H. (© 1992, American Chemical Society. Reprinted from reference 73 with permission.)

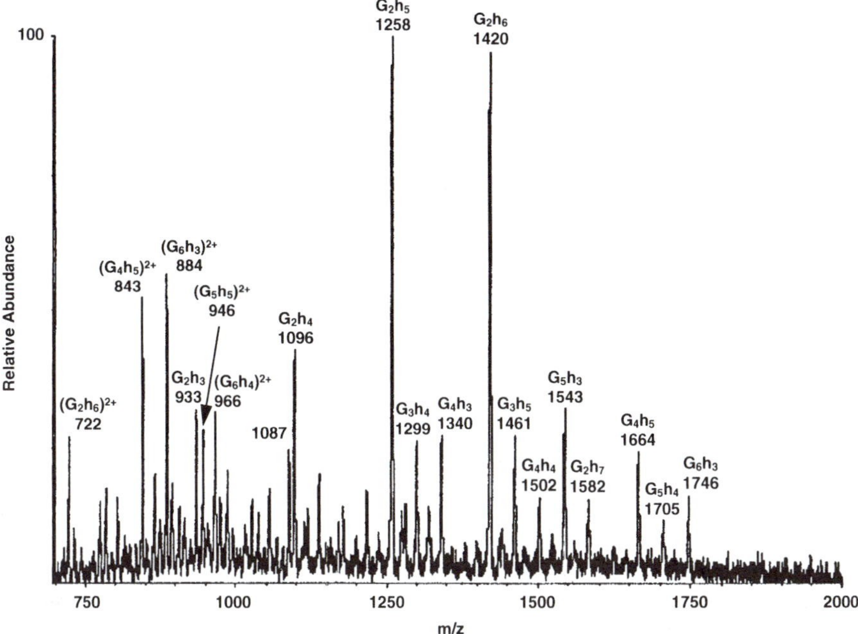

Figure 12.17 Positive-ion ESI-mass spectrum of the various distinct oligosaccharides found on the single N-linked site of ovalbumin after their removal from the glycoprotein by *N*-glycanase. (© 1992, American Chemical Society. Reprinted from reference 73 with permission.)

(Problem 12.16). Note that this spectrum is a presentation of individual, separate oligosaccharides and is not a tandem mass spectrum. Thus, sequences of the residues cannot be obtained.

The next study addresses the question of sequence for a high-mannose glycotype.

Mass Spectral Analysis of the Ribonuclease B High Mannose Glycotype

Ribonuclease B (RNase B) has a number of Asn-linked polysaccharides, where some are of the complex glycotype and others are of the high-mannose glycotype. A tryptic digest was made, and a high-mannose glycopeptide was analyzed by ESI-tandem MS. Figure 12.18 presents the ESI-tandem mass spectrum of the doubly charged species of the glycopeptide, and the structure of the glycopeptide is shown in the figure.[74]

Interpretation is straightforward; however, note that the important high-mass peaks, which provide the Man information, are very low in intensity. This is a consequence of the relatively poor ionization efficiency with protons as the ionizing species.

With either addition of the masses of the individual residues or by simple calculation of the 928^{2+} Da ion, the M species = 1854 Da. Then, 1854 − 1692

= 162 Da (hexose or mannose residue mass), and successive differences yield this species of sugar until m/z 881 is encountered. A GlcNAc is then observed at m/z 881 − 678 = 203 Da. These are a series of Y_n-type ions, as well as the GlcNAc-peptide ion at m/z 678. Lower ion masses are also observed.

PROBLEM 12.17 What are the identities of m/z 441, 522, 603, 684, 765, and 847?

The above N-linked analyses dealt with sequence information; however, no linkage and branching detail could be obtained for the glycoforms, despite the use of tandem mass spectrometry (product-ion spectra). The next section addresses these concerns.

Mass Spectral Structural Analysis of the RNase B High Mannose Glycotype

The high-mannose glycan of ribonuclease B (RNase B) will be revisited here, with the goal of providing a comprehensive understanding of the structure, sequence, linkage, and branching aspects of the glycoforms. Figure 12.19 presents an ESI-mass spectrum of a commercial sample of RNase B.[69] Note the profusion of peaks within each multiply charged ion envelope. Deconvolution[75] of each envelope (Figure 12.19, inset) shows that two separate species are present. The lower mass species is that of the nonglycosylated ribonuclease A

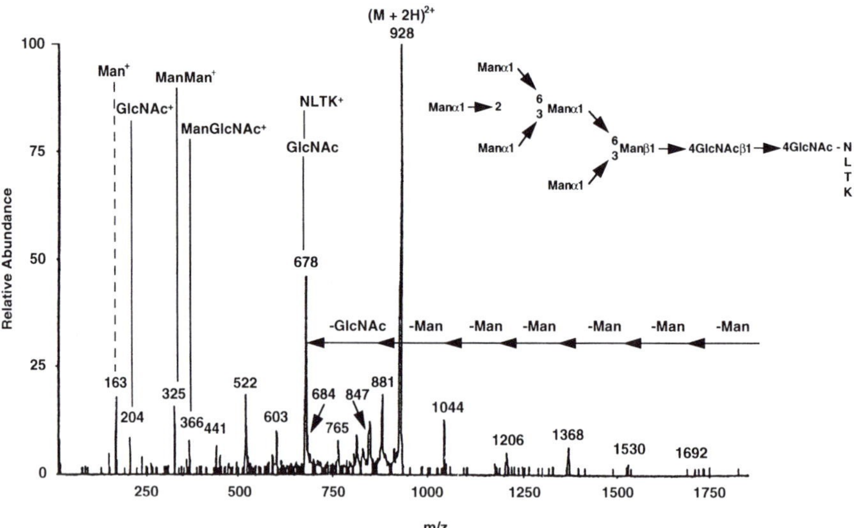

Figure 12.18 The ESI-tandem mass spectrum of one of the glycopeptides isolated from ribonuclease B enzyme. The $(M + 2H)^{2+}$ precursor = 928^{2+}. Inset shows the structure of the glycopeptide where the Manα1–2 signifies a mannose binding on the upper or lower mannose residue of the glycopeptide. (© 1992, American Society for Mass Spectrometry. Reprinted from reference 74 by permission of Elsevier Science, Inc.)

polypeptide (RNase A) contaminant, and the higher mass species is that of ribonuclease B (Figure 12.19, inset). In the main spectrum, each envelope contains an overlap of two distributions of ions, and they differ by one charge. One envelope is highlighted and consists of the 9+ RNase A and 10+ RNase B species.

PROBLEM 12.18 The lower mass species in the deconvoluted spectrum exhibits a regular spacing of signals while the higher mass series exhibits an irregular spacing. For the RNase A envelope, estimate the value of the spacing between the peaks; what structural modification might this signify to the protein?

The RNase B was next subjected to Lys-C protein digestion, and the glycopeptide glycoforms were isolated by RP-HPLC. A glycopeptide that has the high-mannose carbohydrate was isolated and directly infused to provide the ESI-mass spectrum shown in Figure 12.20. Multiply charged envelopes of the glycoforms of the glycopeptide are observed.[69]

PROBLEM 12.19 Label each peak in the 5+ charge state in Figure 12.20 with the proper high-mannose designation. Assume that the polypeptide portion of the glycoforms is 4799.7 Da.

Figure 12.14 provides details as to the various types of high-mannose glycoforms that can be produced. Basically, cellular machinery produces the $Man_9GlcNAc_2$ high-mannose polysaccharide, or M_9, and this unit is susceptible to cleavage by the α-mannosidase exoglycosidase (Table 12.6). The M_9 species can be cleaved at three different terminal Man locations to yield three different

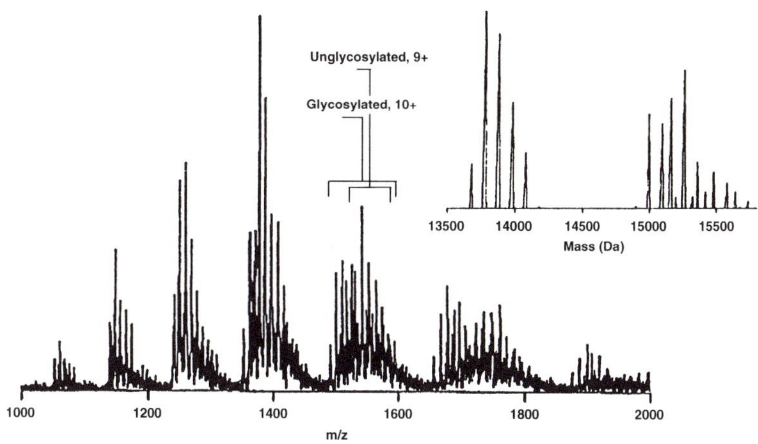

Figure 12.19 The ESI-mass spectrum of a commercially obtained, unprocessed, ribonuclease B sample. The inset shows the deconvoluted, neutral-mass spectrum and exhibits contamination with ribonuclease A enzyme. (© 1994, John Wiley & Sons, Limited. Reproduced from reference 69 with permission.)

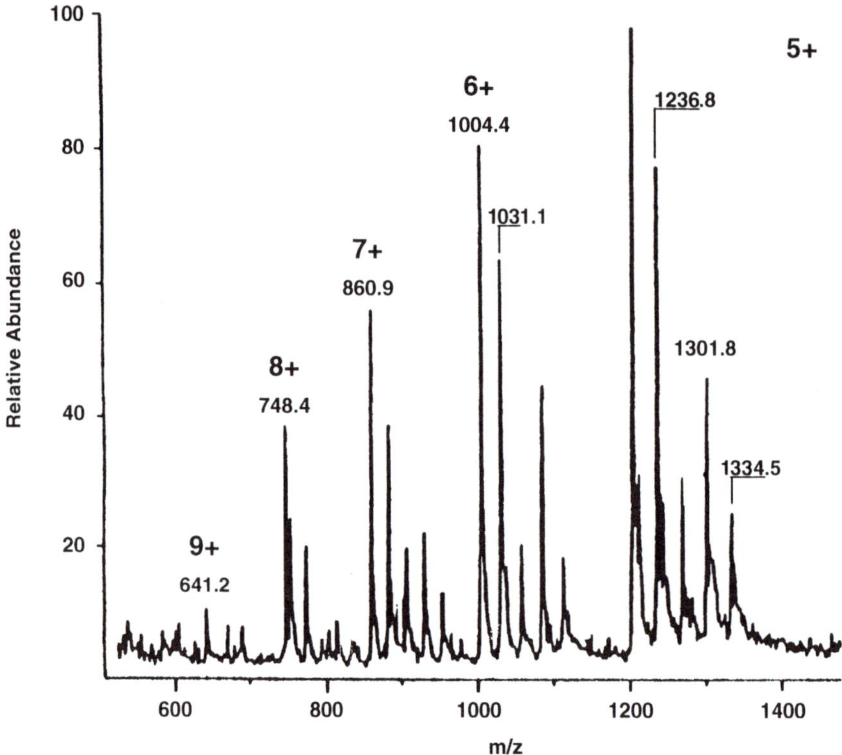

Figure 12.20 The ESI-mass spectrum of a glycopeptide from a Lys-C digest of ribonuclease B. Note the heterogeneity of the glycopeptide, which was isolated by RP-HPLC. (© 1994, John Wiley & Sons, Limited. Reproduced from reference 69 with permission.)

species with eight Man residues (Figure 12.14). These latter three species have a total of nine terminal Man residues, and the α-mannosidase can cleave every one to yield nine M_7 polysaccharides, of which six are different. Therefore, three of the nine structures are redundant. This phenomenon continues, and yields seven different M_6 and six different M_5 glycoforms. The M_4 high-mannose glycoforms are rarely observed (V. Reinhold, 1997, personal communication). Remember, all high-mannose species must retain the core three mannose residues where the central Man is connected to the GlcNAc-GlcNAc diacetylchitobiose disaccharide (Table 12.1).

The identification of each high-mannose glycoform is given in Table 12.14, beneath each structure, and the origin of their nomenclature is shown in Scheme 12.6.[68,69] There are three separate alphanumeric positions, which characterize the structure. First is the total number of Man residues. The middle value is the summation of the carbon-linkage position(s) in each Man residue, not including the C-1 value of 1. The last item is either A, B, or C, and any or all of them are

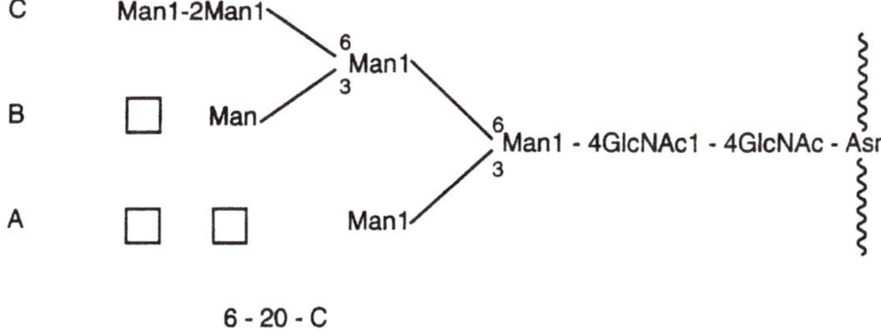

Scheme 12.6 Nomenclature diagram for the high mannose glycoform.

appended when any of the three Man branches have a full complement of mannose residues. Starting from the mannose residue linked to the GlcNAc residue, a full complement for each branch is reached when there are four mannose residues in a series, and each branch has its alphabetic designation appended to the nomenclature label. The boxes show where a mannose residue has been removed.

PROBLEM 12.20 What is the relationship between the six glycoforms that constitute the high-mannose $Man_7GlcNAc_2$? Using Figure 12.14 as a guide, prove that only seven different M_6 glycoforms can be generated from α-mannosidase trimming of the six M_7 glycoforms.

Figure 12.21 provided information on the $Man_{5-9}GlcNAc_2$ glycopeptide glycoforms of RNase B. The glycopeptide glycoform mixture was treated with Endo-H to separate the peptide portion, and the polysaccharide portion was permethylated. Figure 12.21 provides an ESI-mass spectrum of the glycan mixture infused with sodium.[69]

PROBLEM 12.21 Using Table 12.8 as a guide, deduce the identity of each of the mass-labeled peaks in Figure 12.21.

Permethylation, as introduced in Problem 12.21, is attractive for studying polysaccharide structures, because it can provide linkage and branching information when coupled with tandem mass spectrometry. Sodium adduction is usually used, because the adduct increases the ionization efficiency of the otherwise low-proton-affinity permethylated product. The permethylated glycan is largely composed of ether linkages, which are known to exhibit poor ionization efficiency. Tandem mass spectrometry with permethylation also provides information, when compared with underivatized samples, with respect to mass shifts associated with the number of hydroxyl-methoxy species in each residue. Glycosidic cleavages are noted here. However, the positions of interresidue linkages come to light when cross-ring cleavages (A and X species) are observed in the mass spectrum from product-ion analysis. Glycosidic cleavages are easily observed, but cross-ring cleavages provide relatively low-intensity peaks.

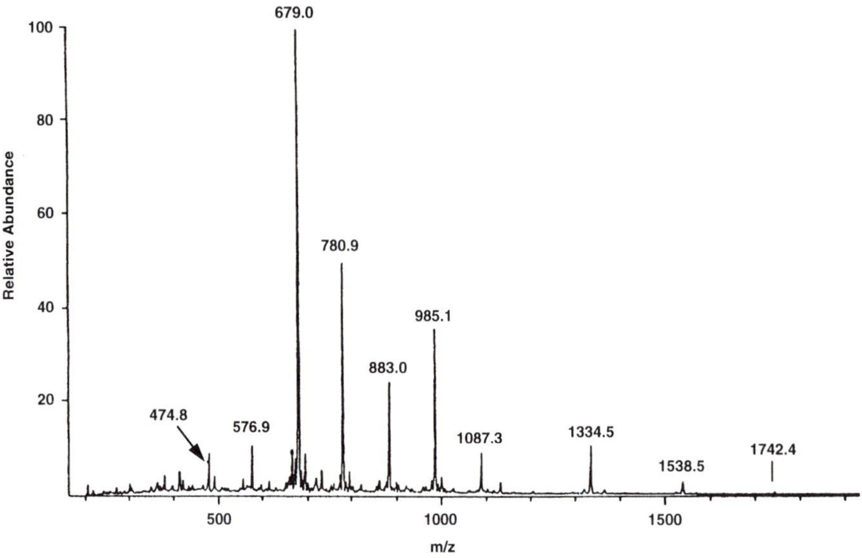

Figure 12.21 The ESI-mass spectrum of the heterogeneous glycan portion of the glycopeptide in Figure 12.20. The glycans were removed by Endo-H deglycosylation, permethylated, and infused into the mass spectrometer with sodium ion by ESI. (© 1994, John Wiley & Sons, Limited. Reproduced from reference 69 with permission.)

However, cross-ring cleavages of sugar residues can be observed under low-energy quadrupole mass spectrometer conditions, unlike the analogous amino acid R-group cleavage CID reactions. The low-intensity cross-ring cleavage fragments are attached to intact residues formed by CID. Table 12.9 shows the ring fragments that can be generated from the four main types of linkages, and A-type ions are produced where the charge is retained as the nonreducing end.[51,53,68,69,76] Hence, the fragment itself comes from a reducing-end residue, and this is attached to the nonreducing-end sugar residue.

Figure 12.22 shows the ESI-product-ion mass spectrum of a specific $Man_8GlcNAc$ glycomer, 8-24-AC.[68,69] Collision-induced dissociation was effected on the disodiated glycoform glycomer at m/z 985.7. The mass-structure insets in Figure 12.22 show the origin of the reducing- and nonreducing-end ions formed by Y-type glycosidic cleavages. As depicted in Scheme 12.3, the reducing-end fragment increases in mass by a proton from the nonreducing-end C-2 hydrogen, and the nonreducing-end mass thus decreases by 1 Da. Each resulting neutral fragment can then adduct to a sodium ion. The $(M + 2Na)^{2+}$ mass (not mass-to-charge ratio) is $985.7 \times 2 = 1971.4$ Da. The singly charged pair of ions: $1729.9 + 241.2 = 1971.1$ Da, and the ion pair: $1525.6 + 445.5 = 1971.1$ Da. The direct addition of the ions generated on either side of a glycosidic bond yields the $(M + 2Na)^{2+}$ mass. Single and doubly charged sodiated ion pairs are listed on the top of the spectrum. Calculation of some of the ion masses from the CID of the $(M + 2Na)^{2+}$ parent ion follows where M = mannose and G = GlcNAc:

Table 12.9 Collision-induced dissociation generation of A fragments

Linkage	Reactant	Product ion	A fragment	M_r, nominal mass of ring fragment attached to R: Underivatized	Methylated
1, 6			$^{0,4}A_n$	60	60
			$^{3,5}A_n$	74	88
1, 4			$^{3,5}A_n$	74	88
1, 3		no fragments			
1, 2			$^{1,3}A_n$	60	74

1(a) $985.7\ (M + 2Na)^{2+} \times 2 = 1971.4$
$1971.4 - 23 = 1948.4 = (M + Na)^+$
$1948.4 - 218.22 = 1730.2$ Da

This is a Y ion that represents the loss of a terminal permethylated mannose residue. The loss is not 219.22 Da because a hydrogen backtransfers to the reducing-end polysaccharide to neutralize the glycosidic oxygen. Sodium ion supplies the single charge.

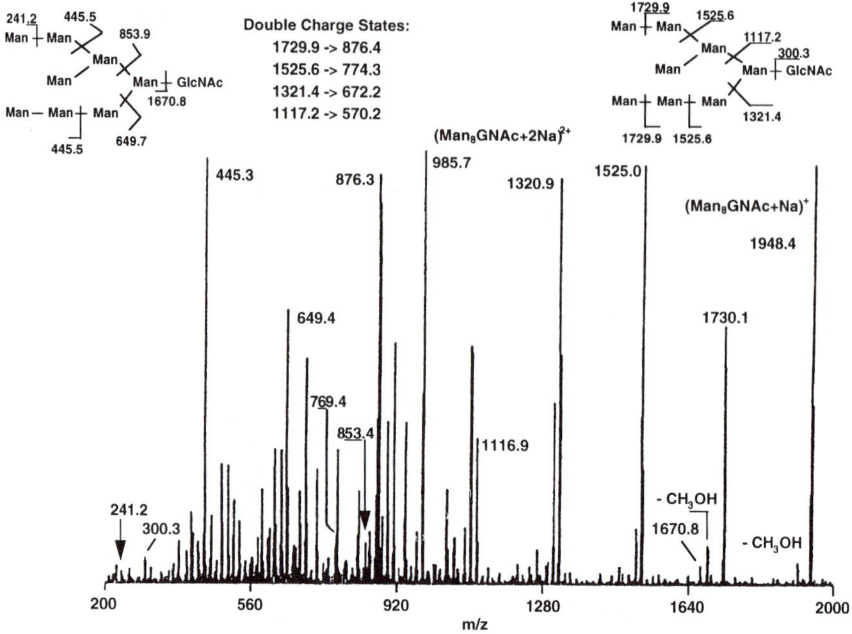

Figure 12.22 The ESI-product-ion mass spectrum of the permethylated 8-24-AC $Man_8GlcNAc$ glycomer. The precursor ion was the disodiated $(M + 2Na)^{2+} = 985.7$ Da. Insets show the structural derivation of the observed ions. (© 1994, John Wiley & Sons, Limited. Reproduced from reference 69 with permission.)

(b) $$1948.4 - 218.22 - 204.22 - 204.22 = 1321.7\text{Da}$$

This represents the loss of the A branch mannose residues.

(c) $$1948.4 - 245.28 - 31(CH_3O) - 1 = 1671.1 \text{ Da}$$

If this were a true B-type cleavage (Scheme 15.2), then the terminal, reducing-end trisubstituted mannose would not lose a hydrogen to GlcNAc. Rather, the ionizing proton attaches to the GlcNAc; therefore, the resulting ion would be 1671.8 Da. The appearance of a 1670.8 Da ion (calculated value of 1671.1 Da) suggests that the Y-type cleavage is operative, where the nonreducing end loses a hydrogen (−1). As Scheme 15.3 shows, an enol and an alcohol form from a Y-type cleavage. Apparently, sodium can attach to either product, and, in this case, the sodium attaches to the nonreducing end:

(d) $$m/z\ 241.2 = 204.22 + 15 + 23 - 1$$

The mannose nonreducing-end leaving group backtransfers one hydrogen to the reducing-end glycosidic oxygen, and Na^+ adducts to the nonreducing-end residue:

(e) m/z 853.9:
$242.2 + 204.22 - 1 = 445.42$, observed in spectrum
$446.42 + [(204.22 - 15) - 1] + (204.22 + 15) = 853.86$ Da

Figures 12.23 and 12.24 show enlargements of certain regions of Figure 12.22 so as to be able to see the low-intensity cross-ring cleavages.[68,69] The methylated column in Table 12.9 can be used as a reference. In Figure 12.23, the lower A-branch, trimannose residue is observed at $(204.22 + 15) + 204.22 + (204.22 - 1) + 23 = 649.7$ Da, and the trimannose structure represents the Na^+ adduct of a neutral, nonreducing-end Y loss.

The reducing-end (Man1–3) of the nonreducing-end fragment loses a proton to the glycosidic-bond oxygen of the trisubstituted mannose. Scheme 15.3 shows that a nonreducing-end product loses a hydrogen to the reducing-end product for Y-ion formation. Sodium can adduct to both neutral fragments, as is observed in the insets to Figure 12.22. All the other ions in Figures 12.23 and 12.24 will be calculated, because important observations can be made with respect to structural, linkage, and glycoform issues. The following information in (2a–h) is presented for observation with respect to linkage and glycoform analysis in Table 12.10.

2(a)

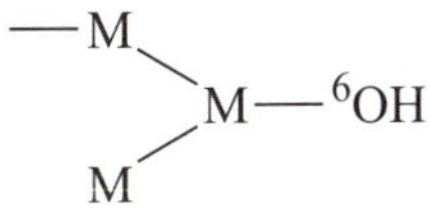

```
—M
   \
    M—⁶OH
   /
  M
```

$$(204.22 + 1) + (204.22 - 15) + (204.22 + 15) + 17 + 23 = 653.7 \text{ Da}$$

(b) **PROBLEM 12.22**

Calculate the mass of the M_5G^{2+} and M_3-^3OH species in Figure 12.23.

(c)

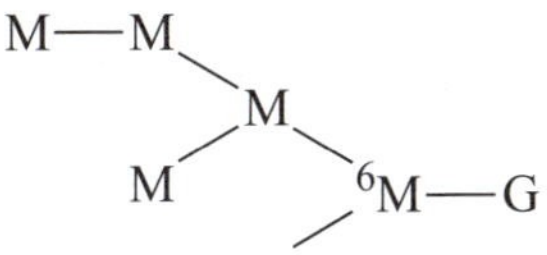

```
M—M
    \
     M
    / \
   M   ⁶M—G
       /
```

Note the subtle difference between this species and that of 665.19^{2+} in Problem 12.22. The difference is a methylene group on the C branch (upper) terminus. Thus:

$$1330.37\ (M + 2Na)^{2+} + 14 = 1344.37 \text{ Da}$$
$$1344.37/2 = 672.18^{2+} \text{ Da}$$

(d)

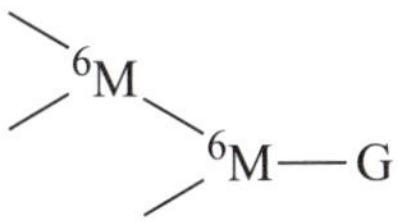

```
 \
  ⁶M
 /  \
     ⁶M—G
     /
```

$$[(204.22 - 15) + 1 + 1] + (204.22 - 15 + 1) + (245.27 + 31) + 23 = 680.71 \text{ Da}$$

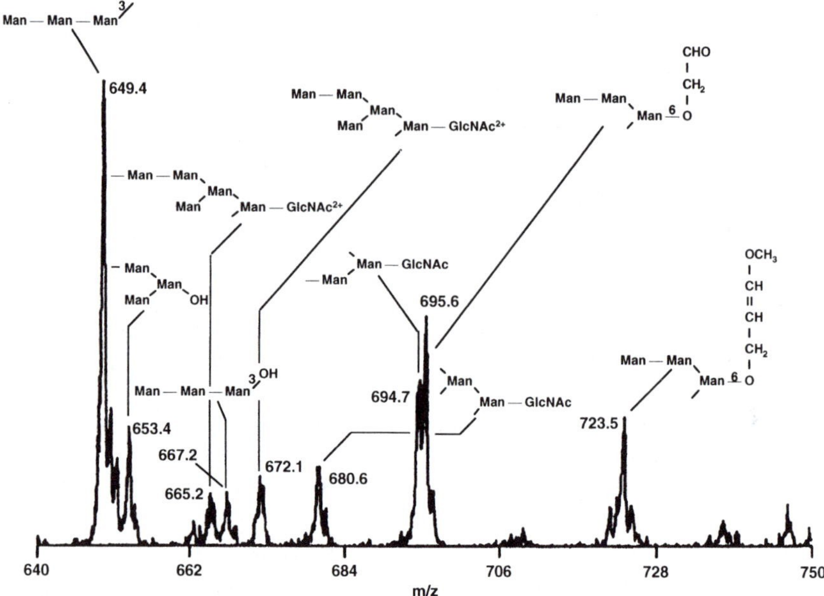

Figure 12.23 Expansion of the m/z 640–750 mass region of Figure 12.22 showing details of cross-ring cleavages. (© 1994, John Wiley & Sons, Limited. Reproduced from reference 69 with permission.)

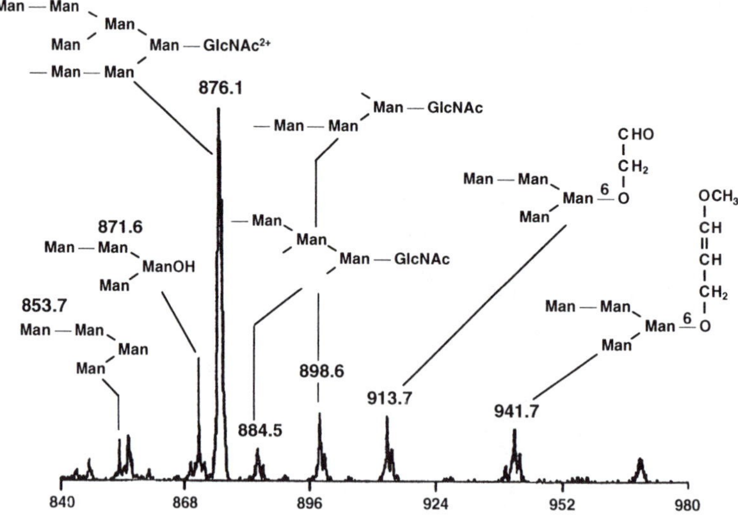

Figure 12.24 Expansion of the m/z 840–980 mass region of Figure 12.22 showing details of cross-ring cleavages. (© 1994, John Wiley & Sons, Limited. Reproduced from reference 69 with permission.)

Table 12.10 Fragment and mass information of the 8-24-AC high-mannose glycoform

	Fragment	Sodiated mass	
		Permethylated	Underivatized
a	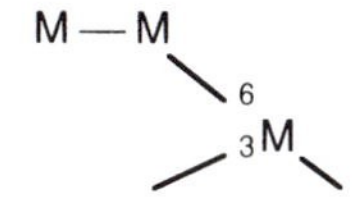 M—M \ 6 3M / \	635.66	509.42
b	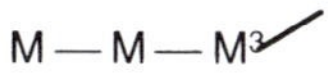 M—M—M3/	649.7	509.42
c	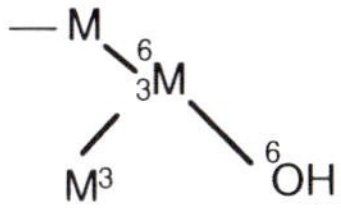 —M \6 3M / \ M3 6OH	653.7	527.42
d	 3OH / M—M—M	667.66	527.42
e	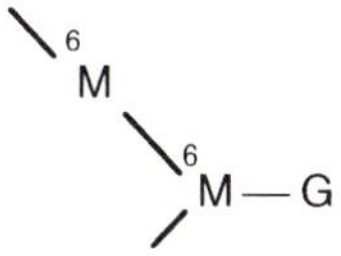 \6 M \6 M—G /	680.71	568.48
f	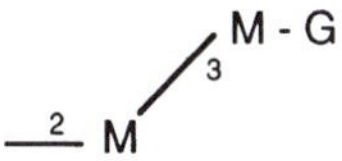 M - G /3 —2— M	694.71	568.48
g	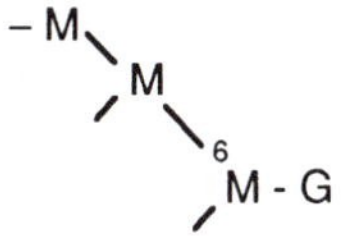 - M \ M / \6 M - G /	884.9	730.61
h	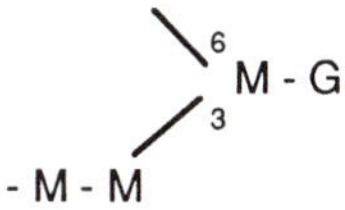 \6 M - G /3 - M - M	898.9	730.61

Note that the two glycosidic bond oxygens on the uppermost M residue both gain a hydrogen from each nonreducing-end dimannose and mannose leaving group.

(e) **PROBLEM 12.23** Calculate the mass value of m/z 694.7 in Figure 12.23.

(f) The next two higher masses (m/z 695.6 and 723.5) in Figure 12.23 are cross-ring fragments of the following:

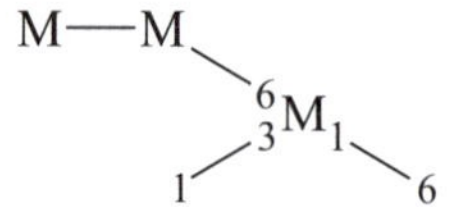

$$\mathrm{M{-}M} \searrow {}^{6}_{3}\mathrm{M}_{1} \searrow 6 \quad (1 \nearrow)$$

This particular fragment is not observed in the spectrum in Figure 12.22; however, it is instructive to calculate its mass in preparation for the cross-ring fragments:

$$(204.22 + 15) + 204.22 + [(204.22 - 15) - 1 + 1] + 23 = 635.66 \text{ Da}$$

Note that this nonreducing-end fragment cannot lose two protons and be electrically neutral. The C-2 of the trisubstituted mannose loses a hydrogen to the M1–6 glycosidic oxygen, and the C-3 glycosidic oxygen gains a proton from the nonreducing-end leaving group.

(g) Cross-ring fragment

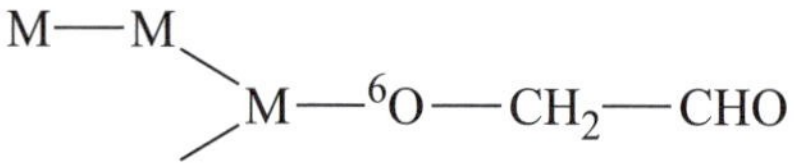

$$\mathrm{M{-}M} \searrow \mathrm{M{-}{}^{6}O{-}CH_2{-}CHO} \quad (\swarrow)$$

$$(204.22 + 15) + 204.22 + [(204.22 - 15) + 1] + 59 + 23 = 695.66 \text{ Da}$$

Note that a mass of 59 Da was added, rather than 60 Da as shown in Table 12.9. This value can be viewed in two ways and originates from m/z 635.66 [see part (f)]. The m/z 635.66 lost a hydrogen from the C-2 atom of the (1–3)(1–6)M1–6 residue in the structure in part (f) to the reducing-end ion fragment. However, this hydrogen is not lost when the ring fragment ($-OCH_2CHO$) is connected to the trimannose residue: 635.66 + 60 = 695.66

The mathematical equation, 636.66 + 59 = 695.66, more accurately reflects the origin of the cross-ring fragment structure.

PROBLEM 12.24 Calculate the mass of the cross-ring fragment in Figure 12.23 that has an unsaturation.

The tetramannose fragment at m/z 853.9 (853.7 in Figure 12.24) was presented in example 2(h).

PROBLEM 12.25 Show the calculations in arriving at the correct mass for the remaining six fragment ions in Figure 12.24.[68,69]

PROBLEM 12.26 Suppose the

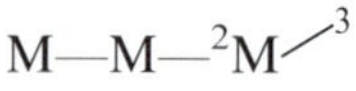

$$\mathrm{M{-}M{-}{}^{2}M} \nearrow 3$$

lower A-branch trimannose residue were a

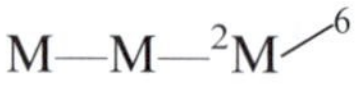

$$\mathrm{M{-}M{-}{}^{2}M} \nearrow 6$$

residue. Would additional peaks be observed in Figures 12.23 and 12.24, and at what m/z value would these be?

PROBLEM 12.27 Calculate the underivatized mass values for all eight fragments in Table 12.10.

Each pair of fragments in Table 12.10 is isomeric when underivatized but is nonisomeric when permethylated. Thus, the permethylated derivatives readily distinguish local domains of glycoforms from each other. The structures in Table 12.10 have implications for distinguishing between the three $Man_8GlcNAc$ glycoforms as shown in Figure 12.14. Neither structure b nor d in Table 12.10 could be present in 8-24-BC (Figure 12.14), because the glycoform does not have a complete A branch as found in the b and d fragments. Structure "a" suggests either 8-24-AC or 8-24-BC, while structure c suggests only 8-24-AC. The incomplete B branch in c has only one mannose residue with a complete nonreducing end, while 8-24-BC and 8-24-AB have two mannose residues on the B branch (complete B branch). A complete B branch does not allow for the central mannose, (1–2)M(1–3), to have a terminal methyl moiety. All three glycoforms show substructures e and f. However, structure g cannot be found in 8-24-AB, because the mannose in the C branch has a terminal methyl group while g does not. Structure h cannot be found in 8-24-BC, because this glycoform has a terminal methyl on the lower A branch, while this moiety is absent in the h fragment. Therefore, the appearance of masses that originate from fragments g and h in Table 12.10 are consistent with the $Man_8GlcNAc$ glycoform 8-24-AC. Without permethylation, the various substructures could not be sorted amongst each pair, and this would not lead to a definitive glycoform analysis. Permethylation even allows for linkage patterns to be delineated among each pair of closely related fragments (Table 12.10), given the established high-mannose motif (Scheme 15.4). The cross-ring fragments provide unambiguous linkage information.

The value of permethylation can also be observed in the following work[77] in the discrimination of glycoforms. The CD2 glycoprotein is located on the surface of the T-cell lymphocyte (Chapter 11), and it functions as a receptor for cell–cell adhesion by binding to the CD58 cell-surface molecule.[78] An example is the CD58 (or LFA-3) surface marker on antigen-presenting cells (APC) (Chapter 11). It has high-mannose glycoforms on an Asn residue in the polypeptide portion. The glycoprotein ribonuclease B also has high-mannose glycoforms. In both glycoproteins, the high-mannose fraction was isolated by endoglycosidase digestion and then directly permethylated. The sodiated $Man_7GlcNAc$ species (m/z 1743.8) from each preparation was subjected to tandem mass spectrometry. In the product-ion mass spectra, the $Man_7GlcNAc \cdot Na^+$ from ribonuclease B exhibited m/z 1103 and 1117, while that of CD2 showed only the m/z 1103 peak for the M_7G glycomer.[77] Figure 12.25 has an analysis of these mass values. For the 7-22-C glycomer:

$$\begin{aligned} M + Na^+ &= 1743.80 \\ & -219.22 \end{aligned}$$

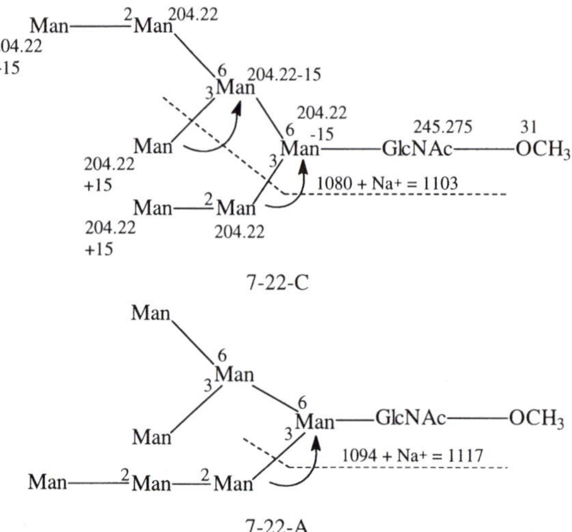

Figure 12.25 Permethylated and sodiated $Man_7GlcNAc$ glycoforms. Both are present in RNase B and only the 7-22-C is found in the CD2 protein.

$$\begin{array}{rr} & -219.22 \\ & \underline{-204.22} \\ & 1101.14 \\ & \underline{+2} \\ m/z & 1103.14 \end{array}$$

The addition of 2 Da originates as the backtransfer of a hydrogen from each of the leaving mannose groups to the main portion of the ion molecule. The two arrows in the schematic of 7-22-C (Figure 12.25) show the transfer route of each hydrogen species. Note that, in addition to 7-22-C, the 7-22-B and 7-22-BC glycomers (Figure 12.14) also can lose the one and two mannose-containing subunits to produce the m/z 1103.14 ion.

The calculation of m/z 1117 is as follows:

$$\begin{array}{rrl} M + Na^+ = & 1743.80 & \\ & -219.22 & \\ & -204.22 & \\ & \underline{-204.22} & \\ & 1116.22 & \\ & \underline{+1} & \text{H backtransfer} \\ m/z & 1117.22 & \end{array}$$

In addition to the 7-22-A glycomer, the 7-18-AB and 7-21-AC species (Figure 12.14) species can also lose a three-mannose residue antenna. Thus, permethylation and tandem mass spectrometry have shown the presence of a different M_7G glycoform on two different glycoproteins.

Another chemical concept exists that can provide even more information in the analysis of linkages in carbohydrate residues. This concept comprises three separate reactions and draws on basic organic chemistry reactions. The combination of the three methods has been labeled by Reinhold[51,66,68,69] as ORM, which means oxidation, reduction, and methylation. The oxidation reaction is effected by adding periodate to the carbohydrate polysaccharide. Adjacent alcohol groups (vicinal diols) react with periodate to cleave the bond between the two carbon atoms and oxidize the alcohol groups to aldehyde moieties (Table 12.11, reaction 1). Reaction 2 in Table 12.11 shows that one compound, a 4,6-linked aldaminitol, has no vicinal diols; therefore, no reaction takes place. A 4-linked aldaminitol has a vicinal diol at carbons 5 and 6. The bond between them is cleaved, and the C-6 group leaves the ring proper. An aldehyde on C-5 remains.

The next reaction in Table 12.11 is reduction with $NaBH_4$ or $NaBD_4$. This was presented in Table 12.4 (reactions 4–6), and this reagent is used in conjunction with cleaving glycopeptides at the amino acid–sugar residue junction. Reaction 3 in Table 12.11 shows the effect that borohydride and borodeuterohydride have on a periodate product. Note that, in the nomenclature, ORM means reduction with $NaBH_4$, and ODM signifies reduction with $NaBD_4$.

The final reaction is (per)methylation and was introduced in Table 12.4 (reaction 8), in the derivatization of sugar residues. Reaction 4 in Table 12.11 shows its effects on oxidized and reduced residues. Pay particular attention to the following facts: in ODM, the two deuterium atoms attached to the oxygen atoms are replaced with methyl groups, and two deuterium atoms remain attached to separate carbon atoms. Reaction 5 in Table 12.11 groups all three reactions together as an ODM sequence, and one deuterium atom remains. Deuterium can play a powerful role in the elucidation of complex polysaccharide structures, as presented below.

As the few examples in Table 12.11 show, different linkages yield different types of products in the periodate reaction. Table 12.12 provides a comprehensive coverage of the topic of ORM and ODM effects on a myriad of sugar residues that highlight details such as linkage(s), hexose, acetylhexosamine, alditol, and open- and closed-ring aldose structures.

The first part of Table 12.12 shows that with a di- or trilinkage, an alditol or aldose cannot be formed, and the ORM/ODM reactions act on internal residues. Note that the arrows in Table 12.12 point to the carbon–carbon bonds that are cleaved by periodate. A 1-linkage cannot produce an aldose or alditol; thus, only vicinal diols react with periodate. A 2-linkage has the potential for either an open- or closed-ring aldose. Note that, in Table 12.12, each form produces a significantly different product. For a 3-linked compound, examples are shown for an alditol and closed-ring aldose. The alditol shows the vicinal diol cleavages between C-4 and C-5, and C-5 and C-6. For the closed ring, 3-linked aldose, the bond between C-1 and C-2 is cleaved. This cleavage leaves an ester functional group after the periodate reaction, where the ester $R^1C(O)O^5CH-$ group contains the carbon atoms as labeled (not shown in Table 12.12). If the acidic

Table 12.11 The ORM/ODM reactions with carbohydrate residues

Method	Mechanism	Reference(s)
1. Periodate oxidation IO_4^-	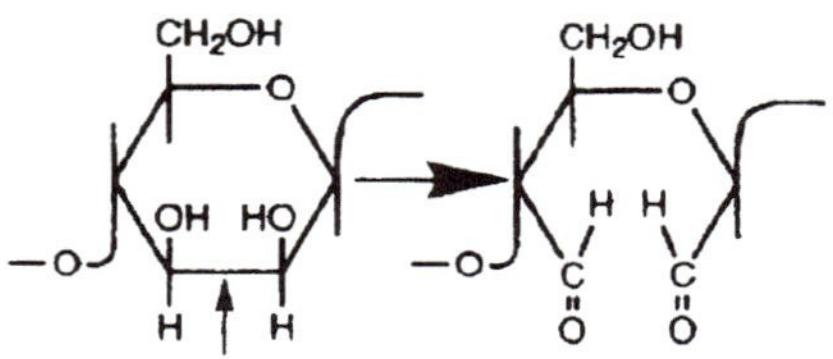	27
2. Periodate—cleaves vicinal alcohol groups; $NaIO_4$	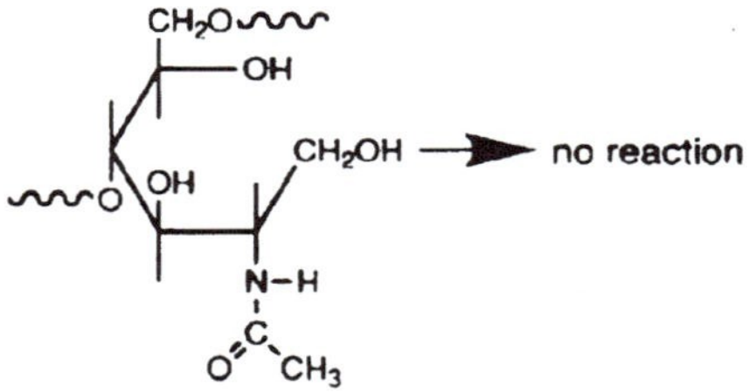 no reaction	28
	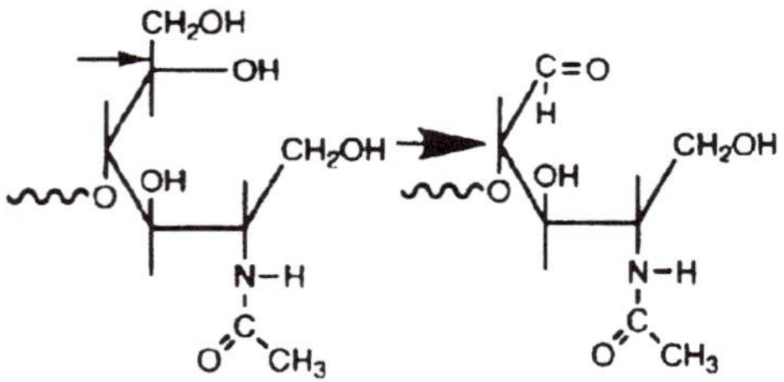	28

3. $NaBH_4$ reduction $NaBD_4$	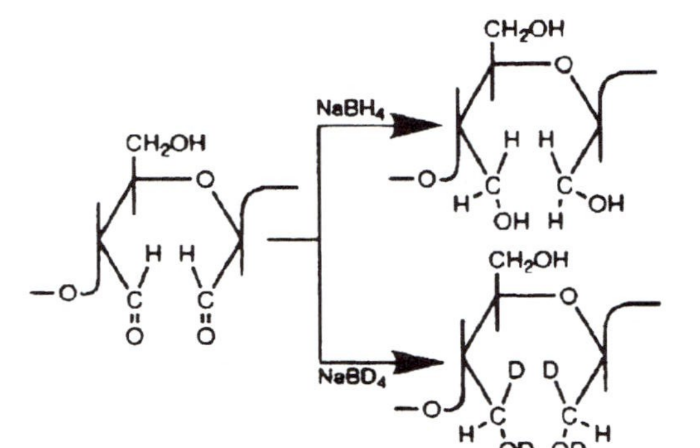 	27
4. Permethylation CH_3I	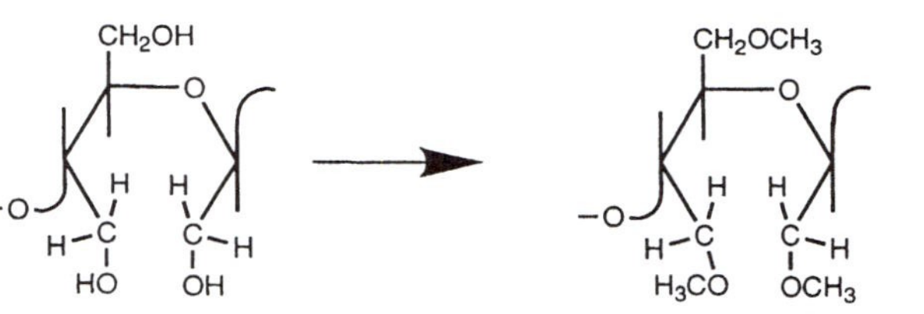 	13, 27

(continued)

Table 12.11 (*continued*)

Method	Mechanism	Reference(s)
or		
H_3C-S-$CH_2^-Na^+$ Methylsulfinyl carbanion		30
5. IO_4^- oxidation NaOH/ $NaBD_4$ reduction Permethylation, CH_3I; ODM	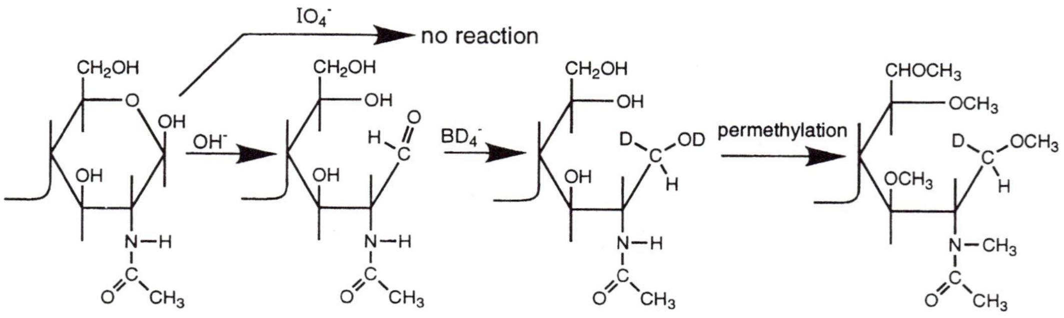	13, 27

condition of periodate cleaved the ester into an acid (formic acid at C-1) and alcohol (at C-5), then most of the remaining sugar residue would be destroyed. Only C-2 and C-4 would be attached to C-3. However, the acidic conditions for the periodate reaction are not conducive for this reaction.[27] Instead, in the subsequent alkaline borohydride reduction reaction, the ester is cleaved to form an alcohol on C-5 and a formic acid leaving group. All four alcohol groups are subsequently methylated. For the 4-linked aldaminitol, only the bond between C-5 and C-6 is cleaved, while a mechanism for the 4-linked closed-ring aldose occurs which is similar to that of the 3-linked closed-ring aldose.

The 6-linked alditol and aldaminitol are essentially destroyed by periodate and yield identical products (Table 12.12). The 6-linked closed-ring residue follows a reaction decomposition pathway similar to the 3- and 4-linked closed-ring residues, where vicinal diols are cleaved, and the borohydride cleaves the ester group.

PROBLEM 12.28 For a mannose residue, draw and calculate the masses of the products of ORM and ODM reactions, in a similar manner to Table 12.12, for the following combination of glycosidic linkages—1,4; 1,2,3; 2-linked alditol—and for an acetylglucosaminitol residue—1,4; 1,3,4; 4-linked, open-ring aldose.

Table 12.13 provides a summary of the hexose and *N*-acetylhexosamine average mass values for different linkage positions in the underivatized and ORM/ODM-treated sugar residues. Open- and closed-ring polysaccharide forms, as well as internal and terminal positions, are also considered in Table 12.13. It is obvious that the derivatization reactions provide a significant measure of mass delineation, including the salient (slight change in mass) yet measurable effect of deuteration.

Some of the di- and trilinked internal hexose linkages can be differentiated. However, note that the discrimination here emphasizes the linkage aspect. Table 12.13 does not discriminate within a sugar class—for example, mannose, glucose, galactose, and so on. Each sugar provides the same mass for each respective entry in Table 12.13. All terminal hexose linkages can be differentiated amongst themselves. A combination of the three masses in Table 12.13 can differentiate between a terminal closed-ring aldose, open-ring aldose, and alditol hexoses, except for the 2- and 4-linked open-ring aldoses and the 3- and 4-linked alditols. Indeed, ORM/ODM is a powerful technique.

For the *N*-acetylhexosamine residue linkages, the discrimination is not quite as good as that of the equivalent hexose species. There are a number of isomeric species in the respective underivatized and ORM/ODM-modified closed-ring situations. However, the open-ring linkages can be completely separated by mass values. This is opposite with respect to the hexose species. Hexose species exhibit no isomers and two pairs of isomers in the closed- and open-ring form, respectively. The two pairs of open-ring hexoses are the 2- and 4-terminal aldoses and the 3- and 4-terminal alditols. Each pair of sugar residues has identical respective masses in all three columns in Table 12.13. The hexosamines have isomers and no isomers in the closed- and open-ring form, respectively.

Table 12.12 The ORM and ODM vicinal alcohol cleavage reactions on carbohydrate residues

Linkage	Residue[a]	M_r[b]	ORM product	Number of additional methyls	M_r (ORM)[b]	ODM product	M_r (ODM)[b]
1,6	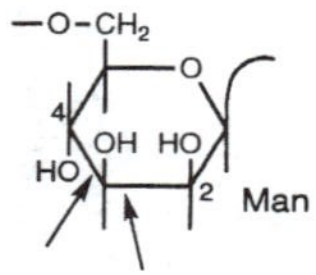	162.142	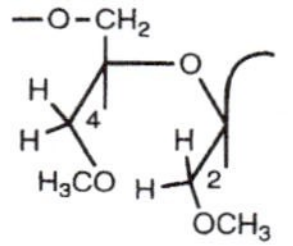	2	162.186	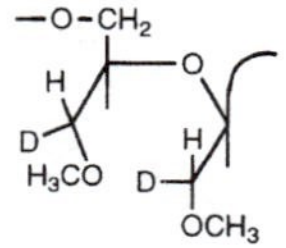	164.198
1,2	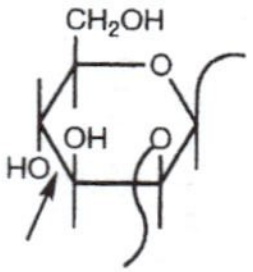	162.142	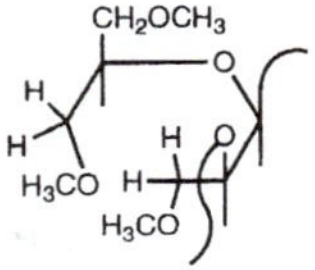	3	206.239	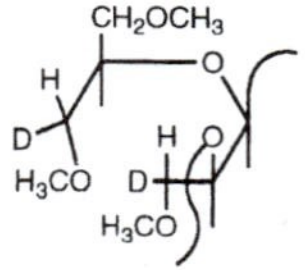	208.251
1,3	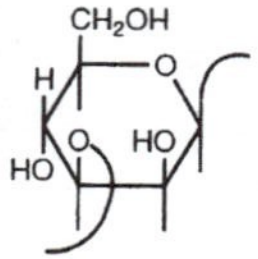	162.142	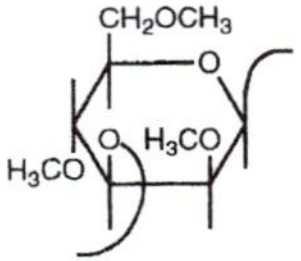	3	204.223	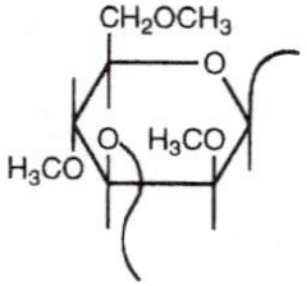	204.223
1,2,6	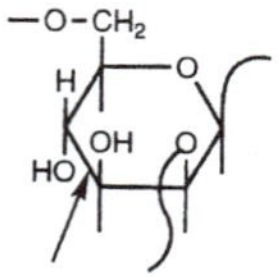	161.134	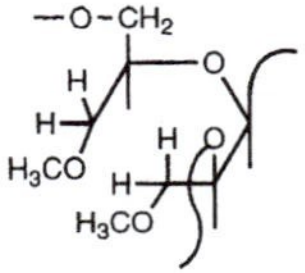	2	191.204	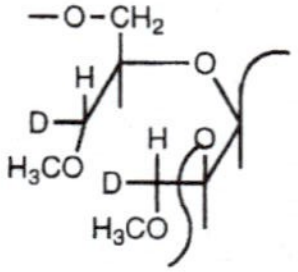	193.22

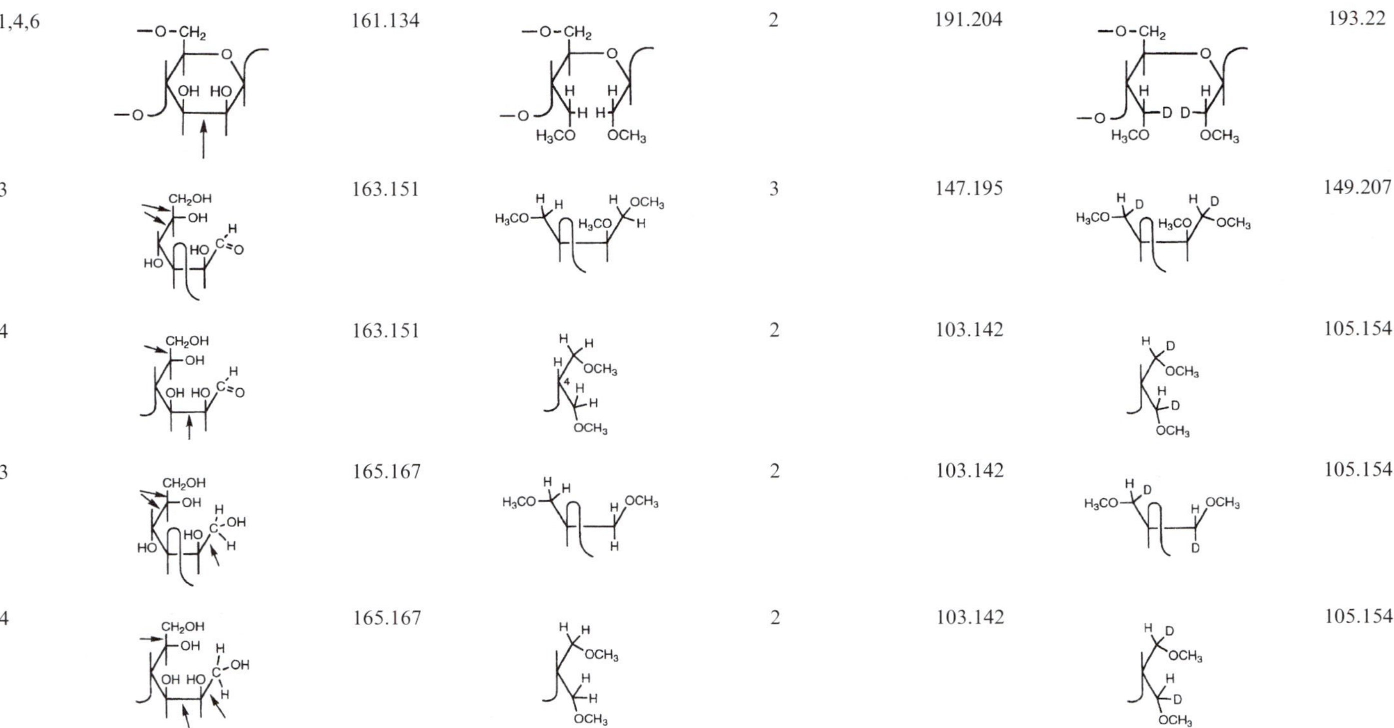

1,4,6		161.134		2	191.204		193.22
3		163.151		3	147.195		149.207
4		163.151		2	103.142		105.154
3		165.167		2	103.142		105.154
4		165.167		2	103.142		105.154

(continued)

Table 12.12 (continued)

Linkage	Residue[a]	M_r[b]	ORM product	Number of additional methyls	M_r (ORM)[b]	ODM product	M_r (ODM)[b]
6		165.167		1	59.088		60.10
1,3		203.195		3	245.276		245.276
6		163.151		1	59.088		60.10
1,3,6		161.134		2	189.189		189.189
1,3,4 1,2,4		161.134	2,6[c] 3,6	2	189.189		189.189

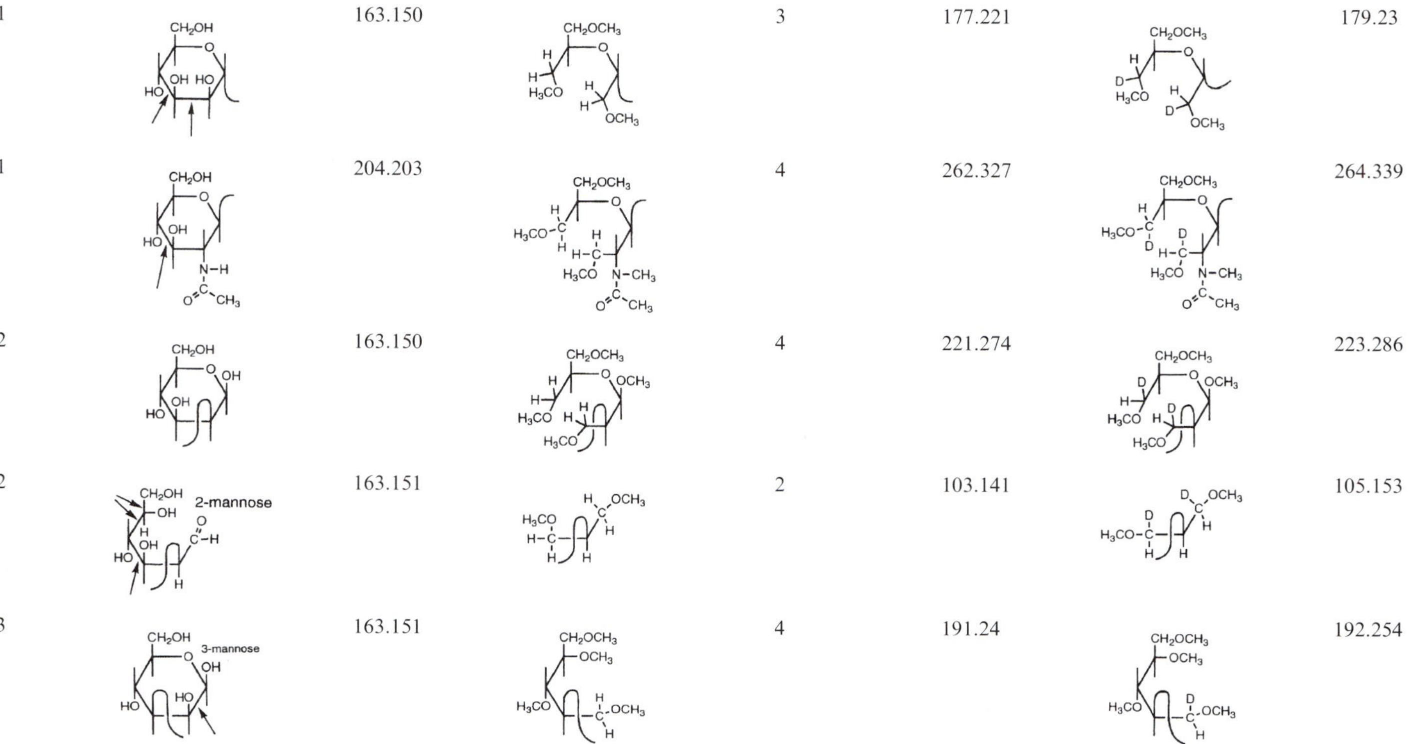

1		163.150		3	177.221		179.23
1		204.203		4	262.327		264.339
2		163.150		4	221.274		223.286
2	2-mannose	163.151		2	103.141		105.153
3	3-mannose	163.151		4	191.24		192.254

(*continued*)

Table 12.12 (*continued*)

Linkage	Residue[a]	M_r[b]	ORM product	Number of additional methyls	M_r (ORM)[b]	ODM product	M_r (ODM)[b]
3	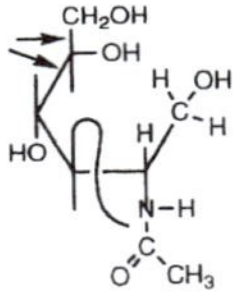	206.219	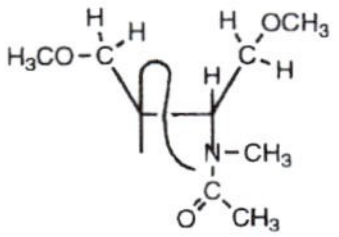	3	188.247	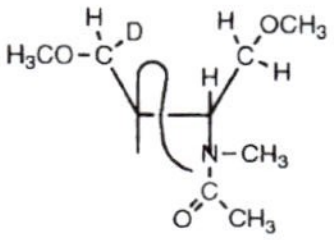	189.253
4	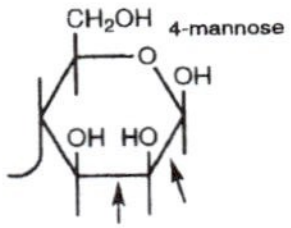 4-mannose	163.151	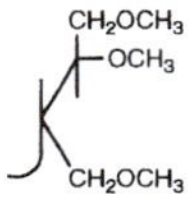	3	147.195	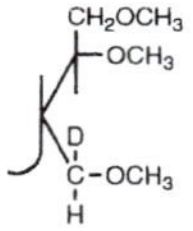	148.201
4	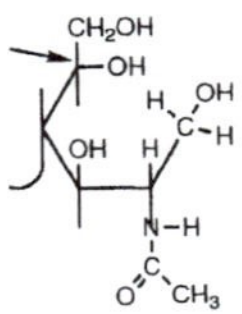	206.219	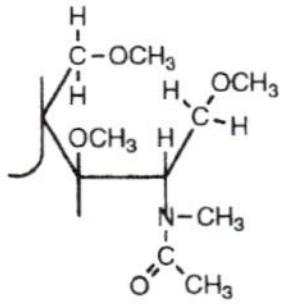	4	232.301	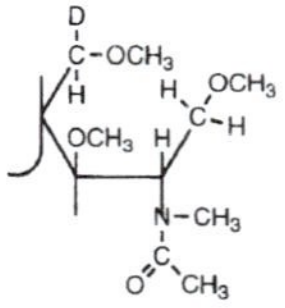	233.31
4	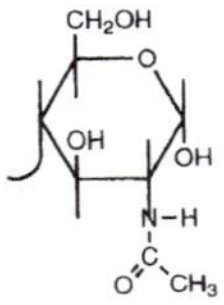	204.203	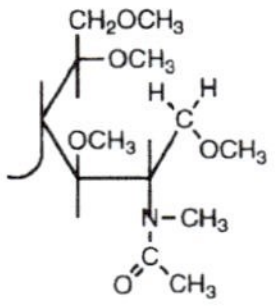	5	276.354	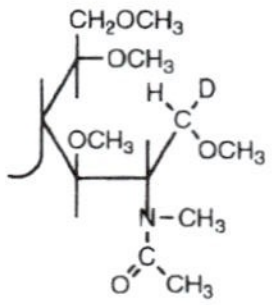	277.36

6	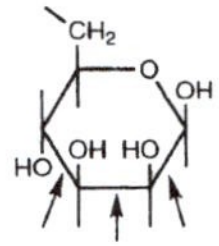	163.151	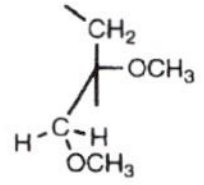	2	103.141	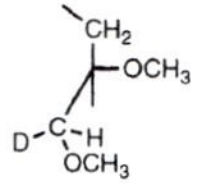	104.147
6	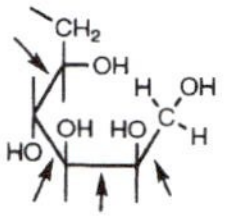	165.167	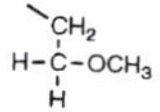	1	59.088	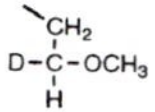	60.1
6	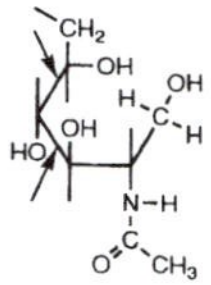	206.219	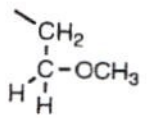	1	59.088	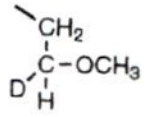	60.1
1,6	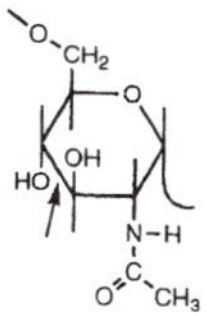	203.195	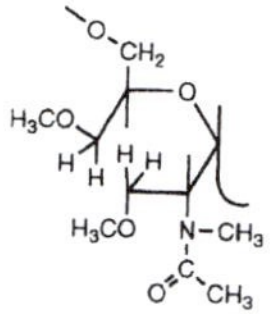	3	247.292	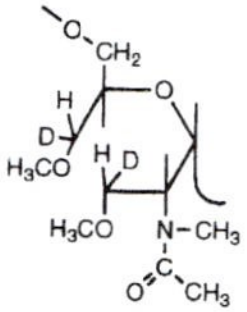	249.304
1,3,6	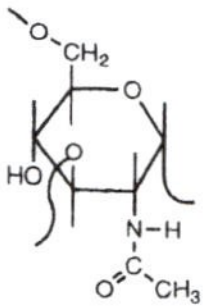	202.195	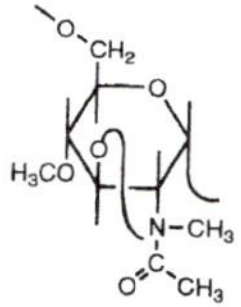	2	230.249	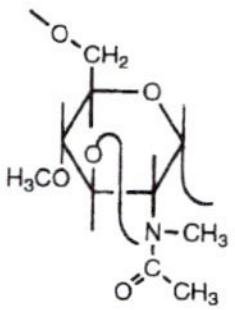	230.249

(continued)

Table 12.12 *(continued)*

Linkage	Residue[a]	M_r[b]	ORM product	Number of additional methyls	M_r (ORM)[b]	ODM product	M_r (ODM)[b]
1,4,6	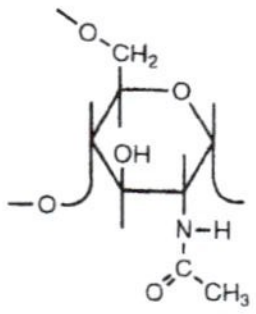	202.195	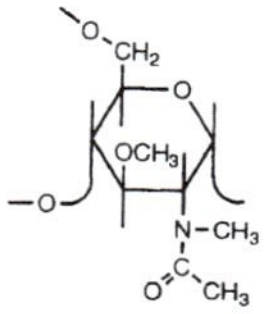	2	230.249	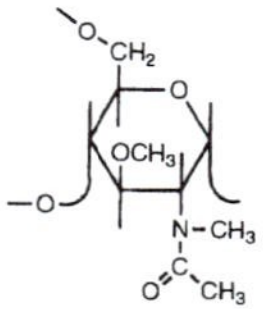	230.249
3	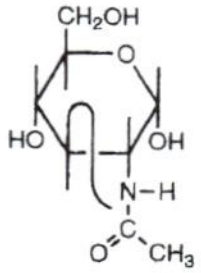	204.203	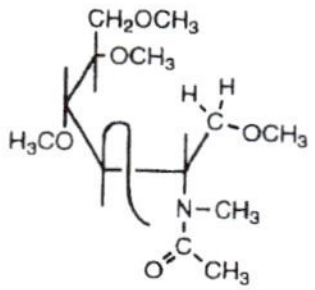	5	276.354	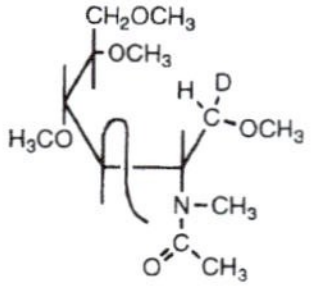	277.360
6	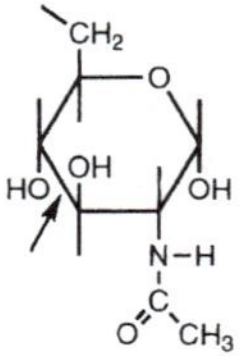	204.219	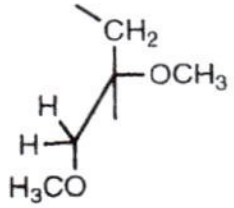	2	103.141	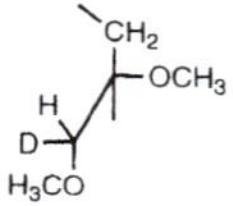	104.148

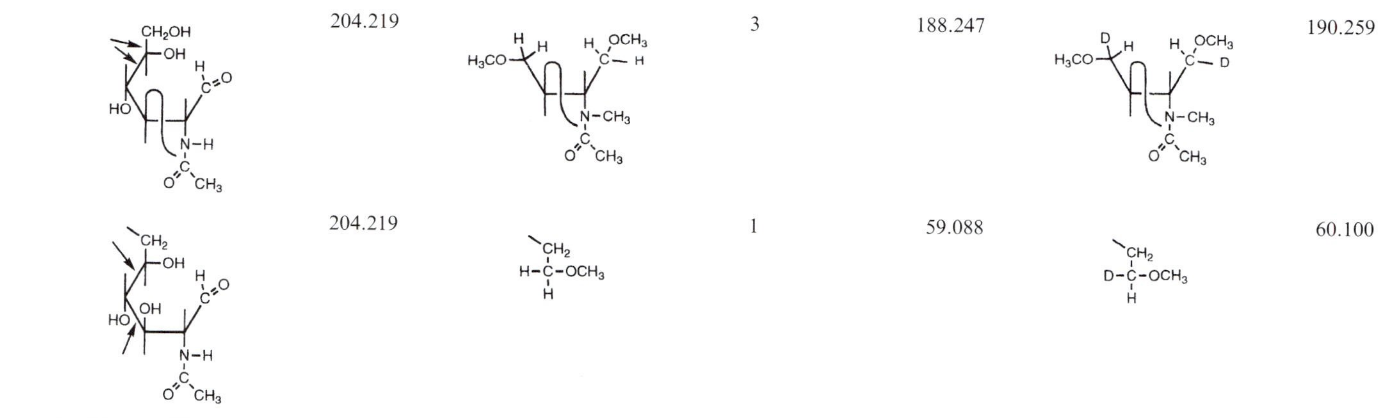

3		204.219		3	188.247		190.259
6		204.219		1	59.088		60.100

[a] Arrow indicates bond cleavage.

[b] M_r, mass in daltons

[c] Indicates the ring carbons that bear a methoxy functional group.

Table 12.13 The ORM/ODM saccharide mass values[a]

Precursor saccharide	Ring position and status	Linkage	Underivatized	ORM (Da)	ODM (Da)
Hexose	Internal	1,2–1,4	162.14	206.24	208.25
		1,3	162.14	204.22	204.22
		1,6	162.14	162.19	164.20
		1,2,3–1,2,4–1,3,4–1,3,6	161.13	189.19	189.19
		1,2,6–1,4,6	161.13	191.20	193.22
	Terminal closed-ring aldose	1	163.15	177.22	179.23
		2	163.15	221.27	223.29
		3	163.15	191.24	192.25
		4	163.15	147.20	148.20
		6	163.15	103.14	104.15
	Terminal open-ring aldose	2	163.15	103.14	105.15
		3	163.15	147.20	149.21
		4	163.15	103.14	105.15
		6	163.15	59.09	60.10
	Terminal alditol	2	165.17	103.14	104.15
		3	165.17	103.14	105.15
		4	165.17	103.14	105.15
		6	165.17	59.09	60.10
N-acetyl-hexosamine	Internal	1,3	203.20	245.28	245.28
		1,4	203.20	245.28	245.28
		1,6	203.20	247.29	249.30
		4,6	203.20	261.32	262.33
		1,3,4	202.20	230.25	230.25
		1,3,6	202.20	230.25	230.25
		1,4,6	202.20	230.25	230.25
	Terminal closed-ring aldose	1	204.20	262.33	264.34
		3	204.20	276.34	277.35
		4	204.20	276.34	277.35
		6	204.20	103.14	104.15
	Terminal open-ring aldose	3	204.22	188.25	190.26
		4	204.22	232.30	234.31
		6	204.22	59.09	60.10
	Terminal aldaminitol	3	206.22	188.25	189.25
		4	206.22	232.30	233.37
		6	206.22	59.09	60.10
Fucose	Terminal	1	147.15	147.20	149.21
NeuNAc	Terminal	2	292.27	288.32	289.33
	Internal	2,8	291.26	361.39	361.39

[a] Select, identical masses in each grouping are underlined.

The next series of investigations shows the potential for the ORM/ODM reactions in the elucidation of structural details in high-mannose polysaccharides. Figure 12.26 presents the ESI-mass spectrum of the high-mannose polysaccharide posttranslational modification portion of a glycoprotein from a plant.[68,69] The polysaccharide was obtained by Endo-H digestion; therefore, a GlcNAc is attached to the trilinked mannose residue. Figure 12.26 shows the methylated, polysaccharide mixture, and various high-mannose species are observed (Table 12.4, reaction 8). All four mass spectral peaks are in the 2+ charge state, with sodium as the ionizing species. The methylation reagent was CD_3I; thus, all hydroxyl and amine methylation sites increase in mass by 3 Da.

In Figure 12.26, m/z 709.0^{2+} is the disodiated cation and represents a $Man_5GlcNAc$ high-mannose species. A question can be raised as to which M5 glycoform(s) is/are represented by the m/z 709.0^{2+} ion. The permethylated structure of the 5-17-C $Man_5GlcNAc$ glycoform is shown in Figure 12.27a, and the shorthand notation used to detail linkages between residues is given in Figure 12.27b. An explanation of each residue is also given in Figure

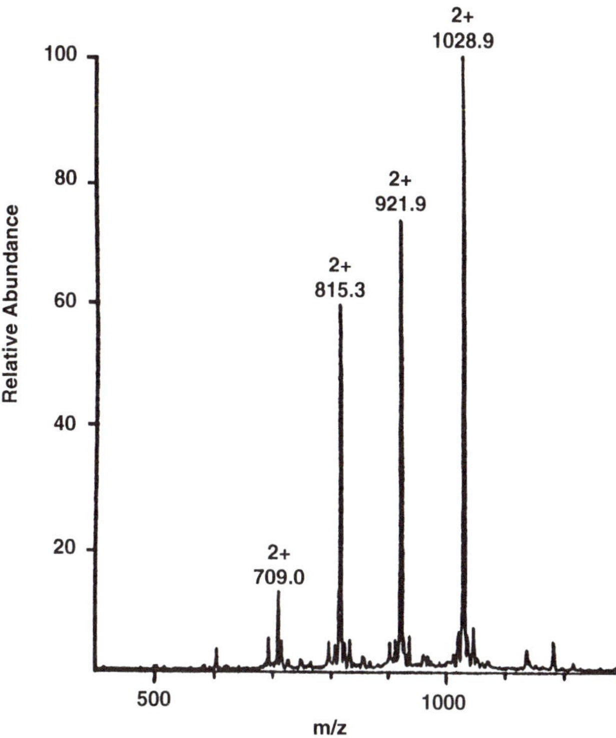

Figure 12.26 The ESI-mass spectrum of the perdeuteriomethylated high-mannose glycan from a plant glycoprotein. The glycan was released from the glycoprotein by Endo-H digestion. (© 1994, John Wiley & Sons, Limited. Reproduced from reference 69 with permission.)

12.27b. Figure 12.27c shows the perdeuteriomethylated structure of 5-17-C; note that the arrow indicates the unmodified methyl moiety on the acetyl group. Complete derivatized structures of the M5 glycoforms in Figure 12.14, such as those in Figure 12.27, will show that they all produce identical masses. Glycoform 5-17-C (Figure 12.14) will be used as an example to represent m/z 709.0^{2+} in Figure 12.26 where (m/z 709.0 × 2) − 46 = 1372 Da. To arrive at this

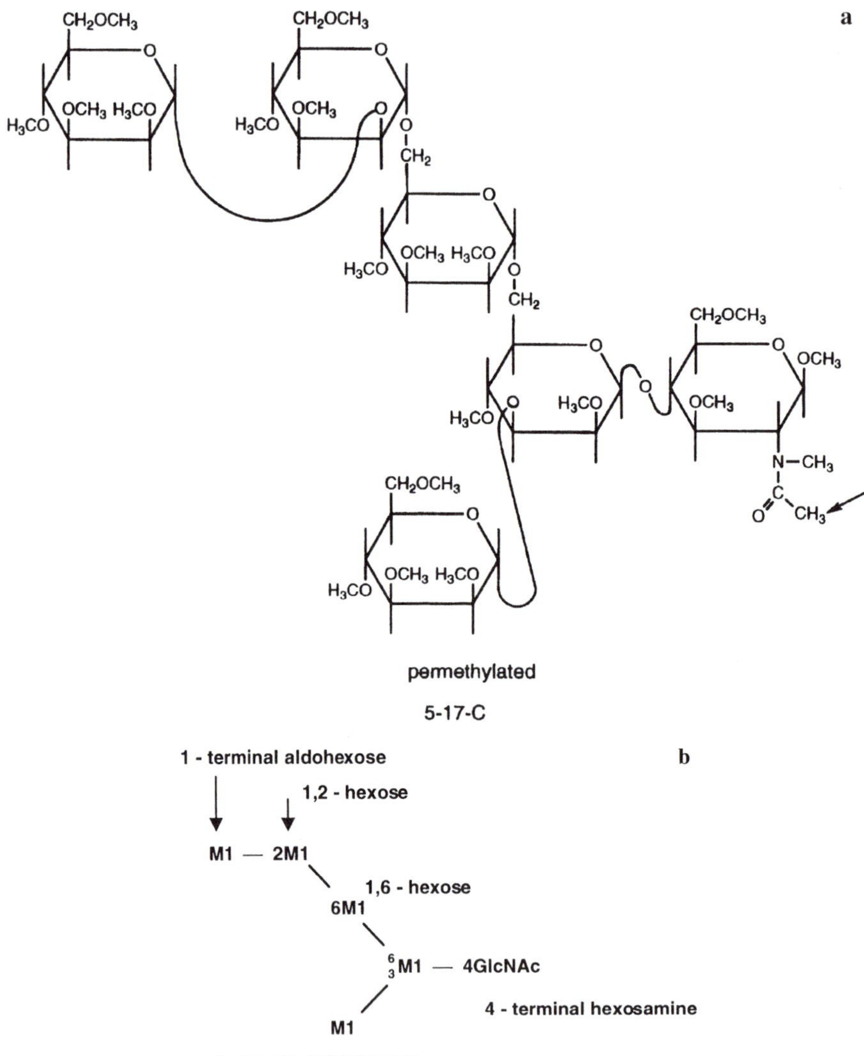

Figure 12.27 Structures of the $Man_5GlcNAc$ 5-17-C high-mannose glycomer. (a) Permethylated structure, where the arrow points to the inherent methyl group (b) shorthand notation of (a), and (c) perdeuteriomethylated structure of (a), where the arrow points to the inherent, unmodified methyl group.

c

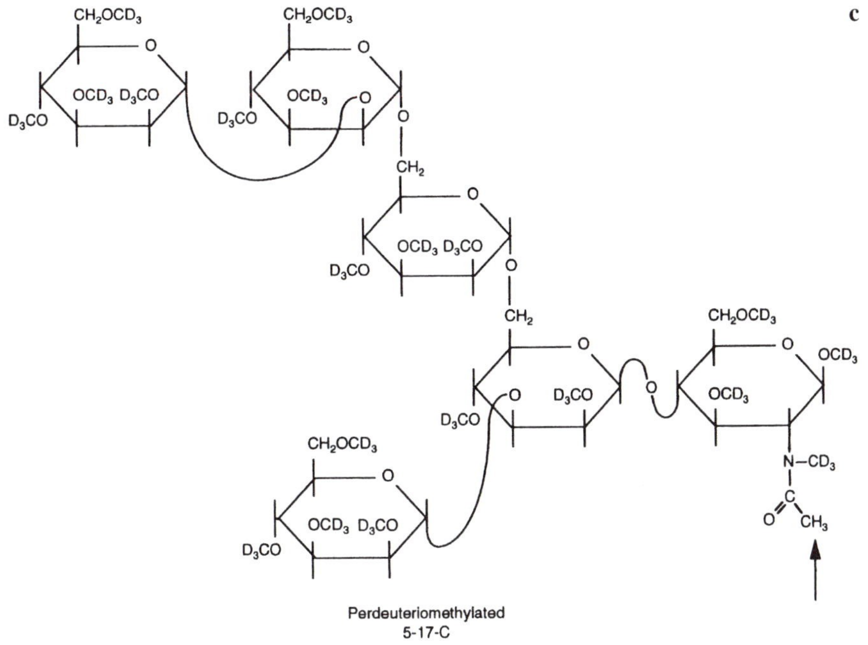

Figure 12.27 (*continued*)

value, Table 12.8a shows that a permethylated $Man_5GlcNAc$ moiety is 1312.46 Da. The equivalent of replacing three methyl hydrogen atoms with three deuterium atoms is 3(2.0141) – 3(1.00797) = 3.018 Da. Figure 12.28c shows that there are 20 CD_3 moieties on the $Man_5GlcNAc$ glycomer and 20 × 3.018 = 60.36. Thus, 1312.46 + 60.36 = 1372.82 Da ≃ 1372 Da; [1372 + 46 (2 Na)]/2 = 709 Da from Figure 12.26. These same numbers are obtained with other $Man_5GlcNAc$ glycoforms.

PROBLEM 12.29 The four peaks in Figure 12.26 represent what series of polysaccharides? Prove this by a comparison of the four mass values; then, calculate and rationalize the three equivalent neutral mass values above m/z 709.0 in Figure 12.26. Is there a preference for which glycoforms are used (Figure 12.14) for each mass value?

The mass values in Figure 12.26 could not be used to differentiate between the $Man_{5-8}GlcNAc$ glycoforms. Figure 12.28, however, provides some discrimination between the glycoforms,[68,69] because the same mixture of polysaccharides used in Figure 12.26 was subjected to ORM (not ODM). The m/z 645.9^{2+} peak in Figure 12.28 yields the following interpretation: (645.9 × 2) – 46 = 1245.8 Da. Thus, each sugar residue in all six M5 glycoforms (Figure 12.14) must be analyzed for ORM mass values. Figure 12.29a provides a detailed example of the 5–18 glycoform. Mass values are appended to each residue (Table 12.13).

Note, in particular, the GlcNAc value. The mass values in Table 12.13, under *N*-acetylhexosamine (3- and 4-linked terminal closed ring), do not include the glycosidic oxygen. This must be added to 276.35 Da and is shown in Table 12.12. The O–C_1 cleavage in GlcNAc is effected by NaOH/$NaBH_4$ (Table 12.4, reaction 4). The *N*-acetylamine moiety prevents IO_4 cleavage of the C_1–C_2 bond; thus, the basic condition of the borohydride merely hydrolyzes the hemiacetal reducing end to an aldehyde on C_1 and an alcohol on C_5. Borohydride then reduces the C_1 aldehyde to an alcohol functionality.

The resulting m/z value of 1202.26 Da (calc) $\neq$ 1245.8 Da (obs) in Figure 12.29a. Figure 12.29b presents shorthand analyses, similar to that in Figure 12.29a, of the five other glycoforms. Only 5–14-B and 5–14 match the observed mass value of 1245.8 Da. Thus, either one or both glycoforms can be generated from the plant cellular machinery for the $Man_5GlcNAc_2$ high-mannose polysaccharide motif in the particular glycoprotein under investigation.

PROBLEM 12.30 In Figure 12.28, the ORM-processed $Man_6GlcNAc$ portion of the high-mannose mixture split into two peaks at m/z 727.0^{2+} and 748.9^{2+}.

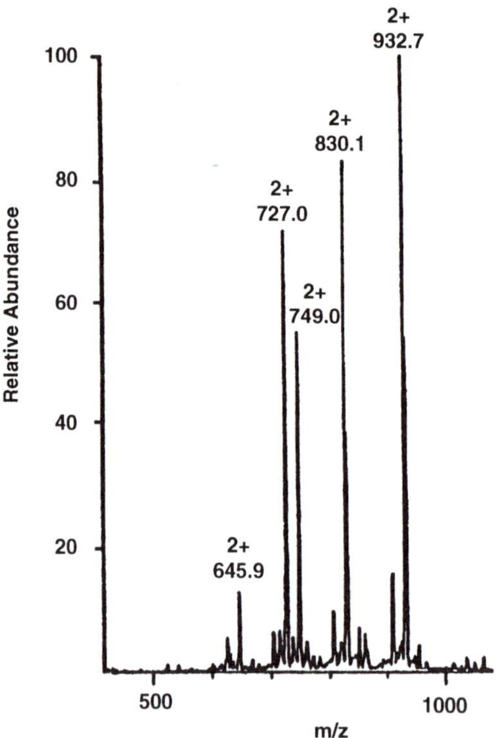

Figure 12.28 The ESI-mass spectrum of the ORM-treated plant glycoprotein glycan mixture from Figure 12.26. (© 1994, John Wiley & Sons, Limited. Reproduced from reference 69 with permission.)

Using Figures 12.14 and 12.29a and b as guides, which M6 glycoform(s) can be attributed to each peak?

PROBLEM 12.31 Repeat Problem 12.30 for m/z 830.1 in Figure 12.28.

PROBLEM 12.32 For the mass at m/z 932.7^{2+} in Figure 12.28, can the $Man_8GlcNAc$ be discriminated by ORM?

Are there any more mass spectrometry/chemical reaction techniques that can be used to differentiate glycoforms? One possibility is ODM, but tandem mass spectrometry of ORM reaction products could yield further clues. Figure 12.30 presents a product-ion mass spectrum of m/z 933.0 = $(M + 2Na)^{2+}$ which was observed as m/z 932.7^{2+} in Figure 12.28, and an analysis of the spectrum follows.[68,69] Given the mass values in Figure 12.30 (inset), reducing-end fragment ions are 1 Da higher in mass, and the nonreducing-end fragment ions are 1 Da less in mass, following typical Y fragmentation (Scheme 15.3). Figure 12.31 provides a schematic of the three glycoforms and the rationale for the observed ions in Figure 12.30:

$$(m/z\ 933.0 \times 2) - 23 = 1843.0 \simeq 1842.4\text{Da (obs)}$$

For the 8-24-AC glycoform we have the following:

$$\begin{aligned}
&m/z\ 1666.68 = 1842.9 - 176.22 \\
&m/z\ 1460.44\ (\simeq 1461.0) = 1842.9 - (177.22 + 20.24) \\
&m/z\ 1254.2 = 1842.9 - (177.22 + 206.24 + 205.24) \\
&m.z\ 1078.0 = 1842.9 - (177.22 + 206.24 + 205.24 + 176.22) \\
&(m/z\ 638.6 \times 2) - 23 = 1254.2\ \text{Da} \\
&(m/z\ 741.5 \times 2) - 23 = 1460.0\ \text{Da} \\
&(m/z\ 844.9 \times 2) - 23 = 1666.8\ \text{Da}
\end{aligned}$$

Glycoform 8-24-BC can form m/z 1666.68 and 1460.44, while 8-24-AB can form m/z 1666.68, 1460.44, and 1254.20. However, only glycoform 8-24-AC can form the unique 1078.0-Da ion. Two or more neutral fragments will not cleave from a glycoform where all neutral fragments have more than one sugar residue. Therefore, the $Man_8GlcNAc$ portion of the high-mannose fraction of the plant glycoprotein consists of only the 8-24-AC glycoform.

Complex Glycotype

The high-mannose and hybrid N-linked glycotypes have been presented, and the complex glycotype is the third important type of glycan found in glycoproteins. Complex types can be found as three basic structures: biantennary (Schemes 12.7 and 12.8), triantennary (Scheme 12.9), and tetraantennary (Scheme 12.10). This glycotype is fairly predictable in basic structure but can have a number of modifications, such as linkage variation, additions (e.g. fucose (Schemes 12.8–12.10), more than one lactosylamine group (Scheme 12.8), variable numbers of sialic acid residues, bisecting biantennary structures, and carbohydrate residue deletions.

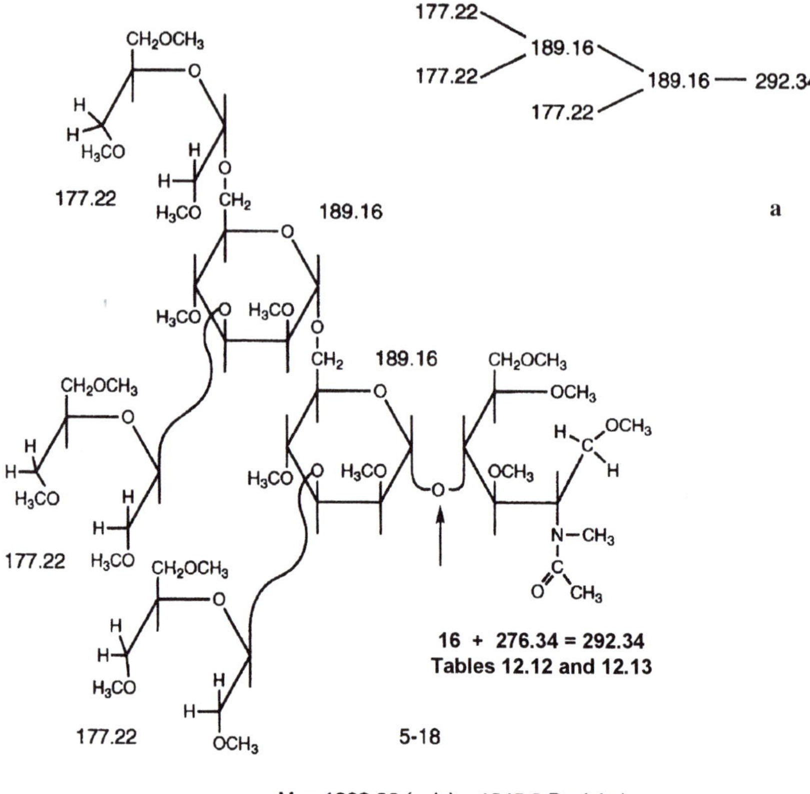

Figure 12.29 Structural details of the high-mannose glycomers after ORM chemistry. (a) Details of the 5–18 glycomer with the inset showing a shorthand notation; (b) shorthand notation of the ORM chemistry on the remaining $Man_5GlcNAc$ high-mannose glycomers.

Some distinctions in Schemes 12.7–12.10 will be highlighted. In Scheme 12.7, note that the linkage between the Neu5NAc and Gal residues can be different (arrows in Scheme 12.7). The GlcNAc residues can display different linkages from the C-1 position to different sites on sugar residues. In Scheme 12.8, the GlcNAc residue in the additional lactosylamine shows a GlcNAcβ1–3Gal linkage, while it is a GlcNAcβ1–2Man linkage for mannose (Scheme 12.7). Fucose is often found in the complex glycotype, and most of the time the residue can be found attached to the reducing-end terminal GlcNAc as a Fucα1–6GlcNAc linkage (Schemes 12.8–12.10). Sometimes fucose is found as a side attachment on one of the antennae residues. Scheme 12.9 shows a fucosylated triantennary glycoform of the complex glycotype with sialic acid residues. The presence of nonreducing-end terminal sialic acid residues is known as “capping” for an antennary structure. There is a change in GlcNAc–Man linkage when comparing the upper and lower antenna of each glycoform in Scheme 12.9, and

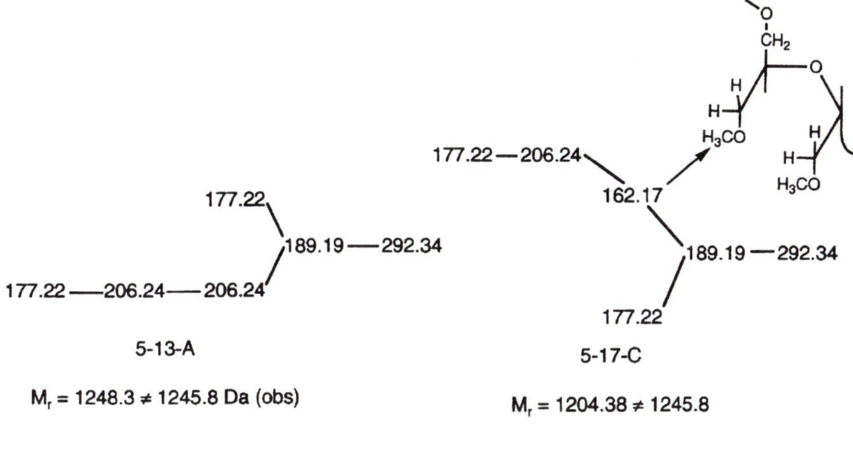

b

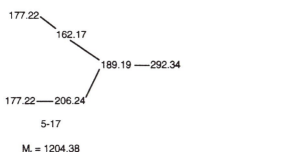

177.22
162.17
189.19 —292.34
177.22—206.24

5-17

$M_r = 1204.38$

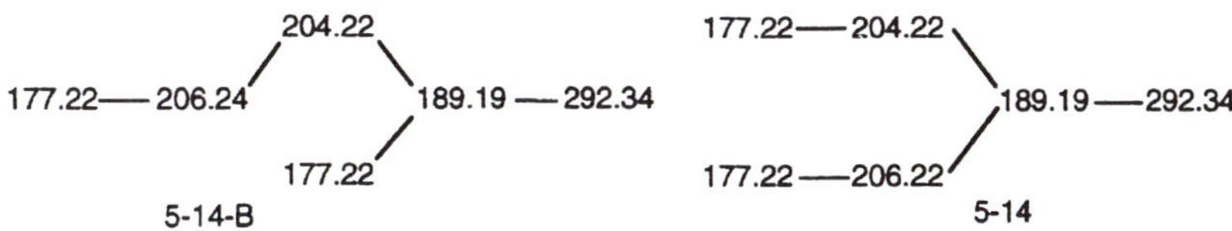

$M_r = 1246.43(\text{calc}) \simeq 1245.8$ Da (obs)

Figure 12.29 *(continued)*

this is marked with arrows. A thought that may come to mind is whether these two structures can be differentiated. This will be addressed below, and the reader can probably arrive at a reasonable method as to how to tackle this problem. Even though this passage is jumping ahead, the reader should be able to calculate, in a very short period of time, exactly what the mass difference would be between the structures in Scheme 12.9 when a certain chemical processing procedure is performed. Scheme 12.10 provides a basic structure for a tetraantennary glycoform; note the same difference in GlcNAc–Man linkage between the antennae as compared to that in Scheme 12.9.

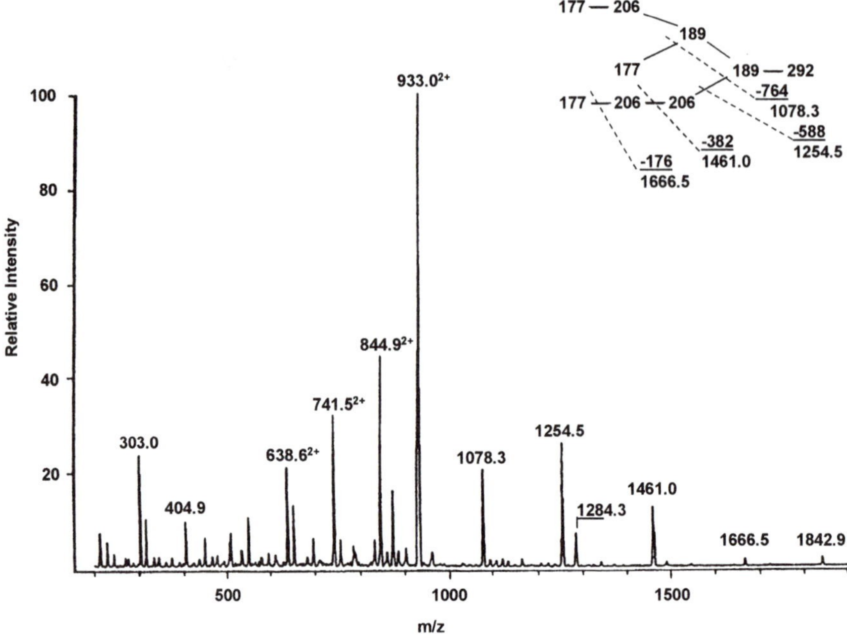

Figure 12.30 The ESI-tandem mass spectrum of $(M + 2Na)^{2+}$ = 933 Da in Figure 12.28. This represents a $Man_8GlcNAc$ glycoform. See text for details. (© 1994, John Wiley & Sons, Limited. Reproduced from reference 69 with permission.)

Table 12.14 provides a partial listing of the many permutations (glycoforms) of the complex glycotype and their mass values. Underivatized permethylated mass values are also listed.

A number of investigations will be presented here that highlight the separation/isolation of the complex glycotype species, and their mass spectral analyses, which usually show a wide heterogeneity in the cellular production of complex glycoforms.

Mass Spectral Analysis of Glycopeptides of Recombinant Tissue Plasminogen Activator

Cells have the capability of providing a wide permutation of progressively different-in-mass glycans on the same amino acid residue in the synthesis of protein. This yields a complex set of different mass species (Table 12.14). Thus, for a constant peptide mass, many different masses can be observed because of the many different attached glycan species. This necessarily reduces the relative intensity of any particular glycoform. Figure 12.32 provides two reversed-phase, high-pressure liquid chromatography (RP-HPLC) mass spectral total-ion chromatograms from the tryptic digestion of recombinant tissue plasminogen activator (r-tPA) glycoprotein.[79] Figure 12.32a shows many peptides, and

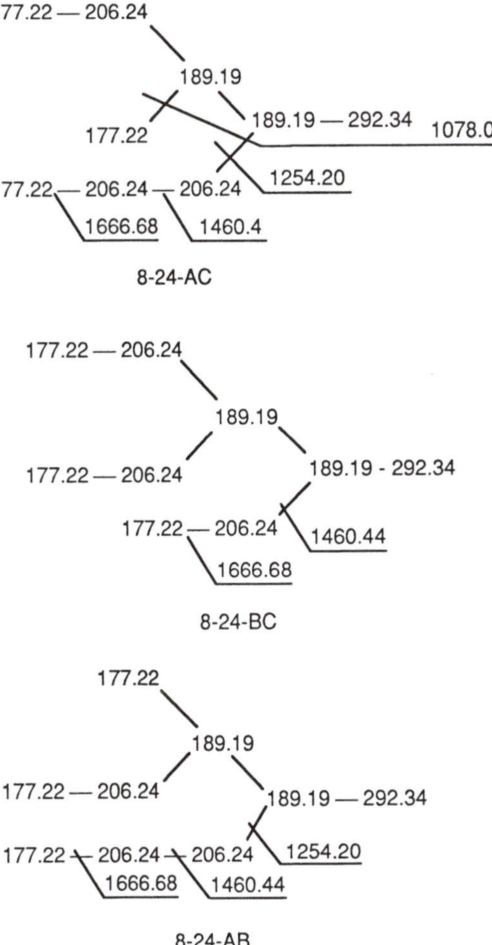

Figure 12.31 Schematic for all three $Man_8GlcNAc$ glycoform and potential (calculated) product-ion values.

three glycopeptides are highlighted. Figure 12.32b is an RP-HPLC of neuraminidase-treated r-tPA. Neuraminidase enzyme cleaves all terminal Neu5NAc residues, and the concept is to reduce the overall amount of heterogeneity in the glycoform distribution.

The notion in LC-MS is that one peak usually corresponds to one compound so that each labeled peak in Figure 12.32 should identify one species. Figure 12.33 shows otherwise.[79] Figures 12.33a and b present the 25–50-min time region from Figures 12.33a and b, respectively. Note that in the three labeled peptide fragments, many ions are observed in a relatively narrow elution window, and these windows are circled in Figure 12.33. In ESI, since gentle ionization processes characterize the method and very little fragmentation

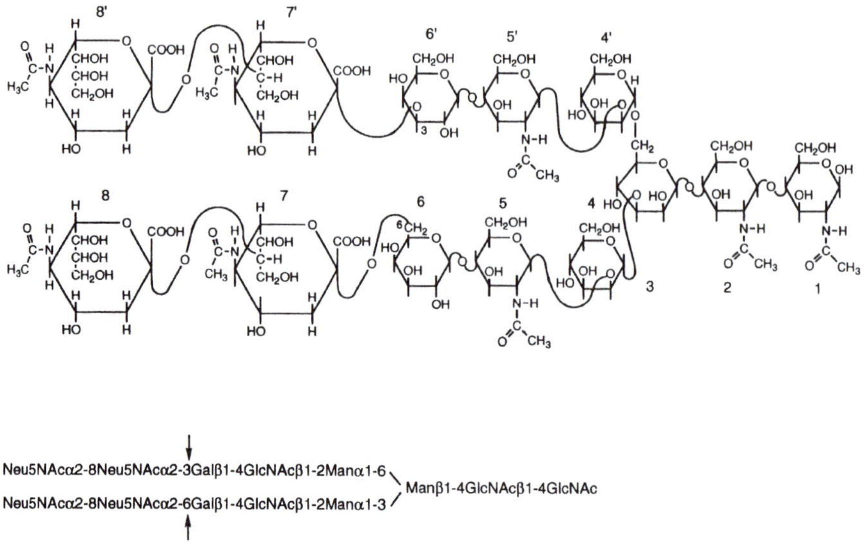
Neu5NAcα2-8Neu5NAcα2-3Galβ1-4GlcNAcβ1-2Manα1-6
Manβ1-4GlcNAcβ1-4GlcNAc
Neu5NAcα2-8Neu5NAcα2-6Galβ1-4GlcNAcβ1-2Manα1-3

Scheme 12.7 Asparagine-linked oligosaccharides: complex biantennary, tetrasialylated glycotype.

occurs with a substance, additional ions are usually multiply charged, separate, intact species in a spectrum. This phenomenon is part of the complex glycan heterogeneity found in Figure 12.33. Many glycoforms such as those listed in Table 12.14 can be found in the narrow elution range in Figure 12.33. Neuraminidase treatment reduces the complexity to a degree from Figure 12.33a to 12.33b. Thus, different-size glycopeptides are found in a very short elution timeframe. The T45 and T17 glycopeptides are of the complex glycotype, while T11 is a high-mannose species that overlaps with T17. It appears that the

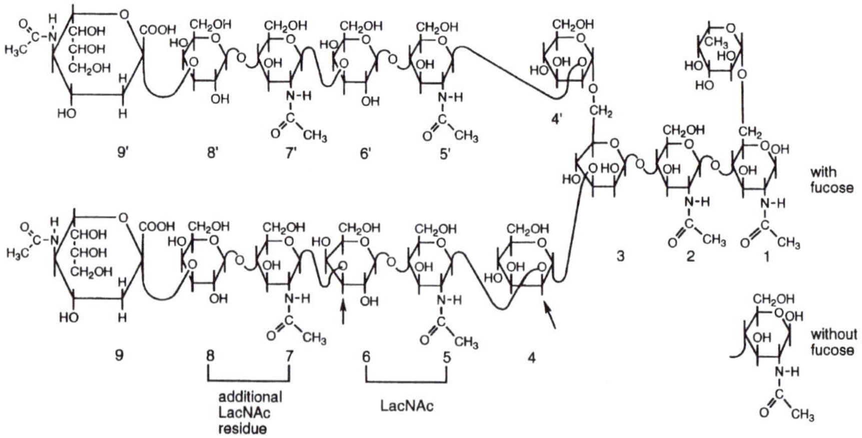
with fucose
without fucose
additional LacNAc residue
LacNAc

Scheme 12.8 Asparagine-linked oligosaccharides: complex biantennary, fucosylated, disialylated, dilactosylamine glycotype.

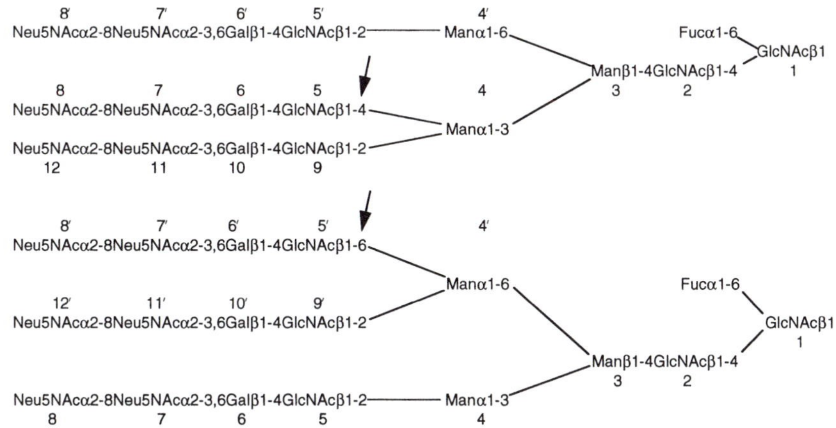

Scheme 12.9 Asparagine-linked oligosaccharides: complex triantennary, fucosylated, hexasialylated glycotype isomers.

glycan heterogeneity has less influence on the magnitude of the RP-HPLC separation than that of the peptide portion. This observation spans the diantennary to tetraantennary mass range, which can be in the thousands of mass units. Thus, separate glycoforms, which span several thousand mass units, can be found in a very narrow elution timeframe. Figure 12.34 shows a view of the T45 glycopeptide region of r-tPA (30–32-min retention time), and the tracings represent the intensity of selected masses (Table 12.14).[79] The legend in the figure explains the various glycoforms, and a number of observations are evident in Figure 12.34a:

1. A greater degree of sialylation produces higher intensity peaks.
2. The intensity increases with decreasing branching structure: that is, biantennary > trianntenary > tetraantennary intensity.

Upon neuraminidase digestion (Figure 12.34b):

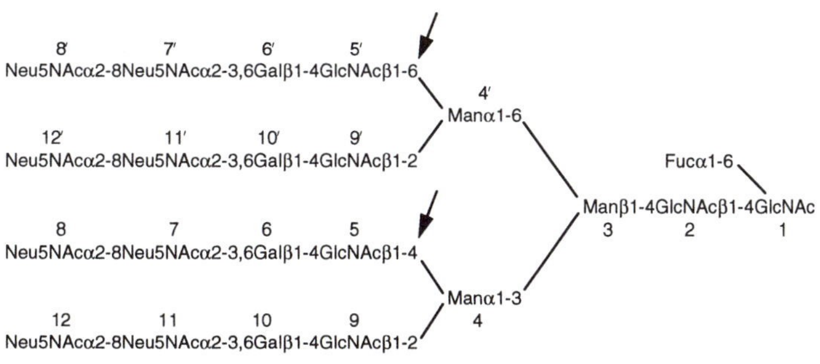

Scheme 12.10 Asparagine-linked oligosaccharide: complex tetraantennary, fucosylated, tetrasialylated glycotype.

Table 12.14 Average mass values of oligosaccharides found in the complex glycotype

Complex carbohydrate	Composition	Number of Neu5NAc[a]					Permethylated[b]; Number of Neu5NAc				
		0	1	2	3	4	0	1	2	3	4
Biantennary	$Gal_2Man_3GlcNAc_4$	1641.50	1932.76	2224.02	2515.26	2806.51	2048.29	2409.68	2771.07	3132.46	3493.85
Biantennary-Fuc	$Gal_2Man_3GlcNAc_4Fuc$	1787.64	2078.91	2370.11	2661.41	2952.61	2222.49	2583.88	2945.27	3306.66	3668.05
LacNAc-biantennary-Fuc	$LacNAcGal_2ManGlcNAc_4Fuc$*	2152.98	2444.25	2735.45	3026.75	3317.95	2671.99	3033.38	3394.78	3756.16	4117.55
$LacNAc_2$-biantennary-Fuc	$LacNAc_2Gal_2Man_3GlcNAc_4Fuc$*	2518.31	2809.58	3100.78	3392.08	3683.28	3121.49	3482.88	3844.27	4205.66	4567.05
Bisecting-biantennary	$Gal_2Man_3GlcNAc_5$	1844.68	2135.91	2427.19	2718.41	3009.69	2293.56	2654.95	3016.35	3377.74	3739.13
Bisecting-biantennary-Fuc	$Gal_2Man_3GlcNAc_5Fuc$	1990.81	2282.07	2573.31	2864.57	3155.81	2467.76	2829.15	3190.55	3551.94	3913.33
Triantennary	$Gal_3Man_3GlcNAc_5$	2006.81	2298.06	2589.31	2880.61	3171.81	2497.78	2859.18	3220.57	3581.96	3943.35
Triantennary-Fuc	$Gal_3Man_3GlcNAc_5Fuc$*	2152.91	2444.20	2735.45	3026.71	3317.91	2671.98	3033.37	3394.76	3756.15	4117.54
LacNAc-triantennary-Fuc	$LacNAcGal_3Man_3GlcNAc_5Fuc^+$	2518.25	2809.54	3100.80	3392.05	3683.25	3121.48	3482.87	3844.26	4205.65	4567.04
$LacNAc_2$-triantennary-Fuc	$LacNAc_2Gal_3Man_3GlcNAc_5Fuc$**	2883.58	3174.87	3466.12	3757.38	4048.58	3570.98	3932.37	4293.76	4655.15	5016.54
$LacNAc_3$-triantennary-Fuc	$LacNac_3Gal_3Man_3GlcNAc_5Fuc^-$	3248.92	3540.21	3831.46	4122.72	4413.92	4020.48	4381.87	4743.26	5104.65	5466.04
Tetraantennary	$Gal_4Man_3GlcNAc_6$	2372.16	2663.41	2954.66	3245.91	3537.16	2947.28	3308.68	3670.07	4031.46	4392.85
Tetraantennary-Fuc	$Gal_4Man_3GlcNAc_6Fuc^+$	2518.31	2809.55	3100.80	3392.01	3683.26	3121.48	3482.87	3844.26	4205.65	4567.05
LacNAc-tetraantennary-Fuc	$LacNAcGal_4Man_3GlcNAc_6Fuc$**	2883.65	3174.89	3466.15	3757.35	4048.60	3570.98	3932.37	4293.76	4655.15	5016.55
$LacNAc_2$-tetraantennary-Fuc	$LacNAc_2Gal_4Man_3GlcNAc_6Fuc^-$	3248.98	3540.22	3831.48	4122.68	4413.93	4020.48	4381.87	4743.26	5104.65	5466.05
$LacNAc_3$-tetraantennary-Fuc	$LacNAc_3Gal_4Man_3GlcNAc_6Fuc$	3614.32	3905.56	4196.82	4488.02	4779.27	4469.98	4831.37	5192.76	5554.15	5915.55

[a]Mass values include the −OH terminal reducing-end moiety and −H nonreducing end.

[b]Mass values include the $-OCH_3$ terminal reducing-end moiety and the $-CH_3$ nonreducing end.

*, **, +, − indicate respective isomers.

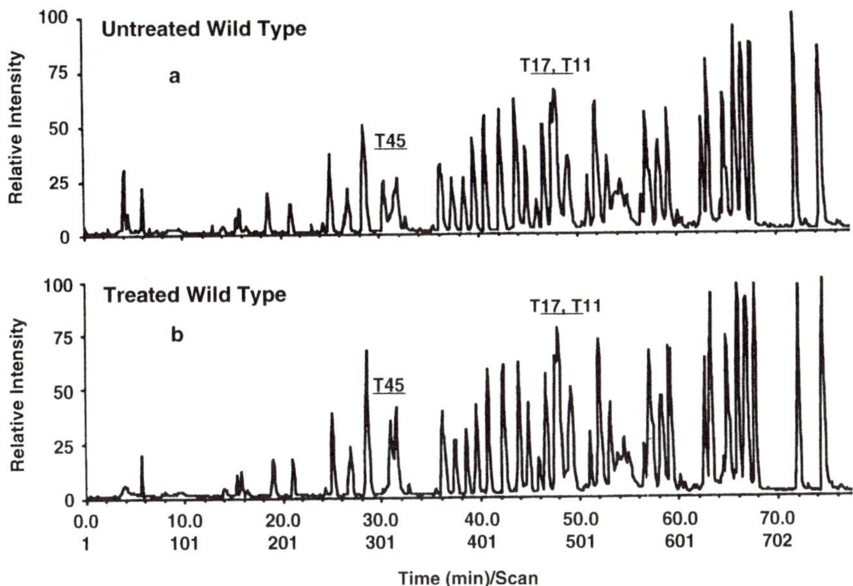

Figure 12.32 The RP-HPLC mass spectral total-ion chromatograms of a tryptic digest of (a) untreated wild type and (b) neuraminidase-treated recombinant tissue plasminogen activator. Glycopeptides are noted by the horizontal bars. (© 1993, American Chemical Society. Reprinted from reference 79 with permission.)

1. Only three main peaks are observed as opposed to eight in Figure 12.35a.
2. Abundance increases as the glycoform mass decreases. It should be kept in mind that there still are multiple compounds (glycoforms) contained in each selected ion current region.
3. The larger glycopeptides elute earlier than the smaller glycopeptides, and the higher mass glycans are relatively more polar than the lower mass glycans. Thus, the more polar, larger polysaccharides elute earlier than the shorter, more nonpolar glycans.

This may seem unusual at first; however, RP-HPLC columns preferentially retain hydrophobic species to a greater extent than polar species. For the RP-HPLC plots in Figures 12.33a and b, the glycopeptides show bands which, upon magnification in Figure 12.34, are clearly shown to be diagonal in nature. The nonglycosylated peptides show a vertical distribution of ions at different retention times (Figures 12.33a and b). Thus, an easy way to spot a glycopeptide is for the distribution of ions to lie on a diagonal, and this is caused by the glycoform heterogeneity (Figure 12.34). This is analogous to sodium dodecyl sulfate polyacrylamide gel electrophoresis (SDS-PAGE). Glycoprotein bands are broad and diffuse in nature, while the protein portion, produced by glycosidase reaction, appears as distinct bands.

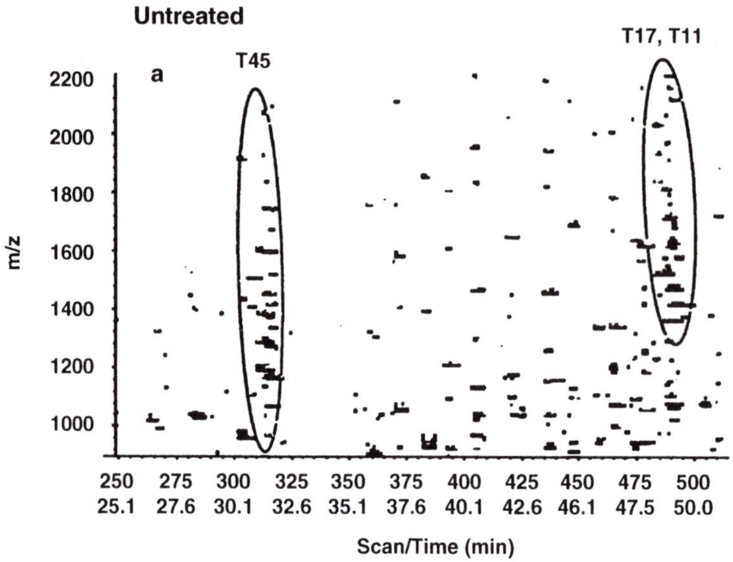

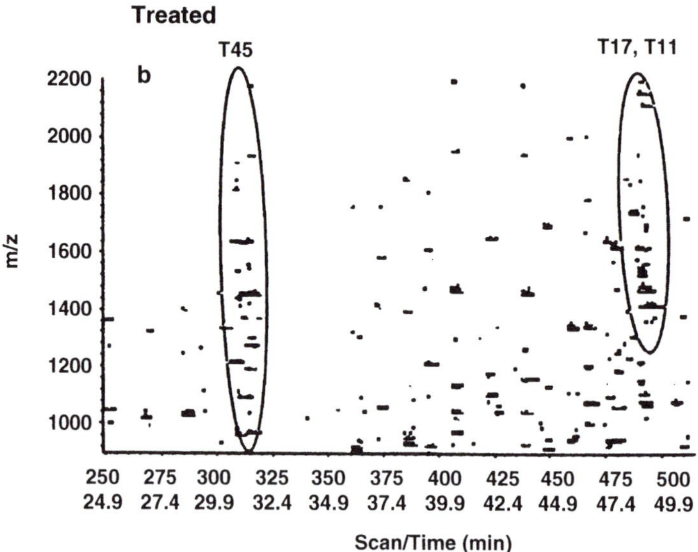

Figure 12.33 Mass-retention time contour plot of the tryptic digest of the (a) untreated wild type and (b) neuraminidase-treated r-tPA. (a) Derived from Figure 12.32a; (b) derived from Figure 12.32b. This plot is equivalent to an overhead view (looking down the *y*-axis or relative *m/z* intensity) with respect to Figure 12.32. The summed intensities of the masses at a particular elution time yields a point on the relative intensity axis in Figure 12.32. This figure resolves each intensity point in Figure 12.32 into its individual *m/z* values. (c) Mass-retention time contour plot of a Thr-103 to N-103 mutant (T103N) of r-tPA. [(a and b) (© 1993, American Chemical Society. Reprinted from reference 79 with permission. (c) (© 1996, CRC Press, Boca Raton, Florida. Reprinted from reference 80 with permission.)

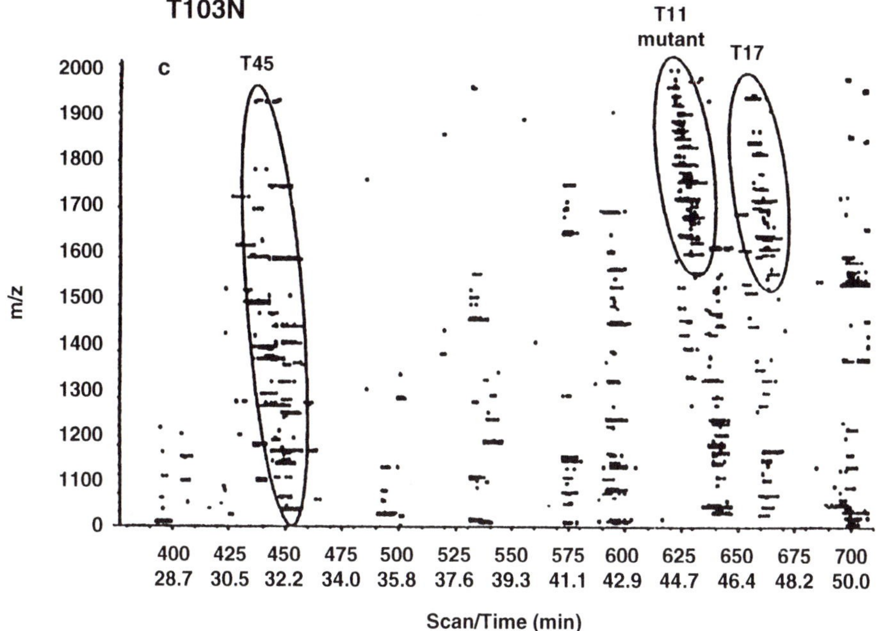

Figure 12.33 *(continued)*

This idea is further reinforced in Figure 12.33c.[80] The tPA was made in a recombinant fashion from Chinese hamster ovary (CHO) cells. The cells were made to undergo site-directed DNA mutagenesis so as to alter amino acid 103 from threonine to asparagine. The Thr-103 residue had no glycosylation, and the idea was to create a site (Asn-103) for potential glycosylation in the T11 fragment by the cellular machinery. The tryptic digest in Figure 12.33c shows that the T11 mutant fragment significantly shifts to an earlier elution time with respect to T17 compared with the T11 fragment in Figures 12.33a and b. The mutated DNA coding region of tPA produced Asn-103, whereby the cellular apparatus glycosylated the newly created amino acid site with complex glycans. The addition of carbohydrate mass decreased the retention time, because the polarity of the mutant is relatively higher than that of the wild type (Figure 12.33a).

Glycotype Specificity of N-Linked Endoglycosidases

Table 12.15 presents a comparison of a number of glycosidases that provides further details to those given in Table 12.5. The endoglycosidases represent a series that displays a measure of selectivity with respect to their relative action on the three N-linked glycotypes.

Amino Acid Residue Consensus Sequence for Glycan Attachment

Glycan moieties are essentially found only on Ser, Thr, and Asn residues. There are a few isolated or unique exceptions to this. It was of interest to see if there was an order or some observed constraints as to which Asn residue(s) in a protein are actually glycosylated. General observations were noted that, in some cases, glycans were attached to Asn residues in the Asn-X-Ser/Thr amino acid sequence.[81] There are many exceptions to this observation; however, a computer investigation (unpublished results presented in reference 82) searched 767 proteins and found 342 glycoproteins that have an Asn-X-Ser/Thr sequence, or so-called consensus sequence or consensus sequon. The X can be any amino acid with the exception of Pro or Asp. Glycosylation at Asn is very rare in prokaryotes, including bacteria; thus, of the 342 entries, 159 eukaryote species (mammals included) were analyzed. Only 49 of the 159 glycoproteins (30.8%) displayed a carbohydrate residue on the Asn in the Asn-X-Ser/Thr motif.[83] Thus, the consensus sequence for establishing glycan presence on an Asn site may be necessary, but it is not sufficient.[81] Many proteins contain this amino acid sequence and are not glycosylated. In this context, the serine and threonine O-linked polysaccharide species have no specific consensus sequence.[34,84]

Mass Spectral Analysis of the Complex Glycotype of LCAT

Human lecithin:cholesterol acyltransferase (LCAT) is an enzyme that effects the transfer of the 2-position fatty acid from phosphatidylcholine to the 3-hydroxyl moiety in blood plasma cholesterol. It is a glycoprotein that consists of five glycosylated Asn residues with complex glycotypes, and three sites that have O-linked glycans. Tryptic digestion was used to effect peptide and glycopeptide production.[62] The five Asn glycosylated sites are found in separate glycopeptides, and the three O-linked sites are distributed in two additional glycopeptides. A number of these glycopeptides were investigated for glycan heterogeneity. One fraction contained the peptide from residues 16–39: AELSNHTRPVILVPGCLGNQLEAK. The protein was carboxymethylated with iodoacetic acid to dissociate disulfide bonds; thus, a Cys residue is CMCys. The average M_r of the peptide is 2618.0 Da, and it is 2617.0 Da when the hydrogen is removed from the glycan site at Asn-20. The ESI-mass spectrum of this particular fraction is shown in Figure 12.35.[62] There are five separate species labeled A–E, and the charge of each peak is found next to the A–E designations.

Thus:

A5 = 1096.7^{5+} = 5478.5; A4 = 1370.4^{4+} = 5477.6; av = 5478.05 Da
B5 = 1126.0^{5+} = 5625.0; B4 = 1406.8^{4+} = 5623.2; av = 5624.1 Da
C5 = 965.3^{5+} = 4821.5; C4 = 1206.2^{4+} = 4820.5; av = 4821.00 Da
D5 = 1038.6^{5+} = 5188.0; D4 = 1297.4^{4+} = 5185.6; av = 5186.8 Da
E5 = 1169.8^{5+} = 5844.0; E4 = 1460.8^{4+} = 5839.2; av = 5841.6 Da

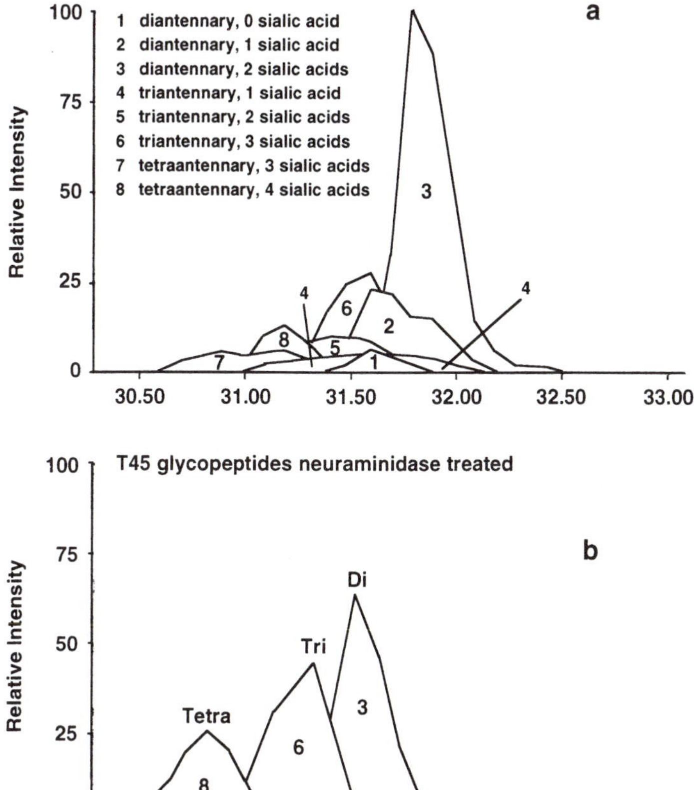

Figure 12.34 Extracted ion profiles (of individual m/z values) of the T45 glycoform region (30–32 min) of Figure 12.33 for the (a) wild type and (b) neuraminidase-treated r-tPA. (© 1993, American Chemical Society. Reprinted from reference 79 with permission.)

Table 12.15 Glycotype specificity of N-linked endoglycosidases

	Oligosaccharide		Bi-	Tri-	Tetra-	
Enzyme	High-mannose	Hybrid		Antennary		Reference(s)
Endo-H	+	+				37, 38, 47
Endo-F_1	+	+/−				37, 38, 47
Endo-F_2	+		+			37, 38, 47
Endo-F_3[a]			+	+		37, 38
PNGase F	+	+	+	+	+	37, 38

[a]Mixture of endo-β-N-acetylglucosaminidases.

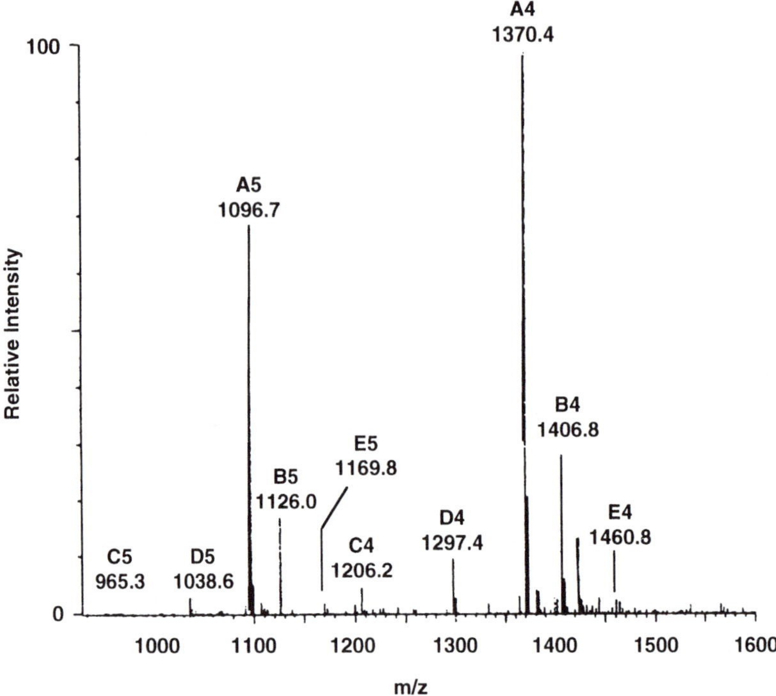

Figure 12.35 The ESI-mass spectrum of a heterogeneous, carboxymethylated, tryptic glycopeptide of LCAT that contains the amino acid residues 16–39. The glycopeptide was isolated by RP-HPLC. (© 1995. Reprinted from reference 62 by permission of Cambridge University Press.)

Subtraction of the mass of the 16–39 peptide residue sequence, 2617.0 Da, from each glycoform glycopeptide yields a respective mass. Note that there are five different types of glycan moieties. The glycan mass value can then be compared to those in Table 12.14 for an identity check. Also, note that 17 Da must be added to each glycan mass for a proper comparison to Table 12.14.

A: $5478.05 - 2617.0 + 17 = 2878.05$ Da $\simeq 2880.61$ trisialo triantennary

B: $5624.10 - 2617.0 + 17 = 3024.1$ Da $\simeq 3026.71$ trisialo fucosylated triantennary

C: $4821.00 - 2617.0 + 17 = 2221.00$ Da $\simeq 2224$ disialo biantennary

D: $5186.80 - 2617.0 + 17 = 2586.8$ Da $\simeq 2589.31$ disialo triantennary

E: $5841.60 - 2617.0 + 17 = 3241.6$ Da $\simeq 3245.9$ trisialo tetraantennary

Note that glycoform B could also be trisialo LacNAc fucosylated biantennary in structure.

PROBLEM 12.33 The ESI-mass spectrum of another glycopeptide fraction from an RP-HPLC separation from a tryptic digest of LCAT is presented in Figure 12.36a.[62] The peptide portion of the glycopeptide contained the 257–276 amino acid residue series MAWPEDHVFISTPSFNYTGR. How many glycoforms are there? Identify them from Table 12.14. Upon *Arthrobacter ureafaciens* neuraminidase digestion of the glycopeptides, Figure 12.36b was obtained. What happened to the glycoforms upon digestion and can they be corroborated in Table 12.14? The products of Figure 12.36b were digested with *Streptococcus pneumoniae* β-galactosidase, and Figure 12.36c was obtained. Explain the data in Figure 12.36c.

Mass Spectral Analysis of Glycans from Bovine Fetuin

The glycoprotein bovine fetuin was reduced with dithiothreitol and alkylated with vinylpyridine in order to cleave the disulfide bonds. This preparation was digested with trypsin, and the peptides and glycopeptides were detected by UV absorbance at 215 nm (Figure 12.37a) and a total-ion chromatogram by RP-HPLC ESI-mass spectrometry (Figure 12.37b).[18] The shaded and scored peaks (peaks 1–7) identify the glycopeptides. These were obtained by subjecting the electrospray eluent to the selected-ion monitoring mode (SIM). The skimmer lens voltage was increased in order to effect "up-front" CID (Chapter 1), such that individual sugar species would fragment from the glycopeptides. These species were detected by their mass—that is, m/z 204 for hexosamine (Figure 12.37c) and m/z 292 for sialic acid (Figure 12.37d) under SIM. Monitoring of these masses throughout the entire chromatographic elution profile assured the detection of the glycopeptide portion of the digest. Peaks 3 and 4 (Figure 12.37) are O-linked and the rest are N-linked. One of these peaks will be investigated.

Glycopeptide Analysis from Bovine Fetuin

Peak 5 in Figure 12.37 consists of a glycopeptide from residues 54 to 85, and the glycan portion is at Asn-81. The peptide sequence is RPTGEVYDIEIDTLETTCHVLDPTPLANCSVR. The ESI-mass spectrum is shown in Figure 12.38a.[18,85]

PROBLEM 12.34 Deduce the saccharide identities of the species marked with symbols in Figure 12.38a.

PROBLEM 12.35 The glycopeptide in Problem 12.34 was digested with *Arthrobacter ureafaciens* neuraminidase exoglycosidase. The resulting glycopep-

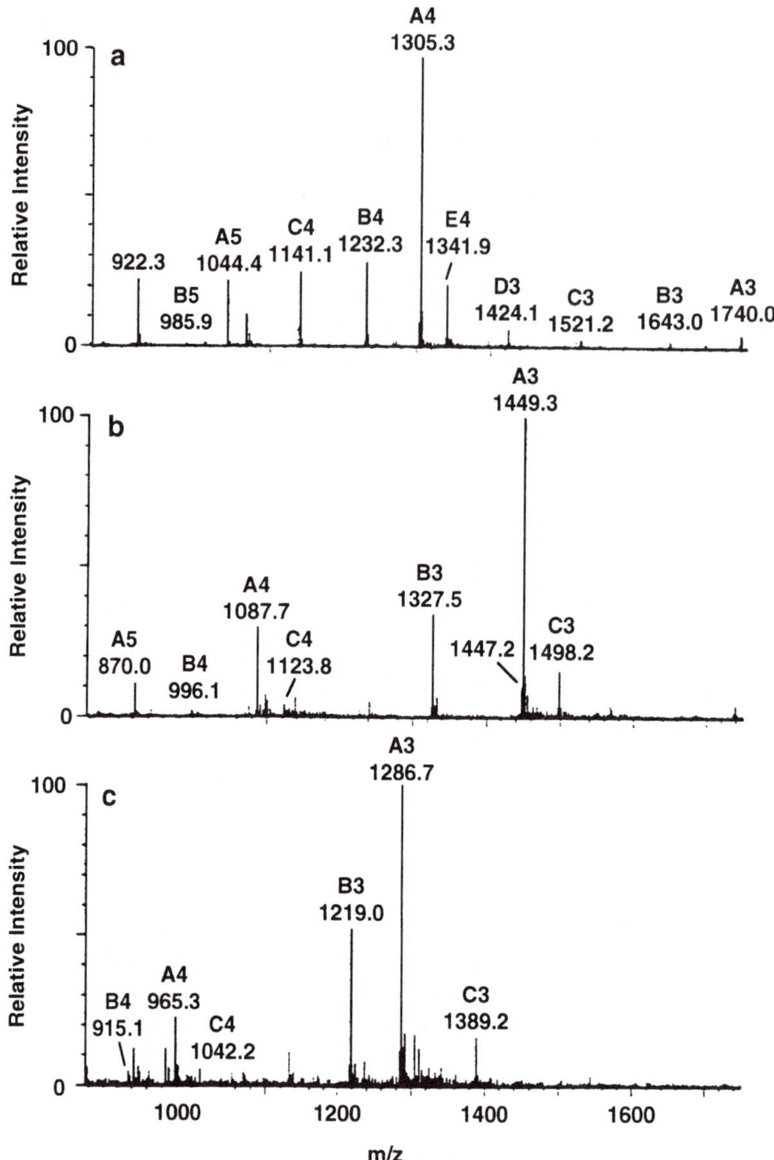

Figure 12.36 The ESI-mass spectra of the 257–276 amino acid tryptic glycopeptide of LCAT isolated by RP-HPLC. (a) Intact glycopeptide, (b) neuraminidase-treated glycopeptide from (a), and (c) β-galactosidase-treated glycopeptide from (b). (© 1995. Reprinted from reference 62 by permission of Cambridge University Press.)

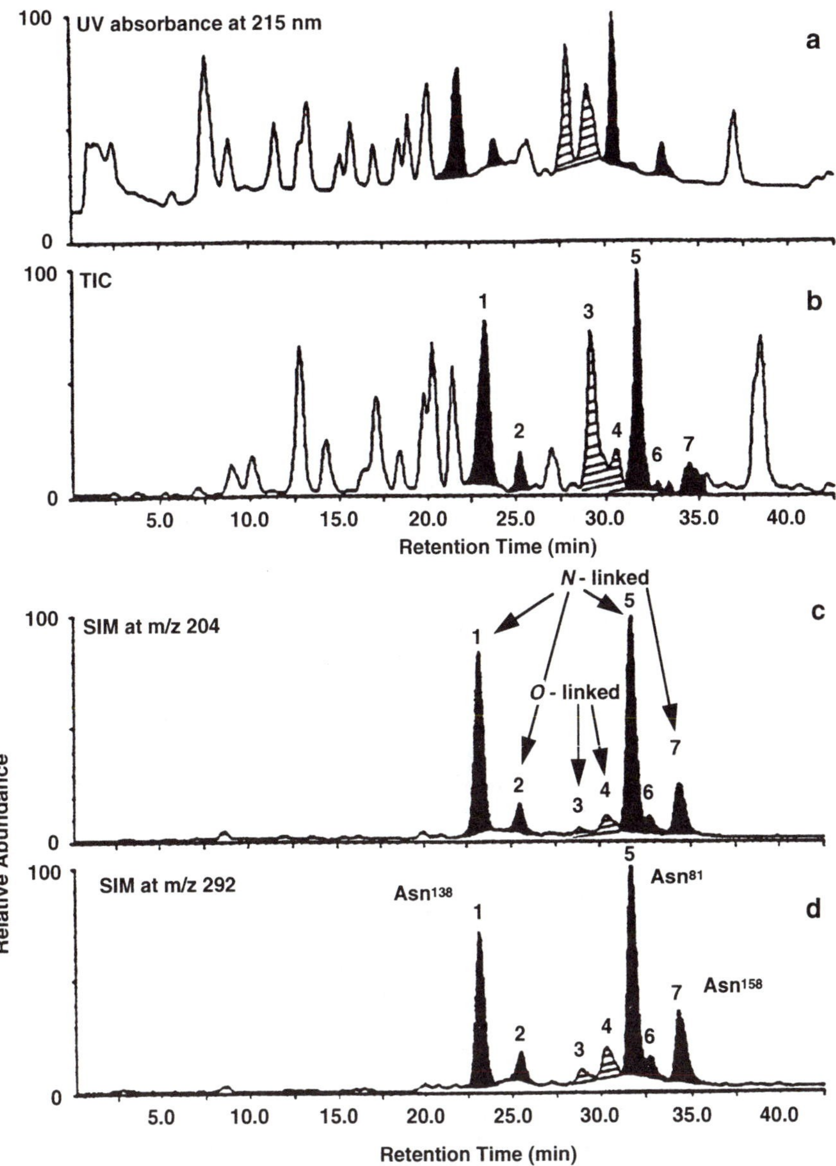

Figure 12.37 (a) Ultraviolet absorbance at 215 nm and (b) the total-ion chromatogram by RP-HPLC-ESI-mass spectrometry of a vinylpyridine-treated, tryptic digest of bovine fetuin. The shaded and scored peaks are the glycopeptide portion of the digest. These peaks were detected by up-front CID where selected ion monitoring (SIM) was performed for (c) m/z 204, hexosamine and (d) m/z 292, sialic acid. Shaded and scored areas identify the N- and O-linked glycopeptides, respectively. (© 1995. Reprinted from reference 18 by kind permission of Elsevier Science-NL, Sara Burgerhartstraat 25, 1055 KV Amsterdam, The Netherlands.)

tide displayed the ESI-mass spectrum shown in Figure 12.38b. Deduce the identity of the two glycoform glycans.

Enzyme Analysis of the Glycopeptides from Bovine Fetuin

The N-linked glycans found in bovine fetuin display some differences from that of Schemes 12.7–12.9. These differences can be found as additional complexity to the basic structures in Schemes 12.7–12.9. This complexity is not obvious, and Scheme 12.11 presents these modifications. These additions are found mainly near the nonreducing end, and the glycoforms include 0–5 sialic acid residues in the indicated positions, as well as a number of permutations of saccharide residue linkages. Most of the antenna galactose residues are 3-linked on the nonreducing end. Most of the antenna-residing GlcNAc residues are 3-linked, as opposed to the 4-link possibility shown in Scheme 12.11. The β-galactosidase action on the desialylated glycopeptide sheds some light on the presence of glycoforms that differ in the type of interresidue linkage.

PROBLEM 12.36 Taking into account that only two glycan species were observed after neuraminidase digestion of bovine fetuin (Figure 12.38b), deduce the identities of the four glycoforms shown in Figure 12.38c. This ESI-mass spectrum was obtained after β-galactosidase digestion of the desialylated glycopeptide, where the enzyme originated from *Streptococcus pneumoniae*. Use Table 12.6 as a guide to solve this problem. It is also useful to know that the specific β-galactosidase preparation also had some peptidase enzyme present, and this cleaves amino acid residues.

Mass Spectral Analysis of the Complex Glycopeptides from gD-2 Protein

Type 2 Herpes simplex virus has a number of proteins, one being a glycoprotein called gD-2. This glycoprotein incorporates itself into the membrane of an infected host cell and is a cell-surface structural element that triggers the immune response. This glycoprotein was produced in a recombinant fashion, r-gD-2, by CHO cells. The glycoprotein was reduced and alkylated with pyridylethylation and was digested with trypsin. No O-linked structures were found but three N-linked glycopeptides were observed (Figure 12.39).[86] Reconstructed ion chromatograms of three key saccharide residues located the three glycopeptides, and these are shown by dotted lines in Figure 12.39. Note that these three glycopeptides, at 32, 33, and 37 min, constitute a relatively minor portion of the tryptic digest (Figure 12.39d). The 33-min peak will be investigated for glycan characterization. The peptide portion was from the amino acids Met-102 to Lys-122, and the sequence is MGDNCAIPITVMEYTECPYNK. The glycan portion resides on Asn-105. The ESI-mass spectrum of the glycopeptide is shown in

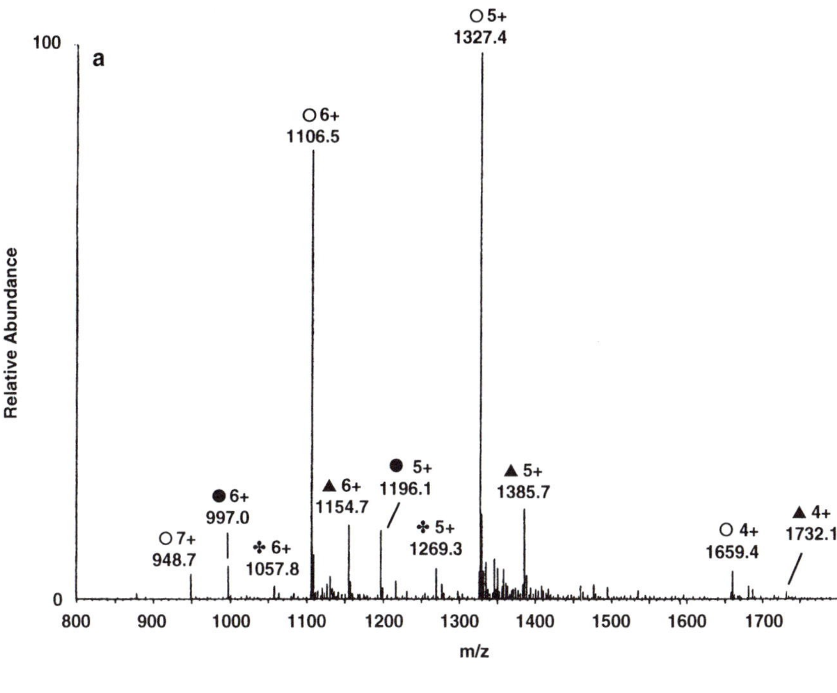

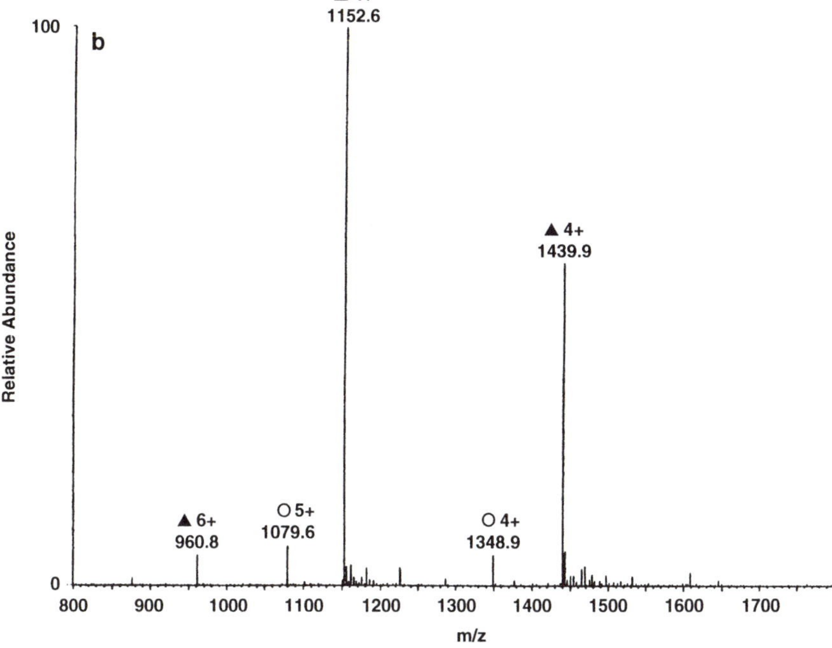

Figure 12.38 The ESI-mass spectra of peak 5 in Figure 12.37. This consists of the heterogeneous glycan portion of the glycopeptide residues 54–85. (a) Heterogeneous mixture of the complete glycans, (b) neuraminidase-treated glycopeptide from (a), (c) β-galactosidase-treated glycopeptide from (b). (© 1995, American Society for Mass Spectrometry. Reprinted from reference 18 by permission of Elsevier Science, Inc.)

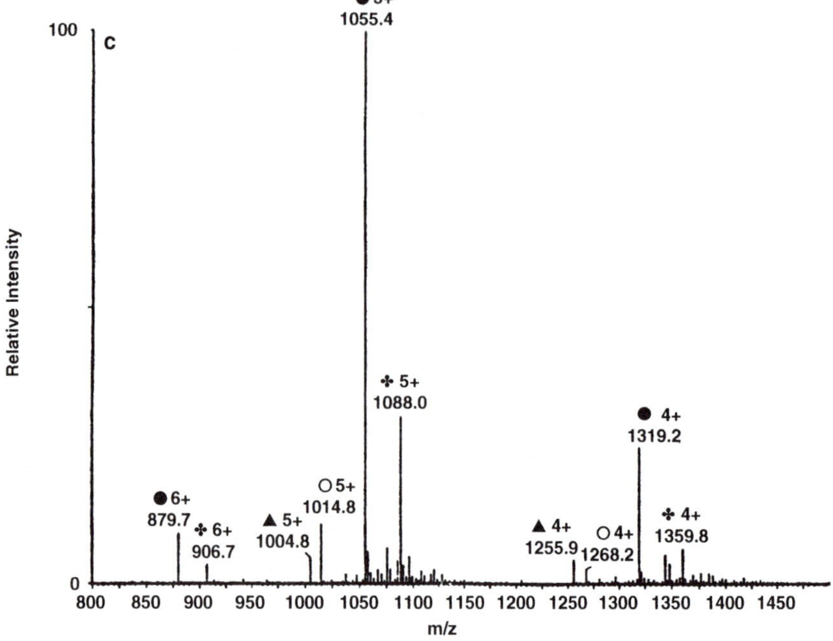

Figure 12.38 *(continued)*

Figure 12.40, and a glycoform distribution can be observed in the 4+ and 3+ charge states.[86]

PROBLEM 12.37 For the m/z 1458.5 and 1580.2 peaks in Figure 12.40, characterize the structures of the two glycoforms.

Glycoform Heterogeneity Analysis of EPO

Recombinant human EPO was considered earlier (see section "Mass Spectral Analysis of a Glycopeptide from Erythropoietin"), and this presentation pro-

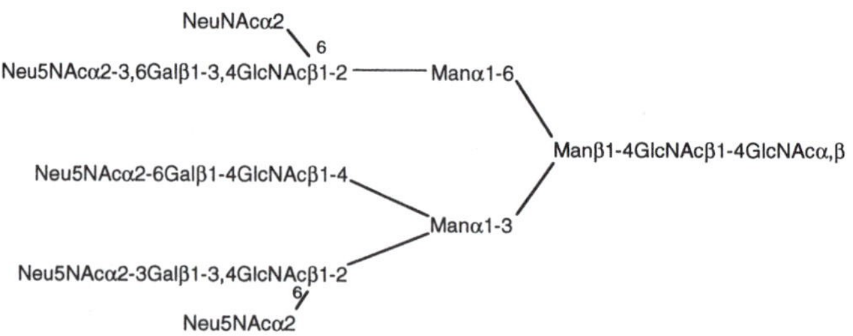

Scheme 12.11 Bovine fetuin triantennary structure.

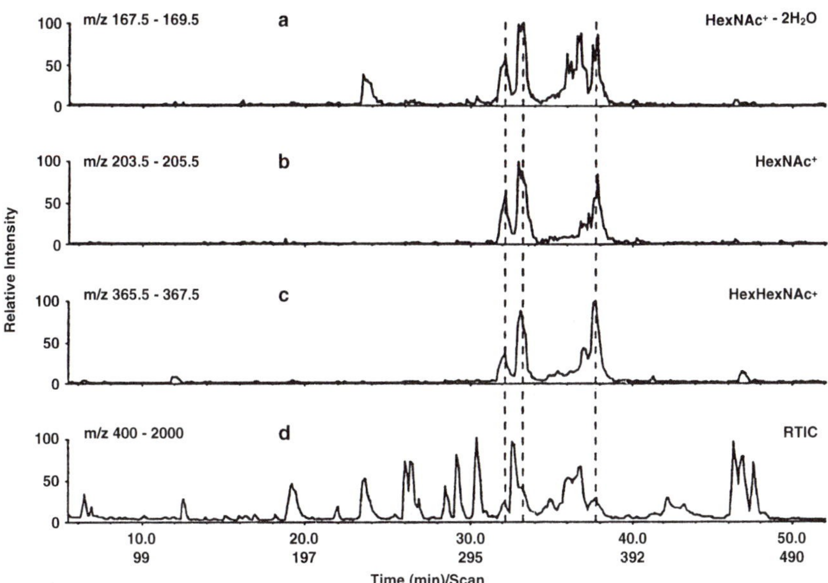

Figure 12.39 The RP-HPLC-ESI-total-ion chromatogram of (d) a reduced, pyridylethylated, neuraminidase-treated, tryptic digest of the gD-2 glycoprotein. Reconstructed ion chromatograms are also shown from the TIC in (d) for (a) *m/z* 168, didehydrated HexNAc, (b) *m/z* 204, HexNAc, and (c) *m/z* 366, HexHexNAc. Three glycopeptides are observed at 32, 33, and 37 min. (© 1996. Reproduced from reference 86 by permission of Humana Press, Totowa, New Jersey.)

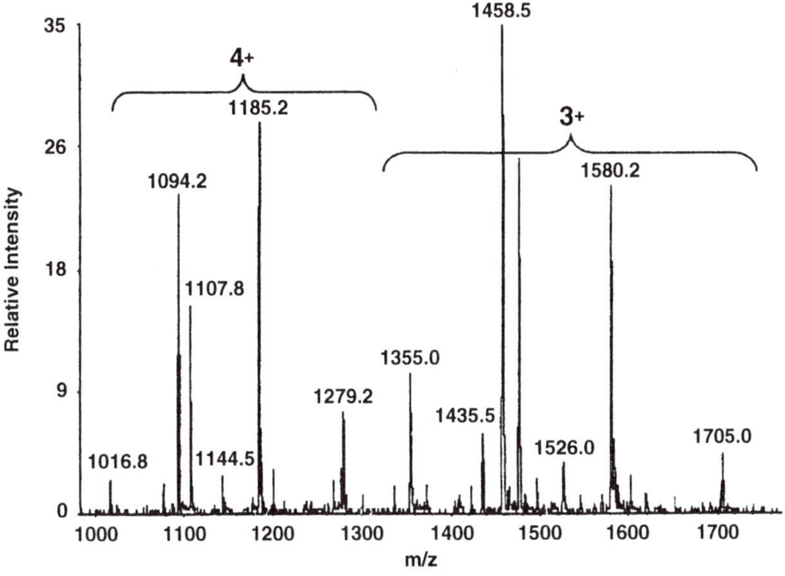

Figure 12.40 The ESI-mass spectrum of the 33-min peak in Figure 12.39 that contains Asn-105. (© 1996. Reproduced from reference 86 by permission of Humana Press, Totowa, New Jersey.)

vides a better appreciation of the glycoform heterogeneity. Figure 12.41a shows a Glu C digest of the unreduced and unalkylated glycoprotein.[87] One O-linked and three N-linked glycans characterize the glycoprotein, and Figure 12.41b provides a detailed observation of the glycopeptide that has an Asn-83 glycan attachment. The mass of the peptide, less a hydrogen, is 4865.7 Da from residues

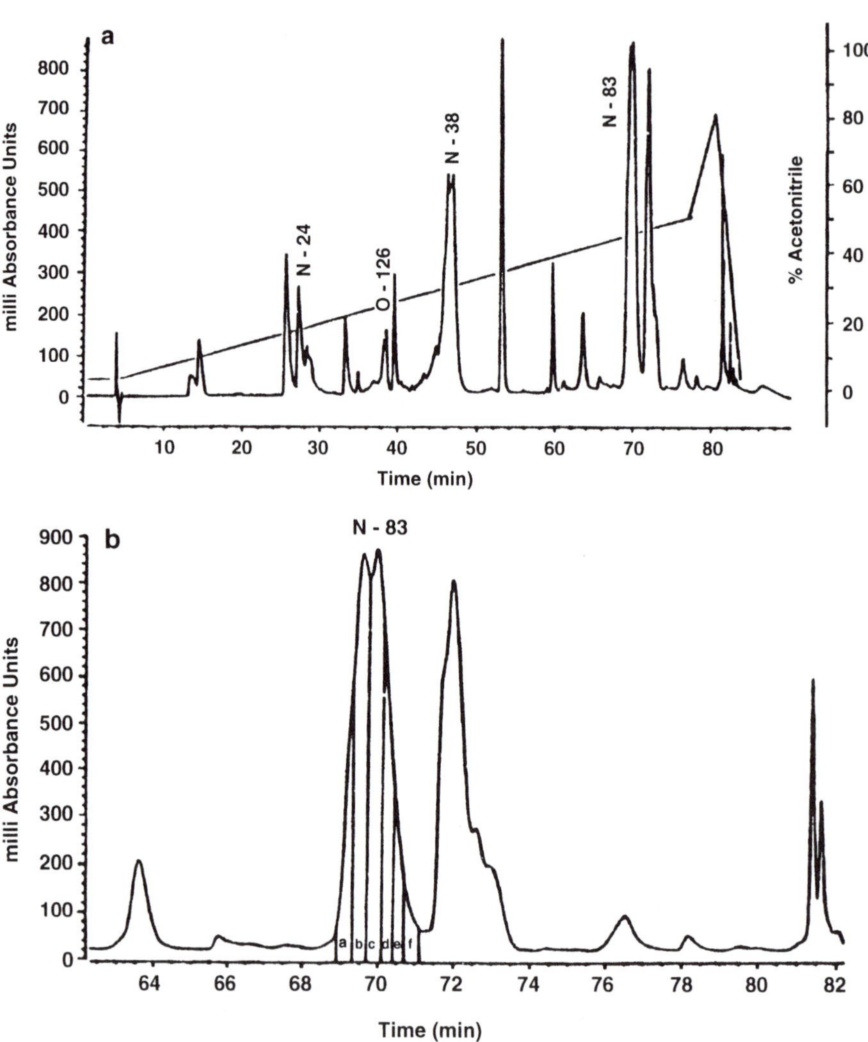

Figure 12.41 The RP-HPLC chromatogram of an unreduced, unalkylated Glu C digest of r-EPO. A C_4 column equilibrated with 0.06% TFA was used. Separation was accomplished by a 1%/min gradient of AcCN at 0.75 ml/min. Detection was at 214 nm. (a) Total chromatogram, (b) expanded portion of (a) to show the Asn-83 glycopeptide elution region. The peak in (b) is delineated into regions labeled a–f. (© 1995, American Chemical Society. Reprinted from reference 87 with permission.)

Ala-73 to Glu-117. Figure 12.41b shows the poorly resolved, doublet RP-HPLC peak split into six elution regions (a–f), and the ESI-mass spectrum of each section is shown in Figures 12.42a–f, respectively. Note the complexity and high degree of glycoforms of each spectrum, and they are primarily 6+ and 7+ species. Note that as elution time increases, a greater distribution of glycoforms is observed. The microheterogeneity can be explained by the diversity of *N*-acetylneuraminic acid groups, the degree of O-acetylation on the alcohol and acid groups of sialic acid (as opposed to the one N-acetylation site), and the degree of LacNAc groups. Permutations of these modifications primarily cause the presence of over 60 glycoforms in Figure 12.42.[87] The general order of elution in increasing time is tetraantennary + multiple LacNAc, tetraantennary + one LacNAc, tetraantennary ± O-acetylation, and triantennary ± O-acetylation. The higher the number of Neu5NAc residues in a given band in Figure 12.41, the earlier that band elutes. These observations follow hydrophilic–hydrophobic interactions of glycan species in an RP-HPLC system, where the more hydrophobic, lower mass glycans elute at a later time than the larger, more hydrophilic glycans.

It is remarkable that enough signal is available for glycan characterization in Figure 12.42. The many individual species present in the Asn-83 elution band (Figure 12.41b) distributes the charge to such an extent so as to significantly lower the absolute intensities of the individual glycoforms.

Chemical Modifications on Complex Polysaccarides

Permethylation derivatization and ORM/ODM analyses will be presented here for further characterization of the complex, N-linked polysaccharide species. The ORM concept, in addition to specific exoglycosidases, provides interresidue linkage information, as was noted in the high-mannose section.

Mass Spectral Analysis of Permethylated Glycans of EPO

The r-human EPO Asn-83 glycopeptide glycan portion is revisited, where, following release by *N*-glycanase, the glycoforms were then permethylated. Figure 12.43 presents the ESI-mass spectrum of the permethylated glycoforms from the Asn-83 glycopeptide;[68,69] note the relatively simpler distribution than that of Figure 12.42. Sodium was infused in the electrospray to effect a more efficient ionization.

PROBLEM 12.38 Figure 12.43 provides labels for the various signals in the ESI-mass spectrum, except that some labels are absent. Provide the glycan identification for the following permethylated signatures: *m/z* 1164.6^{4+}, 1274.8^{3+}, 1544.6^{3+}, 1694.8^{3+}, and 1900.0^{2+}. Some mass values have isomers as given in Table 12.14.

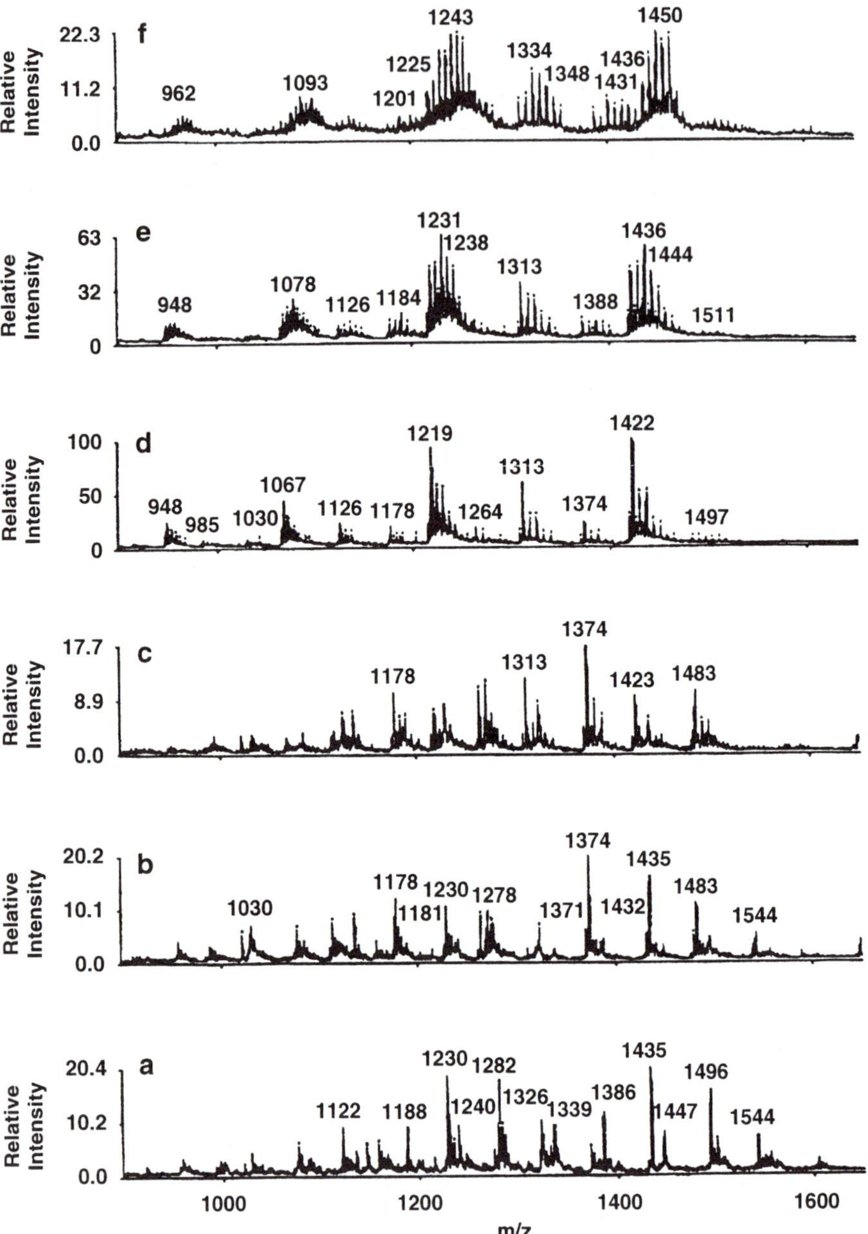

Figure 12.42 (a–f) The ESI-mass spectra of the regions marked a–f, respectively, on the Asn-83 peak in Figure 12.41. (© 1995, American Chemical Society. Reprinted from reference 87 with permission.)

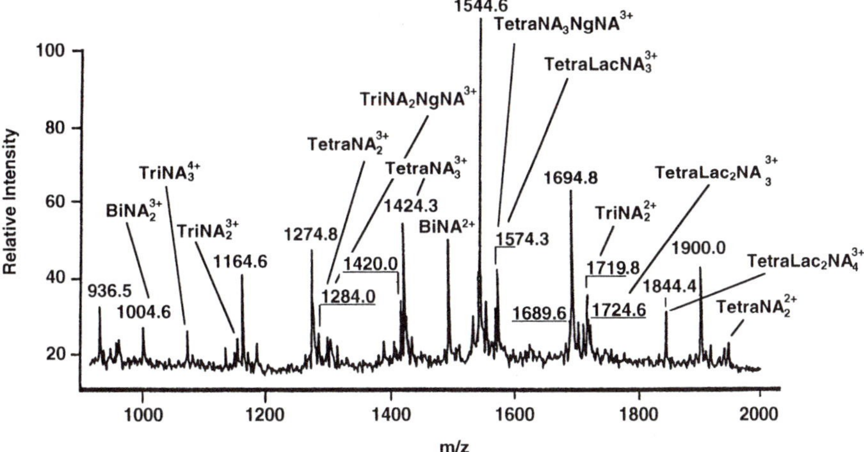

Figure 12.43 The ESI-mass spectrum of the permethylated heterogeneous glycan of the deglycosylated Asn-83 glycopeptide from r-EPO. The ionizing species was sodium ion. See Table 12.1 for abbreviations. (© 1994, John Wiley & Sons, Limited. Reproduced from reference 69 with permission.)

Mass Spectral Analysis of ORM/ODM Treated Glycans of EPO

The N-83 underivatized glycoform mixture was subjected to ODM reaction, and the ESI-mass spectrum shown in Figure 12.44 was effected by sodium ionization.[68,69] All ions in Figure 12.44 have an attached fucose residue. Peaks are labeled according to identity, and mass values shift accordingly with respect to Figure 12.43. The structural and mass value origins will be derived for a few of the ions. Scheme 12.9 shows two structures of a triantennary, fucosylated, hexasialylated complex glycotype. The underivatized and permethylated mass values of both of the fucosylated, trisialylated, triantennary polysaccharide ($TriNA_3Fuc$) structures in Scheme 12.9 are 3026.71 and 3756.15 Da, respectively (Table 12.14). In the ODM-treated glycoform mixture (Figure 12.44), $TriNA_3Fuc$ is observed at m/z 1182.5^{3+} (3478.5 Da). Table 12.16 provides ORM- and ODM-treated structures and mass values for fucose and 2,8-linked *N*-acetylneuraminic acid.

PROBLEM 12.39 Following the format of Table 12.16, provide structure and mass information on the ORM and ODM treatment of 2-linked *N*-acetylneuraminic acid and 4,6-linked *N*-acetylglucosamine residues. Assume that the latter is in the closed-ring form.

Figure 12.41 shows that there are three N-linked glycopeptides in the tryptic digest of r-EPO, and two of them are the N-38 and N-83 glycopeptides. When the glycan portion of each glycopeptide is released, mixtures of glycoforms are observed, and detailed information was presented above for the N-83 glycopeptide. The $TriNA_3Fuc$ glycoforms for N-38 and N-83 are different, and they are derivatives of the structures presented in Scheme 12.9.

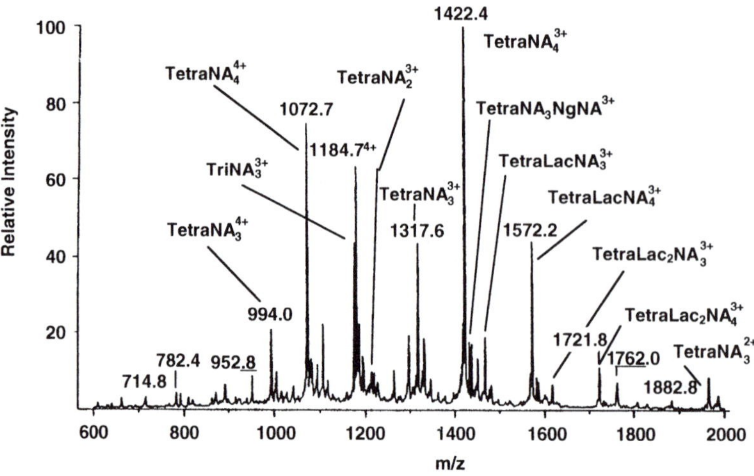

Figure 12.44 The ESI-mass spectrum of the ODM-treated glycan mixture from the deglycosylated Asn-83 glycopeptide from r-EPO. Sodium was the ionizing species, and all ions have an attached fucose residue. (© 1994, John Wiley & Sons, Limited. Reproduced from reference 69 with permission.)

PROBLEM 12.40 (a) The structures of TriNA$_3$Fuc from the N-38 and N-83 glycopeptides are the trisialylated derivatives of those shown in Scheme 12.9, where the NA–Gal linkage is Neu5NAc2–3Gal for all three linkages in both TriNA$_3$Fuc glycoforms. Using Tables 12.12, 12.13, and 12.16, and the information in Problems 12.28 and 12.39, provide detailed information which shows the calculations necessary in order to arrive at the M_r values of the native (untreated) and the ORM- and ODM-treated TriNA$_3$Fuc glycoforms in both the N-38 and N-83 glycopeptides. Assume that the reducing-end GlcNAc is in the closed-ring aldose form. The mass of the N-83 glycoform is higher than that of N-38.

(b) Deduce the identity of the 1184.7^{4+} ion. This can be arrived at by a perusal of Figure 12.45.

(c) Provide calculations to show the ODM mass of 1572.2^{3+} Da for the TetraLacNA$_4$Fuc^{3+} glycoform species in Figure 12.44 where Lac = LacNAc and Tetra = tetraantennary. The NA–Gal linkages are all Neu5NAcα2–3Gal. Calculate the untreated (underivatized) and ORM-processed mass for the tetraantennary glycoform.

(d) As Table 12.14 shows, the TetraLacNA$_4$Fuc glycoform is also a glycomer with TriLac$_2$NA$_4$Fuc. Can both Tri species (Scheme 12.9, 12.12 and 12.13) and the Tetra species (Scheme 12.14) be differentiated amongst each other with respect to mass value? Mass arrangement is similar to that of Schemes 12.12–12.14, and Table 12.16 shows that a 2,8-linked Neu5NAc residue has underivatized, ORM, and ODM mass values of 291.26, 361.39, and 361.39 Da, respectively.

Table 12.16 The ORM and ODM vicinal alcohol cleavage reactions of carbohydrate residues

Linkage	Residue[a]	M_r[b]	ORM product	Number of additional methyls	M_r (ORM)[b]	ODM product	M_r (ODM)[b]
1	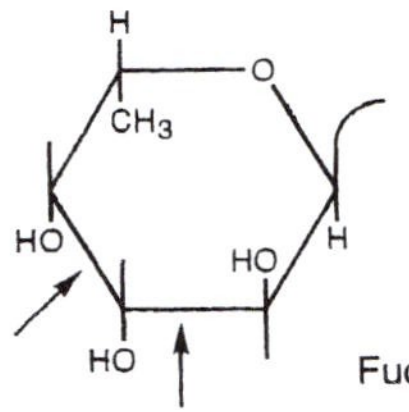 Fuc	147.151	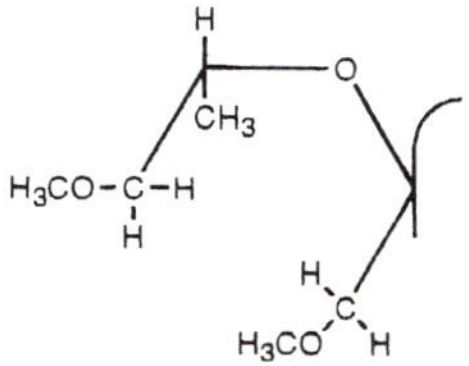	2	147.195	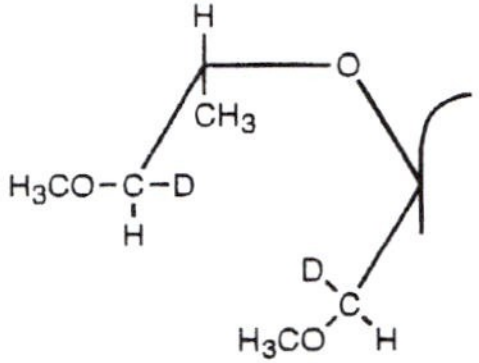	149.207
2,8	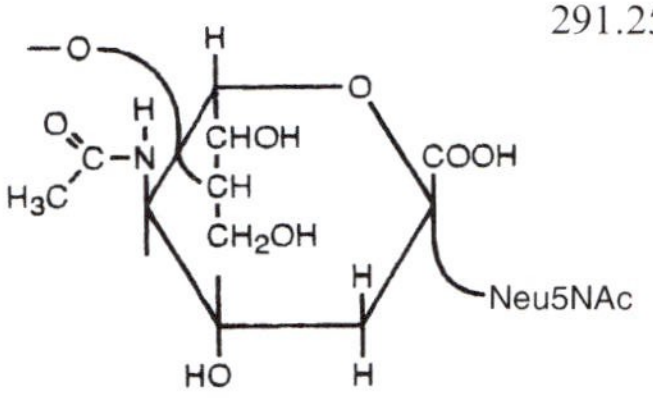 Neu5NAc	291.258	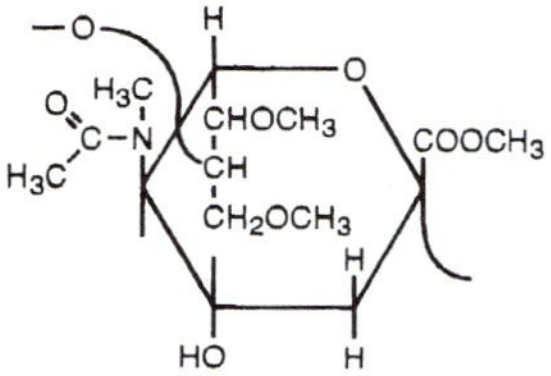	5	361.392	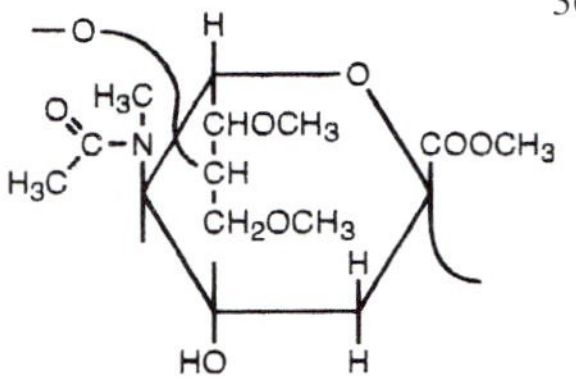	361.392

[a]Arrows indicate bond cleavage.

[b]M_r values in daltons.

A Perspective on Mass Spectral Analysis of Glycans

Despite the wealth of information and biochemical methods available for the mass spectrometry characterization of glycoproteins, the analytical technique still has a long way to go with respect to glycan identification. A great deal of information in this chapter has relied on inference and correlation, as opposed to direct evidence of the presence of a specific sugar residue. In most ESI-MS investigations, the best information obtained consists of hexose/hexosamine presence by virtue of the mass. Sialylation with tetramethylsiloxane of released sugar residues with gas chromatography (GC)-MS detection is one method to definitively identify saccharide presence. The GC retention time can differentiate between the various silylated hexose/hexosamine residues.[10,18,28,29,88,89]

Definitive biological studies of glycan composition can be achieved with specific glycosidase enzymes and nuclear magnetic resonance (NMR).[90] However, relatively significant quantities of sample are usually required for NMR analysis.[76] Certain polysaccharide features can be clarified by MS-methylation analysis and, to a greater extent, by ORM/ODM sample processing. However, Reinhold[51] states that "This series of chemical steps does not identify linkage but occasions a diminished molecular weight that reflects linkage composition," and this provides a cautionary message on the interpretation of ESI-mass spectral results from chemical modification methods in the elucidation of glycan composition.

The relatively limited capability of ESI-mass spectral analysis in the identification of glycans is not surprising given the great deal of structural and stereochemical issues that constitute polysaccharides. The following consitute the major parameters in saccharide identification:[91,92]

ring size; namely, furanose or pyranose

exact sugar identity; this is a stereochemical issue

sugar sequence

type of branching

type of linkage

the anomeric identity—namely, α or β—of each glycosidic linkage

the absolute configuration; namely, the mirror-image D- and L-forms of a sugar

As if these considerations are not daunting enough, Laine estimates that there are at least 50 amino sugars in the biological realm, with approximately 50 more neutral and acidic sugar species.[92] Considering the structural and stereochemical variables available to sugars, a hexasaccharide that contains six different hexoses can theoretically have over one trillion discrete geometries, and, including the L-form of each sugar, over 64 trillion geometries are possible. A nonasaccharide of nine different sugars would generate more than Avogadro's number of isomeric species.

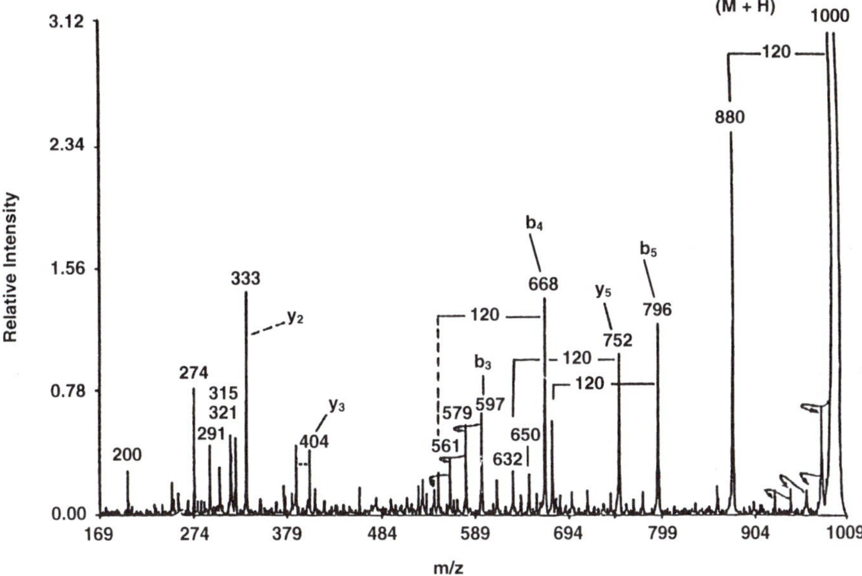

Figure 12.45 The ESI-tandem mass spectrum of the posttranslationally modified FTXAQW peptide from eosinophil-derived neurotoxin. Curved arrows indicate the loss of water (18 Da). The loss of 120 Da from the indicated ions is characteristic of aromatic *C*-glycoside compounds. (© 1996. Reprinted from reference 94 by permission of Academic Press, Inc., Orlando, Florida.)

The importance of ESI-mass spectrometry to the study of glycoproteins, however, cannot be emphasized enough. The ESI-MS technique is very sensitive, and because biological compounds are usually found in very small amounts, ESI-MS capitalizes on this advantage. Even though ESI-MS uses inference and correlation, to an extent, in glycoprotein analysis, the significant amount of current MS research can only improve the understanding of this very important class of biomolecule.

Other Amino Acid Residue Glycan Attachment Sites

Not all carbohydrates in glycoproteins are attached to Ser, Thr, and Asn residues. There are a handful of biological species in which the glycan is attached to another amino acid residue. The enzyme RNase U_s, which depolymerizes RNA in food, is found in human urine.

Eosinophil-derived neurotoxin (EDN) has neurotoxic effects when injected into the cerebellum, and this protein is identical to that of human RNase U_s (from DNA sequence studies), except for a particular amino acid residue at position 7 in the polypeptide.[93,94] Conventional chemical methods could not identify this phenomenon, and a posttranslational modification was suggested for the amino acid at position 7 in EDN. Various proteases were used to generate different N-terminal peptides, and ESI-mass spectral data were collected as follows:

Peptide	Residue number	Sequence	Mass $(M + H)^+$ Observed	Mass $(M + H)^+$ Calculated
1	1–12	KPPQFTXAQWFE	1727.0	1564.8
2	5–10	FTXAQW	1000.4	838.4
3	5–8	FTXA	686.4	524.3
4	6–8	TXA	539.2	377.2

Here, X is the posttranslationally modified amino acid.

PROBLEM 12.41 (a) From the data above, what type and class of compound is the posttranslational modification on the seventh amino acid residue in EDN?

(b) The ESI-tandem mass spectrum of peptide number 2 is presented in Figure 12.45.[94] Deduce the identity of the seventh amino acid residue in the glycopeptide.

Subsequent derivatization and tandem mass spectrometry experiments showed that the hexose was attached to the Trp side-chain indolyl ring by a direct C_1 hexose – C_2 indole ring bond. The NMR techniques confirmed the C_1 hexose – C_2 indole ring bond, and showed that the saccharide is α-mannose.

REFERENCES

1. Gahmberg, C.G.; Tolvanen, M. *Trends Biochem. Sci.* **1996**, *21*, pp 308–311.
2. Varki, A. *Glycobiology* **1993**, *3*, pp 97–130.
3. Paulson, J.C. In *Proteins: Form and Function*; Bradshaw, R.A.; Purton, M., Eds.; Elsevier: Cambridge, U.K., 1990; pp 209–217.
4. Bock, G.; Harnett, S., Eds. *Carbohydrate Recognition in Cellular Function*; Ciba Foundation Symposium No. 145; John Wiley & Sons: New York, 1989, 294pp.
5. Messner, P. *Glycoconjugate J.* **1997**, *14*, pp 3–11.
6. Spellman, M.W. *Anal. Chem.* **1990**, *62*, pp 1714–1722.
7. Zinn, A.B.; Plantner, J.J.; Carlson, D.M. In *The Glycoconjugates: Mammalian Glycoproteins and Glycolipids*; Horowitz, M.I.; Pigman, W., Eds.; Academic Press: New York, 1977; Volume 1, Section 4, pp 69–85.
8. Rademaker, G.J.; Thomas-Oates, J. In *Methods in Molecular Biology: Protein and Peptide Analysis by Mass Spectrometry*; Chapman, J.R., Ed.; Humana Press: Totowa, NJ; Volume 61, Chapter 17, pp 231–241.
9. Plummer, T.H., Jr.; Tarentino, A.L.; Hauer, C.R. *J. Biol. Chem.* **1995**, *270*, pp 13192–13196.
10. Marshall, R.D.; Neuberger, A. In *Glycoproteins, their Composition, Structure and Function*; Gottschalk, A., Ed.; Elsevier Publishing Co.: Amsterdam, The Netherlands, 1972; Volume 5, Part A, Chapter 3, Section 7, pp 322–380.
11. Lee, Y.C.; Scocca, J.R. *J. Biol. Chem.* **1972**, *247*, pp 5753–5758.
12. Morris, H.R.; Dell, A.; Panico, M.; Thomas-Oates, J.; Rogers, M.; McDowell, R.; Chatterjee, A. In *Mass Spectrometry of Biological Materials*; McEwen, C.N.; Larsen, B.S., Eds.; Marcel Dekker: New York, 1990; Chapter 4, pp 137–167.

13. Harvey, D.J. In *Methods in Molecular Biology: Basic Protein and Peptide Analysis by Mass Spectrometry*; Chapman, J.R., Ed.; Humana Press: Totowa, NJ, 1994; Volume 61, Chapter 18, pp 243–253.
14. Takasaki, S.; Mizuochi, T.; Kobata, A. In *Methods in Enzymology: Complex Carbohydrates*; Ginsburg, V., Ed.; Academic Press: San Diego, CA, 1982; Volume 83, Part D, Chapter 17, pp 263–268.
15. Takasaki, S.; Mizuochi, T.; Kobata, A. In *Methods in Enzymology: Complex Carbohydrates*; Ginsburg, V., Ed.; Academic Press: San Diego, CA, 1982; Volume 83, Part D, Chapter 17, pp 263–268.
16. Patel, T.; Bruce, J.; Merry, A.; Bigge, C.; Wormald, M.; Jaques, A.; Parekh, R. *Biochemistry* **1993**, *32*, pp 679–693.
17. Patel, T.P.; Parekh, R.B. In *Methods in Enzymology: Guide to Techniques in Glycobiology*; Lennarz, W.J.; Hart, G.W., Eds.; Academic Press: San Diego, CA, 1994; Volume 230, Chapter 5, pp 57–66.
18. Settineri, C.A.; Burlingame, A.L. In *Modern Chromatographic and Electrophoretic Methods in Carbohydrate Analysis*; Elrassi, Z., Ed.; Elsevier: Amsterdam, The Netherlands, 1995; pp 447–514.
19. Davies, M.J.; Smith, K.D.; Hounsell, E.F. In *Methods in Molecular Biology: Basic Protein and Peptide Protocols*; Walker, J.M., Ed.; Humana Press: Totowa, NJ, 1994; Chapter 17, pp 129–141.
20. Chiesa, C.; O'Neill, R.A.; Horváth, C.G.; Oefner, P.J. In *Capillary Electrophoresis in Analytical Biotechnology*; Righetti, P.G., Ed.; CRC Press: Boca Raton, FL, 1996; Chapter 9, pp 277–430.
21. Ciucanu, I.; Kerek, F. *Carbohydr. Res.* **1984**, *131*, pp 209–217.
22. Marks, G.S.; Marshall, R.D.; Neuberger, A. *Biochem. J.* **1963**, *87*, pp 274–281.
23. Yamashina, I.; Makino, M. *J. Biochem.* **1962**, *51*, pp 359–364.
24. Townsend, R.R. In *Chromatography in Biotechnology*; Horváth, C.; Ettre, L.S., Eds.; ACS Symposium Series 529; American Chemical Society: Washington, DC, 1993; Chapter 7, pp 86–101.
25. Morrison, R.T.; Boyd, R.N., Eds.; *Organic Chemistry*; 2nd Edition; Allyn & Bacon: Boston, MA, 1966.
26. Conn, E.E.; Stumpf, P.K., Eds.; *Outlines of Biochemistry*; 3rd Edition; John Wiley & Sons: New York, 1972; Chapter 2, pp 23–51.
27. Angel, A.-S.; Nilsson, B. In *Methods in Enzymology: Mass Spectrometry*; McCloskey, J.A., Ed.; Academic Press: San Diego, CA, 1990; Volume 193, Chapter 32, pp 587–599.
28. Baenziger, J.; Kornfeld, S.; Kochwa, S. *J. Biol. Chem.* **1974**, *249*, pp 1897–1903.
29. Merkle, R.K.; Poppe, I. In *Methods in Enzymology: Guide to Techniques in Glycobiology*; Lennarz, W.J. Hart, G.W., Eds.; Academic Press: San Diego, CA, 1994; Volume 230, Chapter 1, pp 1–15.
30. Hakomori, S.-I. *J. Biochem.* **1964**, *55*, pp 205–208.
31. March, J., Ed.; *Advanced Organic Chemistry*, 3rd Edition, John Wiley & Sons: New York, Chapter 14, p 629.
32. Zumdahl, S.S., Ed. Chemistry, 3rd Edition, D.C. Heath & Co.: Lexington, MA, 1993; p 1059.
33. Jacob, G.S.; Scudder, P. In *Methods in Enzymology: Guide to Techniques in Glycobiology*; Lennarz, W.J.; Hart, G.W., Eds.; Academic Press: San Diego, CA, 1994; Volume 230, Chapter 17, pp 280–299.

34. Settineri, C.A.; Burlingame, A.L. In *Methods in Molecular Biology: Protein and Peptide Analysis by Mass Spectrometry*; Chapman, J.R., Ed.; Humana Press, Totowa, NJ; Chapter 19, pp 255–278.
35. Maley, F.; Trimble, R.B.; Tarentino, A.L.; Plummer, T.H., Jr., *Anal. Biochem.* **1989**, *180*, pp 195–204.
36. Li, Y.-T.; Li, S.-C. In *The Glycoconjugates: Mammalian Glycoproteins and Glycolipids*; Horowitz, M.I.; Pigman, W., Eds.; Academic Press: New York, 1977; Volume 1, Section 3, pp 51-67.
37. Tarentino, A.L.; Plummer, T.H., Jr., In *Methods in Enzymology: Guide to Techniques in Glycobiology*; Lennarz, W.J.; Hart, G.W., Eds.; Academic Press: San Diego, CA, 1994; Volume 230, Chapter 4, pp 44–57.
38. Tarentino, A.L.; Gómez, C.M.; Plummer, T.H., Jr., *Biochemistry* **1985**, *24*, pp 4665–4671.
39. Gillece-Castro, B.L.; Burlingame, A.L. In *Methods in Enzymology: Mass Spectrometry*; McCloskey, J.A., Ed.; Academic Press: San Diego, CA, 1990; Volume 193, Chapter 37, pp 689–712.
40. Kaartinen, V.; Williams, J.C.; Tomich, J.; Yates, J.R., III; Hood, L.E.; Mononen, I. *J. Biol. Chem.* **1991**, *226*, pp 5860–5869.
41. Shively, J.E. In *Methods in Protein Sequence Analysis*; Jörnvall, H.; Höög, J.-O.; Gustavsson, A.-M., Eds.; Birkhäuser Verlag: Basel, Switzerland, 1991; pp 91–101.
42. Settineri, C.A.; Burlingame, A.L. In *Mass Spectrometry in the Biological Sciences: A Tutorial*; Gross, M.L., Ed.; Kluwer Academic Publishers: The Netherlands, 1992; pp 371–381.
43. Burlingame, A.L. *Curr. Opin. in Biotechnol.* **1996**, *7*, pp 4–10.
44. Li, Y.-T.; Li, S.-C. In *Methods in Enzymology: Complex Carbohydrates*; Ginsburg, V., Ed.; Academic Press: San Diego, CA, 1972; Volume 28, Part B, Chapter 91, pp 714–721.
45. Rohrer, J.; Thayer, J.; Avdalovic, N.; Weitzhandler, M. In *Techniques in Protein Chemistry VI*; Crabb, J.W., Ed.; Academic Press: San Diego, CA, 1995; pp 65–73.
46. Baenziger, J.; Kornfeld, S.; Kochwa, S. *J. Biol. Chem.* **1974**, *249*, pp 1889–1896.
47. Mørtz, E.; Sareneva, T.; Julkunen, I.; Roepstorff, P. *J. Mass Spectrom.* **1996**, *31*, pp 1109–1118.
48. Weber, P.L.; Bramich, C.J.; Lunte, S.M. *J. Chromatogr., A* **1994**, *680*, pp 225–232.
49. Fukuda, M. In *Methods in Enzymology: Complex Carbohydrates*; Ginsburg, V., Ed.; Academic Press: San Diego, CA, 1989; Volume 179, Part F, Chapter 2, pp 17–29.
50. Roberts, G.D.; Johnson, W.P.; Burman, S.; Anumula, K.R.; Carr, S.A. *Anal. Chem.* **1995**, *67*, pp 3613–3625.
51. Reinhold, V.N.; Reinhold, B.B.; Costello, C.E. *Anal. Chem.* **1995**, *67*, pp 1772–1784.
52. Settineri, C.A.; Burlingame, A.L. In *Techniques in Protein Chemistry V*; Crabb, J.W., Ed.; Academic Press: San Diego, CA, 1994; pp 97–104.
53. Domon, B.; Costello, C. *Glycoconjugate J.* **1988**, *5*, pp 397–409.
54. Harvey, D.J.; Bateman, R.H.; Green, M.R. *J. Mass Spectrom.* **1997**, *32*, pp 167–187.
55. Costello, C.E.; Vath, J.E. In Methods In *Enzymology: Mass Spectrometry*; McCloskey, J.A., Ed.; Academic Press: San Diego, CA, 1990; Volume 193, Chapter 40, pp 738–768.
56. Poulter, L.; Burlingame, A.L. In Methods In *Enzymology: Mass Spectrometry*; McCloskey, J.A., Ed.; Academic Press: San Diego, CA, 1990; Volume 193, Chapter 36, pp 661–689.
57. Prome, J.-C.; Aurelle, H.; Prome, D.; Savagnac, A. *Org. Mass Spectrom.* **1987**, *22*, pp 6–12.

58. Peter-Katalinić, J.; Williger, K.; Egge, H.; Green, B.; Hanisch, F.-G.; Schindler, D. J. *Carbohydr. Chem.* **1994**, *13*, pp 447–456.
59. Peter-Katalinić, J.; Ashcroft, A.; Green, B; Hanisch, F.-G.; Nakahara, Y.; Ilijima, H.; Ogawa, T. *Org. Mass Spectrom.* **1994**, *29*, pp 747–752.
60. Harris, R.J.; van Halbeek, H.; Glushka, J.; Basa, L.J.; Ling, V.T.; Smith, K.J.; Spellman, M.W. *Biochemistry* **1993**, *32*, pp 6539–6547.
61. Medzihradszky, K.F.; Settineri, C.A.; Maltby, D.A.; Burlingame, A.L. In *Techniques in Protein Chemistry IV*; Crabb, J.W., Ed.; Academic Press: San Diego, CA, 1993; pp 117–125.
62. Schindler, P.A.; Settineri, C.A.; Collet, X.; Fielding, C.J.; Burlingame, A.L. *Protein Sci.* **1995**, *4*, pp 791–803.
63. Conradt, H.S.; Nimtz, M.; Dittmar, K.E.J.; Lindemaier, W.; Hoppe, J.; Hauser, H. J. *Biol. Chem.* **1989**, *264*, pp 17368–17373.
64. Clogston, C.L.; Hu, S.; Boone, T.C.; Lu, H.S. *J. Chromatogr.* **1993**, *637*, pp 55–62.
65. Raju, T.S.; Lerner, L.; O'Connor, J.V. *Biotechnol. Appl. Biochem.* **1996**, *24*, pp 191–194.
66. Linsley, K.B.; Chan, S.-Y.; Chan, S.; Reinhold, B.B.; Lisi, P.J.; Reinhold, V.N. *Anal. Biochem.* **1994**, *219*, pp 207–217.
67. Reinhold, B.B.; Hauer, C.R.; Plummer, T.H.; Reinhold, V.N. *J. Biol. Chem.* **1995**, *270*, pp 13197–13203.
68. Reinhold, V.N.; Reinhold, B.B.; Chan, S. In Methods in Enzymology: High Resolution Separation and Analysis of Biological Macromolecules; Karger, B.L.; Hancock, W.S., Eds.; Academic Press: San Diego, CA, 1996; Volume 271, Chapter 16, pp 377–402.
69. Reinhold, V.N.; Reinhold, B.B.; Chan, S. In Biological Mass Spectrometry: Present and Future; Matsuo, T.; Caprioli, R.M.; Gross, M.L.; Seyama, Y., Eds.; John Wiley & Sons: Chichester, U.K., 1994; Chapter 3.8, pp 403–435.
70. Chakel, J.A.; Apffel, A.; Hancock, W.S. *LC-GC* **1995**, *13*, pp 866–876.
71. Hunter, A.P.; Games, D.E. *Rapid Commun. Mass Spectrom.* **1995**, *9*, pp 42–56.
72. Kelly, J.F.; Locke, S.J.; Ramaley, L.; Thibault, P. *J. Chromatogr., A* **1996**, *720*, pp 409–427.
73. Duffin, K.L.; Welply, J.K.; Huang, E.; Henion, J.D. *Anal. Chem.* **1992**, *64*, pp 1440–1448.
74. Conboy, J.J.; Henion, J.D. *J. Am. Soc. Mass Spectrom.* **1992**, *3*, pp 804–814.
75. Reinhold, B.B.; Reinhold, N.N., *J. Am. Soc. Mass Spectrom.* **1992**, *3*, pp 207–215.
76. Dell, A.; Reason, A.J.; Khoo, K.-H.; Panico, M.; McDowell, R.A.; Morris, H.R. In *Methods in Enzymology: Guide to Techniques in Glycobiology*; Lennarz, W.J.; Hart, G.W., Eds.; Academic Press: San Diego, CA, 1994; Volume 230, Chapter 8, pp 108–133.
77. Reinhold, B.B.; Reinherz, E.L.; Reinhold, V.N. In *Techniques in Protein Chemistry III*; Angeletti, R.H., Ed.; Academic Press: San Diego, CA, 1992; pp 287–294.
78. Recny, M.A.; Luther, M.A.; Knoppers, M.H.; Neihardt, E.A.; Khandekar, S.S.; Concino, M.F.; Schimke, P.A.; Francis, M.A.; Moebius, U.; Reinhold, B.B.; Reinhold, V.N.; Reinherz, E.L. *J. Biol. Chem.* **1992**, *267*, pp 22428–22434.
79. Guzzetta, A.W.; Basa, L.J.; Hancock, W.S.; Keyt, B.A.; Bennett, W.F. *Anal. Chem.* **1993**, *65*, pp 2953–2962.
80. Guzzetta, A.W.; Hancock, W.S. In *New Methods in Peptide Mapping for the Characterization of Proteins*; Hancock, W.S., Ed.; CRC Press: Boca Raton, FL, 1996; Chapter 7, pp 181–217.
81. Marshall, R.D. *Annu. Rev. Biochem.* **1972**, *41*, pp 673–702.

82. Struck, D.K.; Lennarz, W.J. In *The Biochemistry of Glycoproteins and Proteoglycans*; Lennarz, W.J., Ed.; Plenum Press: New York, 1980; Chapter 2, pp 35–83.
83. Kornfeld, R.; Kornfeld, S. *Ann. Rev. Biochem.* **1985**, *54*, pp 631–664.
84. Greis, K.D.; Hayes, B.K.; Comer, F.I.; Kirk, M.; Barnes, S.; Lowary, T.L.; Hart, G.W. *Anal. Biochem.* **1996**, *234*, pp 38–49.
85. Medzihradszky, K.F.; Maltby, D.A.; Hall, S.C.; Settineri, C.A.; Burlingame, A.L. J. Am. Soc. Mass Spectrom. **1994**, 5, pp 350–358.
86. Hemling, M.E.; Mentzer, M.A.; Capiau, C.; Carr, S.A. In *Mass Spectrometry in the Biological Sciences*; Burlingame, A.L.; Carr, S.A., Eds.; Humana Press: Totowa, NJ, 1996; pp 307–331.
87. Rush, R.S.; Derby, P.L.; Smith, D.M.; Merry, C.; Rogers, G.; Rohde, M.F.; Katta, V. *Anal. Chem.* **1995**, *67*, pp 1442–1452.
88. Laine, R.A. In *Methods in Enzymology: Mass Spectrometry*; McCloskey, J.A., Ed.; Academic Press: San Diego, CA, 1990; Volume 193, Chapter 29, pp 539–553.
89. Hellerqvist, C.G. In *Methods in Enzymology: Mass Spectrometry*; McCloskey, J.A., Ed.; Academic Press: San Diego, CA, 1990; Volume 193, Chapter 30, pp 554–573.
90. Aubert, J.-P.; Biserte, G.; Loucheux-Lefebvre, M.-H. *Arch. Biochem. Biophys.* **1976**, *175*, pp 410–418.
91. McNeil, M.; Darvill, A.G.; Aman, P.; Franzén, L.-E.; Albersheim, P. In *Methods in Enzymology: Complex Carbohydrates*; Ginsburg, V., Ed.; Academic Press: San Diego, CA, 1982; Volume 83, Part D, Chapter 1, pp 3–45.
92. Laine, R.A. *Glycobiology* **1994**, *4*, pp 759–767.
93. Hofsteenge, J.; Müller, D.R.; de Beer, T.; Löffler, A.; Richter, W.J.; Vliegenthart, J.F.G. *Biochemistry* **1994**, *33*, pp 13524–13530.
94. Hofsteenge, J.; Löffler, A.; Müller, D.R.; Richter, W.J.; de Beer, T.; Vliegenthart, J.F.G. In *Techniques in Protein Chemistry VII*; Marshak, D.R., Ed.; Academic Press: San Diego, CA, 1996; pp 163–171.

ANSWERS

PROBLEM 12.1 The m/z 1110.3 appears to be the disodium adduct of the tetrasaccharide-Thr species:

$$(\mathrm{M} - 3\mathrm{H} + 2\mathrm{Na})^- = 1066.94 - 3 + 46 = 1109.94 \simeq 1110.3 \text{ Da (observed)}$$

PROBLEM 12.2 Since Figure 12.2 shows dominant $(\mathrm{M} + \mathrm{Na})^+$ ions with respect to the much weaker $(\mathrm{M} + \mathrm{H})^+$ ions, it seems reasonable that Na^+ is part of these ions.

A straightforward loss is a Neu5NAc. Thus:

$$\text{Ser: } (\mathrm{M} + \mathrm{Na})^+ = 1075.94 \text{ Da}$$

$$1075.94 - 292.26 + 1 = 784.68 \simeq 784.1 \text{ Da}$$

A loss of a sialic acid residue (Neu5NAc) produces a hydrogen transfer to the ion (i.e., 292.26 – 1 = loss of 291.26 or –291.26) in order to neutralize the free oxygen (Scheme 12.3). Since a Na^+ is already present, we have a $(\mathrm{HOR})\mathrm{Na}^+$

ion, which is equivalent to a $(Y + H + Na)^+$ ion. This similar line of reasoning can be found for the other species:

$$\text{Thr: } (M + Na)^+ = 1089.94 \text{ Da}$$
$$1089.94 - 292.26 + 1 = 798.68 \simeq 798.3 \text{ Da}$$

PROBLEM 12.3 A reasonable evaluation is to start with the $(M + Na)^+$ ions again. Therefore:

$$\text{Ser: } (M + Na)^+ = 1075.94 \text{ Da}$$
$$1075.94 - 292.26 + H = 784.68 \text{ Da}$$
$$784.68 - 163.14 + H = 622.54 \text{ Da}$$
$$\text{Thr: } (M + Na)^+ = 1089.94 \text{ Da}$$
$$1089.94 - 292.26 + H = 798.7 \text{ Da}$$
$$798.7 - 163.14 + H = 636.54 \text{ Da}$$

In both species, an $(M + Na)^+$ is assumed and a Neu5NAc loss is postulated as in Problem 12.2. Since the resulting mass is too high, a galactose loss seems reasonable. Remember, its neutral residue mass is now 163.14 Da, not 162.14 Da, because it is a terminal residue. Each sugar loss includes a transfer of a hydrogen to the sodiated species. However, the resulting masses are still too high by 22 Da. It appears that these ions originated from the protonated and not the sodiated form because

$$\text{Ser: } 622.54 - 23 + 1 = 600.54 \text{ Da}$$
$$\text{Thr: } 636.54 - 23 + 1 = 614.54 \text{ Da}$$

and these values are close to the observed values.

Another way of viewing this is by assuming a loss of a Neu5NAcα2–3Gal leaving group of 454.4 Da from the protonated species:

$$\text{Ser: } (M + H)^+ = 1053.94 \text{ Da}$$
$$1053.94 - 454.4 + 1 = 600.54 \text{ Da}$$
$$\text{Thr: } (M + H)^+ = 1067.94 \text{ Da}$$
$$1067.94 - 454.4 + 1 = 614.54 \text{ Da;}$$

These are Y-type ions.

PROBLEM 12.4 Neu5NAcα2-6GalNAc:

$$292.26 + 203.19 = 495.45 \text{ Da}$$

This is a B-type ion. Note that the GalNAc residue picks up a hydrogen from the departing Neu5NAc–Gal disaccharide and is also cleaved from the amino acid group via the B-ion mechanism in Scheme 2.

PROBLEM 12.5 The Neu5NAcα2–6GalNAc seems a likely choice:

$$\begin{array}{rl} 292.26 + 203.19 = 495.45 & \\ -\text{H} & \text{(H goes to Ser)} \\ -\text{H} & \text{for ionization in negative mode} \\ \hline 493.45 & = 493.4 \text{ Da observed} \end{array}$$

This is a $(B_n - 2H)^-$ ion.

PROBLEM 12.6

m/z 987.8:

$$\begin{array}{rrl} (M - H)^- = & 1279.20 & \\ & -291.26 & \\ \hline & 987.94 & \text{This is a Y-type cleavage} \end{array}$$

m/z 696.5:

$$1279.20 - 290.26 - 292.26 = 696.68 \text{ Da}$$

This is analogous to the loss of two NA groups in the first example in this section ("Ser-Thr-Linked Polysaccharide ESI-Mass Spectra").

PROBLEM 12.7

$$\begin{array}{rrl} (M + Na)^+ \text{ calculated} = & m/z\ 1303.2 & \\ & -1 & H^+ \\ \hline & 1302.2 & \\ & +23 & Na^+ \\ \hline (M - H + 2Na)^+ = & 1325.2 & \text{Da} \end{array}$$

PROBLEM 12.8 Provide mass labels for each type of species:

Neu5NAc-Gal-GlcNAc-Fuc-O-polypeptide

$$292.26 + 162.14 + 203.19 + 146.14 + 965.0$$

Summing all five masses yields = 1768.72 = M

$$(M + H)+ = 1769.72$$

This ion is not directly observed, but it may manifest itself as a doubly charged ion: $(1768.72 + 2)/2 = 885.4^{2+}$, and this is observed in Figure 12.7. A loss of a Neu5NAc residue yields the following:

$$1768.72 - 291.26 = 1477.46$$

This is not observed, but it may exist as a doubly charged ion:

$$(1477.46 + 2)/2 = 739.73^{2+}$$

and this is observed.

It seems reasonable to investigate a loss of the Neu5NAc-Gal species:

$$1769.72 - (292.26 + 161.14) = 1316.32\text{Da} = \text{singly charged}$$

$$(1316.32 + 1)/2 = 658.66\text{Da} = \text{doubly charged species}$$

The m/z 658.66 ion appears to be present in Figure 12.7, even though it is not labeled. The ion at m/z 1112.5 is calculated as follows:

$$1768.72 - (292.26 + 162.14) - 203.19 + 1 + H^+ = 1113.13\ \text{Da} \simeq 1112.5\ \text{Da}$$

This ion represents the (Fuc-O-polypeptide + H)$^+$ species. Finally, the ion at m/z 966.4:

$$1768.72 - (292.26 + 162.14 + 203.19) - 146.14 + 1 + H^+ = 967.0\ \text{Da}$$

This represents the polypeptide chain. There are three doubly charged and two singly charged peaks for sequence information purposes.

The data in Figure 12.7 provides information as to the sequence of each of the residues, mainly because each carbohydrate residue has a different mass. Each sugar unit can be traced via successive losses from the intact glycopeptide.

PROBLEM 12.9 There are several ways to calculate the sugar component. The most obvious is to subtract the mass of the intense peak near that of the (M + H)$^+$ signal; however, the mass of the intense peak is not given in Figure 12.8, and only an estimate of this value can be obtained by inspection. Another way is to subtract the a_9^* ion from the a_9 ion. The a_9 ion includes the carbohydrate and the a_9^* does not; therefore, the difference should yield the mass of the carbohydrate.

The carbohydrate can only be located on either Thr^{88} or Thr^{90}; therefore, information on these two positions needs to be obtained. However, the mass of the carbohydrate can also be obtained by subtracting the mass of the entire polypeptide chain from the mass of the glycopeptide, thus:

$$(M + H)^+ = 1649.8\ \text{Da};\ \text{therefore}\ M = 1648.8\text{Da}$$

$$\text{mass of polypeptide} = 1486.7\ \text{-1 (from a Thr hydroxyl)} = 1485.7\ \text{Da}$$

Thus, we have 1648.8 − 1485.7 = 163.1 Da = the mass of a terminal hexose. Now we must determine which Thr residue the hexose is located upon. It is known that the hexose is a mannose residue. Looking at the a and b series, no information can be found for b_3, b_4, b_5 or a_3, a_4, a_5. The b_3^*, b_4^*, and a_4^* provide carbohydrate information, because these ions represent the deglycosylated fragments. Thus, recourse must be sought in the y and z series: y_8, y_9, y_{10}, and z_9. These values overlap Thr^{90}, and y_8-y_{10} are 987.07, 1088.18, and 1187.31 Da, respectively, without taking the carbohydrate into account. These three values are good estimates as to where their mass spectral signals are located by visual inspection. Given this series, z_9 is equal to y_9 − 15 (NH) = 1088.18 − 15 = 1073.18 Da, and this is observed in the spectrum. If a carbohydrate moiety was attached to Thr^{90}, then y_9, z_9, and y_{10} should be significantly higher in mass. By a process of elimination, Thr^{88} must harbor the mannose (hexose) residue.

The mass of the internal VT fragment is 200 nominal mass units. The first two residues, or b_2 = Lys − Ala = 1 + 128.17 + 71.078 = 200 nominal mass units.

PROBLEM 12.10 From Figure 12.9a, we have the following:

$$A2 = 2(1476.3) - 2; \ M = 2950.6 \text{ Da}$$
$$A3 = 3(984.8) - 3; \ M = \underline{2951.4 \text{ Da}}$$
$$\text{av } M_r = 2951.0 \text{ Da}$$

and in Figure 12.9b:

$$A = 1639.7 = M + H)+; \ M = 1638.7 \text{ Da}$$
$$A2 = 2(820.6) - 2; \ M = 1639.2 \text{ Da}$$

The average mass of the intact polypeptide is 1638.95 Da, and this includes the hydrogen atoms on the R-group hydroxyls of Ser-409 and Thr-407. The calculated M_r of the given amino acid sequence = 1655.76 Da. The difference between the calculated and observed polypeptide values is

$$1655.76 - 1638.95 = 16.81 Da \simeq 17 \text{ Da}$$

This information means that one of the amino acid residues underwent a modification in the handling/ESI process, and the result was 17 Da less than the native, calculated value. Table 9.11 shows that only a modification of Q to a pE residue can undergo a 17-Da loss. Thus, the sequence responsible for Figure 12.9b is pEGPPASPTASPEPPPPE. It was given that only two glycosidases were required to completely remove the oligosaccharide; therefore, at least one enzyme must be an endoglycosidase. Table 12.5 shows only one endoglycosidase for an O-linked saccharide, and that is *O*-glycosidase. If the sialic acid is terminal, the first enzyme can be the *N*-acetylneuraminidase, and the second, specific, glycosidase can be the *O*-glycanase from *D. pneumoniae* (Table 12.5). Thus, we have Neu5NAc-Galβ1-3GalNAc-Ser/Thr as the polysaccharide sequence. This same structure can reside on both Thr-407 and Ser-409, and this satisfies the requirement of two Neu5NAc, two Gal, and two GalNAc residues. Figure 12.9a was not required to solve this problem.

PROBLEM 12.11

Letter	Observed mass	M_r (Da)	M_r av (Da)
A	708.2^{3+}	2121.6	2121.4
	1061.6^{2+}	2121.2	
B	916.0^{2+}	1830.0	1830.0
C	805.2^{3+}	2412.6	2412.3
	1207.0^{2+}	2412.0	
E	835.4^{2+}	1668.8	1668.8
F	980.5^{2+}	1959.0	1959.0
G	733.4^{2+}	1464.8	1464.8
	1465.7^{+}	1464.7	

Order the glycopeptides by mass:

Glycopeptide	M_r av
G	1464.8
E	1668.8
B	1830.0
F	1959.0
A	2121.4
C	2412.3

The calculated polypeptide neutral M_r from Glu-117 to Arg-131 = 1465.62 Da. Subtracting a hydrogen from the Ser hydroxyl group yields a mass of 1464.62 Da ($\simeq$ 1464.8 above), and, by adding an ionizing proton, the protonated polypeptide is obtained at 1465.8 Da, as observed in Figure 12.11.

Each combination of difference values yields the following:

G = polypeptide
E − G = 204.0
B − E = 161.2
B − G = 365.2
F − B = 129.0?
F − E = 290.2
A − F = 162.4
A − B = 291.4
C − A = 290.9

It is reasonable to stop calculating differences when the value reaches that of sialic acid.

Simultaneous equations to yield sequences where pp = polypeptide:

E = G + 204 HexNAc−pp
B = 161.2 + E Hex−HexNAc−pp
B = 365.2 + G Hex−HexNAc−pp
F = 290.2 + E NA−Hex−HexNAc−pp
A = 162.4 + F Hex−HexNAc−pp (with NA attached to HexNAc) or Hex−NA−HexNAc−pp
A = 291.4 + B NA−Hex−HexNAc−pp or Hex−HexNAc−pp (with NA attached to HexNAc)

One can be inclined to choose A as the common structure for both simultaneous equations, yet we see in Problems 12.8 and 12.10 that (NA, Hex) HexNAc-pp is a common structural array for an O-linked glycopeptide, while the other two isomers are not observed. The actual structure is NA-Hex-HexNAc-pp.

$$C = 290.9 + A$$

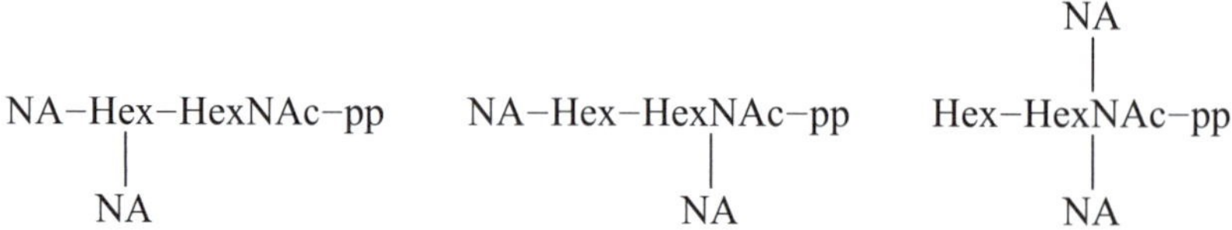

The sections on positive and negative ion mass spectral analyses of two tetrasaccharides dealt with the NA–Hex–HexNAc–pp structure.

The other two do not appear in the previous examples and are generally not present in O-linked polysaccharides.

From the given information, the carbohydrates cannot be placed on a particular Ser residue in the tryptic peptide.

PROBLEM 12.12 Figure 12.46 provides structures of both aldose and alditol derivatives of the heptasaccharide. Arrows indicate the inherent methyl groups. Calculation of the mass of each permethylated residue and their summation yields values of 1544.66 + 23 (Na^+) = 1567.66 Da and 1528.61 + 23 = 1551.61 Da for the alditol and aldose, respectively. Observed ions for the alditol and aldose were 1566 and 1550 Da, respectively (data not shown).

A more straightforward approach in the calculation of the derivatized M_r values is to take into consideration the number of additional methyl groups and reducing-end modification.

Aldose: 1245.1 + 16 + 1 + 19(14 Da) + 23 (Na) = 1551.1 Da

Added to the underivatized M_r is an extra reducing-end oxygen (which was counted as part of the Ser residue in the underivatized analyte) and hydrogen (which is part of the methyl addition on the reducing end), as well as a sodium ion and 19 methylene moieties. Methyl derivatization of an acidic species can be thought of as inserting a methylene group between the proton and oxygen (or nitrogen). There are 19 methylation sites, and, with the addition of a sodium ion, the 1551.1-Da ion was observed (spectrum not provided).

Alditol: 1245.1 + 17 + 1 + 1 + 20(14) + 23 = 1567.1 Da

The 17 Da arises from the Ser oxygen atom plus a hydrogen from the methyl group, and the two individual 1-Da values arise from the addition of a hydrogen on both the C-1 and the methyl attached to the oxygen atom on C-5 on the reducing-end residue. The difference in mass between the derivatized aldose and alditol is 16 Da.

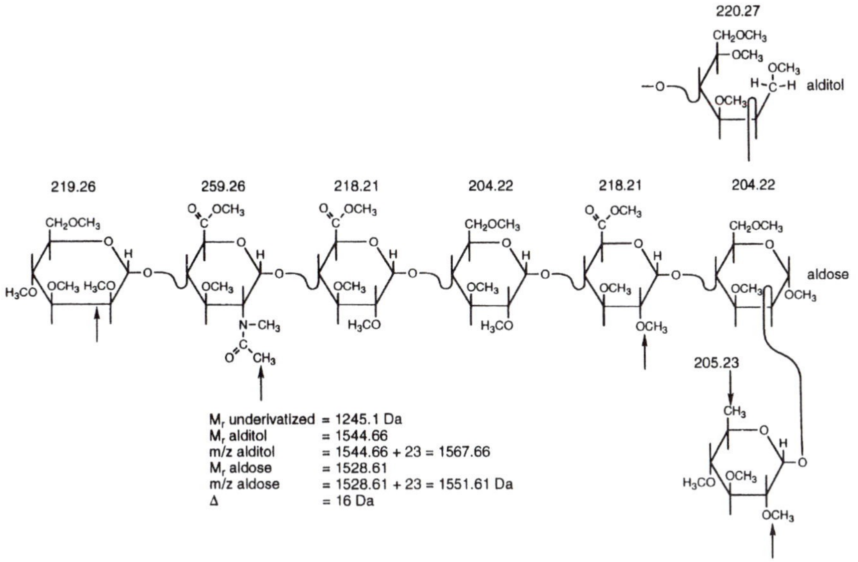
220.27
alditol
219.26
259.26
218.21
204.22
218.21
204.22
aldose
205.23
M_r underivatized = 1245.1 Da
M_r alditol = 1544.66
m/z alditol = 1544.66 + 23 = 1567.66
M_r aldose = 1528.61
m/z aldose = 1528.61 + 23 = 1551.61 Da
Δ = 16 Da

Figure 12.46 The structure of the permethylated alditol and aldose heptasaccharide of the polysaccharide shown in Figure 12.12, including molecular mass calculations.

PROBLEM 12.13 The permethylated forms of each of the following fragments were observed:

Man, nonreducing

Man-GlcUNAc, nonreducing

Man-GlcUNAc-GlcU-Glc, nonreducing

GlcU-Man-Rha, reducing

GlcU-Glu-GlcU-Man-Rha, reducing

GlcUNAc-GlcU-Glc-GlcU-Man-Rha, reducing

PROBLEM 12.14 By subtracting successive deconvoluted mass values from each other, a pattern emerges:

44,857(5)	44,695(3)	44,655(1)	44,249(1)
− 44,695	− 44,492	− 44,492(3)	− 44,087(1)
162	203	163	162

The values in parentheses refer to the error in ± Da.

These peaks or shoulders differ by Hex and HexNAc mass increments; therefore, they must be glycoforms. In addition, they are most likely two groups of glycoforms that represent the high-mannose and hybrid glycotypes.

PROBLEM 12.15 First method: Schemes 12.4 and 12.5 show that high-mannose and hybrid glycotypes are connected to the Asn residue by a GlcNAc, and Table

12.5 provides evidence that Endo-H cleaves a glycan in between the two GlcNAc residues. Thus, 42,869 Da + 203 Da = 43,072 Da. The mass value of 203 Da is used because this can be conceptually thought of as the residue mass being inserted between the H and N atoms of the amide side group of Asn in the polypeptide portion of ovalbumin.

Second method: The observed value from the spectrum is 43,073 Da, and this calculated value is based on the experimentally observed peaks in Figure 12.16 following the format outlined in Chapter 4.

PROBLEM 12.16 From Figure 12.15 and Table 12.8:

$$
\begin{array}{r}
44{,}087(1) \\
-\,42{,}869 \\ \hline
1{,}218 \\
+17 \\ \hline
1{,}235
\end{array}
\quad \text{or} \quad
\begin{array}{r}
44{,}087 \\
-\,42{,}870 \\ \hline
1{,}217 \\
+18 \\ \hline
1{,}235
\end{array}
= \text{Man}_5\text{GlcNAc}_2,\ \text{high mannose (Table 12.8)}
$$

The value of 42,870 for the polypeptide is derived from 43,073 Da (observed value in Figure 12.16) less a GlcNAc residue (203 Da). Continuing in Figure 12.15:

$$
\begin{array}{lrrrrr}
 & 44{,}249 & 44{,}492 & 44{,}655 & 44{,}695(3) & 44{,}857(5) \\
 & -\,42{,}870 & -\,42{,}870 & -\,42{,}870 & -\,42{,}870 & -\,42{,}870 \\ \hline
 & 1{,}379 & 1{,}622 & 1{,}785 & 1{,}825 & 1{,}987 \\
 & +18 & +18 & +18 & +18 & +18 \\ \hline
 & 1{,}397 & 1{,}640 & 1{,}803 & 1{,}843 & 2{,}005 \\
\text{Glycotype:} & \text{high mannose} & \text{hybrid} & \text{hybrid} & \text{hybrid} & \text{hybrid} \\
\text{Glycoform:} & G_2h_6 & G_4h_5 & G_4h_6 & G_5h_5 & G_5h_6
\end{array}
$$

where G = GlcNAc and h = hexose (mannose).

PROBLEM 12.17 These are all doubly charged ions of the higher mass singly charged $(M + H)^+$ peaks:

$$(441 \times 2) - 1 = 881 \text{ Da}$$
$$(522 \times 2) - 1 = 1043 \simeq 1044 \text{ Da observed}$$
$$(603 \times 2) - 1 = 1205 \text{ Da}$$
$$(684 \times 2) - 1 = 1367 \text{ Da}$$
$$(765 \times 2) - 1 = 1529 \text{ Da}$$
$$(847 \times 2) - 1 = 1693 \text{ Da}$$

PROBLEM 12.18 Careful measurement shows approximately a 1.875-mm spacing between the signals. Given a 9.0-mm distance as 500 Da, a value of 98.7 Da is calculated; this suggests a phosphate modification (Chapter 10). The irregular spacings of RNase B are due to both phosphate and carbohydrate posttranslational modifications.

PROBLEM 12.19 Three mass values are given:

$$(1236.8 \times 5) - 5 = 6179;\ 6179 - 4799.7 = 1379.3$$
$$1379.3 + 17\ (\text{OH}) = 1396.3 \simeq \text{Man}_6\ \text{GlcNAc}_2$$
$$(1301.8 \times 5) - 5 = 6504;\ 6504 - 4799.7 = 1704.3 + 17$$
$$= 1721.3 = \text{Man}_8\ \text{GlcNAc}_2$$
$$(1334.5 \times 5) - 5 = 6667.5;\ 6667.5 - 4799.7 = 1867.8 + 17$$
$$= 1884.8 \simeq \text{Man}_9\ \text{GlcNAc}_2$$

Table 12.8 provides the mass matching values. The unlabeled base peak in the 5+ cluster is the $\text{Man}_5\text{GlcNAc}_2$ glycoform, and the low-intensity peak between m/z 1236.5 and 1301.8 must be the $\text{Man}_7\text{GlcNAc}_2$ species.

PROBLEM 12.20 All M7 glycoforms are also glycomers, because they are isomeric. See Figures 12.47–12.49.

PROBLEM 12.21 The most intense peaks are significantly lower in mass than that listed in Table 12.8 under the permethylated M_r column. Therefore, most of the peaks in Figure 12.21 must be doubly charged with Na^+. Also, Endo-H cleaves between the two GlcNAc residues and yields the $\text{Man}_n\text{GlcNAc}$ species; reference is made to the permethylated $\text{Man}_n\text{GlcNAc}$ column in Table 12.8a.

$$m/z\ 474.8 \times 2 = 949.6 = (\text{M} + 2\text{Na})^{2+};\ 949.6 - 46 = 903.6\ \text{Da}$$

The $(\text{M} + \text{Na})^+$ ion, 903.6 + 23 = 926.6 Da, is not observed in Figure 12.21. This is a $\text{Man}_3\text{GlcNAc}$, and is found as 904.013 Da in Table 12.8.

$$m/z\ 576.9 \times 2 = 1153.8;\ 1153.8 - 46 = 1107.8 \simeq 1108.24\ \text{Da} = \text{M}_4\text{G}$$
$$m/z\ 679.0 \times 2 = 1358;\ 1358 - 46 = 1312 \simeq 1312.46\ \text{Da} = \text{M}_5\text{G}$$
$$m/z\ 780.9 \times 2 = 1561.8;\ 1561.8 - 46 = 1515.8 \simeq 1516.7\ \text{Da} = \text{M}_6\text{G}$$
$$m/z\ 883.0 \times 2 = 1766;\ 1766 - 46 = 1720 \simeq 1720.9\ \text{Da} = \text{M}_7\text{G}$$
$$m/z\ 985.1 \times 2 = 1970.2;\ 1970.2 - 46 = 1924.2 \simeq 1925.1\ \text{Da} = \text{M}_8\text{G}$$
$$m/z\ 1087.3 \times 2 = 2174.6;\ 2174.6 - 46 = 2128.6 \simeq 2129.35\ \text{Da} = \text{M}_9\text{G}$$
$$m/z\ 1334.5 = (\text{M} + \text{Na})^+ \simeq 1312.46 + 23 = 1335.46 = \text{M}_5\text{G}$$
$$m/z\ 1538.5 = (\text{M} + \text{Na})^+ \simeq 1516.68 + 23 = 1539.68 = \text{M}_6\text{G}$$
$$m/z\ 1742.4 = (\text{M} + \text{Na})^+ \simeq 1720.9 + 23 = 1743.9 = \text{M}_7\text{G}$$

Note that M_5G, M_6G and M_7G, are observed as the mono- and disodiated species in Figure 12.21.

PROBLEM 12.22

```
—M—M
     \
      M
     / \
    M    M—G2+
        /
```

$$(204.22 + 1) + 204.22 + (204.22 - 15) + (204.22 + 15)$$
$$+ (204.22 - 15 + 1) + (245.27 + 31) + 46 = 1330.37$$
$$1330.37/2 = 665.192 + \text{Da}$$

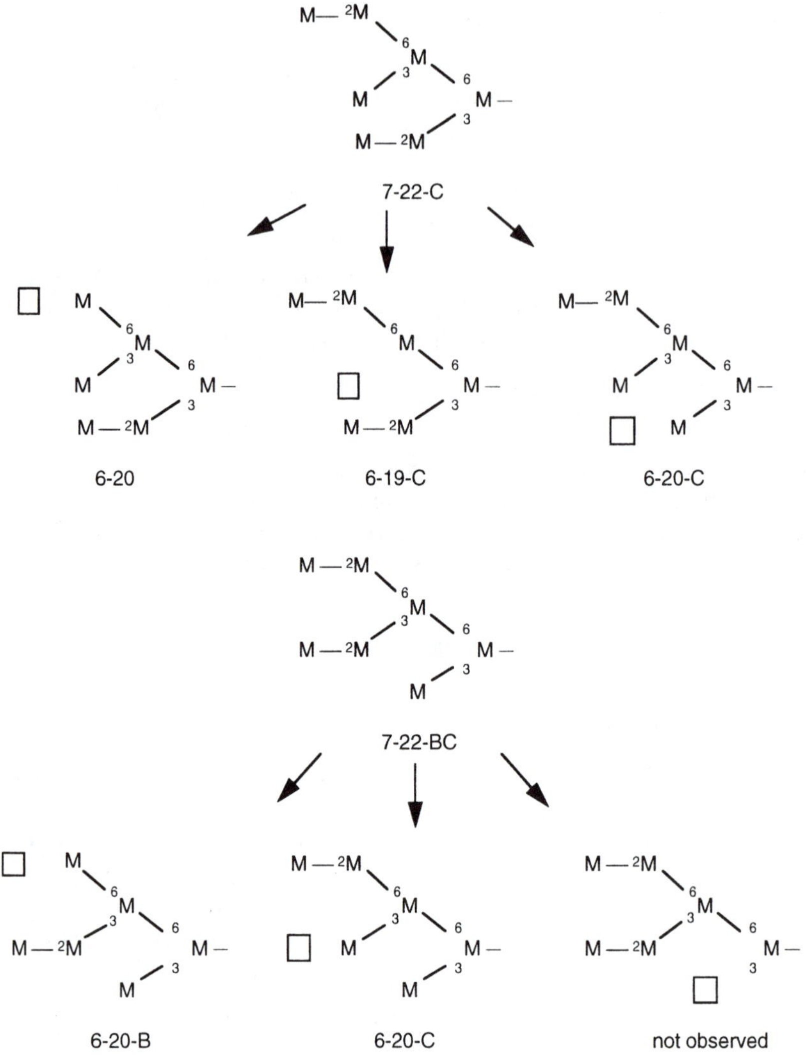

Figure 12.47 Mannose trimming of the 7-22-C and 7-22-BC high mannose glycoforms.

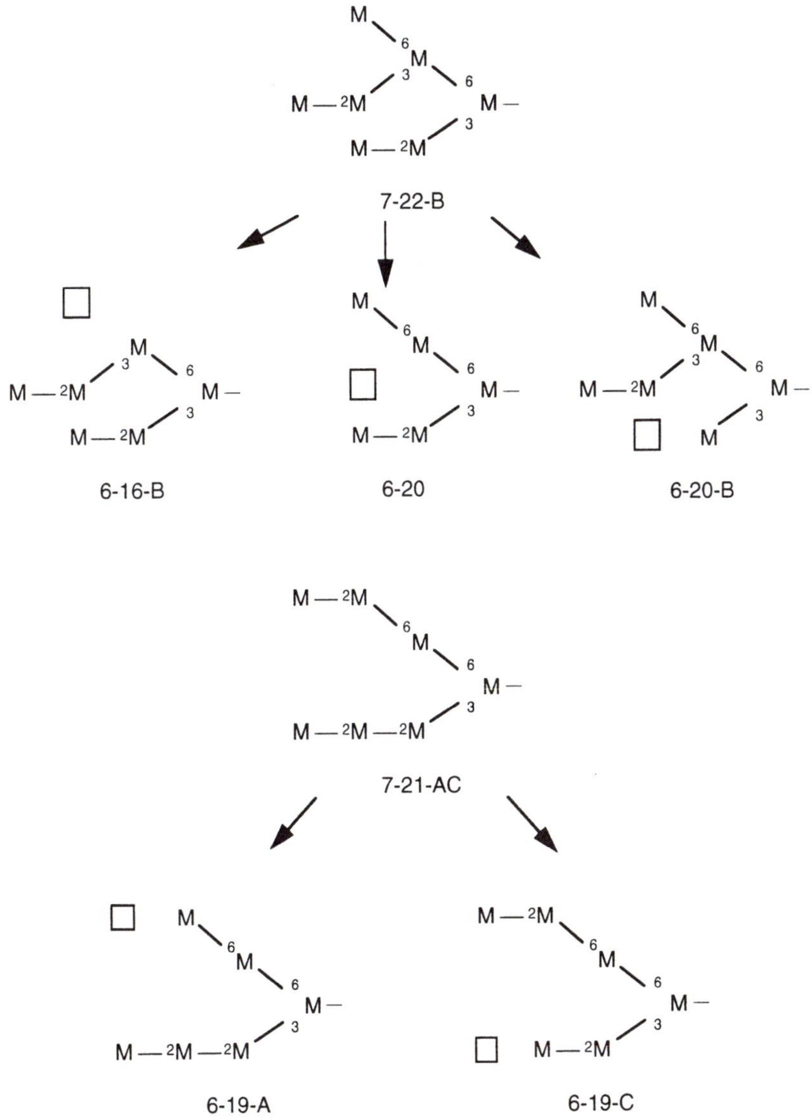

Figure 12.48 Mannose trimming of the 7-22-B and 7-21-AC high mannose glycoforms.

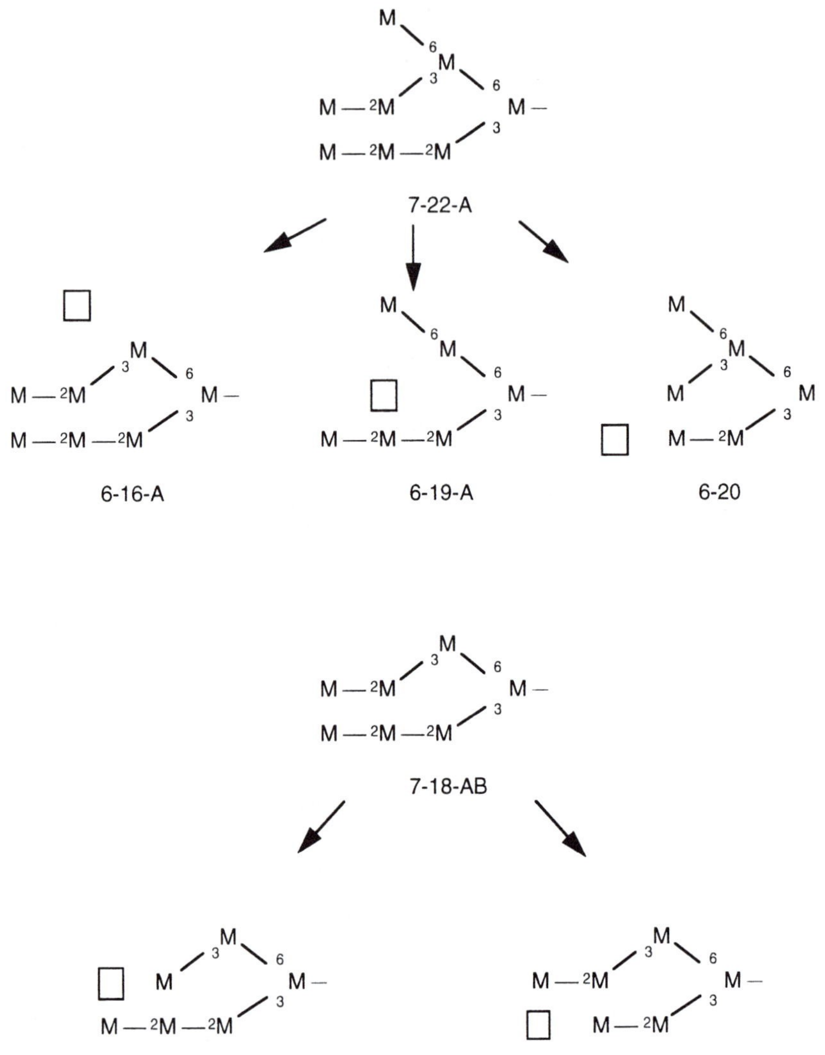

Figure 12.49 Mannose trimming of the 7-22-A and 7-18-AB high mannose glycoforms.

The 31 Da represents a terminal methoxy group on the GlcNAc residue.

M—M—M—^{3}OH

$$(204.22 + 15) + 204.22 + 204.22 + 17 + 23 = 667.66 \text{ Da}$$

PROBLEM 12.23

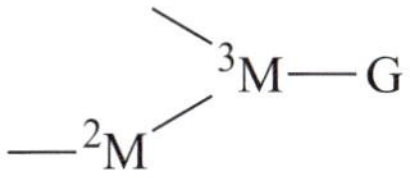

—^{2}M—^{3}M—G

$$(204.22 + 1) + (204.22 - 15 + 1) + (245.27 + 31) + 23 = 694.71 \text{ Da}$$

PROBLEM 12.24

$$635.66 + 88 = 723.66 \text{ Da} \quad \text{or}$$
$$636.66 + 87 = 723.66 \text{ Da}$$

PROBLEM 12.25

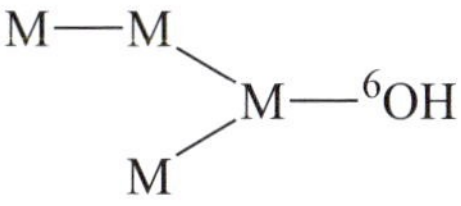

M—M—M—^{6}OH (with M branch)

$$853.9 + 1 + 17 = 871.9 \text{ Da}$$

M—M—M, M—M, —M—M—M—G^{2+}

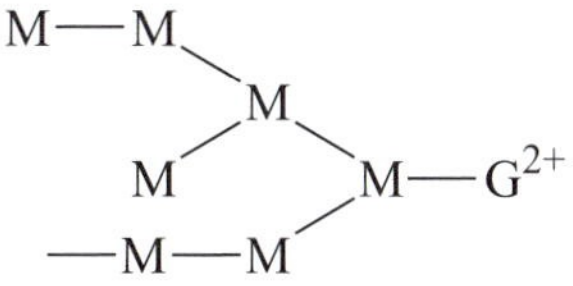

1971.03 = $(M + 2Na)^{2+}$
[1971.03 – (204.22 + 15) + 1] / 2 = 876.4^{2+} Da
reducing end gains the hydrogen

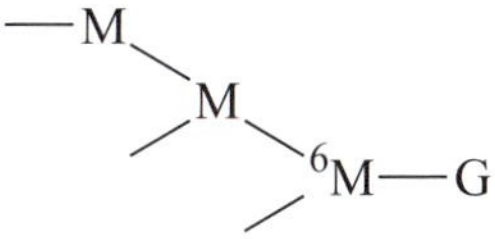

—M—M—^{6}M—G

(204.22 + 1) + [(204.22 – 15) + 1] +
[(204.22 – 15) + 1] + 245.27 + 31 + 23
= 884.9 Da

The 31 Da represents a terminal methoxy group on the GlcNAc residue.

—M—M—$_3$M—G

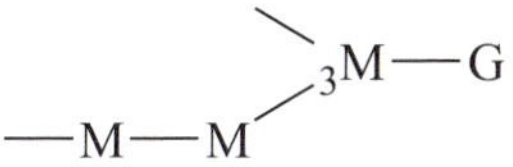

(204.22 + 1) + 204.22 + [(204.22 - 15) +
1] + 245.27 + 31 + 23 = 898.9 Da

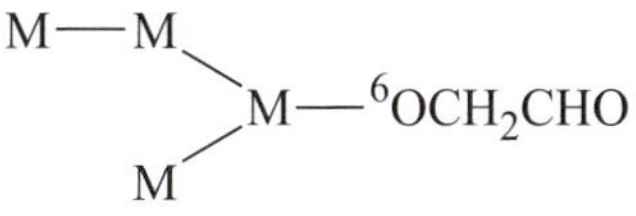

M—M—M—$^6OCH_2CHO$ (with M branch)

853.9 + 60 = 913.9 D
or
854.9 + 59 = 913.9 Da

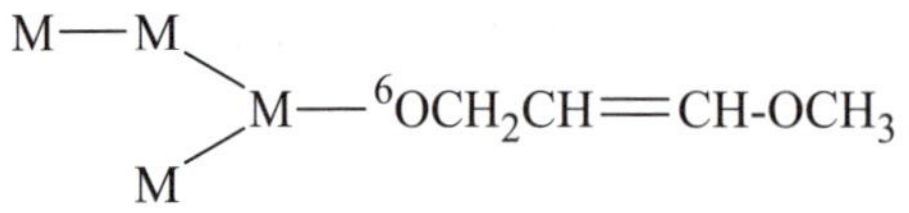

$$M{-}M\backslash M{-}{}^{6}OCH_2CH{=}CH\text{-}OCH_3 \quad / M$$

2(204.22 + 15) + 204.22 + (204.22 – 15) + 23 = 854.88 Da
854.88 + 87 = 941.88 Da

PROBLEM 12.26 Table 12.9 shows that there would be two additional peaks at 60 and 88 Da above 649.7 Da at 709.7, and 737.7 Da, respectively. The spectrum in Figure 12.23 shows essentially no peaks at these values. An M-M-2M6-lower A branch trimannose residue would not produce fragments in the m/z region in Figure 12.24.

PROBLEM 12.27 In the order of the fragments as listed in Table 12.10:

(a) (162.14 + 1) + 162.14 + [(162.14 – 1) + 1 – 1] + 23 = 509.42

(b) (162.14 +1) + 162.14 + (162.14 – 1) + 23 = 509.42

(c) (162.14 + 1) + (162.14 – 1) + (162.14 + 1) + 17 + 23 = 527.42

(d) (162.14 + 1) + 162.14 + 162.14 + 17 + 23 = 527.42

(e) [(162.14 – 1) + 1 + 1] + [(162.14 – 1) + 1] + 203.195 + 17 + 23 = 568.48

(f) (162.14 + 1) + [(162.14 - 1) + 1] + 203.195 + 17 + 23 = 568.48

(g) (162.14 + 1) + [(162.14 – 1) + 1] + [(162.14 – 1) + 1] + 203.195 + 17 + 23 = 730.61

(h) (162.14 + 1) + 162.14 + [(162.14 – 1) + 1] + 203.195 + 17 + 23 = 730.61

All masses are in daltons.

PROBLEM 12.28 Table 12.17 presents the six structures.

PROBLEM 12.29 The difference between the masses in Figure 12.26 are as follows:

1028.9	921.9	815.3
–921.9	–815.3	–709.0
107.0	106.6	106.3

The average value of 106.6 multiplied by 2 charges = 213.2 Da. This value is equal to a permethylated hexose (204.223 Da), where nine hydrogen atoms are replaced with deuterium. The mass values represent perdeuteriomethylated hexose residues in sequence, and the four peaks represent the M5–M8 high-mannose motif.

m/z 815.3:
$(815.3 \times 2) - 46 = 1584.6$ Da

The difference can be due to the average mass value of the atoms used to calculate the Table 12.8 values and experimental uncertainty in the 1584.6-Da value.

m/z 921.9:
$(921.9 \times 2) - 46 = 1797.8$ Da

$Man_7GlcNAc$ from Table 12.8 yields a permethylated value of 1720.90:

$$\begin{array}{l} 1720.90 \\ \underline{+78.45} = 3.018 \times 26 \text{ methyls} \\ 1799.35 \simeq 1797.8 \text{ Da} \end{array}$$

m/z 1028.9:
$(1028.9 \times 2) - 46 = 2011.8$
$Man_8GlcNAc = 1925.128$

$$\begin{array}{l} \underline{+87.52} = 3.018 \times 29 \text{ methyls} \\ 2012.65 \simeq 2011.8 \end{array}$$

For each calculation above, each of the $Man_{5-8}GlcNAc$ glycoforms yields the same respective mass value.

PROBLEM 12.30

$(727 \times 2) - 46 = 1408.0$ Da

$(748.9 \times 2) - 46 = 1451.8$ Da

Table 12.17 ORM/ODM reactions on selected mannose residues

Linkage	Residue	M_r	ORM product	Number of additional methyls	M_r (ORM)	ODM product	M_r (ODM)
1,4	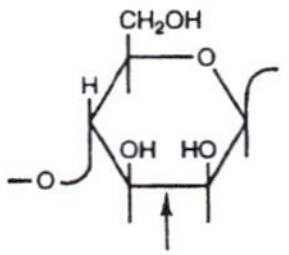	162.142	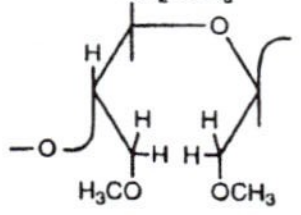	3	206.239	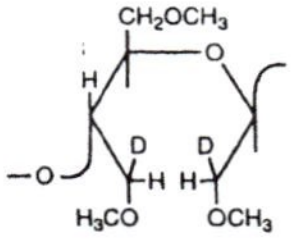	208.251
1,2,3	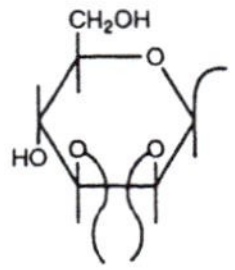	161.134	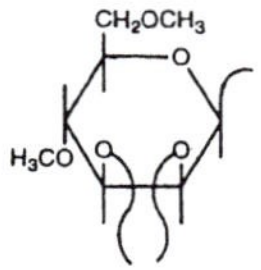	2	189.189	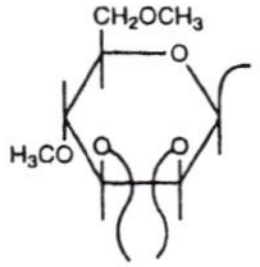	189.189
2	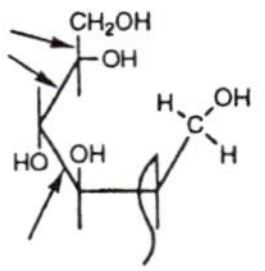	165.167	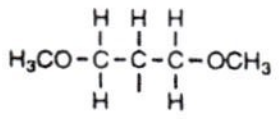	2	103.141	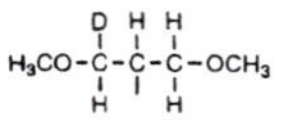	104.147

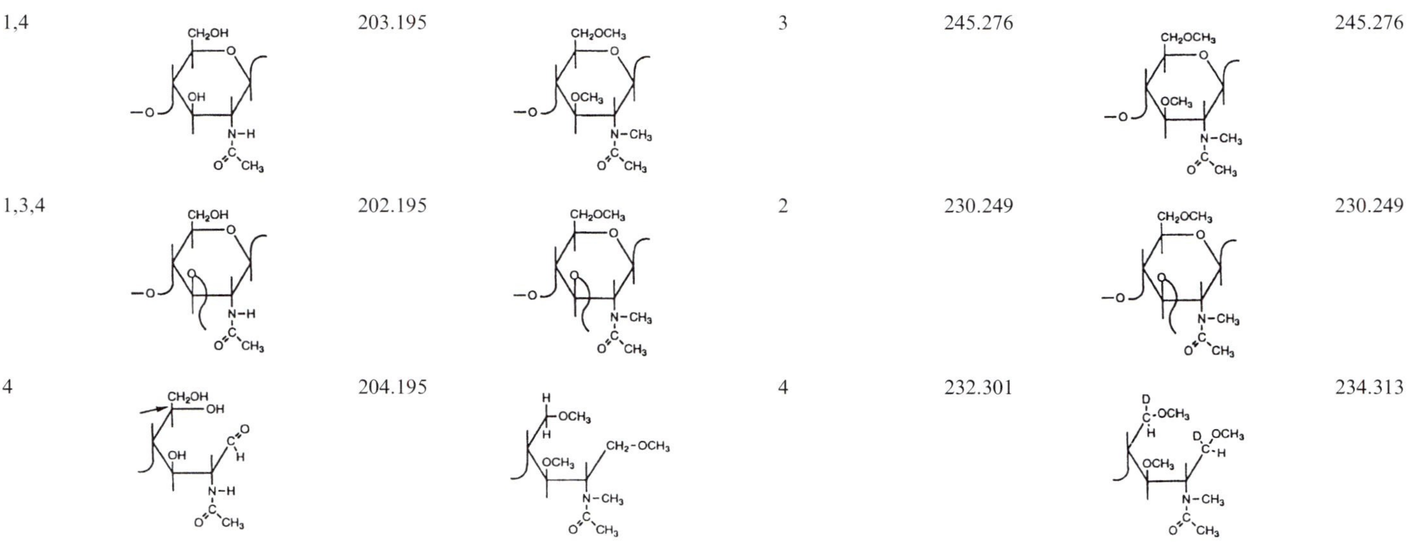

1,4		203.195		3	245.276		245.276
1,3,4		202.195		2	230.249		230.249
4		204.195		4	232.301		234.313

The seven possible glycoforms are analyzed below:

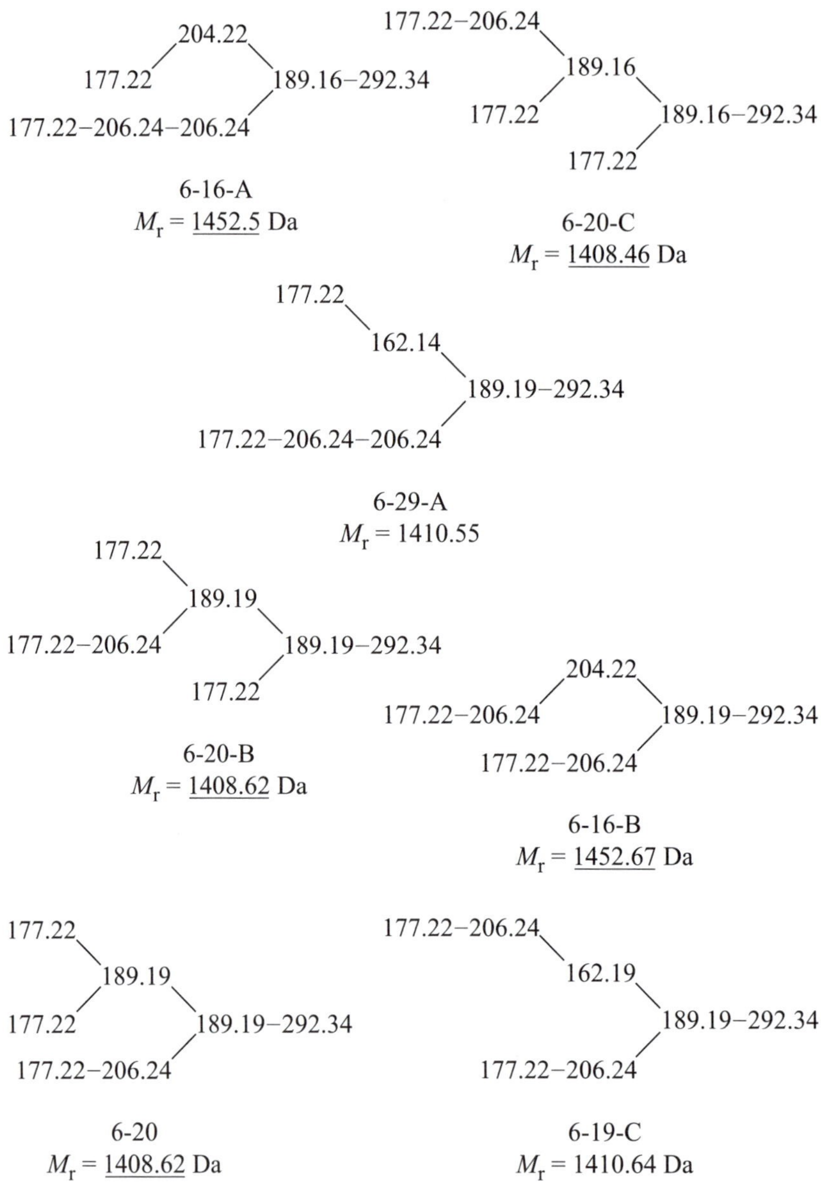

Five compounds are underlined and it appears that m/z 727.0^{2+} could be any combination of 6-20, 6-20-B, and 6-20-C, while m/z 748.9^{2+} could be any combination of 6-16-B and 6-16-A.

PROBLEM 12.31

(830.1 × 2) − 46 = 1614.2 Da (obs)

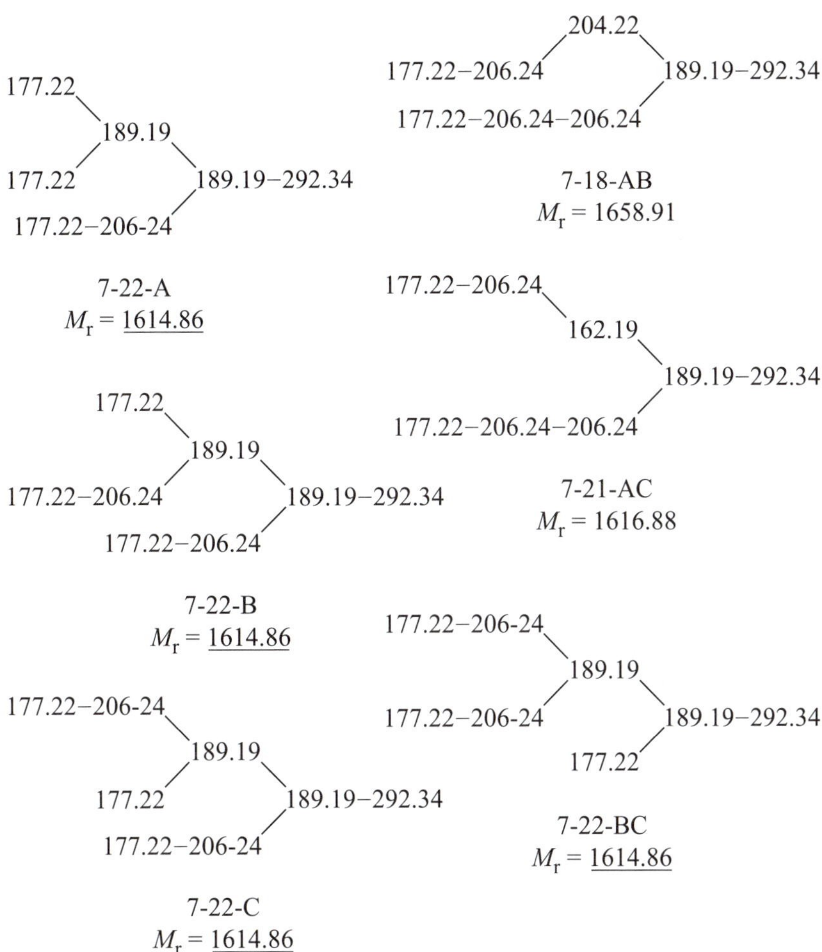

The four underlined structures are observed to fit the observed mass value. Any combination of these values will satisfy the observed m/z 1614.86. Other types of experiments are obviously required to provide further glycoform discrimination.

PROBLEM 12.32 Visual analysis of Figure 12.14 shows that ORM could not differentiate the three glycoforms. The only difference between them is a terminal mannose residue.

PROBLEM 12.33 The M_r of the protein, after subtracting the Asn hydrogen, is 2354.6 Da. The M_r of the five different glycoprotein glycoforms in Figure 12.36a and their identities are as follows:

A5 = 1044.4^{5+} = 5217; A4 = 1305.3^{4+} = 5217.2; A3 = 1740.0^{3+} = 5217; av = 5217.1 Da

B5 = 985.9^{5+} = 4924.5; B4 = 1232.3^{4+} = 4925.2; B3 = 1643.0^{3+} = 4926.0; av = 4925.2 Da

C4 = 1141.1^{4+} = 4560.4; C3 = 1521.2^{3+} = 4560.6; av = 4560.5 Da

D3 = 1424.1^{3+} = 4269.3 Da

E4 = 1341.9^{4+} = 5363.6 Da

Subtraction of the protein mass and adding 17 Da allows comparisons to Table 12.14:

A: 5217.1 – 2354.6 + 17 = 2879.5 $\simeq$ 2880.61 trisialo triantennary

B: 4925.2 – 2354.6 + 17 = 2587.6 $\simeq$ 2589.31 disialo triantennary

C: 4560.5 – 2354.6 + 17 = 2222.9 $\simeq$ 2224.02 disialo biantennary

D: 4269.3 – 2354.6 + 17 = 1931.7 $\simeq$ 1932.76 monosialo biantennary

E: 5363.6 – 2354.6 + 17 = 3026.0 $\simeq$ 3026.75 trisialo fucosylated triantennary or trisialofucosylated LacNAc biantennary

Figure 12.36b represents the mass spectrum of the glycopeptide glycoforms that had their capping neuraminic acid residues removed. Prior to calculations, one should be able to predict the answers. The removal of sialic acid from the five glycoforms in Figure 12.36a yields three species: biantennary, triantennary, and fucosylated triantennary. Three species are observed in Figure 12.36b. Thus, the glycan mass can be found under the zero Neu5NAc column in Table 12.14:

A5 = 870.0^{5+} = 4345.0; A4 = 1087.7^{4+} = 4346.8; A3 = 1449.3^{3+} = 4344.9; av = 4345.57 Da

B4 = 996.1^{4+} = 3983.4; B3 = 1327.5^{3+} = 3979.5; av = 3981.45 Da

C4 = 1123.8^{4+} = 4494.2; C3 = 1498.2^{3+} = 4491.6; av = 4492.9 Da

The method shown below is used to arrive at the amount of sialic acid loss, while the above method identifies the glycoform structure.

B: 3981.45 – 2354.6 + 17 = 1643.85 $\simeq$ 1641.50 biantennary

A: 4345.27 – 2354.6 + 17 = 2007.67 $\simeq$ 2006.81 triantennary

C: 4492.90 – 2354.6 + 17 = 2155.3 $\simeq$ 2152.98 fucosylated triantennary or LacNAc fucosylated biantennary

Subtraction of the mass values (calculated from Figure 12.36b) from that of the respective values of Figure 12.36a mathematically yields the number of sialic acid residues lost:

trisialo triantennary 2879.40
– triantennary 2007.67
871.73/291.26 = 2.99 = 3 sialic acid residues removed where 291.26 is the residue mass of a Neu5NAc carbohydrate

disialo triantennary 2587.70
– triantennary 2007.67
580.03/291.26 = 1.99 = 2 sialic acid residues removed

disialo biantennary 2222.9
– biantennary 1643.85
579.05/291.26 = 1.99 = 2 sialic acid residues lost

monosialo biantennary 1931.7
– biantennary 1643.85
287.85 $\simeq$ 291.26 = 1 sialic acid residue lost

trisialo fucosylated triantennary 3026.0
– fucosylated triantennary 2155.3
870.7/291.26 = 2.99 = 3 sialic acid residues lost

Figure 12.36c represents the loss of terminal galactose residues from the desialylated glycoforms in Figure 12.36b. Schemes 12.7–12.10 show that Neu5NAc species directly link to galactose residues.

A4 = 965.3^{4+} = 3857.2; A3 = 1286.7^{3+} = 3857.1; av = 3857.2 Da

B4 = 915.1^{4+} = 3656.4; B3 = 1219.0^{3+} = 3654; av = 3655.2 Da

C4 = 1042.2^{4+} = 4164.8; C3 = 1389.2^{3+} = 4164.6; av = 4164.7 Da

A: 3857.2 – 2354.6 + 17 = 1519.6

B: 3655.2 – 2354.6 + 17 = 1317.6

C: 4164.7 – 2354.6 + 17 = 1827.1

These values are not listed in Table 12.14; however, the mass values of the A, B, and C glycoforms can be ordered to follow the same trend as in Figure 12.36b; thus:

B peak biantennary 1643.85
−1317.60
326.25/162.14 = 2.01 = 2 Gal lost

A peak triantennary 2007.67
−1519.60
488.07/162.14 = 3.01 = 3 Gal were lost

C peak fucosylated triantennary 2155.3
−1827.1
328.2/162.14 = 2.02 = 2 Gal lost

PROBLEM 12.34 The protein was treated with vinylpyridine; therefore, the two Cys residues were alkylated in the Asn-81 peptide. The S-β-4-pyridylethyl-Cys residue mass is 208.3 Da (Table 9.13). The M_r of the peptide (less the Asn-81 hydrogen) is 3768.097 Da. Subtraction of the peptide mass from the glycopeptide mass values, and adding an hydroxyl group, produces glycan mass values that can be compared to those in Table 12.14:

Shaded circle mass (av):	5975.75 Da − 3768.097 + 17 = 2224.6 disialo biantennary
Asterisk mass (av):	6341.15 Da − 3768.097 + 17 = 2590.1 disialo triantennary
Open circle mass (av):	6633.13 Da − 3768.097 + 17 = 2882.0 trisialo triantennary
Shaded triangle mass (av):	6923.37 Da − 3768.097 + 17 = 3172.3 tetrasialo triantennary

Note the close mass values (Table 12.14) of the tetrasialo triantennary structure (3171.8 Da) and that of the LacNAc$_2$-triantennary-Fuc and LacNAc-tetraantennary-Fuc isomers (3174.89 Da). The experimental values were of satisfactory resolution to be able to distinguish between these species.

PROBLEM 12.35 The neuraminidase cleaves the terminal, nonreducing-end sialic acid. Thus, prior to calculations, one could predict that the two observed species should be a biantennary and a triantennary glycan. This result comes from the answers in Problem 12.34. By cleaving the sialic acid residues from the four glycans, only biantennary and triantennary structures are left (Table 12.14).

Shaded triangle:	5757.3 − 3768.097 + 17 = 2006.2 Da triantennary
Open circle:	5392.3 − 3768.097 + 17 = 1641.2 D biantennary

PROBLEM 12.36 See Table 12.18. The four mass-averaged peaks are:

Shaded triangle	= 5019.3 Da
Open circle	= 5068.9 Da
Shaded circle	= 5272.3 Da
Asterisk	= 5434.8 Da

Four peaks are observed; however, only two desialylated glycopeptides were observed in Figure 12.38b. Working backwards from the two species in

Table 12.18 Abbreviated structures of the four glycoforms of bovine fetuin after neuraminidase and β-galactosidase digestion

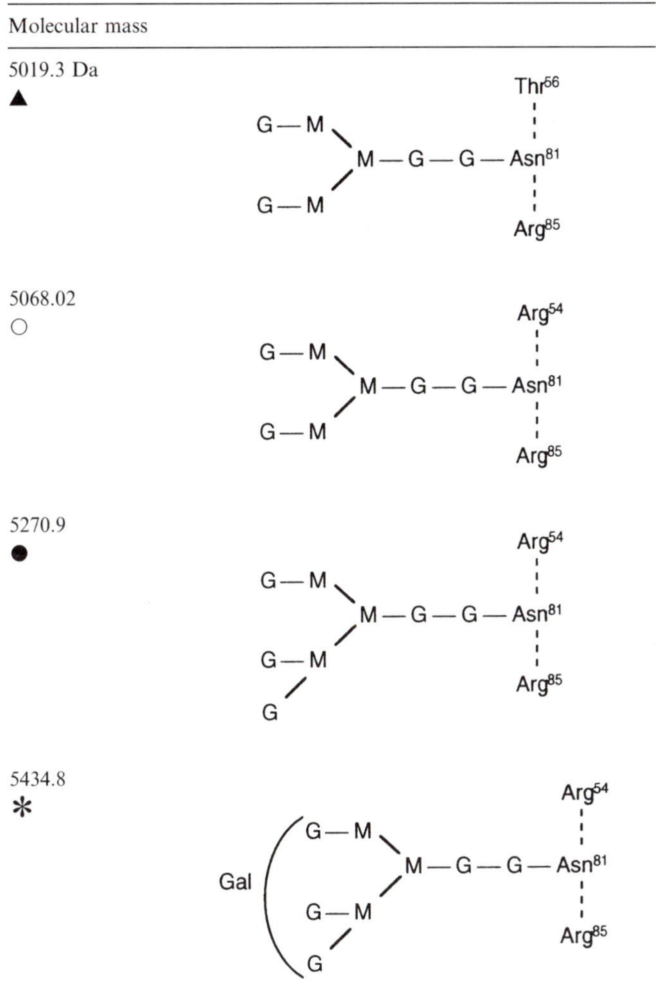

Figure 12.38b by subtracting the galactose residues, two of the four masses should be able to be identified:

3768.097	peptide from Problem 12.34
+ 1641.2	biantennary glycan from Problem 12.35
5409.3	
−17	
5392.3	biantennary glycopeptide from Figure 12.38b
−324.28	2(162.14) = galactose
5068.02	biantennary − 2 Gal

5757.3	triantennary glycopeptide from Figure 12.38b
− 486.42	3(162.14)
5270.9	triantennary − 3 Gal

Thus:

5068.02 = biantennary − 2 Gal $\simeq$ 5068.9 Da from Figure 12.39c
5270.9 = triantennary − 3 Gal $\simeq$ 5272.3 Da from Figure 12.39c

The β-galactosidase enzyme is listed in Table 12.6 as cleaving a Galβ1–4GlcNAc bond; therefore, the 1–4 linkage must be present. There is a higher experimentally observed mass which may have a Galβ1–3GlcNAc linkage (Scheme 12.11): 5434.8 (asterisk average mass in Figure 12.38c) – 5272.3 (shaded circle average mass) = 162.5 Da. Since this difference equals the mass of a hexose, it appears that the higher mass glycoform has both Galβ1–3,4GlcNAc linkages while the two lower mass values have only the Galβ1–4GlcNAc linkage. The lowest experimentally observed mass is m/z 5019.3. This could be due to amino acid removal by the contaminating peptidase. This mass may reflect partial galactose removal, and a question can be raised as to which end of the peptide is affected by the peptidase. By adding back the proper amino acids to 5019.3 Da, the mass of one of the other three peptides should be calculated. Assuming an N-terminal enzyme action, the lowest mass could not come from 5068.9 Da, but possibly by 5272.3 Da − 5019.3 = 253 Da. Table 9.6 shows that this mass equals two combinations of amino acid residues. The Arg–Pro combination is found on the N-terminal side of the Asn-81 glycopeptide. Thus, 5019.3 Da is equal to the triantennary glycopeptide less the Arg–Pro N-terminal amino acid residues. Table 12.18 shows the structures of the four species, where G = glucosamine, M = mannose, and Gal = galactose.

PROBLEM 12.37 The mass of the peptide, less a hydrogen, is 2602.0 Da, and includes the pyridylethylation on both Cys residues.

1458.5^{3+} = 4372.5
− 2602.0
1770.5 + 17 = 1787.5 Da = fucosylated biantennary with no sialic acid groups (from Table 12.14).

1580.2^{3+} = 4737.6
− 2602.0
2135.6 + 17 = 2152.6 Da = fucosylated triantennary or fucosylated LacNAc biantennary with no sialic acid. It is the fucosylated triantennary structure.

PROBLEM 12.38

m/z 1164.6^{4+} = 4658.4 Da
4658.4 − 4(23) = 4566.4 $\simeq$ 4567.0 Da in Table 12.14

Table 12.19 ORM/ODM reaction on two carbohydrate residues

Linkage	Residue[a]	M_r[b]	ORM product	Number of additional methyls	M_r (ORM)	ODM product	M_r (ODM)
2	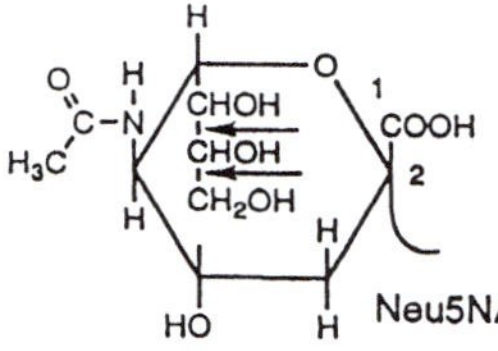 Neu5NAc	292.266	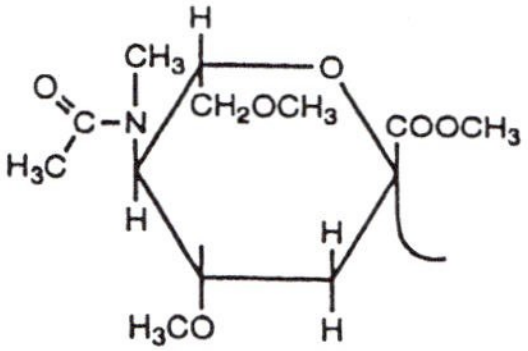	4	288.321	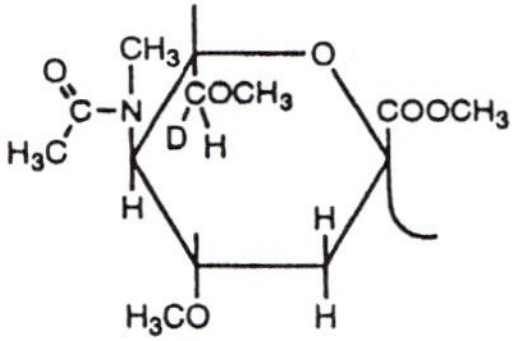	289.327
4,6	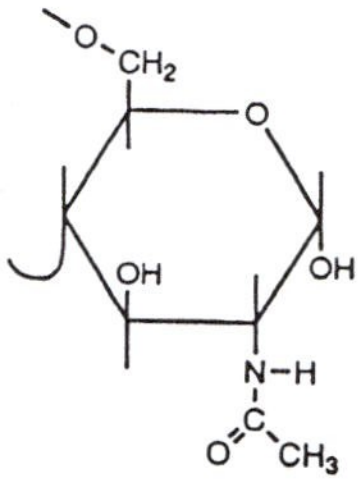	203.195	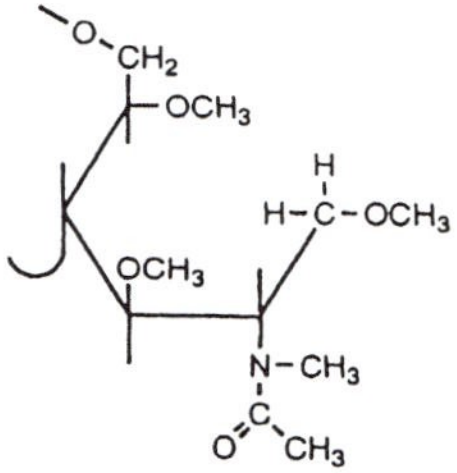	4	261.319	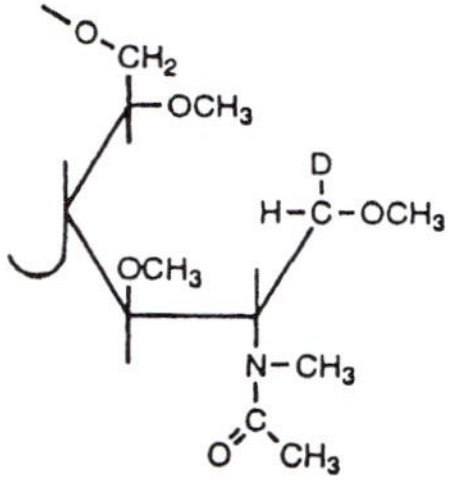	262.325

[a] Arrows indicate bond cleavage.

[b] M_r values in daltons.

Under the central permethylated section of Table 12.14, this mass has three isomers:

$LacNAc_2$-biantennary-Fuc-NA_4
LacNAc-triantennary-Fuc-NA_4
tetraantennary-Fuc-NA_4.

The last species is the correct structure as given in Table 12.4 in reference 87.

$m/z\ 1274.8^{3+} = 3824.4$ Da
$3824.4 - 3(23) = 3755.4 \simeq 3756.16$ Da in Table 12.14
LacNAc-biantennary-Fuc-NA_3
or triantennary-Fuc-NA_3

The last species is the correct one.

$$m/z\ 1544.6^{3+} = 4633.8 \text{ Da}$$
$$4633.8 - 3(23) = 4564.8 \simeq 4567.05 \text{ Da}$$

This is the triply charged ion of $m/z\ 1164.6^{4+}$.

$m/z\ 1694.83+ = 5015.4 \simeq 5016.54$ Da
$LacNAc_2$-triantennary-Fuc-NA_4
LacNAc-tetraantennary-Fuc-NA_4

The latter is the correct species from Table 4 in reference 87.

$$m/z\ 1900.0^{2+} \simeq 1901.08^{2+} \text{ in the last section of Table 12.14.}$$

This is the doubly charged ion of $m/z\ 1274.8^{3+}$.

Tables 3 and 4 in reference 87 provide a comprehensive characterization of Figure 12.43.

PROBLEM 12.39 Refer to Table 12.19.

PROBLEM 12.40 (a) Schemes 12.12 and 12.13 provide the structure and mass value information at the respective residue locations. Note that the listed masses closest to the residue is the native, untreated value, while the middle and outer values represent that of ORM and ODM processing, respectively.

The mass values outlined in a box in both schemes are the only ones that vary from the 4 to 4′ branched glycomer. All other mass values are the same in both glycomers. Note that the glycosidic oxygen in the diacetylchitobiose residue (Table 12.1)—that is, the two reducing-end GlcNAc residues—must be included in the M_r calculations. Problem 12.28 shows that a 1,4-linked GlcNAc does not include the C_1 glycosidic oxygen in its mass, and Problem 12.39 does not include the C_4 glycosidic oxygen in the mass of the 4,6-linked, reducing-end, *N*-acetyl glucosamine residue. Note that the underivatized 4′ and 4-linked $TriNA_3Fuc$ glycoforms are glycomers. The ORM- and ODM-processed glycoforms are not glycomers. The 4 and 4′-branched species differ by only 2 Da for ORM processing and by only 4 Da for ODM processing.

289.33 204.22 245.28 193.22
288.32 204.22 245.28 191.20
292.27 162.14 203.20 161.13
Neu5NAcα2-3Galβ1-4GlcNAcβ1-6

149.21
147.15
4′
289.33 204.22 245.28 Manα1-6 189.19 147.15
288.32 204.22 245.28 189.19 Fucα1-6
292.27 162.14 203.20 161.13
Neu5NAcα2-3Galβ1-4GlcNAcβ1-2 Manβ1-4GlcNAcβ1-4GlcNAc
203.20 16 203.20
245.28 16 261.32
289.33 204.22 245.28 208.25 245.28 16 262.33
288.32 204.22 245.28 206.24
292.27 162.14 203.20 162.14
Neu5NAcα2-3Galβ1-4GlcNAcβ1-2—Manα1-3

M_r underivatized = 3026.78 Da in Table 12.14
M_r ORM = 3469.90 Da
M_r ODM = 3479.97 Da
3479.97 + 3(23) = 3548.97 Da
3548.97/3 = 1183.0^{3+} = 4′-branched $TriNA_3Fuc$ observed in Figure 12.44 = N-83 glycopeptide

Scheme 12.12 The 4′-branched $TriNA_3Fuc$ glycoform from the N-83 tryptic glycopeptide of r-EPO.

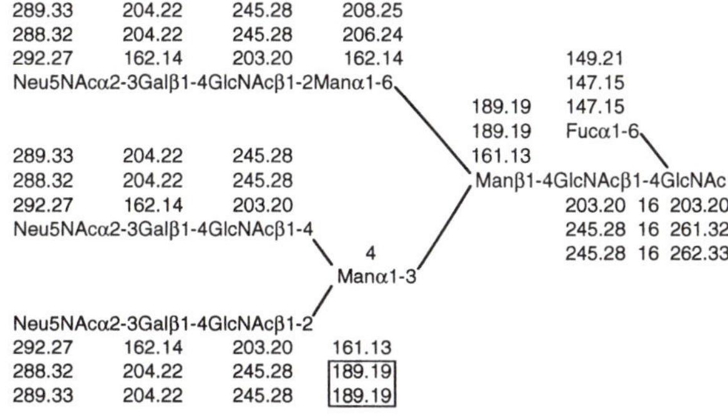

M_r underivatized = 3026.78 Da in Table **12.14**
M_r ORM = 3467.88 Da
M_r ODM = 3475.94 Da
3475.94 + 3(23) = 3544.94
3544.94/3 = 1181.6^{3+} = 4-branched $TriNA_3Fuc$ not observed in Figure **12.44,** therefore must be from the N-38 glycopeptide

Scheme 12.13 The 4-branched $TriNA_3Fuc$ glycoform from the N-38 tryptic glycopeptide of r-EPO.

(b) $$1184.7 \times 4 = 4738.8 \text{ Da}$$

By trial and error, the TetraLacNA$_4$Fuc^{3+} mass of 1572.2^{3+} Da yields $1572.2 \times 3 = 4716.6$ Da (trisodiated)

$$4738.8 \text{ (tetrasodiated)} - 23 = 4715.8 \text{ (trisodiated)} \simeq 4716.6 \text{ Da}$$

(c) Scheme 12.10 can be used as a guide in the solution to this problem, and Scheme 12.14 provides the detailed calculations. The mass value which is closest, in the middle, and most distant from each residue represents the untreated, ORM- and ODM-processed TriLacNA$_4$Fuc glycoform, respectively. An additional LacNAc residue also is found in the glycoform, and would be inserted in one of the antenna on the reducing side of one of the four Neu5NAc residues. Placement of this LacNAc is arbitrary with respect to this exercise, because all four potential glycoforms are transparent with respect to a change in mass under native, ORM- and ODM-processing conditions. However, the position of the LacNAc residue yields different glycomers from a structural point of view.

(d) Schemes 12.15 and 12.16 supply the details with respect to structure and mass values. The positions of the two LacNAc and fourth Neu5NAc residues are arbitrary with respect to this exercise. Both LacNAc moieties can be inserted after the reducing end of any two of the Neu5NAc residues, and the sialic acid has a Neu5NAcα2–8 Neu5NAc linkage (Scheme 12.9).

In Schemes 12.14–12.16, all three underivatized species are glycomers, but the ORM- and ODM-processed glycomers can be distinguished by the change in

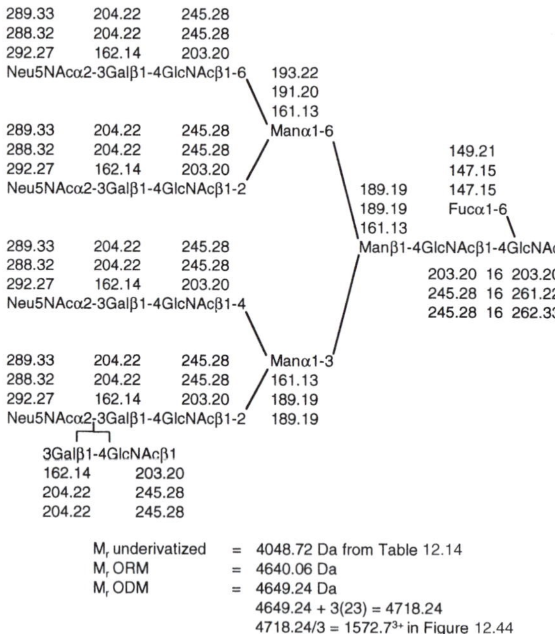

Scheme 12.14 The TetraLacFuc glycoform of the r-EPO N-38 tryptic glycopeptide.

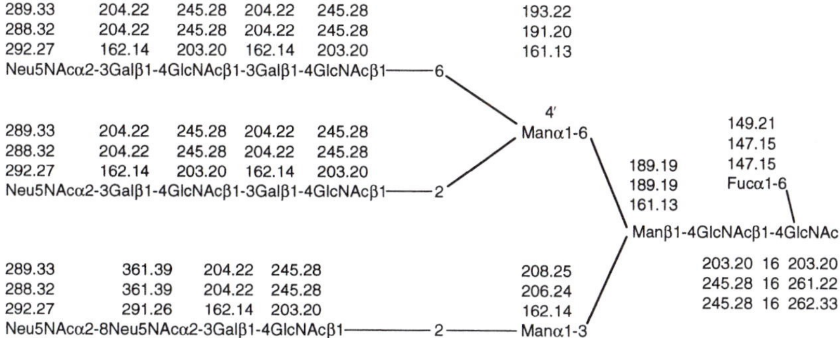

M_r underivatized = 4048.72 Da
M_r ORM = 4730.28 Da
M_r ODM = 4740.36 Da

Scheme 12.15 The 4′-branched $TriLac_2NA_4Fuc$ glycoform.

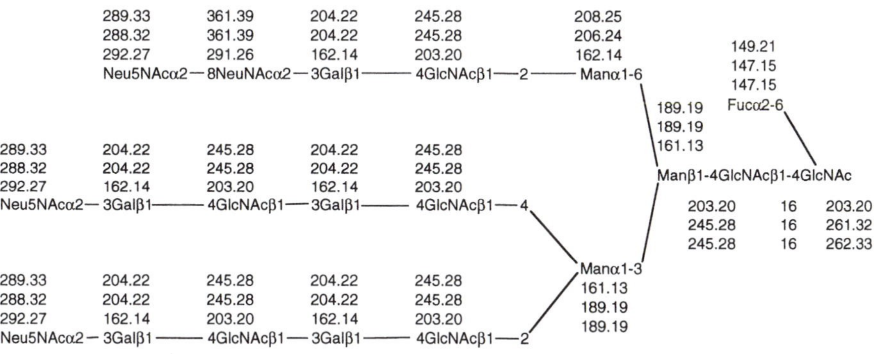

M_r underivatized = 4048.72 Da
M_r ORM = 4728.18 Da
M_r ODM = 4736.33 Da

Scheme 12.16 The 4-branched $TriLac_2NA_4Fuc$ glycoform.

mass. Thus, none of the six glycoforms that result from ORM- and ODM-processing of the underivatized species are glycomers; therefore, they can all be distinguished by different mass values.

PROBLEM 12.41 (a) The difference in mass between the observed and calculated peptides is approximately 162.1 Da, and this strongly suggests that a single hexose saccharide residue is attached to amino acid residue number 7.

(b) The observed b and y ions are presented below:

```
         752       404  333       = y
       F - T - X - A - Q - W
                   |
                  Hex
b =              597  668  796
```

There are two ways to calculate the mass value of X:

1. Subtraction of the mass values of

F-T-X-Hex and F-T-(Hex) = mass of X

$597 - (148.176 + 101.10 + 162.1) = 185.6 \simeq 186.2 =$ tryptophan

2. Subtraction of the mass values of:

$752 - (\text{Hex} + 404) = \text{mass of X}$

$752 - (162.1 + 404) = 185.9 \simeq 186.2 =$ tryptophan

APPENDIX 1

Experimental conditions used to generate electrospray ionization mass spectra

Chapter	Reference(s)	Figure[a]	Method	Solvent gradient or analyte solution[b]	Flow rate[c]	Split	Capillary column[d]	Analyte	ESI sheath flow rate[e]	ESI sheath solvent[b]
4	7 8	3, 18 4, 14	Direct infusion	aq MeOH or AcCN & 5% HOAc	1–10	—	—	Thioredoxin	0.003–0.005	aq MeOH or AcCN & 5% HOAc
4	5	5	Direct	50% aq MeOH, 1 mM NH_4OAc 0.1% HCOOH	2–4	—	—	β-Lactoglobulin	—	—
4	10	7, 10	Direct infusion	10% aq HOAc	1	—	—	Transferrin	25–35 psi	Air flow
4	10	8	Direct infusion	9:2 H_2O:MeOH 0.2% TFA	5	—	—	BSA	25–35 psi	Air flow
4	12	12, 13	Direct RP-HPLC	47.5:47.5:5.0 MeOH:H_2O:HOAc	0.2–1.0		4.6 mm × 25 cm C4	β-Hemoglobins	0.003	Air flow
4	20	17	Direct	0.1% aq HOAc	2.5	—	—	Human growth hormone	—	Air flow
4	22	19, 20	Direct	50% MeOH 5% HOAc	5	—	—	Ovalbumin	—	—
4	9 9	23, 21 9, 6	Direct	50% MeOH 1% HOAc	1–3	—	—	Lysozyme, BSA, horseradish peroxidase	—	—
4	27	27	Direct	70–90% MeOH 1% HOAc	1–2	—	—	Cytochrome c	—	—
4	3	1	RP-HPLC	A = 5% MeOH 2.5 mM NH_4OAc 2.5 mM HOAc B = 50% MeOH gradient elution	40	—	1 mm × 10 cm C18	Bradykinin	—	—

4	11	11	Direct	49.5% MeOH 1% HOAc	1	—	—	Lysozyme	—	—
4	24	24, 26	Direct	1–5% HOAc	1			Ovalbumin	0.003	MeOH or AcCN
4	4	2	Direct	50% aq MeOH 2.5% HOAc	2	—	—	Insulin	—	—
5	10 9	2, 3 4	Direct	1:1 A:B A = H_2O, 0.05% TFA B = 80% aq AcCN & 0.05% TFA	—	—	—	Conotoxin	230 ml/min	Nitrogen gas
5	11	5, 6	Direct	50% aq MeOH 0–5% HCOOH	1	—	—	Bradykinin Insulin	—	—
5	12	7	Direct	50% aq MeOH 0–5% HCOOH	3–5	—	—	Substance P(1–9)	—	—
5	16	9	Direct	50% aq MeOH 1% HOAc	3–5	—	—	Bovine trypsin	—	—
5	19	10, 11	Direct	50% aq AcCN 1% HCOOH	4	—	—	Hemoglobin β-chain	—	—
6	4	1, 5	Direct	50% MeOH 2% HOAc	0.5–2	—	—	Bovine insulin	—	—
6	7	2	Direct	50% AcCN 0.2% HCOOH	3	— —	—	Interleukin-8	85 psi	Gas
			RP-HPLC	gradient of $AcCN:H_2O$ 0.1% TFA	200		2.1 mm × 15 cm C18			
6	12, 13	9, 10	Direct	50% aq MeOH 1% HOAc	1	—	—	Human growth hormone releasing factor	1–5 μl/min	50% aq MeOH 1% HOAc
7	7	3	Direct	10 mM NH_4OAc pH 6.7	1–2	—	—	Myoglobin	600 ml/min	Air

(continued)

Experimental conditions used to generate electrospray ionization mass spectra (*continued*)

Chapter	Reference(s)	Figure[a]	Method	Solvent gradient or analyte solution[b]	Flow rate[c]	Split	Capillary column[d]	Analyte	ESI sheath flow rate[e]	ESI sheath solvent[b]
7	5	4, 10, 11, 12	Direct	± MeOH 1–5% aq HOAc	0.1–1	—	—	Bovine insulin Bovine α-lactalbumin Bovine proinsulin	2–5 μl/min	AcCN
7	2	7, 8	Direct	aq HOAc or 5 mM NH_4OAc pH 6.6	0.5	—	—	Bovine ubiquitin Cytochrome c	—	—
7	11	9	Direct	aq 5% HOAc	0.5	—	—	Melittin	3 μl/min	MeOH
7	10	13	Direct	aq 47% MeOH 6% HOAc	0.5–2	—	—	Bradykinin	—	—
7	15	14	Direct	aq 5% HOAc	0.5	—	—	Cytochrome c	3 μl/min	MeOH
7	16	15	Direct	1:1 MeOH:AcCN 1–5% HOAc	1	—	—	Cytochrome c	3 μl/min	MeOH
7	18	18	Direct	aq 80% AcCN 1–5% HOAc	1	—	—	Bovine hemoglobin	3 μl/min	MeOH or AcCN
8	9	1	Direct	aq 50% MeOH 1 mM NH_4OAc 0.1% HCOOH	2–4	—	—	β-Lactoglobulin	—	—
8	11	6	Direct	0.05–5% aq HOAc or HCOOH	1–5	—	—	Calcitonin	—	—
8	12	4	Direct	20% aq MeOH ±5 mM NH_4HCO_3 pH 8.0	1	—	—	Parvalbumin	—	—
8	15	6	Direct	5 mM NH_4HCO_3	1	—	—	Parvalbumin	—	—
9	14	44–46	RP-HPLC	50% AcCN,0.08% TFA or 1:1 of isopropanol: MeEtOH	50	20:1	1000 mm × 10 cm C18	Yeast SRP protein	—	—

9	123	48, 49	Direct infusion	50% MeOH 2% HOAc	1	—	—	Human pancreatic polypeptide		
	123	51, 52						Corticotropin releasing factor		
9	18 124	S1a–d S5	Direct infusion	90% MeOH 0.1% TFA or 50%MeOH & 5% HOAc	2–10	—	—	Small tryptic-like peptides	—	—
9	125	S1e	CZE	1 mM NH_4OAc				Hemoglobin tryptic peptide		
9	129	S7	RP-HPLC	0–80% AcCN	5	—	50 mm × 7.5 cm	Problem S7	—	—
9	127	S6	HPLC				100 × 15 cm POROS 10-R2	Problem S6		
9 9	27 49	4, 5, 6 7, 33	RP-HPLC packed	0–100% AcCN 0.085% TFA	100	50:1	320 mm × 15 cm C18	Myoglobin Bovine casein	—	—
9	50	53, 54 S4	HPLC & CZE	0–80% HOAc 0.5% AcCN	1–2		50 mm × 70 cm	β-Lacto-globulin	0.002–.006	75/25 MeOH 0.5% HOAc
9	53	9, 55	nano ESI	—	0.03	—	—	Glufibrino-peptide B	—	—
9	55	11–13, 56	Direct infusion	5% HOAc	0.1–1.0	—	—	Cyrochrome c	2–4	MeOH AcCN
9	84	14	RP-HPLC	0–80% AcCN 0.1% TFA	2.5		250 mm× 10 cm C18	Rat cerebellum Calmodulin	—	60:20:20:1 2-MeEtOH: MeOH:H_2O: HOAc
9	58	15	—	—						

(*continued*)

Experimental conditions used to generate electrospray ionization mass spectra (*continued*)

Chapter	Reference(s)	Figure[a]	Method	Solvent gradient or analyte solution[b]	Flow rate[c]	Split	Capillary column[d]	Analyte	ESI sheath flow rate[e]	ESI sheath solvent[b]
9	87	16	Direct infusion	70% AcCn 0.1% TFA	1–5	—	—	Problem 9.6	600	Nitrogen
9	88	17	HPLC packed	50% AcCN 0.05% TFA	40	—	1000 mm× 10 cm ODS II	Problem 9.7 & S2	—	—
9	82	20, 21, 58	Direct infusion	50 mM NH_4HCO_3	—	—	—	r-Hirudin	—	—
9	81	22–26	Direct infusion	80% AcCN 0.1% TFA	2–5	—	—	ACP(65–74)	—	—
9	90	28, 30, 32	Direct infusion	5% HOAc or 50% MeOH & 5% HOAc 50 mM NH_4HCO_3	0.2–0.5	—	—	Substance P; MethGH; hGH, LH-RH	0.0025–0.003	MeOH
9	17	9, 33	Direct infusion	MeOH 0.1% TFA	10	—	—	γ-Endorphin	—	—
9	11	34, 58, S9	Direct infusion		5					
9	109 114	35, 36 38–41	Direct infusion	5% HOAc	0.5–1.0	—	—	Melittin Serum albumins	0.0025–0.005	MeOH
9	116	42								
10	20	1–3	RP-HPLC	0–5 min A 5–100 min B A = aq 0.1% TFA B = AcCN, 0.1% TFA	40	—	1 mm × 10 cm C18	AChR peptide	—	—
10	3 16 19	5, 6 11 15	RP-HPLC	5–50% B 5 min 50–100% B 5 min A = aq 0.1% TFA B = 90% AcCN, 0.1% TFA	4–5	1:7	320mm× 15 cm C18	Small peptides		Air
10	8	7–9	IMAC column	See reference	5			Bovine β-casein	—	—

10	24	12	Direct	aq 50% MeOH 0.2% HCOOH	5	—	—	Fibroblast growth factor receptor tryptic peptide	—	—
10	12	13, 14	Direct	aq 50% MeOH 0.2% HCOOH	5	—	—	Rhodopsin kinase tryptic peptide	—	Nitrogen
10	13	18, 22	RP-HPLC	AcCN gradient 0.1% HCOOH	20		500 mm × 20 cm C18	MAPKK tryptic peptide	—	Nitrogen
10	27	16, 17, 21	RP-HPLC		200	5:195	C18 2.1 mm × 250 mm	gD-2 Tryptic peptide		Nitrogen
10	4	19, 23	Nanospray	48% MeOH 4% NH_4OH or 44% MeOH 12% HCOOH pH $<$ 3.0	20–40 nl/min			Bovine α-casein	—	—
10	18	20, 24	Direct	(+) mode: 47.5% MeOH 5% HOAc; (−) mode: 47.5% MeOH 0.5% piperidine	10	—	—	Small peptides	—	Nitrogen
11	27, 39 23	7, 8 9, 10, 23	RP-HPLC	0–80% AcCN gradient with 0.5% HOAc	1–2 μl/min	—	75 mm× 10 cm C18	Nonapeptides	2–4 μl/min	75% MeOH 0.5% HOAc
11	42, 51	13, 14	RP-HPLC	0.1% TFA first 5 min; 0–40% AcCN in 0.1% TFA for 40 min; then 40–60% AcCN in 0.1% TFA for 10 min	200–400 nl/min		2.1 mm × 3 cm C18	MHC class-II peptides		

(continued)

Experimental conditions used to generate electrospray ionization mass spectra (*continued*)

Chapter	Reference(s)	Figure[a]	Method	Solvent gradient or analyte solution[b]	Flow rate[c]	Split	Capillary column[d]	Analyte	ESI sheath flow rate[e]	ESI sheath solvent[b]
11	40, 71, 75	16, 17	Rp-HPLC	A:2% AcCN, 0.05% TFA B: 80% AcCN, 0.045% TFA; 0–50% B gradient over 15 min	5	1:20	320 mm× 15 cm C18	Tryptic peptides		Nitrogen
12	58	1, 2	Direct	1:1 AcCN:H_2O	3	—	—	Tetrasaccharide	—	—
12	59	3, 4	Direct	1:1 AcCN:H_2O	10	—	—	O-linked polysaccharide	—	—
12	60	7	Direct	50% MeOH 0.1% HCOOH	2	—	—	O-linked glycopeptide	—	—
12	61	8	RP-HPLC	A: 0.1% TFA B: 0.08% TFA 70% AcCN 0–80% B at 4 μl/ min	Post-column addition of 50% AcCN and 0.1% TFA at 2 μl/min	—	1 mm × 100 mm C18	Tryptic glycopeptide	—	—
12	62 62 62	9 36 37	RP-HPL:C	A: 0.1% TFA B: 0.08% TFA 70% AcCN 0–80% B at 4 μl/min	Post-column addition of 1:1 2-Methoxy-ethanol: 2-propanol at 35–40 μl/min; 4–5 μl/min into MS	1:19	1 mm × 250 mm C18	LCAT tryptic glycopeptide	—	—
12	66	11	Direct	70% MeOH 0.5 mM NaOAc	0.75 μl/min	—	—	O-linked glycopeptides of r-EPO	—	—

12	67	13	Direct	60% MeOH 0.25 mM NaOH	0.75	—	—	O-linked glycopeptide	—	—
12	73	16, 17, 18	Direct	30% MeOH 10 mM NH_4OAc or 10 mM NaOAc	2	—	—	Ovalbumin glycopeptides	1.5 L/min	Nitrogen
12	74	19	Direct	50% AcCN 0.05% TFA	2–4	—	—	RNase B glycopeptides	60–65 psi	Nitrogen
12	68, 69	21–25, 27, 29, 31, 44, 45	Direct	Methylated: 70% MeOH 0.5 mM NaOAc Underivatized: 50% MeOH 0.5% HOAc	0.75 μl/min	—	—	Various, see text	—	—
12	79 79 79 80	33 34a,b 35 34c	RP-HPLC	A = 0.5% aq TFA B = 0.05% AcCN; 0–25% B in 50 min then to 60% B by 85 min	0.2 ml/min	1:15	2.1 mm × 250 mm C18	r-tPA Tryptic digest	—	—
12	85 18	38 39	RP-HPLC	A = 0.1% aq TFA B = 0.08% TFA in AcCN Post-column addition of 1:1 2-methoxy-ethanol:isopropanol.	3–5	1:20	1 mm × 100 mm C18	Bovine fetuin glycopetides	—	—

(*continued*)

Experimental conditions used to generate electrospray ionization mass spectra (*continued*)

Chapter	Reference(s)	Figure[a]	Method	Solvent gradient or analyte solution[b]	Flow rate[c]	Split	Capillary column[d]	Analyte	ESI sheath flow rate[e]	ESI sheath solvent[b]
12	86 86	40 41	RP-HPLC		5	1:40	2.1 mm × 250 mm C18	Herpes simplex gD-2 glycopetides	—	—
12	87 87	42 43	Direct	50% aq MeOH 3% HOAc	200–400 nl/min	—	—	r-EPO glycopeptides	35–45 psi	Nitrogen

[a]Figure numbers are given without the chapter prefix.
[b]MeETOH = 2-methoxyethanol.
HOAc = acetic acid.
AcCN = acetonitrile.
TFA = trifluoroacetic acid.
[c]μl/min.
[d]μm i.d. × cm length.
[e]ml/min.

APPENDIX 2

This appendix presents the calculation of the absolute and relative abundances of a fragment of the hemoglobin β-chain that contains the amino acid residues 41–59.

The lowest mass that can be observed from any molecule is its monoisotopic mass, not its nominal mass, because compounds that contain the C, H, N, O, and S elements do not exist at a mass identical to a nominal (integer) mass. In a rigorous fashion, this is true even of compounds below 100 Da[1]. However, compounds with a mass in the low hundreds of daltons have average and monoisotopic masses very close to that of the nominal mass.

The lowest theoretically observable mass of the hemoglobin β-chain fragment is its monoisotopic mass (2057.939 Da, see Problem 3.1), and its absolute abundance is calculated from information in Table 3.1 as follows:

$^{12}C_{93}$	$^{1}H_{135}$	$^{14}N_{21}$	$^{16}O_{30}$	$^{32}S_{1}$
$(0.989)^{93}$	$(0.99985)^{135}$	$(0.9963)^{21}$	$(0.99762)^{30}$	$(0.9502)^{1}$
0.357483	0.979952	0.925109	0.931010	0.9502 = 0.2867

The numbers in parentheses are the abundances of the monoisotopic masses of the elements (Table 3.1), and they are raised to an exponential power, which is equal to the frequency of the respective atom in the molecule. Multiplication of the five resulting numbers—namely, (0.357483) (0.979952) (0.925109) (0.931010) (0.9502)—yields a total absolute abundance of 0.2867 for the monoisotopic species. Thus, the probability or absolute abundance of the presence of the monoisotopic peak in a mass spectrum is 0.2867; however, this value is not on the standard 0–100% mass spectral scale. This value must be adjusted relative to the base isotope peak and is described below.

The next set of calculations are in Table A2.1 of this Appendix and provide abundances of the masses that are approximately one mass unit above the monoisotopic species of the hemoglobin β-chain protein fragment. That is, each isotope molecular formula is identical to the monoisotopic molecular formula except that one atom, one at a time and for each element, is replaced with the respective (A + 1) isotope. Then, the probability of the appearance of each isotope is calculated with the respective one atom substitution. This is achieved by multiplying the five numbers from the exponential expression, for example, for the first entry in Table A.1: (0.369772) (0.979952) (0.925109) (0.931010) (0.9502) = 0.29655 is the absolute abundance or probability of one atom of ^{13}C isotope replacing one ^{12}C atom in an otherwise all monoisotopic element-containing β-chain fragment.

The molecular formula for the amino acid residues 41–59 of the hemoglobin β-chain is $C_{93}H_{135}N_{21}O_{30}S$, as given in Problem 3.1. Table A2.1 presents five different molecular formulae where the molecule can attain an approximately 1-Da increase in mass: $^{12}C_{92}{}^{13}C_{1}{}^{1}H_{135}{}^{14}N_{21}{}^{16}O_{30}{}^{32}S$, etc. The five different molecular formulae originate from the five different (A + 1) elemental isotopes in Table 3.1, because all five elements are found in the β-chain fragment. The basis for the calculations is presented by example, and formal mathematical algorithms can be found in a number of literature sources[2–7]. A comparison of the absolute abundance of each isotope species in Table A2.1 clearly shows that the ^{13}C isotope contributes the most to the absolute abundance, that is, 0.29655(100)/0.33025 = 89.80%. However, the dominant percentage contribution by mass of an element to this hemoglobin fragment is ^{1}H and ^{13}C in the mass defect and isotope shift, respectively. Thus, the probability or absolute abundance (appearance) of either the $^{2}H_{1}$ isotope—namely, $^{12}C_{93}{}^{1}H_{134}{}^{2}H_{1}{}^{14}N_{21}{}^{16}O_{30}{}^{32}S$—or the respective (A + 1) N, O, and S isotope molecular formulae of the hemoglobin fragment are very low compared with the $^{13}C_{1}$ molecular formula isotope, and these values are tabulated in Table A2.1. The masses of the five isotope molecules span a range between 2058.937 and 2058.947 Da, and only 0.010 Da represents the difference in masses between the ^{15}N, ^{14}N and ^{2}H, ^{1}H hemoglobin fragment species within numerical rounding-off in the ten-thousandths dalton place, namely,

$$
\begin{array}{lr}
{}^{2}H - {}^{1}H = 2.014102 - 1.007825 = 1.006277 & 1.006277 \\
{}^{15}N - {}^{14}N = 15.000109 - 14.003074 = 0.997035 & -\underline{0.997035} \\
 & 0.0092\ \mathrm{Da}
\end{array}
$$

Thus, for practical purposes, the five different isotope masses are identical in mass at a distinction level of 0.01 Da. In mass spectrometry terms, this is usually presented as a resolution of 0.01 Da, and this concept is introduced in Chapter 5. Because the ^{13}C isotope species dominates in abundance with respect to that of the other four isotope species, all five isotopes can be considered to contribute to the 2058.944-Da mass.

The 2058.944-Da value does not match either the nominal, monoisotopic, or average mass of the hemoglobin β-chain fragment (Problem 3.1). The average mass is approximately 0.344 Da higher, and there is a distinct difference in mass between the monoisotopic ($^{12}C_{93}{}^{1}H_{135}{}^{14}N_{21}{}^{16}O_{30}{}^{32}S$) and $^{13}C_{1}$-containing

Table A2.1 Calculated abundances of the isotopes that have an approximately 1-Da increase from the monoisotopic mass of hemoglobin β-chain fragment, residues 41–59.

	C	H	N	O	S	Absolute abundance	Mass (Da)
1.	$^{12}C_{92}\,^{13}C$	$^{1}H_{135}$	$^{14}N_{21}$	$^{16}O_{30}$	^{32}S		
	$\frac{93!}{92!\,1!}(0.989)^{92}(0.011)$	$(0.99985)^{135}$	$(0.9963)^{21}$	$(0.99762)^{30}$	$(0.9502)^{1}$		
	0.369772	0.979952	0.925109	0.931010	0.9502 =	0.29655	2058.944
2.	$^{12}C_{93}$	$^{1}H_{134}\,^{2}H$	$^{14}N_{21}$	$^{16}O_{30}$	^{32}S		
	$(0.989)^{93}$	$\frac{135!}{134!\,1!}(0.99985)^{134}(0.00015)$	0.925109	0.931010	0.9502 =	0.0058065	2058.947
	0.357843	0.0198470					
3.	$^{12}C_{93}$	$^{1}H_{135}$	$^{14}N_{20}\,^{15}N$	$^{16}O_{30}$	^{32}S		
	0.357483	0.979952	$\frac{21!}{20!\,1!}(0.9963)^{20}(0.0037)$	0.931010	0.9502 =	0.022359	2058.937
			0.0721479				
4.	$^{12}C_{93}$	$^{1}H_{135}$	$^{14}N_{21}$	$^{16}O_{29}\,^{17}O$	^{32}S		
	0.357483	0.979952	0.925109	$\frac{30!}{29!\,1!}(0.99762)^{29}(0.00038)$	0.9502 =	0.0032761	2058.945
				.0106388			
5.	$^{12}C_{93}$	$^{1}H_{135}$	$^{14}N_{21}$	$^{16}O_{30}$	^{33}S		
	0.357483	0.979952	0.925109	0.931010	0.0075 =	0.0022629	2058.940
						Total = 0.33025	

($^{12}C_{92}{}^{13}C_{1}{}^{1}H_{135}{}^{14}N_{21}{}^{16}O_{30}{}^{32}S$) species. This example has numerous real-life analogies to fractional representations based on average numbers. For example, statistical reduction of numerical datasets that yield averages of 1.2 or 72.9, or 100.000 human beings are routinely obtained. Yet 1.2, 72.9, or 100.000 people do not exist, where in the latter case the .000 implies the possibility of numbers in the ten-thousandths decimal place. There are only 1 or 72 or 100 integer values of human beings when describing the above rational numbers. This is also found in biological molecule molecular masses. For the hemoglobin β-chain fragment, there is no isolated molecule that has a discrete isotope mass of 2059.288 Da (Problem 3.1). This mass value is equal to the average mass of the hemoglobin β-chain fragment.

Continuing, Table A2.2 presents the mathematical permutations and algorithms necessary to arrive at the probability (or absolute abundance) of the (A + 2) molecular isotope peak, which is approximately two mass units above the monoisotopic mass of the hemoglobin β-chain fragment, in the protein fragment molecule. The matrix in Table A2.3 provides a compact format for documenting the various isotope permutations for an approximate 2-Da increase above the monoisotopic mass of the molecule. Each entry in Table A2.3 has an associated abundance or probability value which is calculated in Table A2.2

Remember that the $^{33}S^{33}S$ permutation cannot be used, because the hemoglobin fragment of interest contains only one sulfur atom. Note that each of the individual isotope terms have appeared in Table A2.1, where two separate terms in Table A2.1 are combined to form the respective term in Table A2.3. Thus, the results of the calculations in Table A2.1 can be transferred into Table A2.2.

In a similar manner to the ^{13}C isotope-containing molecule in Table A2.1, the $^{13}C_2$ isotope species also dominates in relative abundance in Table A2.2—namely, 0.1517(100)/0.21846 = 69.44%—with respect to the other possible isotope permutations. There are 16 different molecular isotopes of the hemoglobin β-chain fragment that have a mass approximately 2 Da higher than the monoisotopic mass. The $^{13}C_2$ molecular isotope (2059.947 Da) accounts for only 1/16 or 6.25% of the number of molecular permutations that produce an approximately 2-Da increase from that of the monoisotopic mass species. However, the $^{13}C_2$ isotope-containing molecule commands 69.44% of the presence (absolute abundance) of the 16 protein fragment isotope masses that are approximately 2-Da above the monoisotopic mass. That particular mass at 2059.947 Da is chosen to represent the 16 different, closely spaced isotope masses, and these masses span the range 2059.935–2059.953 Da. A difference of only 0.018 Da contains all 16 isotope masses (Table A2.2). Chapters 5 and 6 show the inability of even modern-day mass spectrometers to isolate many of these molecular isotopes shown in Table A2.2.

These exercises can be continued to solve for the calculated abundances of the hemoglobin β-chain fragment for the approximately 3-, 4-, 5-, (etc.) dalton higher isotope species from that of the monoisotopic mass. The number of molecular isotope permutations increases dramatically as the isotope mass increases from the monoisotopic value, however, only a few of the permutations contribute to a significant extent to the abundance of a particular isotope peak. Each isotope peak of a biological molecule is essentially referred as a $^{13}C_n$ isotope peak, because the ^{13}C, $^{13}C_2$, $^{13}C_3$, (etc.) isotope permutations contribute the most to the abundance or presence of each approximately 1-Da-spaced, observed mass spectral isotope peak. Thus, those particular masses can be calculated by subtracting the atomic weight of a ^{12}C atom and adding a ^{13}C atom for each successive isotope (approximate increase of 1 Da). Diminishing returns take effect, because the calculated abundances for successively higher isotope masses eventually decrease and become negligible. The abundance values of successively higher mass isotopes pass through a maximum and decrease.

For the hemoglobin β-chain fragment, only the $^{13}C_0$ (monoisotopic mass) to $^{13}C_5$ isotope species provide significant abundance values, and these are summarized in reference 8. For the $^{13}C_0$–$^{13}C_2$ isotope species we have the following:

	Absolute abundance	Mass	Relative Abundance %
$^{13}C_0$	0.28670	2057.939	86.8
$^{13}C_1$	0.33025	2058.942	100
$^{13}C_2$	0.21846	2059.946	66.1

A satisfactory comparison is observed between the calculated abundance values listed here and the calculated values listed in Matsuo et al.[8]. Since the abundance distribution of the three masses passes through a maximum at the $^{13}C_1$ isotope species, this can be considered as the most abundant isotope or the base peak (100%) of the entire isotope cluster. Normalizing this distribution, the $^{13}C_1$ peak at 100% yields the $^{13}C_0$, $^{13}C_1$, and $^{13}C_2$ peaks of 86.8%, 100%, and 66.1% abundance, respectively. These calculations were performed on neutral, or zero-charge molecules. In the electrospray ionization positive-ion mode, high-mass molecules are usually charged by the addition of a proton(s), as explained in Chapter 4.

An interesting phenomenon[9] occurs with the ^{13}C isotope species in the example (Table A2.1) in this Appendix. The abundance or probability values of 93 C-12 atoms versus 92 C-12 and one C-13 atom are 0.357 and 0.369, respectively (vide supra). The analogous values for the entire hemoglobin fragment are 0.2867 and 0.2965 for the $^{12}C_{93}$ and $^{12}C_{92}{}^{13}C$ isotopes, respectively. Note that whether the focus is on the carbon atom portion of the molecule or the entire molecule, the relative abundance (or probability of occurrence) increases from the monoisotopic ($^{12}C_{93}$) to the $^{12}C_{92}{}^{13}C$ species. The following shows the point where a series of carbon atoms with one additional ^{13}C atom has a higher probability of occurrence in comparison with the respective monoisotopic series of carbon atoms:

$^{12}C_{89}$, 0.3737	$^{12}C_{90}$, 0.36954	$^{12}C_{91}$, 0.3655	$^{12}C_{92}$, 0.3615
$^{12}C_{88}$, 0.3699	$^{12}C_{89}$ ^{13}C, 0.3658	$^{12}C_{90}$, ^{13}C, 0.3658	$^{12}C_{91}$, ^{13}C, 0.3658

$^{12}C_{93}$, 0.3575
$^{12}C_{92}$, 0.36977

Table A2.2 Calculated abundances of the isotopes that have an approximately 2-Da increase from the monoisotopic mass of the hemoglobin β-chain fragment, residues 41–59

						C	H	N	O	S		Absolute abundance	Mass (Da)
$^{12}C_{91}{}^{13}C_2$	$^{1}H_{135}$	$^{14}N_{21}$	$^{16}O_{30}$	^{32}S	=	0.189186[a]	$(0.99985)^{135}$	$(0.9963)^{21}$	$(0.99762)^{30}$	0.9502	=	0.151724	2059.947
$^{12}C_{92}{}^{13}C$	$^{1}H_{134}{}^{2}H$				=	0.369772	0.0198470	0.925109	0.931010	0.9502	=	0.00600608	2059.950
$^{12}C_{92}{}^{13}C$		$^{14}N_{20}{}^{15}N$			=	0.369772	0.979952	0.0721479	0.931010	0.9502	=	0.0231277	2059.941
$^{12}C_{92}{}^{13}C$			$^{16}O_{29}{}^{17}O$		=	0.369772	0.979952	0.925109	0.0106388	0.9502	=	0.00338875	2059.948
$^{12}C_{92}{}^{13}C$				^{33}S	=	0.369772	0.979952	0.925109	0.931010	0.0075	=	0.00234071	2059.943
$^{12}C_{93}$	$^{1}H_{133}{}^{2}H_2$	$^{14}N_{21}$	$^{16}O_{30}$	^{32}S	=	0.357483	0.000199492[b]	0.925109	0.931010	0.9502	=	0.000058799	2059.953
	$^{1}H_{134}{}^{2}H$	$^{14}N_{20}{}^{15}N$			=	0.357483	0.0198470	0.0721479	0.931010	0.9502	=	0.000452838	2059.944
	$^{1}H_{134}{}^{2}H$		$^{16}O_{29}{}^{17}O$		=	0.357483	0.0198470	0.925109	0.0106388	0.9502	=	0.000066352	2059.951
	$^{1}H_{134}{}^{2}H$			^{33}S	=	0.357483	0.0198470	0.925109	0.931010	0.0075	=	0.000045831	2059.946

$^{12}C_{93}$	$^{1}H_{135}$	$^{14}N_{19}{}^{15}N_2$	$^{16}O_{30}$	^{32}S	=	0.357483	0.979952	0.00267939[c]	0.931010	0.9502	=	0.000830357	2059.935
		$^{14}N_{20}{}^{15}N$	$^{16}O_{29}{}^{17}O$		=	0.357483	0.979952	0.0721479	0.0106388	0.9502	=	0.0002555	2059.942
		$^{14}N_{20}{}^{15}N$		^{33}S	=	0.357483	0.979952	0.0721479	0.931010	0.0075	=	0.000176482	2059.937
$^{12}C_{93}$	$^{1}H_{135}$	$^{14}N_{21}$	$^{16}O_{28}{}^{17}O_2$	^{32}S	=	0.357483	0.979952	0.925109	0.00005876[d]	0.9502	=	0.000018095	2059.949
			$^{16}O_{29}{}^{17}O$	^{33}S	=	0.357483	0.979952	0.925109	0.0106388	0.0075	=	0.000025859	2059.944
$^{12}C_{93}$	$^{1}H_{135}$	$^{14}N_{21}$	$^{16}O_{29}{}^{18}O$	^{32}S	=	0.357483	0.979952	0.925109	0.0559939[e]	0.9502	=	0.0172428	2059.945
$^{12}C_{93}$	$^{1}H_{135}$	$^{14}N_{21}$	$^{16}O_{30}$	^{34}S	=	0.357483	0.979952	0.925109	0.931010	0.0421	=	0.0127025	2059.936
										Total	=	0.21846	

[a] $\frac{93!}{91!\,2!}(0.989)^{91}(0.11)^2$

[b] $\frac{135!}{133!\,2!}(0.99985)^{133}(0.00015)^2$

[c] $\frac{21!}{19!\,2!}(0.9963)^{19}(0.0037)^2$

[d] $\frac{30!}{28!\,2!}(0.99762)^{28}(0.00038)^2$

[e] $\frac{30!}{29!\,1!}(0.99762)^{29}(0.002)$

Table A2.3 Isotope matrix table for an approximately 2-Da increase above the monoisotopic mass for the hemoglobin β-chain fragment

	^{13}C	^{2}H	^{15}N	^{17}O	^{33}S
^{13}C	$^{13}C_2$	$^{13}C^{2}H$	$^{13}C^{15}N$	$^{13}C^{17}O$	$^{13}C^{33}S$
^{2}H	—	$^{2}H_2$	$^{2}H^{15}N$	$^{2}H^{17}O$	$^{2}H^{33}S$
^{15}N	—	—	$^{15}N_2$	$^{15}N^{17}O$	$^{15}N^{33}S$
^{17}O	—	—	—	$^{17}O_2$	$^{17}O^{33}S$
^{18}O	—	—	—	^{18}O	—
^{34}S	—	—	—	—	^{34}S

There is a higher probability of occurrence (higher intensity) for a $^{12}C_{90}$, ^{13}C species than its $^{12}C_{91}$ monoisotopic species.[10] The hemoglobin β-chain fragment lies above the ^{12}C–^{13}C abundance transition such that the probability of occurrence of the $^{12}C_{92}$, ^{13}C isotope is greater than the $^{12}C_{93}$ monoisotopic species. Problem A2.1 is presented for the reader to gain familiarity with the process of calculating the first few abundances of a molecular isotope envelope.

PROBLEM A2.1 Using Tables A2.1, A2.2 and A2.3 as guides, calculate the absolute abundances of the monoisotopic, $^{13}C_1$, and $^{13}C_2$ molecules of bovine insulin $C_{254}H_{377}N_{65}O_{75}S_6$. Also calculate the mass spread in daltons of each ^{13}C cluster. The $^{13}C_3$ molecule—namely, $C_{251}{}^{13}C_3H_{377}N_{65}O_{75}S_6$—represents the true 100% abundance zero-charge peak in the molecular isotope cluster.

REFERENCES

1. McLafferty, F.W.; Tureček, F. *Interpretation of Mass Spectra,* 4th Edition; University Science Books: Mill Valley, CA, 1993.
2. Yergey, J.A. *Int. J. Mass Spectrom. Ion Phys.* **1983,** *52,* pp 337–349.
3. Yamamoto, H.; McCloskey, J.A. *Anal. Chem.* **1977,** *49,* pp 281–283.
4. Hibbert, D.B. *Chemom. Intell. Lab. Syst.* **1989,** 6, pp 203–212.
5. Grange, A.H.; Brumley, W.C. *J. Am. Soc. Mass Spectrom.* **1997,** *8,* pp 170–182.
6. McCloskey, J.A. In *Methods in Enzymology;* McCloskey, J.A., Ed; Academic Press: New York, 1990; Volume 193, pp 882–886.
7. Beynon, J.H., Ed., *Mass Spectrometry and its Applications to Organic Chemistry*: Elsevier Publishing Co.: Amsterdam, The Netherlands, 1960.
8. Matsuo, T.; Sakurai, T.; Matsuda, H.; Wollnik, H.; Katakuse, I. *Biomed. Mass Spectrom.* **1983,** *10,* pp 57–60.
9. Fenselau, C. *Anal. Chem.* **1982**, *54*, pp 106A–116A.
10. Johnstone, R.A.W.; Rose, M.E. *Mass Spectrometry for Chemists and Biochemists*, 2nd Edition; Cambridge University Press: Cambridge, U.K., 1996; Chapter 12, p442.

ANSWERS

PROBLEM A2.1 The monoisotopic mass of bovine insulin is calculated as given in Problem 3.3, and is equal to 5729.601 Da.

The absolute abundance of the monoisotopic mass is calculated as follows:

$^{12}C_{254}$	$^{1}H_{377}$	$^{14}N_{65}$	$^{16}O_{75}$	$^{32}S_{6}$
$(0.989)^{254}$	$(0.99985)^{377}$	$(0.9963)^{65}$	$(0.99762)^{75}$	$(0.9502)^{6}$
0.0602363	0.945015	0.785884	0.836346	0.736021

Multiplication of the five terms yields 0.02754 for the absolute abundance of the monoisotopic mass.

Tables A2.4 and A2.5 present the calculations for the approximately 1- and 2-Da mass increase (^{13}C and $^{13}C_2$, respectively) from the monoisotopic mass of the insulin molecule. Table A2.4 yields an absolute abundance of 0.088093 for the approximately 1-Da mass increase (^{13}C peak), with an isotope mass range of 5730.598–5730.607 Da and 0.009 Da mass spread. Table A2.5 provides an absolute abundance of 0.15198 for the $^{13}C_2$ mass, with an isotope mass range of 5731.595–5731.613 Da and a mass spread of 0.018 Da.

Calculation of the absolute abundance of the $^{13}C_3$ isotope cluster (not shown) would allow the determination of the relative abundances[4] of the ^{12}C, $^{13}C_1$, and $^{13}C_2$ isotopes. This entails the construction of a $^{13}C_3$ table of isotopes (not shown).

Table A2.4 Calculated abundances of the isotopes that have an approximately 1-Da increase from the monoisotopic mass of bovine insulin

	C	H	N	O	S	Absolute abundance	Mass (Da)
1.	$^{12}C_{253}{}^{13}C$	$^{1}H_{377}$	$^{14}N_{65}$	$^{16}O_{75}$	$^{32}S_6$		
	$\frac{254!}{253!\ 1!}(0.989)^{253}(0.011)$	$(0.99985)^{377}$	$(0.9963)^{65}$	$(0.99762)^{75}$	$(0.9502)^6$		
	0.170172	0.945015	0.785884	0.836346	0.736021 =	0.077797	5730.604
2.	$^{12}C_{254}$	$^{1}H_{376}{}^{2}H$	$^{14}N_{65}$	$^{16}O_{75}$	$^{32}S_6$		
	$(0.989)^{254}$	$\frac{377!}{376!\ 1!}(0.99985)^{376}(0.00015)$					
	0.0602363	0.0534486	0.785884	0.836346	0.736021	0.0015575	5730.607
3.	$^{12}C_{254}$	$^{1}H_{377}$	$^{14}N_{64}{}^{15}N$	$^{16}O_{75}$	$^{32}S_6$		
	0.0602363	0.945015	$\frac{65!}{64!\ 1!}(0.9963)^{64}(0.0037)$				
			0.189707	0.836346	0.736021 =	0.0066475	5730.598
4.	$^{12}C_{254}$	$^{1}H_{377}$	$^{14}N_{65}$	$^{16}O_{74}{}^{17}O$	$^{32}S_6$		
				$\frac{75!}{74!\ 1!}(0.99762)^{74}(0.00038)$			
	0.0602363	0.945015	0.785884	0.023893	0.736021 =	0.00078670	5730.605
5.	$^{12}C_{254}$	$^{1}H_{377}$	$^{14}N_{65}$	$^{16}O_{75}$	$^{32}S_5{}^{33}S$		
	0.0602363	0.945015	0.785884	0.836346	$\frac{6!}{5!\ 1!}(0.9502)^5(0.0075)$	0.0013204	5730.600
					0.034857 Total =	0.088093	

Table A2.5 Calculated abundances of the isotopes that have a nominal 2-Da increase from the monoisotopic mass of bovine insulin

						C	H	N	O	S		Absolute abundance	Mass (Da)
$^{12}C_{252}{}^{13}C_2$	$^{1}H_{377}$	$^{14}N_{65}$	$^{16}O_{75}$	$^{32}S_6$	=	0.239428[a]	0.945015	0.785884	0.836346	0.736021	=	0.109458	5731.608
$^{12}C_{253}{}^{13}C$	$^{1}H_{376}{}^{2}H$	$^{14}N_{65}$	$^{16}O_{75}$	$^{32}S_6$	=	0.170172	0.0534486	0.785884	0.836346	0.736021	=	0.00440010	5731.611
$^{12}C_{253}{}^{13}C$	$^{1}H_{377}$	$^{14}N_{64}{}^{15}N$	$^{16}O_{75}$	$^{32}S_6$	=	0.170172	0.945015	0.189707	0.836346	0.736021	=	0.0187796	5731.601
$^{12}C_{253}{}^{13}C$		$^{14}N_{65}$	$^{16}O_{74}{}^{17}O$		=	0.170172	0.945015	0.785884	0.023893	0.736021	=	0.00222252	5731.608
$^{12}C_{253}{}^{13C}$				$^{32}S_5{}^{33}S$	=	0.170172	0.945015	0.785884	0.836346	0.0348568[b]	=	0.00368433	5731.604
$^{12}C_{254}$	$^{1}H_{375}{}^{2}H_2$	$^{14}N_{65}$	$^{16}O_{75}$	$^{32}S_6$	=	0.0602363	0.00150748[c]	0.785884	0.836346	0.736021	=	0.0000439282	5731.613
	$^{1}H_{376}{}^{2}H$	$^{14}N_{64}{}^{15}N$			=	0.0602363	0.0534486	0.189707	0.836346	0.736021	=	0.000375974	5731.604
	$^{1}H_{376}{}^{2}H$	$^{14}N_{65}$	$^{16}O_{74}{}^{17}O$		=	0.0602363	0.0534486	0.785884	0.023893	0.736021	=	0.0000444956	5731.611
	$^{1}H_{376}{}^{2}H$		$^{16}O_{75}$	$^{32}S_5{}^{33}S$	=	0.0602363	0.0534486	0.785884	0.836346	0.0348568	=	0.0000737610	5731.606
$^{12}C_{254}$	$^{1}H_{377}$	$^{14}N_{63}{}^{15}N_2$	$^{16}O_{75}$	$^{32}S_6$	=	0.0602363	0.945015	0.0225447[d]	0.836346	0.736021	=	0.000789984	5731.595
		$^{14}N_{64}{}^{15}N$	$^{16}O_{74}{}^{17}O$		=	0.0602363	0.945015	0.189707	0.023893	0.736021	=	0.000189907	5731.602
		$^{14}N_{64}{}^{15}N$	$^{16}O_{75}$	$^{32}S_5{}^{33}S$	=	0.0602363	0.945015	0.189707	0.836346	0.0348568	=	0.000314814	5731.597
$^{12}C_{254}$	$^{1}H_{377}$	$^{14}N_{65}$	$^{16}O_{73}{}^{17}O_2$	$^{32}S_6$	=	0.0602363	0.945015	0.785884	0.000336733[e]	0.736021	=	0.0000110874	5731.609
			$^{16}O_{74}{}^{17}O$	$^{32}S_5{}^{33}S$	=	0.0602363	0.945015	0.785884	0.023893	0.0348568	=	0.0000372575	5731.604
$^{12}C_{254}$	$^{1}H_{377}$	$^{14}N_{65}$	$^{16}O_{74}{}^{18}O$	$^{32}S_6$	=	0.0602363	0.945015	0.785884	0.125751[f]	0.736021	=	0.00414055	5731.605
$^{12}C_{254}$	$^{1}H_{377}$	$^{14}N_{65}$	$^{16}O_{75}$	$^{32}S_5{}^{34}S$	=	0.0602363	0.945015	0.785884	0.836346	0.195663[g]	=	0.00732654	5731.597
										Total	=	0.15189	

[a] $\frac{254!}{252!\,2!}(0.989)^{252}(0.011)^2$ [b] $\frac{6!}{5!\,1!}(0.9502)^5(0.0075)$ [c] $\frac{377!}{375!\,2!}(0.99985)^{375}(0.00015)^2$ [d] $\frac{65!}{63!\,2!}(0.9963)^{63}(0.0037)^2$

[e] $\frac{75!}{73!\,2!}(0.99762)^{73}(0.0038)^2$ [f] $\frac{75!}{74!\,1!}(0.99762)^{74}(0.002)$ [g] $\frac{6!}{5!\,1!}(0.9502)^5(0.0421)$

APPENDIX 3

Molecular formulae and molecular mass information for peptides and proteins

Protein/peptide	Molecular formula	Mass value: Nominal	Mass value: Monoisotopic	Mass value: M_r average	Reference[a]	Sequence[b], remarks
Thyroid releasing hormone, TRH	$C_{16}H_{22}N_6O_4$	362	362.17	362.4	1, 2, 3	pEHP-NH_2
Leu-enkephalin	$C_{28}H_{37}N_5O_7$	555	555.34	555.6	4, 5–7	YGGFL
Met-enkephalin	$C_{27}H_{35}N_5O_7S$	573	573.3	573.7	7, 8	YGGFM
Dynorphin A (1–6)	$C_{34}H_{49}N_9O_8$	711	711.37	711.82	2, 9	YGGFLR
Kemptide, dephosphorylated	$C_{32}H_{61}N_{13}O_9$	771	771.47	771.92	10, 11, 12	LRRASLG
β-Casomorphin, bovine	$C_{41}H_{55}N_7O_9$	789	789.4	789.9	2, 13	YPFPGPI
Angiotensin III	$C_{46}H_{66}N_{12}O_9$	930	930.51	931.11	8	RVYIHPF
Oxytocin	$C_{43}H_{66}N_{12}O_{12}S_2$	1,006	1,006.44	1,007.2	12, 14–16	CYIQNCPLG-NH_2
Angiotensin II	$C_{50}H_{71}N_{13}O_{12}$	1,.045	1,045.53	1,046.2	2, 5, 17	DRVYIHPF
Arg[8]-vasotocin	$C_{43}H_{67}N_{15}O_{12}S_2$	1,049	1,049.45	1,050.23	2, 16, 18	CYIQNCPRG-NH_2
Bradykinin	$C_{50}H_{73}N_{15}O_{11}$	1,059	1,059.56	1,060.22	8, 16, 17, 19, 20–22	RPPGFSPFR
Acyl carrier protein, ACP (65–74)	$C_{47}H_{74}N_{12}O_{16}$	1,062	1,062.53	1,063.18	2, 14	VQAAIDYING
Substance P(1–9)	$C_{52}H_{77}N_{15}O_{12}$	1,103	1,103.59	1,104.27	5, 8	RPKPQQFFG
Dynorphin(1–9)	$C_{52}H_{84}N_{18}O_{11}$	1,136	1,136.7	1,137.36	2, 23	YGGFLRRIR

Gramicidin S	$C_{60}H_{92}N_{12}O_{12}$	1,140	1,140.70	1,141.5	<u>19</u>, 22, 24, 25	LFPVOrnLFPVOrn, cyclic
Luteinizing hormine-releasing hormone, human LH-RH	$C_{55}H_{75}N_{17}O_{13}$	1,181	1,181.57	1,182.3	<u>2</u>, <u>8</u>, 26, 27	pEHWSYGLRPG-NH_2
Cyclosporin A	$C_{62}H_{111}N_{11}O_{12}$	1,201	1,201.84	1,202.6	12, <u>28</u>, <u>29</u>	
Angiotensin I	$C_{62}H_{89}N_{17}O_{14}$	1,295	1,295.7	1,296.5	5, <u>19</u>, <u>26</u>, 30, 31, 32	DRVYIHPFHL
Substance P(1–12)	$C_{63}H_{98}N_{18}O_{13}S$	1,346	1,346.6	1,347.6	<u>2</u>, 21, 33	RPKPQQFFGLM-NH_2
Vasoactive intestinal contractor peptide (1–12)	$C_{61}H_{88}N_{18}O_{22}$	1,424	1,424.6	1,425.5	<u>2</u>, 34	HSDAVFTDNYTR
Conotoxin	$C_{55}H_{80}N_{20}O_{18}S_4$	1,436	1,436.48	1,437.6	<u>2</u>, <u>12</u>	ECCNPACGRHY SC-NH_2
Human fibrinopeptide A	$C_{63}H_{97}N_{19}O_{26}$	1,535	1,535.68	1,536.58	<u>8</u>, 12, 16	ADSGEGDFLAEGG GVR
Glu[1]-fibrinopeptide B	$C_{66}H_{95}N_{19}O_{26}$	1,596	1,596.67	1,570.59	<u>35</u>, 36	EGVNDNEEGFFSAR
Gramicidin D = 85% gramicidin A Gramicidin B = Trp^{11}-Phe^{11} Gramicidin C = Trp^{11}-Tyr^{11}	$C_{82}H_{119}N_{17}O_{15}$	1,581	1,581.91	1,582.95 (D)	<u>37</u>, 38, 39 <u>38</u> 38	HCO-VGALAVVVWLWLW-$NHCH_2CH_2$-OH
Dynorphin A(1–13)	$C_{75}H_{126}N_{24}O_{15}$	1,602	1,603.0	1,604.0	<u>2</u>, <u>19</u>	YGGFLRRIRPKLK
Bombesin	$C_{71}H_{110}N_{24}O_{18}S$	1,618	1,618.81	1,619.87	<u>2</u>, 18	pEQRLGNQWAVGHL M-NH_2

(continued)

Molecular formulae and molecular mass information for peptides and proteins (*continued*)

Protein/peptide	Molecular formula	Mass value: Nominal	Mass value: Monoisotopic	Mass value: M_r average	Reference[a]	Sequence[b], remarks
Somatostatin, growth hormone release inhibiting factor	$C_{82}H_{111}N_{21}O_{20}S_2$	1,636	1,636.72	1,637.91	8, 15	AGCKNFFWKTFTSC (disulfide link C3–C14)
Neurotensin	$C_{78}H_{121}N_{21}O_{20}$	1,671	1,671.91	1,672.95	9, 16, 19	pELYENKPRRPYIL
α-Endorphin, human β-lipotropin(61–76)	$C_{77}H_{120}N_{18}O_{26}S$	1,744	1,744.91	1,746.0	19, 21, 40	YGGFMTSKSQTPLVT
Ribonuclease S-peptide(1–15)	$C_{73}H_{118}N_{24}O_{24}S$	1,746	1,746.84	1,747.96	41, 42	KETAAAKFERQHMDS - NH_2
Renin substrate angiotensinogen(1–14), porcine	$C_{85}H_{123}N_{21}O_{20}$	1,757	1,757.92	1,759.04	2, 20	DRVYIHPFHLLVYS
Renin (human) substrate angiotensinogen(1–14)	$C_{83}H_{122}N_{24}O_{19}$	1,758	1,758.93	1,760.03	2	DRVYIHPFHLVIHN
γ-Endorphin	$C_{83}H_{130}N_{19}O_{27}S$	1,857	1,857.9	1,859.1	2	YGGFMTSEKSQTP LVTL
Ribonuclease, S-peptide(1–21)				2,166	42, 43	
Bovine insulin, α-chain	$C_{97}H_{151}N_{25}O_{34}S_4$	2,337	2,337.97	2,339.68	19, 44	
Kaplan peptide	$C_{103}H_{166}N_{30}O_{33}$	2,350	2,351.2	2,352.6	45	SAISLDGEKVDFNBFR GRAVK
Adrenocorticotropic hormone, human ACTH(18–39) (CLIP)	$C_{112}H_{165}N_{27}O_{36}$	2,463	2,464.19	2,465.71	16, 46	RPVKVYPNGAEDEAE AFPLEF

Bovine insulin, α-chain (oxidized)	$C_{97}H_{151}N_{25}O_{46}S_4$	2,529	2,529.91	2,531.67	16, 19, 47	GIVEQC(SO$_3$H)C(SO$_3$H) ASVC(SO$_3$H)SLYQL ENYC(SO$_3$H)N
Melittin (bee venom)	$C_{131}H_{229}N_{39}O_{31}$	2,843	2,844.75	2,846.51	19, 25, 46, 48–51	GIGAVLKVLTTGLPAL ISWIKRKRQQ-NH$_2$
Bovine insulin, β-chain	$C_{157}H_{232}N_{40}O_{41}S_2$	3,396	3,397.67	3,399.95	19, 44	FVNQHLCGSHLVEAL YLVCGERGFFYTPKA
β-endorphin, β-lipotropin(61–91), bovine, ovine	$C_{155}H_{250}N_{42}O_{44}S$	3,434	3,435.9	3,438.0	2, 40	YGGFMTSEKSQTPLV TLFKNAIIKNAYKKGE
β-endorphin, β-lipotropin(61–91), human	$C_{158}H_{251}N_{39}O_{46}S$	3,461	3,462.89	3,465.0	2, 52 (Y,E)	YGGFMTSEKSQTPLV TLFKNAIIKNA (Y,H)KKG(E,Q)
Glucagon	$C_{153}H_{224}N_{42}O_{50}S$	3,480	3,481.6	3,483.8	19, 24, 41, 46, 53–56	HSQGTFTSDYSKYLD SRRAQDFVQWLMNT
Bovine insulin, β-chain (oxidized)	$C_{157}H_{232}N_{40}O_{47}S_2$	3,492	3,493.64	3,495.95	19, 41, 46	FVNQHLC(SO$_3$H)GSHL VEALYLVC(SO$_3$H)GE RGFFYTPKA
Adrenocorticotropic hormone, ACTH(1–39), human	$C_{207}H_{308}N_{56}O_{58}S$	4,536	4,538.26	4,541.1	2, 9, 48, 57, 58	SYSMEHFRWGKPVG KKRRPVKVYPNGAED ESAEADPLEF
Human parathyroid hormone(1–44)	$C_{225}H_{366}N_{68}O_{61}S_2$	5,058	5,060.7	5,063.9	46, 50	SVSEIQLMHNLGKHL NSMERVEWLRKKLQ DVHNFVALGAPLAPR

(continued)

Molecular formulae and molecular mass information for peptides and proteins (*continued*)

Protein/peptide	Molecular formula	Mass value			Reference[a]	Sequence[b], remarks
		Nominal	Monoisotopic	M_r average		
Transforming growth factor-αTGF-α				5,500	59, 60	50 residues
Bovine insulin, α- and β-chains	$C_{254}H_{377}N_{65}O_{75}S_6$	5,727	5,729.6	5,733.6	17, 19, 21, 24, 25, 44, 61–65	
Porcine insulin	$C_{256}H_{381}N_{65}O_{76}S_6$	5,771	5,773.6	5,777.6	12, 25, 53, 65	
Human insulin, α- and β-chains			5,804.6	5,807.6	2, 12, 66	
Cerato-ulmin				7,619	16, 67	
Insulin-like growth factor, IGF-1				7,649	16, 34	
Bovine ubiquitin	$C_{376}H_{624}N_{106}O_{119}S$	8,556	8,560.57	8,565.01	9, 19, 34, 37, 66, 68, 69	
Bovine proinsulin				8,681	16	
Porcine proinsulin	$C_{398}H_{617}N_{109}O_{123}S_6$	9,079	9,083.4	9,089.4	12, 24	
apo-ACP, acyl carrier protein				10,422.7	16, 70	
holo-ACP				10,763.1	70	
Ribonuclease S-protein(22–124)				11,534	42	
Thioredoxin (E. coli)				11,673.4	9, 12, 18, 34, 71, 72, 73	
Rabbit parvalbumin				11,976.7	74	
Porcine cytochrome c				12,231	16	
Bovine cytochrome c				12,231	16, 19	

Cytochrome c (horse heart)				12,359.9	20, 25, 66, 69, 75–79	104 residues
Ribonuclease B, deglycosylated				13,682	16,80	
Bovine pancreas ribonuclease A, RNase A				12,682.3	16, 42, 51, 64, 80, 81, 82, 83, 84	
Bovine ribonuclease S				13,700	42	
Bovine α-lactalbumin				14,175.0	9, 69, 72, 74, 81, 85	
Recombinant angiogenin				14,268.1	86	
Lysozyme (chicken egg)	$C_{613}H_{951}N_{193}O_{185}S_{10}$	14,289	14,295.81	14,305.16	17, 19, 20, 21, 25, 34, 62–64, 72, 81, 87–89	
Bovine ribonuclease A, $(pyridylethyl)_8$				14,531	80, 82	
Bovine α-lactalbumin, carboxymethylated (6 per molecule)				14,647.4	16, 18, 81	
Lysozyme (human milk)				14,692.8	25	
Ribonuclease B, bovine $Man_6GlcNAc_2$				15,061	26	
Human hemoglobin, α-chain				15,126.4	69, 90, 91, 92, 93	
Recombinant interleukin-2α*, IL-2α				15,547.2	94, 95, 96	133 residues
Human hemoglobin, β-Hafnia				15,858.2	91	
Human hemoglobin, β-chain				15,867.2	69, 90, 91, 93, 97, 98, 99	

(continued)

Molecular formulae and molecular mass information for peptides and proteins (*continued*)

Protein/peptide	Molecular formula	Mass value			Reference[a]	Sequence[b], remarks
		Nominal	Monoisotopic	M_r average		
Human hemoglobin, β-Quebec-Chori				15,879.3	91	
Hemoglobin, δ-chain				15,924.2	16, 93, 100	
Bovine calmodulin				16,791.5	16, 101	
Horse apomyoglobin (heart/skeletal muscle)	$C_{769}H_{1212}N_{210}O_{218}S_2$	16,932	16,940.96	16,951.53	<u>19</u>, 31, <u>40</u>, 63, 102	
holo-Whale myoglobin (skeletal muscle)				17,199.91	16, 34, 102	
Recombinant interleukin-1α*, IL-α				17,646	16	
Dimer of porcine proinsulin	$C_{796}H_{1234}N_{218}O_{246}S_{12}$	18,158	18,166.8	18,178.8	24	
Bovine β-lactoglobulin B				18,277.1	16, <u>19</u>, <u>103</u>, <u>104</u>	162 residues
Bovine β-lactoglobulin A				18,363.1	13, 16, 49, 64	
Human interferon, α-2 α-1				19,238	<u>96</u> 16	165 residues
Trypsin inhibitor (soybean)				20,090.9	9, 25	
Recombinant interleukin-6				20,907.3	16, 34	
Recombinant porcine somatotropin				21,802	16	
r-Bovine somatotropin				21,816	16	
Recombinant human growth hormone, hGH				22,125	16	

Methionine, human growth hormone				22,256.2	9, 33, 57	
Prolactin				22,660.0	16, 34	
α-Casein, dephosphorylated				22,974	16, 18	
Bovine trypsin, β-form				23,293	9, 26, 72, 105, 106	
Bovine trypsin, α-form				23,309.6	16, 83	
Bovine trypsin, Ψ-form				23,329.0	16, 83	
Porcine trypsin				23,463.3	72, 106	
Trypsinogen, bovine				23,981.1	9, 16, 18, 62, 64, 72	
Bovine chymotrypsin				25,234	9, 16, 81	
Chymotrypsinogen, (bovine)				25,656	16, 18, 107	
Subtilisin (Carlsburg)				27,288	16, 18	
Subtilisin BPN				27,534.0	16, 64	
Staphylococcal enterotoxin B	$C_{1275}H_{1945}N_{328}O_{386}S_{10}$	28,333	28,346.97	28,365.17	<u>108</u>	239 residues
Carbonic anhydrase, bovine				29,024	9, 16, 18, 25, 64, 109	
Thioesterase II, rat mammary, N-acetylated(1–261)				29,230	16, 18	
Thermolysin				34,334	16, 18	
Porcine pepsin				34,584	16, 110	326 residues
Lactate dehydrogenase, rabbit				36,500	16, 18	
Alcohol dehydrogenase (yeast,N-actyl)				36,748	16, 18	

(continued)

Molecular formulae and molecular mass information for peptides and proteins (*continued*)

Protein/peptide	Molecular formula	Mass value			Reference[a]	Sequence[b], remarks
		Nominal	Monoisotopic	M_r average		
Recombinase A (E. coli)				37,842	16, 18	
Alcohol dehydrogenase (horse liver)				39,830.0	9, 16, 107	
Actin (rabbit muscle)				41, 862	16, 18	
Plasminogen-activating inhibitor, PAI				42,769.9 Na^+ impurities	16, 34	
Horseradish peroxidase, HRP, dephosphorylated				43,261	16, 62	
Ovalbumin				44,300	9, 16, 18, 25, 34	
Bovine serum albumin, BSA				66,430.3	9, 62, 64, 67, <u>111</u>	583 residues
Human serum albumin, HSA				66,436.1	16, 67, <u>99</u>, 112	
apo-Bovine serum transferrin, apo-BST				78,030	16, 67	
apo-Human serum transferrin, apo-HST				79,557	16, 67	
Lactate dehydrogenase				140,000	69	4 subunits
Immunoglobulin G, IgG				149,190	16, 26	
Glucose isomerase				172,400	26	
Catalase				236,230	26	

[a]Underlined reference provides the amino acid residue sequence.
[b]Orn, ornithine; references in this column provide sequence information.

REFERENCES

1. Johansson, I.M.; Huang, E.C.; Henion, J.D.; Zweigenbaum, J. *J. Chromatogr.* **1991**, *554*, pp 311–327.
2. Sigma Chemical Co. catalogue, St. Louis, MO, 1996.
3. Jones, W.M.; Manning, J.M. *Biochem. Biophys. Res. Commun.* **1985**, pp 933–940.
4. Metzger, J.W.; Eckerson, C. In *Microcharacterization of Proteins*, Kellner, R.; Lottspeich, F.; Meyer, H.E., Eds., VCH Publishers: New York, 1994; Volume 2, pp 167–187.
5. Starrett, A.M.; DiDonato, G.C. *Rapid Commun. Mass Spectrom.* **1993**, *7*, pp 12–15.
6. Tomer, K.B. *Mass Spectrom. Rev.* **1989**, *8*, pp 483–511.
7. Mück, W.M.; Henion, J.D. *J. Chromatogr.* **1989**, *495*, pp 41–59.
8. Ashcroft, A.E.; Derrick, P.J. In *Mass Spectrometry of Peptides*; Desiderio, D.M., Ed.; CRC Press: Boca Raton, FL, 1991; Chapter 7, pp 121–138.
9. Smith, R.D.; Loo, J.A.; Edmonds, C.G.; Barinaga, C.J.; Udseth, H.R. *Anal. Chem.* **1990**, *62*, pp 882–899.
10. Schroeder, W.; Covey, T.; Hucho, F. *FEBS Lett.* **1990**, *273*, pp 31–35.
11. Till, J.H.; Annan, R.S.; Carr, S.A.; Miller, W.T. *J. Biol. Chem.* **1994**, *269*, pp 7423–7428.
12. Biemann, K.; Martin, S.A. *Mass Spectrom. Rev.* **1987**, pp 1–76.
13. Hunt, D.F.; Shabanowitz, J.; Moseley, M.A.; McCormack, A.L.; Michel, H.; Martino, P.A.; Tomer, K.B.; Jorgenson, J.W. In *Methods in Protein Sequence Analysis*; Jörnvall, H., Höög, J.-O.; Gustavsson, A.-M., Eds.; Birkhäuser Verlag: Basel, Switzerland, 1991; pp 257–266.
14. Schnölzer, M.; Jones, A.; Alewood, P.F.; Kent, S.B.H. *Anal. Biochem.* **1992**, *204*, pp 335–343.
15. Aplin, R.; Anderson, K. *Proc. 43rd ASMS Conference on Mass Spectrometry and Allied Topics* **1995**, p 1276.
16. Smith, R.D.; Loo, J.A.; Loo, R.R.O.; Busman, M.; Udseth, H.R. *Mass Spectrom. Rev.* **1991**, *10*, pp 359–451.
17. Larsen, B.S.; McEwen, C.N. *J. Am. Soc. Mass Spectrom.* **1991**, *2*, pp 205–211.
18. Ashton, D.S.; Beddell, C.R.; Cooper, D.J.; Green, B.N.; Oliver, R.W.A. *Org. Mass Spectrom.* **1993**, *28*, pp 721–728.
19. Stephenson, J.L. Jr.; McLuckey, S.A., *Anal. Chem.* **1997**, *69*, pp 281–285.
20. McEwen, C.N.; Larsen, B.S. *Rapid Commun. Mass Spectrom.* **1992**, *6*, pp 173–178.
21. Murata, H.; Takao, T.; Shimonishi, Y. *Rapid Commun. Mass Spectrom.* **1994**, *8*, pp 205–210.
22. Wang, G.; Cole, R.B. *Org. Mass Spectrom.* **1994**, *29*, pp 419–427.
23. Lee, E.D.; Henion, J.D.; Covey, T.R. *J. Microcolumn Sep.* **1989**, *1*, pp 14–18.
24. Zubarev, R.A. *Int. J. Mass Spectrom. Ion Processes* **1991**, *107*, pp 17–27.
25. Loo, J.A.; Udseth, H.R.; Smith, R.D. *Anal. Biochem* **1989**, *179*, pp 404–412.
26. Loo, J.A.; Edmonds, C.G.; Smith, R.D.; Lacey, M.P.; Keough, T. *Biomed. Environ. Mass Spectrom* **1990**, *19*, pp 286–294.
27. Schally, A.V. *Science* **1978**, *202*, pp 18–28.
28. Husek, A. *J. Chromatogr. A* **1997**, *759*, pp 217–224.
29. Havliček, V.; Jegorov, A.; Sedmera, P.; Ryska, M. *J. Mass Spectrom. & Rapid Commun. Mass Spectrom.* **1995**, pp S158–S164.
30. Bruins, A.P.; Covey, T.R.; Henion, J.D. *Anal. Chem.* **1987**, *59*, 2642–2646.
31. Schwartz, J.C.; Syka, J.E.P.; Jardine, I. *J. Am. Soc. Mass Spectrom.* **1991**, *2*, pp 198–204.

32. Meng, C.K.; McEwen, C.N.; Larsen, B.S. *Rapid Commun. Mass Spectrom.* **1990**, *4*, pp 151–155.
33. Edmonds, C.G.; Loo, J.A.; Fields, S.M.; Barinaga, C.J.; Udseth, H.R.; Smith, R.D. In *Biological Mass Spectrometry*; Burlingame, A.L.; McCloskey, J.A., Eds.; Elsevier Science Publishers B.V.: Amsterdam, The Netherlands, 1990; pp 77–100.
24. Jardine, I. In *Methods in Enzymology: Mass Spectrometry*, McCloskey, J.A., Ed; Academic Press: San Diego, CA, 1990; Volume 193, Chapter 24, pp 441–455.
35. Covey, T.R. In *Biological and Biotechnological Applications of Electrospray Ionization Mass Spectrometry*; Snyder, A.P., Ed; American Chemical Society: Washington, DC, 1996; Chapter 2, pp 21–59.
36. Hail, M.; Lewis, S.; Zholl, J.; Jardine, I.; Whitehouse, C. In *Current Research in Protein Chemistry: Techniques, Structure and Function*; Villafranca, J.J., Ed.; Academic Press: San Diego, CA, 1990; Chapter 10, pp 105–116.
37. Dienes, T.; Pastor, S.J.; Schürch, Scott, J.R.; Yao, J.; Cui. S.; Wilkins, C.L. *Mass Spectrom. Rev.* **1996**, *15*, pp 163–211.
38. Kleinekofort, W.; Avdiev, J.; Brutschy, B. *Int. J. Mass Spectrom. Ion Processes* **1996**, *152*, pp 135–142.
39. Biemann, K. In *Mass Spectrometry of Biological Materials*; McEwen, C.N.; Larsen, B.S., Eds.; Marcel Dekker: New York, 1990; Chapter 1, pp 3–24.
40. Zhang, X.; Dillen, L.; Vanhoutte, K.; Van Dongen, W.; Esmans, E.; Claeys, M. *Anan. Chem.* **1996**, *68*, pp 3422–3430.
41. Caprioli, R.M.; DaGue, B.; Fan, T.; Moore, W.T. *Biochem. Biophys. Res. Commun.* **1987**, *146*, pp 291–299.
42. Ogorzalek Loo, R.R.; Goodlett, D.R.; Smith, R.D.; Loo, J.A. *J. Am. Chem. Soc.* **1993**, *115*, pp 4391–4392.
43. Goodlett, D.R.; Ogorzalek Loo, R.R.; Loo, J.A.; Wahl, J.H.; Udseth, H.R.; Smith, R.D., *J. Am. Soc. Mass Spectrom.* **1994**, *5*, pp 614–622.
44. Bruce, J.E.; Anderson, G.A.; Smith, R.D. *Anal. Chem.* **1996**, *68*, pp 534–541.
45. Dobberstein, P.; Muenster, H. *J. Chromatog, A* **1995**, *712*, pp 3–15.
46. Dongre, A.R.; Somogyi, A.; Wysocki, V.H. *J. Mass Spectrom.* **1996**, *31*, pp 339–350.
47. Loo, J.A.; Ogorzalek, R.R.; Light, K.J.; Edmonds, C.G.; Smith, R.D. *Anal. Chem.* **1992**, *64*, pp 81–88.
48. Weigt, C.; Meyer, H.E.; Kellner, R. In *Microcharacterization of Proteins*; Kellner, R.; Lottspeich, F.; Meyer, H.E.; Eds.; VCH Verlagsgesellschaft mbH: Weinheim, Germany, 1994; Chapter V.3, pp 189–205.
49. Covey, T.R.; Bonner, R.F.; Shushan, B.I.; Henion, J. *Rapid Commun. Mass Spectrom.* **1988**, *2*, 249–256.
50. Smith, R.D.; Loo, J.A.; Barinaga, C.J.; Edmonds, C.G.; Udseth, H.R. *J. Am. Soc. Mass Spectrom.* **1989**, *90*, pp 53–65.
51. Loo, J.A.; Edmonds, C.G.; Udseth, H.R.; Smith, R.D. *Anal. Chim. Acta* **1990**, *241*, pp 167–173.
52. Zhu, X.; Desiderio, D.M. *Mass Spectrom. Rev.* **1996**, *15*, pp 213–240.
53. Fenselau, C.; Yergey, J.; Heller, D. *Int. J. Mass Spectrom. Ion. Phys.* **1983**, *53*, pp 5–20.
54. Fenselau, C. *Anal. Chem.* **1982**, *54*, pp 106A–116A.
55. Yergey, J.A. *Int. J. Mass Spectrom. Ion Phys.* **1983**, *52*, pp 337–349.
56. Nguyen, D.N.; Becker, G.W.; Riggin, R.M., *J. Chromatogr.* **1995**, *705*, pp 21–45.
57. Loo, J.A.; Edmonds, C.G.; Smith, R.D. *Anal. Chem.* **1993**, *65*, pp 425–438.
58. Mirza U.A.; Chait, B.T. *Anal. Chem.* **1994**, *66*, pp 2898–2904.

59. Spear, K.L.; Sliwkowski, M.X. In *Techniques in Protein Chemistry II*; Villafranca, J.J.; Ed.; Academic Press: San Diego, CA, 1991; Chapter 22, pp 233–240.
60. Derynck, R.; Roberts, A.B.; Winkler, M.E.; Chen, E.Y.; Goeddel, D.V. *Cell* **1984**, *38*, pp 287–297.
61. Busch, K.L. *Spectroscopy* **1994**, *9*, pp 21–22.
62. Chapman, J.R.; Gallagher, R.T.; Barton, E.C.; Curtis, J.M.; Derrick, P.J. *Org. Mass Spectrom.* **1992**, *27*, pp 195–203.
63. Dobberstein, P.; Schroeder, E.; *Rapid Commun. Mass Spectrom.* **1993**, *7*, pp 861–864.
64. Chowdhury, S.K.; Katta, V.; Chait, B.T. *Rapid Commun. Mass Spectrom.* **1990**, *4*, pp 81–87.
65. Edmonds, C.G.; Loo, J.A.; Barinaga, C.J.; Udseth, H.R.; Smith, R.D. *J. Chromatogr.* **1989**, *474*, pp 21–37.
66. Li, Y.; Hunter, R.L.; McIver, R.T.,Jr. *Int. J. Mass Spectrom. Ion Processes* **1996**, *157/158*, pp 175–188.
67. Feng, R.; Konishi, Y.; Bell, A.W. *J. Am. Soc. Mass Spectrom.* **1991**, *2*, pp 387–401.
68. Winger, B.E.; Hein, R.E.; Becker, B.L.; Campana, J.E. *Rapid Commun. Mass Spectrom.* **1994**, *8*, pp 495–497.
69. Edmonds, C.G.; Loo, J.A.; Ogorzalek Loo, R.R.; Smith, R.D. In *Techniques in Protein Chemistry II*; Villafranca, J.J., Ed.; Academic Press: San Diego, CA, 1991; Chapter 47, pp 487–495.
70. Chapman, J.R.; Gallagher, R.T.; Mann, M. *Biochem. Soc. Trans.* **1991**, pp 940–943.
71. Loo, J.A.; Quinn, J.P.; Ryu, S.I.; Henry, K.D.; Senko, M.W.; McLafferty, F.W. *Proc. Natl. Acad. Sci. U.S.A.* **1992**, *89*, pp 286–289.
72. Barber, M.; Green, B.N. *Rapid Commun. Mass Spectrom.* **1987**, *1*, pp 80–83.
73. Herrera, A.E.R.; Lehmann, H. *Biochim. Biophys. Acta* **1974**, *336*, pp 318–323.
74. Hu, P., Buckel, S.D.; Whitton, M.M.; Loo, J.A. *Proc. 43rd ASMS Conference on Mass Spectrometry and Allied Topics* **1994**, p 322.
75. Krishna, R.G.; Wold, F. In *Methods in Protein Sequence Analysia*; Imahori, K.; Sakiyama, F.; Eds.; Plenum Press: New York, 1993; pp 167–172.
76. Andrien, B.; Boyle, J.G. *Spectroscopy* **1995**, *10*, pp 42–44.
77. Lee, T.D.; Shively, J.E. In *Methods in Enzymology: Mass Spectrometry*; McCloskey, J.A., Ed.; Academic Press: San Diego, CA, 1990; Volume 193, Chapter 19, pp 361–375.
78. Eshraghi, J.; Chowdhury, S.K. *Anal. Chem.* **1993**, *65*, pp 3528–3533.
79. Pugh, D.J.; Woolfitt, A.R.; Bateman, R.H. *Proc. 43rd ASMS Conference on Mass Spectrometry and Allied Topics* **1995**, p 657.
80. Caprioli, R.M.; Whaley, B.; Mock, K.K.; Cottrell, J.S. In *Techniques in Protein Chemistry II*; Villfranca, J.J.; Ed.; Academic Press: San Diego, CA, 1991; Chapter 48, pp 497–510.
81. Edmonds, C.G.; Smith, R.D. In *Methods in Enzymology: Mass Spectrometry*, McCloskey, J.A., Ed; Academic Press: San Diego, CA, 1990; Volume 193, Chapter 22, p 412–431.
82. Smith, D.L.; Zhou, Z. In *Methods in Enzymology: Mass Spectrometry*; McCloskey, J.A., Ed.; Academic Press: San Diego, CA, 1990; Volume 193, Chapter 20, pp 374–389.
83. Loo, J.A.; Edmonds, C.G.; Smith, R.D. *Science* **1990**, *248*, pp 201–204.
84. Loo, J.A.; Edmonds, C.G.; Ogorzalek Loo, R.R.; Udseth, H.R.; Smith, R.D. In *Experimental Mass Spectrometry*; Russell, D.H., Ed.; Plenum Press: New York, 1994; Chapter 7, pp 243–286.

85. Yan, L.; Tseng, J.-L.; Fridland, G.H.; Desiderio, D.M. *J. Am. Soc. Mass Spectrom.* **1994**, *5*, pp 377–386.
86. Monegier, B.; Clerc, F.F.; Dorsselaer, A.V.; Vuilhorgne, M.; Green, B.; Cartwright, T. *BioPharm* **1990**, *3*, pp 26–35.
87. Cody, R.B.; Tamura, J.; Musselman, B.D. *Anal. Chem.* **1992**, *64*, 1561–1570.
88. Gallagher, R.T.; Chapman, J.R.; Mann, M. *Rapid Commun, Mass Spectrom.* **1990,** *4*, pp 369–372.
89. Thibault, P.; Pleasance, S.; Laycock, M.V.; MacKay, R.M.; Boyd, R.K. *Int. J. Mass Spectrom. Ion Processes* **1991**, *111*, pp 317–353.
90. Wada, Y.; Tamura, J.; Musselman, B.D.; Kassel, D.B.; Sakurai, T.; Matsuo, T. *Rapid Commun. Mass Spectrom.* **1992**, *6*, pp 9–13.
91. Ferrige, A.G.; Seddon, M.J.; Green, B.N.; Jarvis, S.A.; Skilling, J. *Rapid Commun. Mass Spectrom.* **1992**, *6*, pp 701–711.
92. Shackleton, C.H.L.; Witkowska, H.E. In *Mass Spectrometry: Clinical and Biomedical Applications*; Desiderio, D.M., Ed; Plenum Press, New York, 1994; Volume 2, Chapter 4, pp 135–199.
93. Summerfield, S.G.; Buzy, A.; Jennings, K.R.; Green, B. *Eur. Mass Spectrom.* **1992**, *2*, pp 305–322.
94. Lahm, H.-W.; Stein, S. *J. Chromatogr.* **1995**, *326*, pp 357–361.
95. Johnstone, R.A.W.; Rose, M.E. *Mass Spectrometry for Chemists and Biochemists,* 2nd Edition; Cambridge University Press: Cambridge, U.K., 1996; Chapter 12, p 442.
96. Stein, S. *BioPharm* **1989**, *2*, pp 30–37.
97. Witkowska, H.E.; Bitsch, F.; Shackleton, C.H.L. *Hemoglobin* **1993**, *17*, pp 227–242.
98. Sanderson, K.; Andrén, P.E.; Caprioli, R.M.; Nyberg, F. *J. Chromatogr., A* **1996**, *743*, pp 207–212.
99. Light-Wahl, K.J.; Loo, J.A.; Edmonds, C.G.; Smith, R.D.; Witkowska, H.E.; Shackleton, C.H.L.; Wu, C.C. *Biol. Mass Spectrom.* **1993**, *22*, pp 112–120.
100. Oliver, R.W.A.; Green, B.N. *Trends Anal. Chem.* **1991**, *10*, pp 85–91.
101. Hu, P.; Ye, Q.; Loo, J.A. *Anal. Chem.* **1994**, *66*, pp 4190–4194.
102. Zaia, J.; Annan, R.S.; Biemann, K. *Rapid Commun. Mass Spectrom.* **1992**, *6*, pp 32–36.
103. Covey, T.R.; Shushan, B.I.; Thomson, B.A.; Henion, J.D. *Proc. 37th ASMS Conference on Mass Spectrometry and Allied Topics* **1989**, pp 558–559.
104. Turula, V.E.; Bishop, R.T.; Ricker, R.D.; de Haseth, J.A. *J. Chromatogr. A* **1997**, *763*, pp 91–103.
105. Chowdhury, S.K.; Chait, B.T. *Biochem. Biophys. Res. Commun.* **1990**, *173*, pp 927–931.
106. Ashton, D.S.; Beddell, C.R.; Green, B.N.; Oliver, R.W.A. *FEBS Lett.* **1994**, *342*, pp 1–6.
107. Meng, C.K.; Mann, M.; Fenn, J.. *Z. Phys. D: At. Mol. Clusters* **1988**, *10*, pp 361–368.
108. Marrack, P.; Kappler, J. *Science* **1990**, *248*, pp 705–711.
109. Beu, S.C.; Senko, M.W.; Quinn, J.P.; Wampler, F.M.III; McLafferty, F.W. *J. Am. Soc. Mass Spectrom.* **1993**, *4*, pp 557–565.
110. Winger, B.E.; Light-Wahl, K.J.; Ogorzalek Loo, R.R.; Udseth, H.R.; Smith, R.D. *J. Am. Soc. Mass Spectrom.* **1993**, *4*, pp 536–545.
111. Hirayama, K.; Akashi, S.; Furuya, M.; Fukuhara, K., *Biochem. Biophys. Res. Commun.* **1990**, 173, pp 639–646.
112. Meloun, B.; Morávek, L.; Kostka, V. *FEBS Lett.* **1975**, *58*, pp 134–137.

GLOSSARY

Numbers in square brackets refer to the chapter where the term is initially used.

aldaminitol An alditol with an amide group. [12]
alditol Linear six-carbon sugar where all carbons have the hydroxyl functional group. All carbons thus have a saturated oxygen species. [12]
amino acid An organic molecule that has a carbon as the central atom, with the following four groups: a hydrogen atom, an amino group, a carboxyl group, and an organic side group. [9]
amino acid residue An amino acid without an amine hydrogen and the hydroxyl group from the carboxyl group. This depiction represents how an amino acid occurs in a protein.
ampholyte An amphoteric compound that can exist as an anion or cation, depending on the solution pH. [2]
amphoteric compound A compound that can exist in a net positive, negative, or neutral state, depending on the pH. [2]
arithmetic average mass An experimentally derived average mass, M_{expt}. This is slightly lower in mass than the average mass (M_r). [4]
average mass The weighted mass of all the isotopes of a particular atom or molecule. This takes into account the different masses and relative abundance of each mass. [3]
capillary isoelectric focusing (CIEF) An ampholyte mixture of polyamines and polycarboxyl compounds and the analyte fill the entire column and are allowed to separate in different static pH zones under an electric field. Thus, the pH increases or decreases across the capillary column, and the ends of the capillary are placed in low- and high-pH solutions. The compounds and analyte finally stop moving in the capillary at their pI value. One end is connected to an ESI source and all components are eluted and electrosprayed with the aid of a salt buffer and positive pressure. [2]
capillary isotachophoresis (CITP) This technique uses the same instrumentation as CZE except for a different electrolyte scheme. For the separation of a mixture of cation analytes, a highly mobile (fast-moving) cationic compound (leading electrolyte) is placed

at the beginning of the column. The analyte is then introduced and a slow-moving cationic electrolyte (trailing electrolyte) is added. Buffering anions are contained in each of the three applications. The analytes separate between the two electrolytes and all bands move to the detector electrode. [2]

capillary zone electrophoresis (CZE) A capillary column is filled with a buffer electrolyte, and an electric field is applied across the capillary. Ion analytes migrate toward the detector end of the capillary. [2]

charge-site-remote fragmentation This phenomenon is based on the observation that for a stable charge in a molecule, the farther away a bond cleavage occurs from the charged (e.g., protonated) site, the relatively higher in intensity is that product ion with respect to product ions that fragment closer to and retain the charged site. [9]

CID-MS See up-front CID. [1]

collision-activated dissociation (CAD) See collision induced-dissociation.

collision-induced dissociation (CID) This procedure can occur in tandem-in-time or tandem-in-space mass spectrometers. A particular ion (precursor ion) is selected and stored in an ion-trap or Fourier transform (tandem-in-time) system, or only the precursor ion is allowed to pass through a linear mass-analyzing element (quadrupole or sector) and into a second non-mass-analyzing element in a tandem-in-space instrument. The precursor ion is caused to travel repeatedly in the tandem-in-time system, where it continuously experiences collisions with the collision gas (usually helium). In a tandem-in-space system, the ion travels from one end to the other end in a non-mass-analyzing element. This element is filled with gas (usually argon) and the precursor ion experiences collisions. The precursor ion in both cases breaks apart into lower mass fragments, and this process is called CID. [1]

complex glycotype A polysaccharide that can have many linear combinations of different saccharide units. These structures can be biantennary (two chains connected to a common linking oligosaccharide–Asn structure), triantennary, or tetraantennary. [12]

conventional mass spectrum A record of the masses of ions and their relative intensities that enter the mass spectrometer system. [1]

Edman degradation This is a procedure that sequentially, and starting from the N-terminal protein position, removes an amino acid, one at a time, by derivatizing the N-terminal residue, cleaving the residue, and identifying it by RP-HPLC. The cycle is repeated for each subsequent N-terminal amino acid residue. This degradation process is ineffective if the N-terminus is not in the free amine state. [9]

electroosmotic flow (EOF) The ion flow in the liquid under the influence of an applied voltage in CE. [2]

Electrospray The process of electrostatically passing and/or forcing (via mechanical and pressure procedures) a liquid stream through a capillary orifice. The resulting mist consists of electrically neutral drops. [1]

electrospray ionization (ESI) The process of passing a liquid through a capillary tube with a 4–5 kV electric field at the tip; charged droplets are formed which desorb adducts of the neutral analyte and ionized reagent molecule. [1]

Endo-F Cleaves the glycosidic bond between the first two sugar residues attached to the peptide, and this occurs for all carbohydrates attached to an Asn residue. [12]

endoglycosidase An enzyme that cleaves bonds from internal carbohydrate residues as opposed to that of terminal residues. [12]

exact mass See monoisotopic mass. [3]

exoglycosidase Cleaves terminal (nonreducing end) sugar residues. [12]

glycoform Rarely do cells make a glycoprotein molecule that exists in only one biochemical structure. Usually, the same polypeptide template experiences multiple sugar

residues in polysaccharide sequence and amino acid residue attachment sites. The collection of these molecules, which differ in the amount and amino acid residue attachment site(s) of the saccharide/polysaccharide species, are labeled glycoforms. [12]
glycomer Glycoform molecules differing only by individual isomeric sugar residues, and also includes aspects such as linkage and branching features. [12]
glycoprotein A protein that contains carbohydrate monomer/polymer groups. The carbohydrate portion is bound to a protein, primarily through serine, threonine, or asparagine residues. [12]
glycoprotein heterogeneity Glycoform distribution at a certain glycoprotein amino acid residue. [12]
glycotypes The three types of Asn-linked oligosaccharides: high-mannose, hybrid, and complex polysaccharides. [12]
high-mannose glycotype Term that describes a polysaccharide that can consist of up to nine mannose and two *N*-acetylglucosamine residues in a branched structure. The high-mannose polysaccharide attaches to a polypeptide by the Asn residue. [12]
hybrid glycotype A polysaccharide that can have many linear combinations; however, the antenna polysaccharides characterize both high-mannose and complex glycotypes. [12]
immobilized metal-ion affinity column (IMAC) A capillary column that has a bound metal ion. This metal ion is selected for its preferential binding to an analyte or group of analytes of interest in a sample. [10, 11]
ionspray This uses the same experimental conditions as ESI except a pneumatically assisted (compressed air-assisted) component is added in order to facilitate the entrainment and dispersion of the liquid stream. [1]
isotope An atom that can be found as more than one mass. The collection of atoms are identical—that is, are of the same element and hence have the same number of protons (atomic number)—except the different masses are due to the different number of neutrons. For example, the two most common naturally occurring carbon masses are ^{12}C, carbon-12, which was given an arbitrary mass of 12.0000 Da by the International Union of Pure and Applied Chemistry and the International Union of Atomic Weights, and ^{13}C, carbon-13, which has a mass of 13.003354 Da. The additional 1.00335 value is due to an extra neutron. [3]
isotope shift Differences between the average and monoisotopic masses of a compound. [3]
kinase Catalyzes the addition of a phosphate moiety to the hydroxyl side group of a Ser, Thr, or Tyr residue. [10]
major histocompatibility complex (MHC) A glycoprotein that is produced inside a cell, it accepts an 8–25 amino acid peptide from internal protein degradation, and the complex is transported to the cell surface. The MHC complex presents the peptide for scrutiny by circulating cytotoxic T-lymphocytes. There are type-I and type-II MHC molecules, which are formed by different biochemical pathways, and the MHC I and II units bind peptides that are 8–12 and 13–25 amino acid residues in length, respectively. The MHC I molecules are found in tissue cells, and MHC II complexes are found in mobile white blood cells. [11]
mass defect This refers to the difference in mass of the nominal mass and any isotope of the compound. [3]
mass spectrum See conventional mass spectrum. [1]
microheterogeneity See glycoprotein heterogeneity. [12]
molecular mass See monoisotopic mass. [3]
monoisotopic mass The lowest mass isotope of an atom or compound. [3]

M_r Symbol for average mass, average molecular mass, average chemical mass, or relative molecular mass. [3]

multiply charged species The addition of charges to a neutral analyte molecule. These charges can take the form of ionized species such as H^+, Na^+, and K^+. [4, 5, 8]

neutral-loss mass spectrum This spectral scan can be accomplished on a tandem-in-space mass spectrometer, where the first and third mass analyzers are scanned at the same rate. There is a numerical mass difference or offset between the two analyzers. This mass difference is set to equal the mass of a structural moiety on one or more precursor ions that can fragment upon CID. The scan displays the precursor ions that fragmented upon CID to release the moiety. [10]

***N*-glycanase** Cleaves all N-linked glycoproteins at the Asn–carbohydrate linkage. [12]

nominal mass The whole number value of an atom's or compound's mass. [3]

ODM ORM, except the reagent $NaBD_4$ replaces $NaBH_4$. [12]

***O*-glycanase** An enzyme that releases a polysaccharide from an O-linked Ser or Thr residue. [12]

oxidation, reduction, and methylation (ORM) An organic chemistry series of reactions used in polysaccharide analysis that cleaves the bond between vicinal diols and results in two methyl ether groups. Reduction usually takes place with $NaBH_4$. [12]

peptide bond The covalent bond formed when two amino acids join by a dehydration reaction. See amino acid residue. [9]

phosphatase Catalyzes the removal of a phosphate moiety from the side group of a Ser, Thr, or Tyr residue. [10]

phosphoprotein A polypeptide chain that contains one or more H_2PO_3 moieties attached to the side group oxygen of either a Ser, Thr, or Tyr residue. [10]

phosphotransferase The most common type of protein kinase. [10]

pI Isoelectric point of a molecule. This is equal to the pH of a solution at which the compound does not move (electrically neutral) in an electric field. [2]

PNGase F See N-glycanase. [12]

polypeptide A relatively short series of amino acid residues compared with a protein; a "small" protein.

posttranslational modification (PTM) This is a molecular addition, modification, or deletion to a polypeptide/protein chain by the cellular apparatus after the messenger RNA has completed the synthesis of the amino acid chain. [9]

precursor ion (formerly parent ion) An ion found in the conventional mass spectrum. [1]

principal peak See monoisotopic mass. [3]

product ion (formerly daughter ion) An ion formed from the fragmentation or CID of a precursor ion. [1]

product-ion spectrum (formerly daughter-ion spectrum) The mass spectral record of fragments produced from the collision-induced dissociation (CID) process from a precursor ion. [1]

proline effect If the residue in the sequence X-Pro is of a charged nature—for example, Arg, Asp, or Glu—the ion that results from the cleavage of the bond is of a low intensity. If the X residue is neutral, then an intense signal is observed. These mass spectral observations can define the presence of proline, and the mass spectral peak intensity observations are operative for both ···X and Pro··· peptide fragments. [9]

protein A substance made up of a covalent, linear combination of amino acids. The amino acid residues are bound by a dehydration reaction between an amine from one amino acid residue and a carboxyl group from another residue. [9]

recombinant technology The DNA technology used when a protein is desired in significantly greater amounts then that obtained from a conventional purification and isolation method. The DNA responsible for production of the protein is spliced from the native host genomic material and inserted into the genome of another cell that has the capacity to make significant amounts of the desired protein. [12]

residue See amino acid residue. [9]

resolution Defines the relative separation of two closely spaced mass spectral peaks, where resolution $= R = M/\Delta m$. M, experimentally measured mass of a peak; Δm, width in daltons of the peak or width between overlapping peaks. [5]

resolving power See resolution. [5]

reverse aldol condensation Known as peeling and occurs on polysaccharides. Base-catalyzed reactions can degrade the terminal sugar, and this process is repeated, which causes the destruction of the entire polysaccharide. [11]

reversed-phase, high-pressure liquid chromatography (RP-HPLC) A method used to separate mixtures of compounds. It is also used to concentrate individual species. A capillary column is coated with a nonpolar stationary phase and a polar liquid passes through the capillary. The liquid phase can be a mixture of different aqueous and non-aqueous solvents. An important class of stationary phases that are coated on the inside of the capillary column are silicon derivatized with octadecylsilane, and this is abbreviated as C_{18}. [2]

sample-stacking In the CZE, CITP, and CIEF methods, when the conductivity of the analyte is lower than the main buffer-solvent in the capillary, the analyte "focuses on itself," and thus concentrates itself into a narrow band. This is referred as sample-stacking. [2]

sheath flow An independent liquid or gas stream that flows in a collinear fashion about the central liquid analyte electrospray emission. [1]

skimmer voltage The voltage difference that causes "up-front" CID or CID-MS. This voltage difference is manifested by a number of physical devices located near the vacuum region of the mass spectrometer or just after the pinhole inside the mass spectrometer in the reduced pressure transition region. [1]

up-front CID This technique was developed on a quadrupole mass spectrometer, where the sample that is entering the vacuum experiences a relatively high voltage to effect fragmentation. Even if a sample is composed of a number of precursor ions, they all fragment, and this results in a mixture of product ions. A conventional mass spectrum is then taken of all the mass fragments. [1, 10]

weighted average mass Average weighted mass; see average mass. [3]

Index